WYXIE

by DICK OSGOOD

WONDERLAND

(An Unauthorized 50-Year Diary of WXYZ Detroit)

Bowling Green University Popular Press
Bowling Green, Ohio 43403

Acknowledgements

Material from this book was used in slightly different form in *Horizon* magazine and the *Detroit News* Sunday magazine. Permission to reprint was granted by *Horizon* and the *Detroit News*.

Permission to publish the *Lone Ranger* pictures was granted by Lone Ranger Television, Inc., a subsidiary of Wrather Corporation. "Lone Ranger" (T.M.) and Tonto are trademarks of Lone Ranger Television, Inc.

Library of Congress Catalog Card No.: 81-82501

ISBN: 0-87972-186-3: Clothbound
0-87972-187-1: Paperback

CONTENTS

II
Stray Thoughts On A Picket Line

Dedicated to the Memory of
FRAN STRIKER

Dick Osgood
(Courtesy of The Henry Ford Museum, Dearborn, Michigan)

Foreword

THIS IS THE STORY of one station—WXYZ—but the general history of broadcasting is reflected in it—from scientific experiment and a rich man's hobby, through the era of great creativity, to what we have today. The phasing out of radio network programming is a part of it, as are the rise of the disc jockey, the fixed television quiz show and payola, the revolution of music from house band, remotes, staff pianists, organists and singers, to rock and roll records, Top 40, "underground" music to MOR.

WXYZ was the home of *The Lone Ranger*, *The Green Hornet* and *Sergeant Preston*. Among its alumni are Douglas Edwards, Fred Foy, John Hodiak, Betty Hutton, Martha Scott, George Seaton, Danny Thomas and Mike Wallace; also network presidents, station managers, managers and executives now scattered from New York to Los Angeles and from Florida to the state of Washington. Writers, producers, directors and the great stars of every field of sports and entertainment have visited it for interviews and appearances. It was the scene of the revival of Chuck Berry and the discovery of Stevie Wonder. Soupy Sales, Don Ameche, Herbert Marshall, Betty Clooney, Johnny "Scat" Davis and Lou Gordon all worked there. It lifted the city in which it is located to fourth place in the nation as a production center, and it started a national network.

But it is more than all this. Unknown men and women have dedicated their lives to it, found and lost mates in it. It has been the cause of divorce. And death. It has been wife, master, mistress, "a puzzlement" and an obstinate bitch. It has been known to sap the creative power of genius and leave it dry. Engineers, singers, sound men, musicians, arrangers, actors, writers, announcers, newscasters, disc jockeys, salesmen and hard-headed business executives have devoted their waking days and nights to the sound of it. Sometimes the sound was good; sometimes it was obnoxious; once, more than a hundred people made a co-operative effort to silence it.

It has weathered poverty, panic and payola. It has survived five changes of ownership, a dozen changes in management and several major shifts in policy. A house divided, it has yet stood firm. It has been a beacon attracting outstanding talent from across the land. It has made fortunes, caused breakdowns and driven men and women to drink.

The story as told in this book is as close to the truth as it can be. If there are errors, they are honest errors of memory. Recollections are not always sharp after forty years, or even ten. Often it has been possible to combine the fragments of faded recall into a whole and bring it into focus. Unintentional mistakes have been corrected as one version of an incident has been matched against another. Sometimes one memory has been outnumbered by two or three. Occasionally it has been necessary to compare reminiscences and choose those which seemed younger or sharper or more positive. Motives, emotions and integrity have been considered.

What is told here appears to be the truth.

D.O.

January 1981

"A poor man is better than a liar."
Proverbs 19; 22.

Part One
Silver Was A Plough Horse

1
The Galloping Wonder

PEOPLE JUST COULD NOT believe it. When it was explained to a lady at the World's Fair in St. Louis in 1904, she cried. "It's a fake! You connect what you send with what is received by a silk thread just so you can say it's wireless."

Visitors stood at the base of the demonstration tower, looking up, trying to SEE the messages go out into the air. They could see nothing and there were no wires. News was sent from that tower to a nearby newspaper by Morse code; this was the wireless telegraph.

The inventor had been working on it since his college days at Yale. He wrote the thesis that earned his degree while soldiering in the Spanish-American War. This was the first time the public had witnessed his marvel. He was Dr. Lee De Forest and he won the Fair's Grand Prize for establishing wireless communication between St. Louis and Chicago, three hundred miles away.

Two years later, De Forest invented a kind of glass tube which, in end effect, amplified, or increased, the sound. He called it "the Audion tube." Men who tried to sell stock in it were indicted for fraud.

In summing up his prosecution, the District Attorney said, "Let me tell you, gentlemen of the jury, these men are preying on the minds of simple people. This device is without merit. It is a piece of glass which has been built in the form of a lamp, not to perform scientific wonders, but to sell stock."

The men were sent to prison.

Telephone companies were not interested: a conversation sent out into the air could be picked up by anybody; there would be no privacy. Dr. De Forest's vision took a different direction. He saw a day coming when, by means of his invention, opera could be brought into every home.

"Some day," he predicted, "news, and even advertising, will be sent out to the public on the wireless telephone."

To prove it, in 1910, he sent the singing voice of Enrico Caruso vibrating into the air. Still, for twelve years, his invention rested in limbo. To survive, he had to take a job with a telephone company in California. When he tried to borrow $125 to save his patents from expiring, nobody would lend it to him. Finally, at the end of 1921 he went to Germany and experimented with phonofilm—a kind of talking motion picture. Six months later, he returned to the United States and was surprised to find himself a hero.

Somehow his scientific marvel, the wireless telephone, had escaped from the laboratory and the constriction of limited imagination, and had finally caught the fancy of a few men of influence. One of these, the publisher of *The Detroit News*, **James E. Scripps, had been so impressed with the demonstration of a local inventer,** Thomas E. Clark, that he had given him a check for $1000 with which to carry on his experiments. In 1919, James' son, William, acquired a transmitter and set up an experimental sending station called 8MK. In Pittsburgh, a vice president of

Westinghouse, H.P. Davis, and a research engineer, Dr. Frank Conrad, broadcast the returns in the Harding-Cox Presidential election over equipment called KDKA.

During the first eleven months of 1921 factories were producing an average of 5,000 tubes a month for the Radio Corporation of America. Then suddenly public interest was ignited and broadcasting jumped into favor overnight. On December 30, the RCA production schedule was increased to 40,000 tubes a month. Six months later, monthly production had jumped to 200,000 tubes.

"Simple people" were adding new words to their vocabularies: amplifier, detector, static, condensor, balancer, booster and induction coil. The new field was winning the attention of investors. Broadcasting stations were springing up like crabgrass. A conference was called in Washington to determine how the Department of Commerce could apply its existing power of regulation to the new development, and to recommend legislation to Congress which would expand that control.

"We have witnessed in the last four or five months," said Secretary of Commerce Herbert Hoover, "the most astounding thing that has come under my observation of American life. This department estimates that today over 600,000 persons possess wireless telephone receiving sets, whereas there were less than 50,000 such sets a year ago."

Mr. Hoover also commented, "One of the problems that enters into this whole question is—who is to support the sending stations?"

In its Sunday edition of April 16, 1922, *The Detroit Free Press* devoted four full pages to a Radio Show that would open in that city the following Thursday.

The J.L. Hudson department store entered an ad which read: "Radio has been called an 'influence for democracy.' People in towns and villages and those in desolate country places, may share alike in its wide sphere of influence. Concerts by world-famous artists, sermons by great religious leaders, and all the features of life in metropolitan centers, are carried through space to every place that a man calls his home."

Profesor T.E. Reynolds of the University of Detroit predicted that radio would keep country kids down on the farm.

People were still struck with wonder. Benjamin F. Bailey, Ph.D. and Professor of Electrical Engineering at the University of Michigan wrote: "When I sit with a pair of wireless phones on my ears and hear a clear, distinct voice speaking to me out of the unknown, or hear what might be the voice of an angel singing in the sky, I am lost in wonder at the marvel that the mind of man has accomplished. *The thing is impossible; but there it is.*"

He added: "When I examine the apparatus required to work this miracle, my wonder is only increased. A few turns of wire, a few sheets of tinfoil and paper, a piece of crystal and a fine wire touching it and a telephone receiver comprise the wonder-working outfit."

In its Sunday edition, *The Detroit News* observed that the "growth of interest and demand for equipment" had "not been equalled in any other industry."

The Free Press agreed. "Ninety days ago," it commented, "even two months ago, radiophony had received but scant attention; but now everybody is interested, far more so than they were years ago in the automobile and earlier in the bicycle, because the auto and bicycle cost real money, and a radio set, complete including installation by a so-called expert, may be had for from $20 to $30."

It was predicted that the radiophone receiving instrument would be as popular as the phonograph. People could even build their own sets. All you needed to know was how to put together those few turns of wire, the tinfoil, the piece of crystal, the fine wire and the telephone receiver. The rush demand for "How to —" books had begun at the Detroit Public library six months before and now nearly 350 copies were being taken out as fast as they were returned.

The big Radio Show opened on Thursday, April 20, 1922. Several thousand people braved a storm to get to it that night. Mayor James Couzens opened it officially, speaking from a radio station. One hundred exhibits crammed eight thousand square feet of the hall on the fourth floor of the General Motors Building. It was the first radio show of any size in the United States. Two which had preceded it in New York and Chicago were only half as big.

"DETROIT RADIO HUB OF NATION" headlined the *Detroit News* the next day. "City's Manufacturers at Show Amazed as Orders Pour in from Everywhere."

Forty companies in Detroit were manufacturing complete receiving sets; sixty concerns were making parts; twenty Detroit factories were building cabinets to put the sets in; one hundred wholesalers and retailers were engaged in marketing the new product. Dealers came to the show from all parts of the Middle West.

Six booths were devoted to radio batteries alone. There was a stall devoted exclusively to radio books and magazines. RCA displayed its latest invention, "the new General Electric frequency amplifying transformer" which did away with the aerial. One exhibitor concentrated on receiving sets for canoes. Speeches were made. The inventor, Thomas E. Clark, paid tribute to "the late James E. Scripps, founder of *The Detroit News,* as one who foresaw the future of wireless telephony and liberally financed experiments."

The Scripps' experimental station 8MK had changed its identification to WWJ, and WWJ had a display in the show as "the pioneer newspaper station of the world."

The show lasted four days. Among the thousands who flocked to it had to be George Harrison Phelps, who, like radio, was himself a galloping wonder.

2
The Man Who Went Nowhere But Up

GEORGE HARRISON PHELPS had come a long way from the country kid he had been at Miller's Falls, Massachusetts, the son of a grain dealer. He had grown up with *Sink or Swim, Do or Die* and all the other books by Horatio Alger, and that philosophy had rubbed off on him. His favorite saying was "hitch your wagon to a star." George was always looking ahead and up, and anything that was new caught his interest.

What was new when he graduated from Cornell Univesity with a Bachelor of

George Harrison Phelps of WGHP

Arts degree in 1903 was the automobile and that was the star he hitched his wagon to. No horses and grain business for him. He opened what some said was the first automobile service garage, in Worcester, Massachusetts.

When he was 23 and had a chance to become assistant manager of the Buick office in Boston, he sold his garage and moved to the city that dominates New England. With one foot firmly planted on the Buick rung, he now had the other one free to feel upwards for the next one. His searching toe found an opportunity to move up from assistant manager to manager and he took advantage of it without hesitation even though it meant changing companies.

He left Buick to manage the automotive concern called Everett, Metzger and Flanders. In 1910 or early 1911 E.M. and F. was taken over by the Studebaker Corporation financed by J.P. Morgan & Co., and that led to the next step up for George Harrison Phelps. Only six years after he had sold his garage in Worcester, the fast-moving young man from Miller's Falls owned the New York branch of Studebaker. He was not yet 30. The automobile business had caught a comer.

He had met a girl named Laura Stephens in Ithaca while attending Cornell. They were now married and had two sons. In New York, Phelps joined the right clubs, moved on the edge of the theatrical world which he loved and became acquainted with important visitors from Detroit. These last carried the word back to the heart of the new industry that the man running Studebaker in New York was a live one. Soon the word reached Dodge.

In 1914, when Dodge was incorporated, a man was needed to help organize a sales department and direct the company's advertising. The offer went to George Harrison Phelps. He sold his New York Studebaker business and moved with his family to Detroit.

The Phelps family bought a home on Edison Avenue. Among their neighbors were Edsel Ford, Fred Haynes of the Dodge Company, and John Dodge himself. George lost no time firmly weaving himself and his family into the social fabric of Detroit. They joined the Detroit Athletic Club, affectionately referred to as the D.A.C. He was elected President of the Players.

The Players was—and is—a unique and exclusive club dominated by professional and advertising men. Its members dedicate themselves to the production each month of a play or a batch of one-acters. They own a perfectly equipped little theater, build their own scenery, and hire a man to make them up. Only men can join, if invited.

Presumably it was for The Players that Mr. Phelps with M.I. Pitkin, wrote a short two-act play called "The Pearl Thief." He also wrote a book of "impressions" aimed at improving the technique of automobile salesmen. It was called "GO." *Automobile Topics* declared that any man would be improved for having read it, especially the two types of salesmen—those who knew it all and those who didn't know enough.

It was neither a play nor a book, however, that brought fame and fortune to George Harrison Phelps. It was his application of the single word "dependable" to the Dodge automobile. Phelps made that word so synonymous in advertising Dodge that for several years the word "Dependable" alone, printed in white against a background of cobalt blue was all that was required on a billboard to remind the masses of the car it stood for, and because of the connection in the public mind, Dodge prospered.

So did the Phelps. They moved from Edison Avenue to a fancy house on Bishop Road in Grosse Pointe Park. They joined The Detroit Club and The Detroit Yacht Club; and they did a great deal of entertaining.

In February, 1922, George Harrison Phelps made an important announcement: he was leaving Dodge to open his own advertising agency and was taking the Dodge account with him. By April, he had turned a fine old three-story house at Rowena and John R into an attractive place of business, and one month short of his 39th birthday, his hair thinning a bit, his face full and rather handsome, his manner bright and aggressive, he hosted a housewarming. As a reporter from *Automobile Topics* wrote later, it was "all that a housewarming should be"—complete with good liquor smuggled across the Detroit River from Canada. Phelps' old friend, De Wolf Hopper, trained in from New York for the occasion and recited his famous "Casey at the Bat."

As the host moved among his guests with his usual amiability, he had the air of a man in possession of a secret. Everyone knew that Saturday night in April, 1922, that Phelps had, locked in the safe of his new enterprise, a contract with the Dodge Brothers.

"George is the only man I know," one guest commented, "who will have made one million dollars in a single year."

It was rumored that one clause in the new contract was quite unheard of then in the advertising business: it was to run for five years. Only three men at the housewarming, other than Phelps himself, knew the real secret. One was Fred Haynes, President and General Manager of Dodge since the brothers' death two years before. Another was Phelps' brother-in-law, Bill Laurie, who had moved over from Chicago to work with him and who senior ladies today recall was "the best-looking man in Grosse Pointe." The third was Charley Matheson, another Dodge bigwig with a shrewd mouth, a square face, rimless eyeglasses and a pearl stickpin.

The contract was non-cancellable.

Ten days after his housewarming, George Harrison Phelps went to the big Radio Show in the General Motors Building with Bill Laurie and Max Hunt. Hunt was with the Hutchinson Steamship line which sailed the Great Lakes. They went from the Show to Charlie Fisher's Uptown Bridge Club, a speakeasy two blocks away. (Charlie Fisher's is now Fisher 666, a respectable restaurant on Lothrop.)

It was late afternoon. The general atmosphere of Charlie Fisher's was one of dark red mahogany, leather, amber lights and cigar smoke. The men began talking with friends along the bar about the wonders of wireless sound.

"I read in *The Free Press* that the Seeandbee is being equipped with radio by RCA," said a man named Roberts who had jowls and pale, thyroid eyes.

The Seeandbee which sailed the Great Lakes was the largest freshwater passenger ship in the world.

"We're putting radio telephone sets on all seventeen of our ships," said Max Hunt.

"You mean that people on the ship will be able to talk with people on the shore by radio?" Roberts marvelled.

"Right through the air without any wires," Hunt declared.

"Amazing," said Roberts, shaking his jowls.

Bill Laurie clucked his tongue.

A new voice came from behind them.

"That's *nothing.*"

They turned to see a man sitting at a table with a young blonde. He was wearing a carnation. He said, "There's a piece in tonight's *News* about some scientist who proposes linking movies to the radiophone gadget and sending *pictures* through the air."

The men at the bar gave him a long look.

"Impossible," said Max Hunt at last.

The men at the bar turned back to their drinks.

Four months after Phelps opened his agency, the A.T.& T. station in New York, WEAF (now NBC) broadcast the first radio commercial. It was a ten-minute speech extolling the advantages of owning one's own home. The speech was paid for by a real estate company.

Eight months after Phelps' housewarming and the Detroit Radio Show, the first agency to buy advertising on radio purchased time on WEAF. William F. Rankin, whose name the agency bore, delivered a talk on the subject "Advertising and Its Relation to the Public." Rankin was testing the power of this new means of communication. From that broadcast he received 25 letters and telephone calls. One of them resulted in a new account worth half a million dollars.

Rankin had no further doubt about the power of radio and was one of the first to put his clients on the air. For Goodrich Rubber he took an ordinary singer and using a gimmick made him one of the first idols of broadcasting, "The Silver Masked Tenor."

This was advertising and show business, and Phelps loved both.

"What do you think about advertising on the radio?" Fred Haynes of Dodge asked at one of Phelps' parties.

"Controversial," said Phelps. "There's a great deal of doubt about the ethics of direct advertising on a public medium."

"I'm inclined to go along with Herbert Hoover," said Bill Laurie.

Hoover had said the quickest way to kill broadcasting would be to use it for direct advertising.

"*You* know why they put programs on the air," said Phelps. "It's to stimulate the sale of receiving sets. And they're doing it. The sale of receiving sets last year amounted to $136,000,000. When I opened the agency last year there were only fifty broadcasting stations. Do you know how many there are today? Fourteen hundred."

Bill Laurie clucked his tongue, shaking his head.

"They had a gold rush in the 1800s," he said. "And now they've got a radio rush."

"No question about it," Phelps laughed. "Radio is the coming thing."

He wanted to join the rush and he could afford to. His agency already had branches in New York, London, Paris, Berlin and California—wherever Dodges were sold.

Radio was in a state of chaos. The government had not yet established an agency to control it. Applicants got licenses to operate from the Department of Commerce. In the beginning all stations were assigned to one of the only two frequencies, but now, because of the rush, this was no longer sufficient. Following the urgent recommendation of another National Radio Conference, Herbert Hoover opened up the spectrum and assigned each applicant a definite frequency. There were still not enough to go around. Several stations in different parts of the country operated on the same frequencies.

The Department of Commerce was not allowed to deny any applicant a license to broadcast; it had no authority to restrict an operator as to what band he broadcast on. It could not prevent one station from taking over the frequency of another. Piracy was rampant.

Into this turbulent sea plunged George Harrison Phelps—for fun. Through a friend in Washington he secured a Class B license and a wave-length far enough removed from other Detroit stations to eliminate the possibility of serious interference.

He found a man described later by an expert as "more of a mechanic than an engineer" to build the station. His name was Harold Elliott. Elliott had a friend in Chicago licensed with Western Electric whose equipment was standard. Certain pieces of it somehow got sidetracked from Chicago to Detroit and found their way into what had been the carriage house behind Phelps' agency. Even William Laurie who was to manage the new enterprise for his brother-in-law, referred to it 45 years later as "a bootleg station." The vast majority of radio stations at that time were no better.

On Saturday night, October 10, 1925, three-and-a-half years after he opened his agency, five months after he had been photographed cruising down the Detroit River aboard his new 45 mph, 9-passenger Hacker-Dolphin Speedster, and after a series of test broadcasts which had left everyone satisfied that the new station could be heard, George Harrison Phelps went on the air with his own radio station. He gave it his initials. It was station WGHP.

3
Rich Man's Plaything

There were now six radio stations in Detroit including WGHP. Scripps and *The Detroit News* were operating WWJ; E.D. Stair and *The Detroit Free Press* had WCX with its studios in the Book-Cadillac Hotel; WJR was owned by the Jewett Radio and Phonograph Company; WWI belonged to The Ford Motor Company for private use; and the Detroit Police Department had its own station with the appropriate call letters KOP. The police department was pioneering the use of radio in police automobiles. There was also Station WDYC owned by the Detroit Yacht Club which restricted itself to relaying messages to its members as they sailed their boats and drove their cruisers on Lake St. Clair and the Detroit River.

The most popular programs were George Lamp and his *Red Apple Club* on WCX and a big show broadcast Wednesday nights by the Burroughs Adding Machine Company on WJR. The busiest musician in the city had to be Gene Goldkette. With different groups, Goldkette was on the air morning, noon and night, bouncing between WCX and WJR. The two stations were approaching a merger.

WGHP lured Bernice Rayburn, "statuesque queen of the ivories," away from Mr. Stair's WCX to become staff pianist. No radio station then could exist without one. Wherever holes appeared in the programming—and that was often indeed—they were plugged up by staff pianists who played ragtime and light classics, accompanied singers and vamped their way through hours of emergency.

Three weeks after its opening, the new station WGHP earned this comment in *The Detroiter:* "The programs that have thus far been presented by this station have made a decidedly favorable impression with radio audiences in these parts."

Unfortunately "these parts" did not extend far enough. So a new transmitter with twin towers was built some twenty miles north of the city near Mt. Clemens. Even then, engineer Alvin Schaub recalls, they were "overdoing their output at one time and an inspector made them cut their power."

The radio advertising policy of George Harrison Phelps was ultra-conservative, restricted to a mere listing of his clients at the beginning and end of each program. The station was really little more than a plaything. Neither Phelps nor Bill Laurie paid much attention to it.

They were more interested in what was going on at the Dodge Brothers and had reason to be concerned. The Dodge account continued to be the main course on the Phelps meal ticket and non-cancellable or not the contract found itself unexpectedly in a chilling tray.

Horace and John Dodge had been dead for five years but the Dodge Company, under Fred Haynes' management, had prospered mightily, but there was a problem with the two widows. Mrs. Horace Dodge and Mrs. John Dodge had become equal partners upon the death of their husbands in 1920. In five years they had lost interest in the business. After all, what lady would not prefer a few more millions in cash to a factory? So, in April, 1925—six months before George Harrison Phelps opened his radio station, the widows decided to sell.

General Motors offered them $134,000,000. A New York banking company, Dillon, Read & Co., offered $152,000,000 in cash. The Dillon, Read offer was accepted. Clarence Dillon immediately re-organized the company and sold new stock in it for a profit of $13,000,000, besides retaining control of the enterprise. Fred Haynes stayed where he was for the time being but Dillon, Read & Co. put a man named Edward G. Wilmers in charge. Mr. Wilmers had revived the Goodyear Rubber Company from a sinking spell but knew almost nothing about marketing automobiles and even less about George Harrison Phelps.

It was immediately obvious to Mr. Phelps that Dodge was no longer precisely the same company with which he had signed a non-cancellable contract.

Two things happened in 1927 which affected Station WGHP—one in New York, the other in Washington.

In January, a New York outfit calling itself United Independent Broadcasters, Inc. headed jointly by the concert impresario Arthur Judson and a pioneer broadcaster named J. Andrew White, announced its intention of forming a network of stations to facilitate selling time to advertisers who needed more than one city. It also promised to supply the sixteen stations which would be included in the network with programs. United would pay each station $500 a week for ten hours of time.

Before this announcement could translate itself into action, however, the Columbia Phonograph Company set up a branch to serve as the sales agency of United. This branch, called the Columbia Phonograph Broadcasting System, soon figured it had stepped into a losing proposition and was dissolved. All its stock went to United Independent Broadcasters. Most of the stock of the floundering company was now purchased by an enterprising young Philadelphian, William S. Paley, and his family.

The Paleys had saved their fading cigar business (La Palina Cigars) by advertising over Station WCAU in Philadelphia and had purchased time on the network now threatened with untimely demise. William S. Paley, 27 years old, with his family purchased this new and troubled network.

The station chosen to be the CBS outlet in Detroit was George Harrison Phelps' WGHP.

Meanwhile in Washington, Congress passed the Radio Act of 1927 which created the Federal Radio Commission to bring order out of chaos. The license of every radio station then operating was automatically cancelled with the passing of this legislation. The Commission was given authority to issue new licenses, assign frequencies, and enforce the amount of power to be used by any station, and the hours it would be allowed to broadcast. One hundred and fifty stations for which there was found to be no legitimate place were squeezed off the air.

WGHP survived.

The studios in the garage on the alley at Rowena and John R proved to be inadequate. New studios were installed in the 15th floor penthouse atop the Maccabees Building (now the School Center Building) at Putnam and Woodward Avenue near the then new public library and the Detroit Institute of Arts.

The psychology of the Dodge widows' sell-out and the change of management did not help the Dodge Company. In 1927 it skipped a dividend.

At this time a man was moving up in the automobile business named Walter P. Chrysler. He had come from General Motors and taken over the old Maxwell Company. Early in 1928 he introduced the DeSoto. He needed an additional plant and wanted another car; it was always wise to have several makes going at one time in case one turned out to be a lemon.

In the spring of 1928 Clarence Dillon arranged a meeting with Chrysler and offered him the Dodge Company. Chrysler agreed to take over Dodge *if* 90% of the

Studio WGHP, Maccabees Building, 1926, Courtesy Ray Schaub

stockholders gave their consent. Dillon, Reed & Co. had to do some fast work on the open market to accumulate enough stock (and votes) to meet this requirement.

At 5:30 one afternoon in the early summer of 1928, K.T. Keller of the Chrysler Corporation sat at one end of an open telephone line in Detroit for one hour before he heard the treasurer of the company, E.E. Hutchinson, say on the New York end, "We have bought the Dodge Brothers."

This was bad news for George Harrison Phelps. A close associate of Walter P. Chrysler was the automotive advertising expert from whom Phelps had wrested the Dodge account when he opened his own agency, Theodore MacManus.

At the same time, the non-cancellable five-year contract expired.

Thinking back over this expiration, William Laurie commented: "We really hoped that they would renew it but we soon found out that they weren't going to. So then the thing got to be pretty gloomy-lookin'."

The "thing" was Station WGHP.

Laurie continued: "The thing, instead of paying off, kept losing and his [Phelps'] losses got bigger and bigger. It cost $125,000 a year to run and finally he just couldn't put up the money to run it."

Bill Laurie found himself trying to peddle a losing proposition. Leo Fitzpatrick of WJR wanted to buy the station but was unable to raise the money. Detroit Edison turned it down. Relief came from another state.

Major J. Andrew White who had figured in the formation of the Columbia Broadcasting System was impressed with the success of a station in Toledo. It was owned by two men in the steel and oil business—J. Harold Ryan and George B. Storer. Their Toledo station, WSPD, took its name from a gasoline called Speedeen which they sold in a chain of filling stations. Major White persuaded Ryan and Storer to form an Ohio corporation and put in a bid for the CBS outlet in Detroit whose owner was known to be having financial difficulties. The Ohio corporation leased WGHP for $200 a month with an option to buy, and took over management of the station on October 1, 1928. Although his name remained on the glass entrance to the station, George Harrison Phelps was no longer active in the broadcasting game. His advertising agency, however, continued to prosper in spite of the loss of the Dodge account.

Several changes were made in the staff of WGHP but the real hero was its new Chief Engineer, Henry Grossman. Grossman had to take on the maddening task of trying to decipher the unorthodox equipment which had been installed by Harold Elliott. In the end, Grossman tore the whole thing down and rebuilt the works according to Western Electric standards. He also obtained a permit from the City of Detroit and in the spring of 1930 built two transmitting towers on top of the Maccabees Building.

Ray Schaub, engineer hired by Grossman, remembers George B. Storer as "a fine young man with the spark of a Wall Street banker."

As Schaub described him, "He meant business. We gave him results under the engineering head of Henry Grossman. Grossman later was hired by CBS as engineering chief of the New York division."

The most obvious change to WGHP listeners under the Storer regime was in the programming and new advertising policy. Paul H. Bruskie wrote about it in the slick little weekly, *Detroit Saturday Night:*

"WGHP now sells time aggressively and in almost any amount. The little fifty-word laudations that the announcer edges in between canned cacophony cost the business houses mentioned between five and six dollars in lots of twenty or more."

This was a local reflection on what was beginning to occur all over the country. The good-will-service-to-receiving-sets approach was giving way to the hard sell. The radio sales pitch had arrived.

The "canned cacophony" was another new development. There was a national controversy at this time (1929) over the use of phonograph records in broadcasting. Recording companies were threatening to sue radio stations that played their product. No less unhappy were the recording artists themselves who were thus forced to perform by proxy without pay. The Storer regime, however, went along with the records.

On Wednesday, October 23, 1929, the first symptom of what was to settle into the nation's economic arteries like chronic blood poisoning burst like a boil on the stock market. This day of minor panic was followed by "Black Thursday" when spectators gathered in the Wall Street sector with true Manhattan optimism to watch victims of "the crash" jump from high windows. Five days later the avalanche of selling orders hit a new high and the hopes of speculators sank to new depths of despair.

The people of America soon realized the poison was not confined to the boil. The entire national economy was sick for many reasons, each of which had its own complications understood only by experts who survived to apply hindsight. Whatever the causes, whoever was to blame, the United States of America came down with that ten-year sickness referred to as The Great Depression.

As the nation collapsed, two shrewd and immensely successful Detroit theater men with coast-to-coast reputations found themselves wallowing in several million dollars in cash. They had made a most extraordinary deal and were rumored to be shopping around for a radio station.

4
This Was Show Biz

John Kunsky's family had come from Bohemia. He loved life and enjoyed people. He was short, had a large nose, and his face worked its way into permanent lines of happiness. He was shrewd but sentimental.

George Washington Trendle's father and mother both came from Germany. He was tall, had a pink complexion, keen blue eyes and always looked as though he had just had a shower and put on a clean shirt.

Kunsky and Trendle met when very young at the Caille Perfection Motor

Company owned by two brothers, Adolph who was "kind of dumb" and Arthur, the one with the brains. They manufactured slot machines in a little factory near the Burroughs Adding Machine plant in Detroit.

When John Kunsky first worked for them he earned ten dollars a week wrapping asbestos around pipes. Then he became a "machine fixer." Trendle got a part time job at the plant doing some bookkeeping, then he was made purchasing agent.

Kunsky was enterprising. He bought a Caille Brothers slot machine on the side and made money with it. He bought a second machine, and pretty soon had quite a number of them—all making money.

In 1905 he made a trip to New York City to investigate a new idea in entertainment. In a little store on 14th Street he saw his first movie.

Kunsky returned to Detroit and with Arthur Caille purchased a projector, rented a store and became the first man in Michigan to show motion pictures commercially. He called his first nickelodeon The Casino. Soon he had a second called The Star, and hired young Trendle to work nights on his books. Trendle was studying law.

Movies soon graduated from stores into theaters. Caille and Kunsky built the third movie theater ever erected. The first was in Pittsburgh; the second in Cincinnati; then theirs in Detroit at 34 Monroe Street. It was called The Royal.

They also built the Hippodrome on lower Woodward Avenue. It was a three-story complex with movies on the first floor, freaks on the second, and a menagerie on the third. It was the only failure in Kunsky history, and it had been Caille's idea.

George Trendle graduated from the Detroit College of Law in 1908 at age 24. He was a handsome young man with full lips, a firm chin, and brown wavy hair which he parted near the middle. He wore rimless eyeglasses that fastened to his nose and had a string that went over his right ear.

He practiced law for twelve years. Finally, Trendle abandoned all his clients to devote full time to Caille and Kunsky and when that partnership broke up he "handled their divorce papers."

Kunsky now invited Trendle to go on the payroll of the Kunsky Theatrical Enterprises and help full time with the theaters. They helped organize the First National Exhibitors Circuit, one of whose purposes was to press producers into making movies of better quality. There were 40 exhibitors in this group, each from a different section of the country. They contracted among themselves not to encroach upon one another's territory.

This same group purchased property near Hollywood, built a studio, and created a production company known as First National Pictures, Inc. Eventually, this property was taken over by the Warner Brothers.

In 1917, Trendle observed that the neighborhood where Kunsky had his theaters was deteriorating.

"Mr. Kunsky," he said, "I think we should move our business a little to the north. I believe that if we built a nice theater somewhere near Grand Circus Park we'd made money with it."

"Go ahead," said John Kunsky.

Trendle managed to get a 99-year lease on a choice piece of property opposite the southern side of the park. On it, they erected a building which contained offices, stores and a handsome theater which they named the Madison. It opened on the 1st of February in 1918. In its first year the Madison turned a profit of $400,000.

Next they built a theater on the north side of the park and called it the Adams, followed by a third around the corner on Broadway which they named the Broadway-Capitol.

George W. Trendle developed into quite a showman. He convinced the aging Edna Wallace Hopper to have a movie made of her face being lifted. The film was first shown in the Kunsky theaters and did such a sensational business it was booked all over the country. As a result, Edna Wallace Hopper—never much of a talent—was offered half interest in a cosmetics company and made a fortune.

Kunsky and Trendle introduced "talkies" in Detroit even before movies were made with sound. They hired actors to wear rubber-soled sneakers and move around quietly behind the screen of their Lafayette Theater and invent dialogue to match the action in the movie.

To Trendle "it was a mish-mash." But it made money.

Later, the Kunsky theaters actually did introduce talkies in Detroit. New York exhibitors were afraid to book "The Jazz Singer" even though it starred Al Jolson. They felt the Warner Brothers who produced it were demanding too high a price for their "experiment." In Detroit, seeing ahead again, George Trendle dickered. He persuaded the Warners to accept a percentage of the take above a reasonable guarantee. So, in 1927, Kunsky and Trendle in Detroit broke the dam and sent the talkies racing to the sea of success.

Kunsky Theaters kept up with the growth of the city. Whenever a new population center developed a new Kunsky theater would be built there. The fame of the profitable Kunsky operation spread throughout the country. A Chicago theatrical concern known as Balaban and Katz built a huge downtown movie palace in Detroit and named it the Michigan. This invasion was contrary to the agreement among the First National Exhibitors Circuit.

To Trendle it was a fighting circumstance. About the consortium of forty he said: "They were afraid of sticking their noses into something. They said why don't you arbitrate the matter? Well, I said, our agreement is spelled out right here in this contract. What's there to arbitrate?

"Well, they didn't know."

"I said, 'How about Sid Kent? He's the President of Paramount Pictures. He's your good friend. And I think he's honest; I'm perfectly satisfied to arbitrate this whole thing before Sid Kent'."

So they arbitrated. The result was that John Kunsky got the Michigan Theater and Balaban and Katz was given a 25% interest in the entire Kunsky chain of theaters.

Acquisition of the Michigan Theater opened up an opportunity for an old friend of Kunsky and Trendle from the bygone days at Caille Brothers. Howard O. Pierce grew up with George Trendle and was hired at Caille Brothers to replace him when he was promoted into the front office. Pierce was now working in the Kunsky

George W. Trendle outside of the theatres he managed for Kunsky.

Howard Pierce

Theaters, and he wanted to try his hand at producing some of the stage shows. To break him in, Kunsky let him book the various acts that appeared in his theaters.

Once Pierce bumped into a critic on *The Detroit News,* Al Weeks.

"Hey," said Weeks. "There's a 9-piece orchestra playing a dance date out in Ann Arbor. It's good."

"Yeah?" said Pierce. "Who is it?"

"Kid by the name of Fred Waring. Waring's Pennsylvanians."

"Never heard of 'em," said Howard Pierce. He went to Ann Arbor to watch and listen.

Remembering the occasion Howard Pierce remarked: "They were pretty good. So I said to Fred, 'What numbers could you play on a stage?' Well, he played the 'Song of India.' 'Well—what other numbers? Don't you have some others?' 'Yes,' he said, 'we've got a couple others.' So I booked 'em into the Madison.

"The opening day they had a matinee. There were these boys I had set up on the stage and they were grinning at the audience and they looked—well they looked like toy soldiers—big blobs of red on their cheeks. I had to show 'em how to make up but they were young and they were good and that was their first engagement on a stage. Then, of course, they got 'em on the Publix Circuit."

The most important theaters on the Publix Circuit were the Michigan in Detroit, the Paramount in New York, and one owned by Balaban and Katz in Chicago. Howard Pierce was given his chance to stage units for this circuit. A unit consisted of several acts put together in a miniature musical, elaborately staged, beautifully lighted, and dressed in special new costumes.

Trendle appreciated them. "Balaban and Katz produced one unit a month in Chicago and one was staged in New York by the Paramount. *We* produced *two* a month. They all played over the whole circuit. Howard produced ours by himself. They were the best stage shows of any I ever saw."

John Kunsky who enjoyed having fun was leaving more and more work to his General Manager. Trendle established policy. Above all, he insisted that all entertainment in the Kunsky theaters be clean family entertainment. Kunsky theaters prospered mightily.

Then Sam Katz took another chance in Detroit. As Trendle commented on it: "So Sam Katz made a deal to put on in the United Artists Theater in Detroit the same kind of stunt he'd pulled at the Michigan. I called him up and reminded him of what had happened before. Katz was then trying to be made president of Paramount and he didn't want Zukor to know that he was antagonizing one of his biggest customers—us."

Adolph Zukor was unquestionably one of the most important executive leaders of the motion picture industry, let alone top man of Paramount Pictures and it was true, the 16 Kunsky Theaters in Detroit were among Paramount's most important customers.

Sam Katz decided to go easy. The United Artists Theater was to be built just one block away from the Michigan. Katz called Kunsky. Trendle happened to be in Kunsky's office and took the call.

"Wouldn't you and Kunsky like to have the United Artists in your circuit?" Katz asked.

"No," said George Trendle. Never again would he trust anybody; he was fed up. Then he had an idea. He asked, "How would you like to buy the whole Kunsky circuit?"

"Well," said Katz in Chicago, "that's a thought."

"Think about it," said Trendle. He put down the phone. A wry grin twisted his lips and he rubbed his hands together, looking at Kunsky. Kunsky's eyes were wide and his mouth hung open.

"Sell our theaters?" said Kunsky with disbelief, his voice soft with sentiment.

"Now's the time," said Trendle cold as ice and happy as a six-foot penguin. "Our—your—theaters are making over $900,000 a year. They'll have to pay top price and Sam's in a spot where he doesn't want to make enemies."

The two men waited.

Great changes were taking place in the world of entertainment. Flesh-and-blood stock companies all over the country were dying as a result of the movies; but this new thing called radio was *free*. The movies which had just discovered sound and were only now learning to talk, viewed radio as a towering monster in horror make-up.

Furthermore, the changeover from silents to talkies was expensive business for the average theater owner. The cost of sound equipment had already forced 240 theaters in small cities to close. By comparison, top men in radio seemed to be sitting pretty. They even predicted that radio would eventually dominate all show business. George W. Trendle was inclined to agree.

Everybody liked radio, from Herbert Hoover who was now President and had already spoken on the air 19 times, to a spiritualist in Chicago who declared that even the dead used radio to communicate with the living.

One day Howard Pierce stood in the entrance of the Madison Theater Building when Trendle stepped off the elevator.

"Howard," he said, "why don't we go into the broadcasting business?"

"Okay," said Howard, laughing. "How do we do that?"

"First," said Trendle, "we have to get a wave-length."

"Well," said Howard, "where do we get a wave-length?"

"I don't know," Trendle admitted. "You can dig one up."

He was serious.

Trendle's telephone conversation with Sam Katz brought results. Trendle went to John Kunsky with an offer.

"George," said Kunsky, "I won't sell any theaters unless you agree to go into some other business with me."

"I'll go into some other business with you," Trendle agreed. "But only if it's on a 50-50 basis."

"I'll agree to that," Kunsky said after a moment.

"We ought to take Howard Pierce in with us."

"For how much?"

"I'll take ten percent less if you'll take ten per cent less," said Trendle. "Forty per cent for you; forty per cent for me; and twenty per cent for Howard."

"Fair enough," said Kunsky. "Do you have a business in mind?"

"Let's get our money out of the theaters first."

So negotiations started.

Meanwhile, Howard Pierce was looking for a wave-length. He belonged to The Players and cornered George Harrison Phelps.

"George," said Pierce. "I hear you're gettin' out of the radio business."

"Takes too much of my time. And money," Phelps admitted. "It's just a hobby."

"Maybe we might be able to make a deal," Pierce suggested. "George Trendle has kind o' asked me to look around for a wave-length, and you've got one you don't want."

Phelps laughed. "Sorry, Howard. I just sold it to George Storer."

"Sold it? When did you do that?"

"Yesterday."

Newspapers reported that the Kunsky theaters were sold for $6,000,000, 72% of which was in Paramount stock which was then selling at $75 a share. When the stock market fell like Humpty Dumpty a short time later, Paramount plunged with it to $13 a share. But, according to the newspaper reports, Kunsky and Trendle were left sitting high and secure, and practically alone on the top of the wall because their deal with Paramount-Publix called for a guaranteed price on the stock—that price, $75 a share.

George Trendle's story was different, and probably right. He said that for everything except the Madison Theater, which they kept, they took $4,000,000—half of it in cold cash, the other half in notes to be paid off at the rate of $20,000 a year.

Either way, as the Great Depression spread its poison throughout the body of the nation, John Kunsky and George Trendle were scouting around Detroit for a radio station—cash in hand.

On April 25, 1930, eighteen months after George B. Storer purchased WGHP from George Harrison Phelps for $40,000, he sold it to Kunsky, Trendle and Pierce for $250,000.

5

The Last Word in Radio

July 1, 1930. President Hoover was predicting a good year. Ex-President Calvin Coolidge urged Americans to have more faith in themselves. The fiscal year had ended the day before. Washington was broke and Congress blamed. An Associated Press tabulation showed that 4,818,371 people now lived in Michigan. The state was $4,000,000 in debt. Detroit was reckoned as the fourth ranking city in the nation with more than a million-and-a-half inhabitants. The city was emerging from a hot spell. In suburban Birmingham, prohibition was being fought by a group of ladies; and a tunnel was being built under the Detroit River to connect the United States with Canada.

The real news was lost on the lower half of page 13 in the *Free Press*, confined to Pat Dennis' column in the *Times*, and did not appear at all in *The Detroit News. The Free Press* topped its little 26-line, one-column story with the headlines:

STATION WXYZ OPENS
TUESDAY WITH PROGRAM.

The program would commence at 8:30 P.M., said *The Free Press*, with the playing of a new march composed for the occasion by Carl Rupp and Julian Webster. Rupp was conductor of the WXYZ Studio Orchestra. Representative Robert H. Clancy was scheduled to make the "inaugural address."

Pat Dennis, Radio Editor of *The Detroit Times*, called this first official program of WXYZ, sound unheard, "a good bet."

The Detroit News ignored the event completely. *The News* listed the programs of WWJ, its own station, twice on one page—first, haughtily isolated at the top, and again, as a magnanimous gesture of integration, mixed in with the program listing of other stations lower down. WWJ was not calling any more attention than necessary to the competition. In fact, in the 8:30 P.M. spot, it did not even list the inaugural program of WXYZ but listed instead something called "Savino Tone Pictures." It did, however, generously suggest that at 9:30 (after the special initial program was over) listeners tune to WXYZ to hear an abbreviated form of Saint-Saens' opera "Samson and Delilah" which would reach Detroit through WXYZ from the Columbia network.

The Station WXYZ referred to was none other than WGHP.

From the moment Kunsky, Trendle and Pierce had come into possession of Station WGHP in April, they had embarked upon a carefully plotted advertising campaign to promote the station, based on their long and successful experience in the theaters.

To begin with, George W. Trendle foresaw the slogan: "The Last Word in Radio," and in keeping with this idea he wanted to change his call letters to the last four letters in the alphabet. The "W" was a sure thing. Getting the XYZ was not so easy.

Trendle "found that these call letters were held by the Army and the Navy and we couldn't have 'em; so I went to the Navy in Washington and had a long talk with the man in charge.

"He said, 'Well, Trendle, if you can get the Army to do this job for you, we'll okay it. But,' he says, 'you can never get them to do it.'

"So then I went over to the Army and talked with the fellow over there and he said to me, 'Well, if you can get the Navy to do this thing, we'll be willing. But,' he said, 'I'm sure they won't.'

" 'Well,' I said, 'will you put it in writing that if the Navy is agreeable, you're agreeable?'

"So the guy wrote me a letter. I took it over to the Navy, and I said, 'Here you are, my friend.'

"He looked rather dumbfounded. But he okayed it."

Trendle then obtained a similar letter from the Navy and submitted both with his application for a license to the Federal Radio Commission.

Thus it was that on July 1, 1930, an advertisement, three columns wide and about ninety lines deep appeared in all three Detroit newspapers:

TODAY! DETROIT'S GREAT NEW RADIO STATION! WXYZ! THE LAST WORD IN RADIO! New personalities in the radio entertainment field! John H. Kunsky and George W. Trendle, names known to all as the foremost in the Detroit theater and entertainment field and whose tireless efforts have made Kunsky Detroit Theaters predominant among the great theaters of the country, with Howard O. Pierce, present Detroit's newest radio station, WXYZ, bringing you the greatest programs on the air!
Through WXYZ, Detroit hears and enjoys such personalities as Chic Sale, Julia Sanderson, Frank Crummit, *The Major and the Senator*, Carl Rupp and his Michiganders, Norman Brokenshire, Ted Husing, Evangeline Adams, the Jesse Crawfords, Will Osborne, *Mr. and Mrs.*, David Rose, *Arabesque*, *The Crocket Mountaineers*, *True Story Hour* and others too numerous to mention.
Station WXYZ, fittingly named "The Last Word in Radio" will bring you the best in entertainment.
Always dial Station WXYZ—and enjoy the most from your radio.

And there was more. It was an ad worthy of a Cecil B. DeMille epic from Paramount.

Among the programs listed for WXYZ on that official opening day, July 1, 1930, were Ida Bailey Allen, O'Cedar Mop Time, the Army Band, Ann Leaf (organist), *The Couple Next Door,* Kaltenborn Edits the News, Anheuser Busch, and, of course, the inevitable Studio Orchestra.

Two days later, a newspaper reported that Carl Rupp and his Orchestra (WXYZ) was "the only network program originating in Detroit." Rupp was heard through CBS in New York from Detroit at dinner time on Mondays and Fridays. So popular was he already that his time would be extended the following week from half to three-quarters of an hour. Thus, for the record, Carl Rupp was the first WXYZ-created program to go national.

There would be others. Once a rich man's plaything, the station was now in the hands of dynamic showmen. George Washington Trendle, one-time newsboy, was about to surround himself with some dynamic executives and talented writers, musicians and performers and emerge as a national force in the industry.

But not without a struggle.

The new company was known as the Kunsky-Trendle Broadcasting Corporation. Mr. Kunsky's name came first but he was not the President. He was the Vice-President and Howard Pierce was named Secretary and Director of Broadcasting. The President of the Company was Trendle. The law school graduate with the full lips and the wavy brown hair was now 46 years old, balding, but clear-eyed and always fresh as a newly plucked carnation. He was tall and carried his weight well. And he didn't have to throw it around; everybody knew he was the boss; there was never any doubt of that. He was quiet and soft spoken, but he inspired the proper fear. He was well organized, considered first things first, but always had something cooking on the back burners and in the oven. With the love for work that dominated his life, he settled down to the creation of a distinctive radio station.

According to Howard Pierce, "George was the brains. George was the aggressive one. Let's not say he dominated. He just put the ideas into John Kunsky's head."

Trendle backed up Pierce's statements: "I just put an X where I wanted him to sign his name and that he did and went about his business. He never interfered."

From the beginning, Station WXYZ (as is true of most broadcasting stations) was divided into two camps. One, made up of talent and creative personnel, worked in the studios on the two top floors of the Maccabees Building, always referred to by Mr. Trendle as "out at the station" because they were three or four miles north of his headquarters. The other was the executive and sales branch of the organization housed downtown in the Madison Theater Building.

Although Howard Pierce worked as Program Director, his office was in the Madison with Kunsky and Trendle. Radio was a whole new thing and Howard was obviously not at home in it, but he had been great in the theaters and Trendle thought he would work out.

Also in the Madison were Barney Kilbride from the theaters now in charge of personnel and the books, and Fred Roach writing publicity and the ads. And there was a lean, active salesman named Garde Wallace who was outstanding. Owen Uridge was Sales Manager. He had come with the station. Trendle was not too sure about Uridge.

Trendle was leaning back at his desk mulling over his staff when a sharp little female face peered at him around the frame of his office door and lit up dimly with a vague little smile. Five little fingers appeared in a kind of wave. Then the apparition disappeared.

This mysterious little woman who also occupied an office in the Madison Theater Building was certainly not part of the staff yet seemed in some obscure way to have some connection. Trendle knew the connection but to most of his employees she was an enigma. A tiny little blonde, she went about her eerie business wearing gloves and out-of-date black satin dresses. Her movements were slow and she seemed to float rather than walk. She was Maude Weston. Madame Weston.

According to Pierce: "She was the best astrologer I ever heard!" Did Trendle confer with her?

According to Pierce: "Yes. He did."

He had faith in her?

"Yes. Yes. He did."

Did Mr. Trendle remember Madame Weston?

George W. Trendle.
He said it was his baby—his, and only his alone!

Howard Pierce

"Yes."

Did he consult her occasionally?

"Well, she had her office in the Madison Theater Building. And Howard Pierce swore by her. I talked to her a couple of times. But I never had a hell of a lot of faith."

Did she ever influence any of his decisions?

"No. No."

When Trendle turned his mind to his employees "out at the station," invariably the first one he thought of was Alice Gorham. Alice had come to WXYZ from Fred Grennell's advertising agency. She was writing newscasts and publicity. He had noticed immediately that she had a very nice figure. She had vitality; she was "smart as a whip." Her hazel eyes were wide and innocent and bright as buttons and her smile was ready and open. She had gusto and she was happy. Occasionally the bright eyes and the happy smile became a mask hiding a firm obstinacy when she had made up her mind about something and somebody tried to oppose her. But she was feminine. Trendle liked Alice Gorham.

He had a new Chief Engineer at the studio named Irving Bell. Reception of the station's signal was still not good in some parts of the city and Trendle wanted it improved.

According to Ray Schaub, Bell improved it. "Bell discovered that the Maccabees Building was a reinforced structure from the fifth floor up. This made for poor radiation. He built what is known as a counterpoise, consisting of a network of wires strung around the top of the building between the towers. Things like this are not being done these days but Bell's counterpoise really helped our signal."

Jack Bundy, the Studio Manager, and Fred Jenks, carrying the title Program Director, both did announcing. Other announcers were Ed Smith, Harry Tanner, Howard Butler and Del Sharbutt. The station's first sportscaster was George Sutherland; he was especially good on the hockey games. Carl Rupp was musical director and Ruth Ward was staff pianist. Another pianist, Verne Willard, filled in. Marguerite Werner supplied organ programs from one of the theaters. She was a big, handsome woman with deep dimples, reddish hair and a soft voice.

The real prize out at the station, as far as Trendle was concerned in the beginning, was a young writer named James Jewell. Show business was in his blood, bones and brain. He had written dramatic material in New York for John Golden, the theatrical producer, for such stars as Alla Nazimova and Olga Petrova, and for Paramount Pictures out on Long Island when Emil Jannings and Lya de Putti were making movies there. And he had done a bit of acting. He had run his own marionette show with a summer off-shoot of the Jessie Bonstelle stock company, made his own marionettes. He had left Bonstelle and come to WXYZ through his brother Ralph who played cornet in the studio orchestra.

Trendle didn't care much for Jewell's announcing but he and Howard Pierce liked his theater background and lost no time naming him Dramatic Director of WXYZ. Jewell had thick, straight dark hair, green eyes, a little mustache and had a habit of wrinkling his forehead in such a way that horizontal lines formed above the outside ends of his eyebrows giving him a rather ferocious look. Like Owen Uridge, he had come with the station.

George Sutherland, the sportscaster, went downtown to talk with Trendle one day in the fall of 1930 about a man he thought would be worth hiring and who might come to Detroit. His name was Harold True.

"Have him come to see me," said Trendle.

Harold True had started in radio at WTAM in Cleveland in 1922. He had turned down offers from George Storer, both in Toledo and in Detroit, but he came to see George Trendle.

He was profane and Trendle did not like profanity. But he knew right away that Harold True was his kind of man. He was lean and he was hard and he was aggressive and fearless. His eyes were black and snappy; his hair was red, straight and inclined to be unruly. He was neat, clean-cut, and had a ready, uncomplicated laugh. But the main thing was his voice. It expressed the man. It was vibrant and clear and forceful. Trendle hired him at once and put him on the news.

Trendle might have put him to singing, but True never told Trendle that he could sing. "I never told anybody I could sing because they were only paying singers five bucks a program. This was also true of the actors."

The country was sinking deeper and deeper into the Great Depression. There had been demonstrations by the unemployed in Seattle, Los Angeles and Chicago. Bread lines on the Bowery in New York were attracting 2,000 hungry people every day. In early November 6,000 unemployed men took up positions in New York to sell surplus apples. The idea spread to other cities.

Luxury items, except radio, were especially hard hit—among them, the theater. Actors found the going tough. New York actors considered the theater their important occupation. They did radio acting, as time allowed, on the side. More and more time allowed. Only the biggest hits managed to eke out a road tour. More than half the New York theaters closed down. Actors had to take what they could get. This was as true in Detroit as it was in New York. That's why James Jewell could line up players for "five bucks a program."

Jewell's first experiences were interesting. "The first radio drama I did was a mystery show that I pulled out of the trunk. I had written one-act plays. I had probably thirteen, fourteen, fifteen in my trunk. And I figured this would last. But of course it was used up in two weeks. The third week, Howard Pierce came to me and said. 'Well—now what?' "

Jewell started writing a detective series. His brother's cornet teacher had been a police inspector. Another inspector lived across the street. They had been Jim's heroes as he grew up. He now got ideas from them. And he borrowed from the classics.

According to Jewell, "There was a series on the air called the *Columbia School of the Air*. After one week I put on Poe's *The Gold Bug*. Well, immediately after we had done our show, the Columbia School of the Air came on from CBS with—*The Gold*

James Jewell

Harold True (from collection of the late Howard Pierce)

Bug. The same thing! And if that wasn't enough, the following week we put on another Poe story feeling sure they would not put on another Poe, and by God if we didn't do the same thing again! With *The Pit and the Pendulum*."

It was during this period that Jewell claims to have made radio history.

According to him, "The first RADIO show that was ever adapted to THEATER—of course, theater shows were adapted to radio—was one we did in the first series of programs I did for WXYZ.

"Jessie Bonstelle played the lead character in the stage production. Her company did it for a curtain raiser for *Bird in the Hand*, a John Drinkwater play; and I'd be happy to show you the critique by Len Shaw (*Detroit Free Press*). He was a great critic. And I never swelled up so big nor did I ever feel so important as when I read him where he said it was a shame that the heavy and ponderous play by Drinkwater had to follow this beautiful little thing. Mine!"

Yes, George Trendle was very well pleased with Jim Jewell. What Jewell had to have now was material. They'd find it.

In general, Trendle felt, the programming was not up to the standard attained in their theaters. Even with the Depression and the jobless, good men were hard to find. He needed new announcers; and the downtown staff could stand strengthening. At the moment the man he was least pleased with was the Sales Manager.

6

It Was The Place. It Was The Time. He Was The Man.

Owen Uridge knew at once that this man was no victim of the Depression. His suit was of blue flannel and it fit, might even have been tailored. His shoes looked comparatively new and they were polished. His shirt was white and his tie harmonized. The top coat he tossed over a chair when he came in looked soft, warm and expensive. From the moment he entered the office, H. Allen Campbell dominated the interview.

"I'd like to make you a sportin' proposition if you—. You do need a salesman?"

"We can use a salesman," Uridge admitted. "We pay 15% commission here—no drawing account."

"That's what I expected," said Campbell easily. "It just so happens I can't work on that kind of deal. I believe I can prove that I'm able to sell. I'd like to make you a sportin' proposition. Never mind the 15%. I'd like to go to work here for four weeks. But I want $200 a week. For four weeks. Now if I am not able to make a thousand dollars a week in four weeks, you don't have to pay me the two hundred. In four weeks I'll make up my mind if I want to stay in radio. And you naturally are goin' to have the same privilege of sayin', 'You're no damn good and we don't want you'."

Uridge sat back in his chair. There wasn't a trace of eagerness or anxiety on the face of H. Allen Campbell. Uridge frowned a little as though he couldn't quite make it out.

"So," said Campbell as though to make it easier. "You're goin' to gamble $800.

H. Allen Campbell

I'm makin' you a sportin' proposition."

Uridge suddenly looked over Campbell's head. His eyes widened and he stood up. There, behind Campbell, stood George Trendle.

"Mr. Trendle," said Uridge. "This is — er —."

"Allen Campbell, Mr. Trendle," said Campbell. He stood up smiling and shook hands with Trendle. Trendle moved across the office to drop some papers on Uridge's desk. His back still turned, he said, "I just happened to overhear something about a sporting proposition."

He dropped the papers and turned to face Campbell.

"What's the proposition?"

"Ask Mr. Uridge," said Campbell, picking up his top coat, his strong teeth bright in the grin. "I've finished and I'm leavin'.'

Almost any man who has ever worked has sooner or later found himself in a political squeeze. In the politics of government the average citizen—usually, and unwisely, of course—feels remote—like a man in the back row of a stadium watching a football game. The players on the field may be trying to kill one another but the distant spectator watches it all impersonally.

When a situation of conflict develops between two forces in a business, on the other hand, the spectator may suddenly find himself, as in a bad dream, down there on the field, both teams on the offensive—and himself the football.

H. Allen Campbell was not one to be turned into a football and he was escaping from a bad dream. When he first went to work for *The Detroit Times*, William E. Anderman under whom he worked was the Advertising Director. Then Anderman was promoted to publisher and a man named Darmstetter replaced him as Campbell's immediate boss and the situation developed. For some reason, Darmstetter thought young Campbell was the publisher's pet and he began to ride him. Campbell was in display advertising, manager of real estate sales. The Depression had made business tough enough without working in the shadow of a man's hatred.

The Detroit Times was a Hearst newspaper and in 1930 a man in the Hearst organization could move from one paper to another if there was an available opening and he was well recommended. So young Campbell went to Bill Anderman and told him he wanted to move to another paper. Opportunities developed on *The Baltimore Sun* and*The Atlanta Georgian.*

Campbell was uniquely fitted for either city. He was from the South. He had a peculiar clipped southern accent, very different from the conventional drawl. It did not dominate his speech, only showed up occasionally, but it was there contributing subtly to the drive of his personality. This, plus his northern training made him a natural, besides which he had ease, charm, optimism and a dynamic selling ability.

But Mrs. Campbell, who was in poor health, did not want to leave Detroit.

"I'm leavin' *The Detroit Times,*" Campbell insisted. "But if you're concerned about leavin' Detroit, I'll explore another thing."

What he had in mind was radio. He had never known George W. Trendle nor

John Kunsky nor Howard Pierce but like everybody else, he knew their reputation in the theaters. So he had called WXYZ and asked for the Sales Manager. Owen Uridge had been very pleasant on the phone, had told him to drop in.

The night following his interview, Campbell received a telegram asking him to go to the Madison Theater Building the next morning and see George Trendle. The telegram was from Trendle.

When he walked into the reception room the next day, everybody seemed to have been alerted.

"Mr. Campbell," said the grey-haired Adelaide Werner at the switchboard. "You're to see Mr. Trendle."

"Well, maybe so," said Campbell. "I have a wire from him. But I've sort of a military background and my interview yesterday was with Mr. Uridge. So I would like to see Mr. Uridge first—or at least let him know that I am here."

Ad Werner plugged into Uridge's office, announced Campbell, listened, then looked at Campbell and said, "No, Mr. Campbell, he says you are to see Mr. Trendle."

Campbell made a note that George W. Trendle was a man who went over his employees' heads.

The two shook hands and Trendle motioned him to a chair. As Campbell put his brown hat and top coat over another chair, Trendle approved the brown suit Campbell was wearing, the nicely polished tan shoes, the white shirt, the dark red tie. This man had a good face, a strong face with strong teeth, shrewd blue eyes, rather small—and too much wavy brown hair.

"Now," said Trendle after Campbell had sat down smiling, and waiting for Trendle to speak. "Will you repeat this sporting proposition?"

Trendle got up from his desk halfway through Campbell's repetition of the offer he had proposed to Uridge. He walked up and down. He stopped when Campbell had finished and stood looking at him. Suddenly he stuck out his hand and said, "Shake. You got a deal."

The two shook hands again.

"Can you start right away?"

"Yes," said Campbell.

The next morning he talked with Owen Uridge. Overnight he had picked out a half dozen accounts he had sold on the newspaper which he thought he might be able to lure into radio. The trick in 1930 was to woo advertisers away from print into the new medium. Money was very, very tight because of the Depression; it was not going to be easy, he knew. Besides, this was a new field.

"What are you endeavoring to really sell in radio?" he asked Uridge. "I mean—are you tryin' to sell 15-minute programs? Are you tryin' to sell 'em on a weekly basis? Once a week? Three times a week? Six? Or are you tryin' to sell spot announcements? I mean, what is your preference? What are you tryin' to build up on this station?"

"Well," Uridge told him, "I suppose—maybe—uh—the 15-minute program—

uh—half-hour programs would be more impressive to Mr. Trendle than anything else."

"What half-hour programs do you have to sell?"

Uridge couldn't think of any, except, maybe, The Early Birds in the morning.

"You know," Campbell said. "I had hoped you'd say that what this station needs is money, that you want it any way you can get it and get the station on the road."

Uridge looked uneasy. Campbell became even more decisive.

"It seems to me," he said, "that spot announcements is the way that will get results for people I've been accustomed to dealin' with."

"Well," said Uridge defensively, "I'm not going to stop you from selling announcements."

"We're goin' to have to show them results," Campbell went on. "We're goin' to have to have ideas. We're goin' to have to have copy. We're goin' to have to have a L-LOT of things."

One of the first customers Campbell called on was Harry Hyman, President of United Shirt Stores. When Campbell walked in, Hyman was slouched over his desk, looking very depressed.

"Hi, Allen," he said sadly. "What can I do for you?"

Campbell laughed. "Harry, you can't do one goddamn thing for me in the mood you're in. I'm not even goin' to talk to you."

Without a smile, Hyman shrugged, one hand upward to indicate a chair. Campbell stood there smiling down at him broadly.

"I came over here to talk about some advertising that I think would help you," he said. "But until I can get you out from under the desk I don't want to talk with you."

"Listen, Allen, things are tough. I've had to take myself off a $50,000 drawing account as President to satisfy the stock holders. I'm off salary 'til things get better."

"I'm comin' back to see you tomorrow," Campbell promised, laughing. "You haven't got anything better to do. What the hell—business is lousy."

He went in the next day all primed.

"Harry, can you get a lot o' shirts—white shirts? And can you get 'em cheap?"

"I can get all the merchandise in the world," said Hyman. "The trick is to sell it."

"Well, I have an idea for you. It's an old one, but it's timely."

"What's the idea?"

"You sell one shirt for $1.99 and two for $2.00—somethin' o' that kind. You add a second shirt for a penny."

"Where does radio come in?"

"We tell the world. You put on a spot announcement every three hours through the day, six days a week."

Hyman was thinking about it; he wasn't turning it down.

"And Harry," said Campbell. "That's a hell of a lot o' spot announcements. You owe 'em some entertainment. You owe 'em a half-hour show."

"What would be the show?"

"Russ Morgan," said Campbell. "He's new here and he's got one hell of a band. And we'll throw in some singers."

H. Allen Campbell went back to the Madison Theater Building with an account worth around $800 a week—unheard of in those times on local stations. Some say the Russ Morgan show was the first half-hour commercial program on WXYZ. The United Shirt account was supposed to have been a very short one—two or three weeks. It stayed on the air for over a year.

In the meantime, H. Allen Campbell had made his first sale in radio and George W. Trendle had found his man—if he could keep him.

Campbell brought in two or three other new accounts that first week.

"We like what you're doing," said Trendle.

The second week he brought in still more orders for spot announcements.

"You're building up a pretty sizable commission," Trendle observed.

Campbell reminded him he was not on commission.

"You're going to get a commission. Don't worry about that," Trendle promised. "I've gone over the figures and you're over your $200 salary already. How do you like radio?"

"I made a four-week deal with you and I'm not goin' to tell you until four weeks."

The fourth week ended on a Saturday. Trendle called him into his office.

"All right," Campbell," he said. "How about radio?"

"I like it."

Trendle closed the door of his office. He walked back to his desk and sat down. He said, "We're going to have a sales meeting in ten minutes and at this meeting I want to announce that you're our new Sales Manager."

"Mr. Trendle, this comes a little fast," Campbell said after a moment. "I'm not concerned about gettin' together on salary arrangements. But—I have some other things in mind."

Trendle waited—cold, calm, and interested.

"You are a man with a big reputation," Campbell continued. "I'm sure you have a lot o' friends and I'm afraid that you might be a person who would want to suggest friends that you would want me to hire for sales jobs."

Trendle's expression didn't change. Campbell went on.

"You've made it pretty clear that you want a new Sales Manager, but what are you goin' to do with Uridge? I don't want to cut his throat."

"Uridge is a good technical man," said Trendle. "I'm going to send him out to the station and make him Studio Manager. In fact, I'll give him a raise."

Trendle leaned back in his chair and studied Campbell.

"Anything else?" he demanded.

"There's one other thing," said Campbell. "If I'm the Sales Manager and you come in to our Saturday meetings and conduct the meetings, I'm not the guy you want. I'll tell you anything you want to know, but when you come in over my head to talk to the salesmen—. First of all, they're not goin' to tell you a damn thing; they're goin' to clam up tight."

Still no change in Trendle's expression. He waited.

"Anything else?"

"No, sir."

"All right," said Trendle. "I don't know anything about sales. I don't want to hire any salesmen. And I don't want to be in those meetings."

Trendle took Campbell into the meeting that morning and presented him as the new Sales Manager.

Thinking back over it Campbell was amazed: "As a young punk with not a hell of a lot of experience, I don't know where I got the courage to talk to him the way I did."

The year 1931 that began so well with the acquisition of H. Allen Campbell was to end in a threat of catastrophe because of the making of an extremely unwise decision—so unwise that forty years later no one would admit he had made it.

7

No Runs; Four Hits; One Error; And Everybody Left On.

A red light was shining over the door. Studio A was on the air. It was the Russ Morgan show for United Shirts. Suddenly the door crashed open and two men scuffled their way out of the studio. One man broke loose and ran down the hall to the stairway and down the stairway to the men's john, the other man right after him. Inside, their excited voices bouncing from one tile wall to another, the two men faced each other, anger in their eyes.

"What's the matter with you?" demanded Russ Morgan.

"You son-of-a-bitch, you *know* what's the matter," yelled Harold True.

Russ Morgan, in the middle of the program, on the air, had hollered at his drummer, disobeying the prime rule of radio: there must be no sound whatever in a live studio except what is to be heard on the receiving sets. The announcer of the show was Harold True and he had yanked Morgan off the conductor's platform and dragged him out of the studio.

"Don't push me around," yelled Morgan in the john.

"You think you're tough, you son-of-a-bitch, but I'm just as tough as you are and I'm not one goddam bit afraid of you! When you're on the air you don't holler at your drummer!"

Russ Morgan had been discovered by Howard Pierce.

Reminiscing about it later Pierce remembered: "We had a theater called the State and we had stage shows there. Morgan had a 9-piece stage band. So I came in one morning to supervise, and there was a young feller down in the pit rehearsing a special number and he had his coat off. He was in his vest. So I listened to him for a while and he did a darn good job.

"So the time went on and we had the station and I hired him. And I fired him. And I hired him and I fired him."

Another contemporary recalls, "Morgan was tough. Great musician, though."

WXYZ made four major acquisitions in 1931. Irving Bell was Chief Engineer. His assistant was Gordon Heatley. A friend of the Heatley family was a young man from Canada, short with a round face, named John Tierney. Tierney had been first man "on the line" centering pistons at Essex automobile.

He wanted to get into radio; so he quit Essex not long before it went out of business, and attended a radio school, expecting to go to sea. But there was no ship work, so he went to see Gordon Heatley and was hired at WXYZ, February 24, 1931. He remained with the station for over 41 years—even tempered, ever smiling and helpful.

To H. Allen Campbell goes credit for a major acquisition. He was in Indianapolis trying to land an account. Wherever he went, he listened to radio. He sat in his Indianapolis hotel room listening to a program of poetry. He thought the man had a great voice and listened to hear if his name would be given. It was—at the end of the program. Campbell called the station and asked if he might talk with Brace Beemer. The deep, vibrant voice he had just heard on the radio came on the phone.

"Mistuh Beemer," said Campbell. "My name is Allen Campbell and I'm Sales Manager of Station WXYZ in Detroit. I heard your program a few minutes ago and I'd like to talk to you. Can you come down to my hotel? I could have something to offer you."

Beemer came to the hotel and met Campbell. He had the looks to go with the voice. He was six-feet-four or five, solidly built, had a round face, healthy complexion and a thoroughly masculine manner.

Campbell hired him immediately to come to Detroit on a four-week basis, which was standard procedure in those early days.

When Beemer arrived in Detroit, Campbell arranged for him to audition for Trendle and Pierce. In the middle of the cluster of offices in the Madison Theater Building was a kind of board room, carpeted in green, with mahogany chairs and a huge mahogany table. There was no window in the room but there was a fireplace and the fireplace was wired to the Maccabees Building. The control room could connect that fireplace with any studio.

Campbell sent Beemer out to Harold True at the studios and called Trendle and Pierce to the board room. Campbell didn't pretend to be a talent expert, but this was different; he had hired Beemer and they had not heard him.

When Trendle and Pierce were ready, Campbell called True at the station and gave the go ahead. Beemer did a half hour of news, poetry, a little acting, sight reading, commercial announcements—the works.

It was the first time in two days that Trendle and Pierce had seen each other, and a few moments after Beemer's voice began to come from the fireplace, the partners began to talk. They talked all the time Beemer was auditioning. When Beemer ended his audition it was like an interruption to their conversation.

"Well, Howard," said Trendle. "What do you think?"

"Well," said Pierce. "I didn't think he was so hot."

Trendle agreed. "I'm not too impressed either."

Campbell's face was red. Each minute of the half hour had found him angrier than the minute before. He did not raise his voice. He spoke in a normal tone. But his speech became more and more clipped and decisive. A little of the south crept into it.

"Gentlemen," he said. "I hope that you will not feel that I am too much out of orduh, but I'm havin' a difficult time controlling my tempuh. I hired this man. I believe he has one of the fahnest voices I've ever heard but for one half hour you men have sat there yakkin'—to the extent that *I* couldn't even hear the audition. Ah didn't need to because Ah've already heard him. *But*—this is no way to judge a man."

Neither Trendle nor Pierce spoke. It was their turn to have red faces. Campbell got up and returned to his office.

Brace Beemer stayed at the station. Permanently. Almost.

Another of the year's four major human acquisitions was to the credit of Barney Kilbride. Barney was a nice little man with a thin face and spectacles.

One day a 19-year-old boy named Jimmy Riddell showed up to apply for a job. He had pink cheeks, clear blue eyes and a wide, toothy smile. He was bright, happy-looking and Barney Kilbride hired him as mail boy.

The other two human assets acquired in 1931 came to the station through James Jewell.

One was Ted Robertson, a mellow, middle-aged man, when I taped him, whose memories of the station seemed to be wrapped in cotton. In 1931 he was a juvenile actor with straight blonde hair, blue eyes and a lot of fire. He had worked with Jessie Bonstelle's summer company which toured Michigan, acting at night and helping Jewell with his puppets in the afternoon.

As he remembered it: "I used to stand around and watch and Jewell'd say to me, 'Would you like to help us?' And so I got in and he taught me how to work the puppets and do the voices.

"Well, he left. He disappeared. And the reason was, he got a job at WXYZ."

When Ted Robertson returned to Detroit at the end of the summer tour, he couldn't get a job. The grip of the Depression was getting tighter. At the beginning of the year 6,000,000 people had been unemployed. Millions more were out of work now.

Finally, Ted Robertson saw mention in a newspaper of the Jewell Players at WXYZ and he was reminded of the puppet man. He called WXYZ and Jewell offered him a job helping Bert Djerkiss on sound effects.

"I can't pay you anything," Jewell had to admit. "But if you would like to work—."

Such were the times that Ted would. And did.

Meanwhile, Jewell had been forming his company of actors. He was aiming at a half-hour drama every night. He had been buying scripts from a Don Becker in Cleveland. Harold True had known him at WTAM.

True remembered another man who had been program director of WTAM for a while and who could also write. He was no longer in Cleveland and they started to trace him. True recalled that he had come from Buffalo, New York. And that's where they found Fran Striker.

He was now operating a one-man syndicate, belting out scripts which he sold to radio stations wherever he could. He was collecting royalties that ranged from two to ten dollars depending upon the power and coverage of the station using them. Jewell got him on the phone.

"This is the situation here," Jewell explained. "WXYZ ranks third in Detroit. It's under new management and it is aggressive. I am Dramatic Director here and I'm organizing a radio stock company and I need scripts. I want six cops-and-robbers stories a week. Can you provide them?"

"I sure can," said Fran Striker.

"What would it cost us?"

"How powerful is your station?"

"A thousand watts."

There was a short silence on the Buffalo end of the line. Then: "How's $4 per half-hour script for first performance rights in the Detroit area?"

"I'll call you back," Jewell promised. He knew he would have to talk first with George Trendle.

It was a deal. Fran Striker began sending to WXYZ his *Warner Lester, Manhunter* series, his *Dr. Fang* series and *Thrills of the Secret Service.*

The year ended in catastrophe.

When Campbell had customers primed for radio advertising, he needed programs. For programs he had to go to Howard Pierce. He felt the programs Pierce suggested would not deliver an audience. Trendle remained loyal to his old friend.

"Howard Pierce can produce anything."

As the weeks went by, the conflict grew. Campbell had one advertiser ready for the air and had been waiting for Howard to come up with a good program idea. Nothing happened. He went to Pierce.

"Have you done anything on this particular idea that I talked to you about?"

"No. I haven't."

"When can I get somethin' on it? I can sell it if you can come up with a program."

Pierce became impatient.

"Allen Campbell," he said, "you don't get these things overnight. Takes time. You got to write 'em. You got to get the people."

Behind Campbell's half smile was a hard, unyielding determination.

"Mr. Trendle tells me that you're the greatest producer in the world," he said. "I mean, you built all those beautiful stage shows for the theaters."

"Yes," said Pierce. "I did. But damned if I ever produced nineteen in a day!"

Pierce became more and more defensive. When the local feud drew toward a climax the Columbia Broadcasting System was sucked into the fray.

"You know, Howard," Campbell said one day. "I have a news account that I can sell on this station—52 weeks—non-cancellable—6 days a week. Now that's important money."

"Well," said Pierce who had come to dislike young Campbell. "Why don't you do it?"

"You can't give me the time for it," said Campbell. "CBS is in there."

The next time Trendle talked with Pierce about programs, Pierce said, "It isn't a matter of programs, George. It's a matter of time. Campbell complains that he doesn't have the time to sell. It's the network. The network grabs all the best time."

Being on a network did present problems. In 1931 a local station could establish a program in a given time spot and if the network wanted that time for one of its national programs, it would pre-empt the time; the local station would be obliged to either cancel its program or move it. Either way, it risked losing a sponsor.

And WXYZ was beginning to make money. It was reasonable to expect, especially with Campbell in sales, that the station would make more of it if it could count on time. At the end of 1931, WXYZ and the Columbia Broadcasting System ended an association which had lasted from the days of George Harrison Phelps.

Trendle told us it was Campbell who persuaded him to cancel CBS and called it "the only mistake Campbell ever made." Campbell blames Howard Pierce for loss of the network. Pierce, when asked, without a moment's hesitation, said the decision to drop CBS was George Trendle's. Since Trendle made all major decisions, this would appear to be reasonable.

Campbell disclaimed all involvement: "It was silly as hell. A round-robin. I didn't even know they were gettin' to that point at all. Now then! Here I am, Sales Manager, left with—zero."

At the outset, it looked like catastrophe. But such was the caliber of all concerned both downtown and out at the station that the greatest burst of creativity WXYZ has ever known was ignited.

Out of the dilemma rode radio's greatest hero.

8

Bedlam

The first days without a network were chaos. No programs were transcribed then; everything was live. Phonograph records were still not used on major stations except in the early morning and occasionally late at night. Announcers, singers, musicians, actors and sound men fell over one another in confusion.

A number of combinations were appearing together for the first time—soloists became singers in quartets and trios, group singers became soloists. Actors announced and announcers acted. One never knew five minutes ahead what he would be called upon to do next. The four staff pianists—Ruth Ward, Eve Gladstone, Henry Finney and Harry Wismer (not the sports-caster)—had quarter-hour fills; Finney and Wismer had orchestra programs besides; and to their horror, all four learned they were to do one program every day as a 4-piano team!

The studio lay-out added to the confusion. When WXYZ lost the network, there were only two studios, one control room and the transmitter. Now, with all the additional activity, more space was needed. An art school which had occupied half of the penthouse was ousted, and carpenters, electricians and engineers added to the bedlam as they converted that space into two large additional studios. In one of these, adding to the uproar, a large pipe organ was installed. It came from a theater and with it to the studios came Marguerite Werner. Marguerite and Adelaide Werner of the downtown switchboard were sisters-in-law. Marguerite's husband, Eduard, was musical director of the Michigan Theater and conducted important special programs on the station. Adelaide's husband, Franz, was head of WXYZ's music library. He had China-blue eyes, a large bulbous nose, spoke with a strong Austrian accent, and had lost two or three fingers fighting for the Kaiser in World War One.

The music library was a large square room on the relatively peaceful 14th floor below the studios. There was less traffic but more noise in the news room next door. Small offices had been partitioned off around its perimeter. The noise was created by the newswire machinery. Two men prepared newscasts for the announcers—Rollie Hoover, an easy-going man whose black eyes looked upon the turbulent radio world around him with chronic amusement, and Henry Alexander, a nervous little man with a ready laugh.

Beyond the newsroom and a bank of six elevators, three on each side of the hallway, were the offices at the front of the building of Owen Uridge, James Jewell, Harold True, Alice Gorham, a secretary and a receptionist.

The excitement was really upstairs. When they were finished, the four studios formed a square-shaped U around the control room and transmitter with a hallway extending across the top of the U. To get to the organ studio at the bottom of the U, it was necessary to walk through the studio between it and the front hallway unless one descended to the floor below by the front stairway, walked the length of the 14th floor and climbed back up to the penthouse by the back stairs. Often there was no time for such a detour and since all studios were used for many things, there was much traffic through them even when they were on the air. It was awkward.

One group might be interrupted in the middle of rehearsal by another group piling into the same studio at the last minute before airtime, as silently as possible. An organ program might give way to an orchestra program in the same studio. It was fantastic.

The only unhurried, unexcitable persons were, as always, the engineers—Al and Ray Schaub,John Alan, Larry LaDue, Roy Gardner, Irving Bell, Jack Tierney and the others. Nothing roused them ever from the relaxed state of calm for which radio engineers are noted. Tierney did bounce a bit in an emergency, but he usually met disaster with a happy grin.

One day, a pert and vivacious singer, Leta Wales, was rehearsing with a small orchestra in one studio. Leta had recently returned from a tour of the Orpheum circuit. Vaudeville was dying, if not already dead, and a theater man had introduced her to Howard Pierce who gave her a job. The song she was rehearsing was in the wrong key for her but time was, as usual, of the essence, and she was doing her best without asking the musicians to transpose. A tall young man with wavy light brown hair walked through the studio and remarked, "Miss Wales, that sounds putrid."

"Well, thank you so much!" said the diminutive Leta.

It was Brace Beemer. As he walked on through and out of the studio, Leta recalled a fortune teller who had invaded the railroad car reserved for the vaudeville bill with which she had been travelling in Texas. Leta had always loved tall, dark, brown-eyed, black-haired men. But the fortune teller had said, "You are going to meet a tall, blonde, blue-eyed man who's going to change your life completely."

Leta Wales became the second Mrs. Brace Beemer, retired from singing, and took over three sons from the first wife.

Beemer was now doing a poetry program called *The Wandering Vagabond.* Bromley House, another announcer, sang on it; Pete Van de Van played the accordian; and there was a gypsy violin by Marion Raviski. Anne Campbell, from a newspaper, was doing a morning show; there was an opera singer from Chicago, Lois Johnson; and Russell Neff who sang vigorous tenor, light opera style. Russ also announced. Harold True had brought him on from Cleveland. Another Russ—Russell Wood—directed and made arrangements for the vocal groups including the Frohne Sisters and Spartan Singers. Hank Finney composed arrangements for the orchestras.

Among the announcers now: True, Beemer, Neff, House, Herbert Green, Garnett Garrison (who later taught at the University of Michigan). John Prosser (later manager of WKAT at Miami Beech) and Del Sharbutt.

One day with brisk efficiency, Thelma Watson ushered two young men into the office of George Trendle. They represented the Detroit Federation of Musicians. Buddy Fields and Raymond Meurer were members of the Board. A situation had developed between the union and the Kunsky-Trendle Broadcasting Corporation. It was ticklish. An outgoing president of Local 5 had proposed forming a non-union orchestra at the station, and this traitorous suggestion had come to light at union headquarters.

As Fields and Mueurer settled into chairs, George Trendle leaned back comfortably behind his oversize desk and calmly sized up his visitors. It was the first time he had met the pink-faced Raymond Meurer, but he was aware of him as a young lawyer-musician who played banjo and sang with an orchestra out at the station and recalled that he had even won a little court case for him; but he was not about to dole out any flattery and pleasantries at this particular meeting, and he gave no sign of recognition.

As a matter of fact, Trendle had been only mildly interested in the proposition of the deposed union President; he accepted unions; but there was no point in making it easy. At the moment, he had little to lose and they had much to gain. In these strenuous times, he was confident he could find plenty of hungry musicians who would jump for jobs on his station—union or no union.

"Mr. Trendle," Ray Meurer began, "you have a reputation of being a very smart man and we thought you *were* a smart man."

"You thought I was a smart man?" Trendle stiffened a little.

"Yes," Meurer continued. "But there is someone taking advantage of you who hasn't a thing to lose. But you've got much to lose."

"I have?" said Trendle.

Meurer's eyes were steady and his voice poured out like oil. He had a way of talking that somehow reminded one of a mother cooing to a baby through protruding lips, and he broadened vowel sounds occasionally which seemed peculiar for a native of Michigan.

"You need the union, Mr. Trendle," he continued. "At this toime you are not on a network, but looking ahead to the future you would never be able to get into any type of situation with any network if you did not have union musicians. It would just simply be catastrophic."

"For the musicians," said Trendle, smiling.

"And for the station, Mr. Trendle," Muerer insisted.

"The networks are already unionized. You can imagine the situation if you wanted to negotiate with a network. You might even want to feed programs to a network—as you did with Carl Rupp. Now, if you are a non-union station—."

There was no threat in the friendly, deliberate manner of Raymond Meurer, but the friendliness seemed a little phony.

"I am not against working with the union," Trendle admitted. "But some of your rules are cock-eyed as applied to radio."

"How do you mean?" asked Buddy Fields.

"I have a staff orchestra of 17 men," Trendle explained. "Now if I want to break that down and do some programs with fewer men—seven or eight men—I can't do it. I have to pay for full orchestras."

"Something of that koind might be negotiated, Mr. Trendle," ventured Meurer.

Something of the kind was negotiated. The first contract signed with a union at WXYZ was with the musicians.

As Raymond Meurer described it: "The Board of Directors sort of worked out a situation to make me what we called a contractor. He was responsible for the hiring of the musicians and worked with the leader and the union and the station—in effect, a koind of steward. I began to handle the labor situation for the radio station."

Meurer and his brother Henry moved their law offices into the Madison Theater Building.

"Jim, we're in a peck o' trouble."

Howard Pierce had dropped into Jewell's office.

"The little kid show that comes from Webster Hall."

"What about it?" Jewell asked without interest.

"Well, the man who's been runnin' that show has been picked up by the police on a complaint from one of the mothers. He abused one of those kids. He's been arrested. Naturally, whatever happens, we can't have him back on that show. And Jim, I need that little show. You were with the Detroit Recreation Department with your marionettes. You must know all about kids."

"Me!" exclaimed Jewell. "Howard, I'm up to my ears. I'm doing a half-hour drama every night of the week."

"But this is just a little ad lib thing, Jim."

So Jewell fell heir to a kid show. He changed the name of it to *Kiddie Karnival.* And he was thrilled by the talent of some of the children. After three broadcasts, he went downtown to meet Trendle and Pierce.

"Some of these kids are just too good!" he enthused. "There's a little Betty Hutton and her sister; she's got more pep! and there's a girl with a beautiful voice—Joan Leslie. They have to be *seen.* I want to put on a show with the best of these kids in a theater."

"Are they really that good?" Trendle asked.

"Mr. Trendle, I could put on a good show once a week with these kids."

"Every week?"

"Every Sunday."

"I could probably get you the Regent Theater," said Pierce.

Jewell named the big Sunday show *Children's Theater of the Air.* He wrote sketches for it. The kids acted in them, and sang and danced. Jewell was as fierce a perfectionist with the youngsters as he was with his adult actors. Only the best survived.

The talent out at the station was excited and happy in the chaos of creativity; but two men sat in their offices in the Madison Theater Building facing unpleasant facts. Both soon discovered the mistake of losing CBS.

George W. Trendle was pouring money into additional programs which in the end simply did not stack up against a network. The station had lost prestige and with it—business. And that's what challenged H. Allen Campbell.

"Mr. Trendle," he said, "WJR has us at a disadvantage. They have 10,000 watts; we have 1,000. We can't compete. We have WOOD—WASH in Grand Rapids. Why don't we take in some other stations out through the state?

"Are you talking about a network?" Trendle asked.

"Our own network," said Campbell. "I believe we might make some money with it. Certainly we could compete with WJR. We could do even bettuh."

"Look into it," said Trendle.

H. Allen Campbell took a swing through Michigan. He called on station managers and owners in Battle Creek, Kalamazoo, Jackson, Flint and Bay City. He proposed forming a network. They would pay for the telephone lines connecting them with WXYZ but in exchange they would be getting free programs as network members, *and* they would be paid for their "time" when WXYZ sold a sponsor who wanted his message to reach through the state. This would not only put WXYZ in competition with WJR but give the out-state stations an edge on their competition as well.

Campbell's heavy magnetism and his calm, decisive way of talking won over these out-state men immediately. They were not as desperate as WXYZ at the time, but they were all feeling the Depression.

Times were even worse in '32 than they had been, and they would get still worse.

In January, the American Legion had pushed a drive to find a million jobs under a five-days-a-week plan. Even the reserved J.P. Morgan went on the radio to support it. During the year, Congress passed and President Hoover signed a bill appropriating more than $2 billion for relief. There were hunger marches. People were eating leftover food from restaurants, stale bread from bakeries and spoiled fruit from grocery stores. This was the year when 273,000 people lost their homes through mortgage foreclosures.

Radio suffered less than other enterprises; it cost nothing to listen to radio, and that was excuse enough to buy a set. People were not buying new automobiles, washing machines, electric stoves or waffle irons, but 4,000,000 families managed to purchase radio sets between 1930 and 1932.

People ate at home and listened to the radio. Hotels installed radio sets in their lobbies where guests would gather to listen to programs, even coming in off the streets to hear favorites. In the homes of unemployed fathers and husbands, adults in families stayed up late listening; there was no job to get up for in the morning. Restaurants and movies suffered disastrously. Party-ing declined noticeably; it was harder to buy clothing and old clothes kept people at home. Radio found an eager family audience.

Some people behind the mikes were not having it much better. Ted Robertson, the new sound boy who was not being paid anything at WXYZ, finally ran out of money. He was living with an aunt and uncle. In the middle of one afternoon the phone rang. The aunt answered it. It was James Jewell.

"Where's Ted?" he wanted to know.

"He's right here," said the aunt.

"Why isn't he here?" Jewell demanded. "Why hasn't he come to the studio?"

"Frankly, Mr. Jewell," said the aunt, "he doesn't have carfare, and I feel I've loaned him all I can."

"Could you lend him twelve cents to come down tomorrow?" Jewell asked after a moment. "And we'll discuss the situation."

The next day Ted borrowed carfare.

"Look," said Jewell. "Things are difficult but I think I can muster some money."

What he managed was $6.25 a week. It was better than nothing. Ted worked again.

Jewell was now directing six half-hour dramas a week and in one of them—*Warner Lester, Manhunter*—he was playing the title role. A versatile actress from stock companies, Ruth Dean Rickaby, was his leading lady. She was a short woman around forty with a thick body, black hair and brown eyes. There was a third "running character" (in every story) named Axford. He was played by an old actor, Jim Irwin. They were very popular.

Jewell moved the *Kiddie Karnival* from Webster Hall to the Colonial Department Store in the heart of the city. He appeared in it as a clown in full make-up and costume.

Describing the circumstances, Ted Robertson explained: "It meant every day he had to go down to the Colonial Department Store and somebody had to drive him down there and he did the show and when he finished he had to come back to the studio. So he was Jimbo-Jambo, the Happy Clown, and I became Tadpole Teddy."

Kids in the store were rounded up and put on the air. Sometimes with startling results.

Howard Pierce explained one: "One day, Jim was askin' questions and he had this little kid sittin' on his knee.

"Said, 'What's your name?'

"S'he 'My name is John.'

"S'he, 'That's fine. John *what*?'

" 'John.'

" 'Well I know—but what's your last name?'

"And the innocent, wide-eyed little boy said, 'What's it to ya, you son-of-bitch?' "

George W. Trendle sat in his office with Howard Pierce. WXYZ had now been without a national network for several months. Locally, its programs compared well with any other local programs in the city, but network programs won the listeners.

"I think Jimmy's doin' a great job with *Warner Lester*," said Pierce.

"Cops and Robbers have always been box office, Howard," said Trendle. "And I've been thinking of something else that always did well for us in the theaters. We never did bad business with a good western. In fact, we did good business with bad westerns. I've got a character in mind, Howard. He hasn't taken definite shape yet, but —."

Later, Trendle called in Thelma Watson, tall, pale and forty with cinnamon eyes.

"A memo," said Trendle. He looked at his calendar. "Send copies to Pierce, Jewell, True, Roche and Owen Uridge. Here's the memo: Arrange to be in my office at 11 A.M. Friday to discuss a new program."

9
The Jigsaw Controversy

The hero riding around in George W. Trendle's head eventually developed into *The Lone Ranger*. It began in his mind and it ended with his approval. He deserved much of the credit he took for the creation of radio's greatest hero, but the whole story, the true story, of the evolution of this historic program has never been told.

There have been three main sources of controversy in this puzzling bit of radio history: George W. Trendle at the head of executive forces: the studio group "out at the station," headed by James Jewell; and a lone writer 250 miles away in Buffalo named Fran Striker.

Even now after 40 odd years, emotions flare up in discussing what has become an extremely touchy subject. One of the first with whom we talked was the late Harold True. When we asked him to tell us what he knew about how *The Lone Ranger* came into being, his black eyes caught fire and his voice sharpened with rage. Surprisingly, he banged a table with his fist and exclaimed, "*The Lone Ranger* was *my* idea!"

At the time, Owen Uridge was going to move from WXYZ to WJR and Harold True was the Studio Manager. George Trendle, according to True, "came out to the Maccabees Building" every Thursday night to talk over the programming. One Thursday night late in 1931, he told True they were going to drop CBS and be an independent station.

"Mr. Trendle," said True, "that's going to take an awful lot of money and a lot of people to fill all that time."

"I'm aware of it," Trendle replied. "But Pierce and Campbell seem to think we can make more money if we don't have CBS taking our time away from us."

True was thinking out loud. "Singers—musicians—."

"And I want some new dramatic shows," said Trendle.

According to True, he at this time told Trendle of an idea he had had when working in Cleveland at WTAM of "a sort of modern Robin Hood." He said he called Jewell in for this conversation and he said that at that meeting he suggested Fran Striker. But, said True, Jewell wrote the first *Lone Ranger* scripts. Both Trendle and Jewell have confirmed this.

Unemotionally, James Jewell explained to us that the half-hour dramas he was producing six nights a week in 1932 were independent, individual dramas, each one complete in itself. They were connected only by the overall title: *Hunters of Men.*

One night on this series, Jewell says, he did a western story in which a man "with a shadow hanging over him" had to disguise himself. He came back to the town where he was in trouble, to clean it up. And to disguise himself, he wore a mask covering the *lower* half of his face.

Both the Thursday night meeting with True and the production of the Jewell western occurred some time before Trendle sent the memo setting up the first official morning conference to discuss a new program.

Fran Striker, in Buffalo, knew nothing of WXYZ's plans to create a western hero until he received a letter from James Jewell on the 28th of December, 1932. In this letter, Jewell suggested that Fran write up "three or four wild west thrillers" using as the central figure a lone cowboy. Jewell suggested that he put in his scripts all the "usual hokum" of rustlers, the heroine on the railroad tracks, the fight on top of box cars, Indian bad men, two-gun bank robbers, etc. etc. etc.

Five days later, Jewell received a reply from Striker saying the wild west stuff appealed to him a lot and that some scripts would arrive in Detroit that week.

Fran had become interested in the west when he was a boy. A bachelor uncle—Uncle Joe—had lived with the family. He had spent some time in the west and told great tales about it and Fran had been fascinated by them.

Striker had already written a series of radio dramas called *Covered Wagon Days*. He had been trying to peddle this series to his syndication customers since 1930. Apparently the sample scripts he mailed to WXYZ were revisions of some of these scripts. In a letter to Jewell, he suggested the name of their new program be *Covered Wagon Days.*

Fran's hero in this series was a happy-go-lucky cowboy who laughed and sang through his adventures. This conception persisted through his earliest scripts. But this was not the hero George Trendle had in mind. Trendle's hero was to be a dedicated upholder of justice. Sober perfection. The idol of children.

Considerable light is shed upon the misunderstanding that grew into controversy by the file of letters that passed between Jewell and Striker, and by a review of his own career by Striker himself. As an imaginative man, he might sometimes exaggerate a little for the sake of telling a good story but in vital matters he was absolutely without guile. There never lived a more fundamentally honest man than Fran Striker.

Not until after the series was on the air did Fran Striker have any direct contact with Trendle. Quite properly his sole contact with WXYZ in the beginning was the station's Dramatic Director, James Jewell. And it is easy to understand from Striker's correspondence why he believed himself, as author, to be creating this new hero. He did not know that the constant demand for changes in the scripts he was mailing to Jewell was, in fact, coming indirectly from Trendle. He believed this new program was evolving from his *Covered Wagon Days.* He may even have believed at this time that he owned the series, for he was being paid for it just as he was being paid for *Warner Lester*—a fee of $4 per performance.

Fran Striker's son, Robert, says today: "As far as any of the evidence is concerned in terms of letters or anything like that, the only thing they were paying on any of these scripts was royalties; dad's writing something and they're only paying him for the right to use it. So, on the basis that they were still paying royalties at the time he was writing a show called *The Lone Ranger*, that show existed before Trendle bought any rights whatsoever to the show per se.

"Dad would adapt anything for anybody. He would kill himself trying to please people—to make it suit their wishes."

This is true. At the very time Detroit was engaged in a frenzied effort to create a western, the Dramatic Director of the Yankee Network in Boston was using Striker's *Warner Lester* series to fight a popular national program, *Eno Crime Club. Eno Crime Club* was broadcast on Tuesday nights. To steal its thunder the Boston director, Richard E. Osgood, was having Striker expand his half-hour *Lester* series into hour-long stories with a climax in the middle so strong that the first half hour could be broadcast Monday nights and leave listeners so interested they would be compelled to tune back to it Tuesday nights rather than switch to *Eno Crime Club.*

So revising scripts was nothing new to Fran Striker; unknown to anyone in Detroit, he was doing it at the same time for Osgood in Boston. In fact, he was so busy on these rewrites that he fell behind in production of new stories for Jewell's other dramas and received frenzied letters, urging him to catch up and get ahead.

The times were frenzied at WXYZ. As Ted Robertson recalled it: "There were times—I have to tell ya—that because of the mails, the scripts did not arrive—and—uh—Jewell wrote while we rehearsed. He sometimes wrote those scripts—was still writing them when we went on the air."

Such was the modest, eager nature of Fran Striker that he offered no alibis.

Instead, he apologized and promised to try harder; and he did his best to keep his promise. He thanked Jewell for suggesting changes and expressed vigorous enthusiasm for the new western series.

Trendle at this time knew nothing of Fran Striker other than that he syndicated scripts from Buffalo and WXYZ was using some of them; Trendle had no idea yet that Striker was involved with the new series.

As Trendle remembered: "I first had Jimmy Jewell write a couple of stories and they weren't very good and I told him so. And gradually they got better. And then I found out that he wasn't writing them at all. They were being written in Buffalo by Fran Striker."

Jewell was the legitimate go-between. It was crucial, as Director of the dramatic department, that he find satisfactory scripts. When his own scripts failed to suit Trendle, it was natural to turn to the author of the immensely popular *Warner Lester* series. Ideas that came out of story conferences in Detroit were passed on to Striker in Buffalo in letters written by Jewell. It is reasonable to presume, too, that during the meetings in Detroit some of the ideas expressed in Striker's letters might have been proposed by Jewell. And there is no doubt that Jewell himself contributed ideas of his own.

In one of his letters, Striker wrote, "The thought struck me as I was writing the play that you might organize a Lone Ranger Club for Boys and ask them to write in for their membership cards."

The Lone Ranger Safety Club was not formed until two years later, and then the idea came from a sponsor.

The great white horse *Silver* was derived from Trendle's own love and knowledge of horses. He was a good rider, himself. At one of the meetings someone suggested an Arabian steed. Trendle turned that down: "too small." He insisted on a "beautiful, high-stepping, big, white horse."

It has been said the idea for the silver bullet came from Howard Pierce and this may well be true; he says it was. On the other hand, Fran's happy-go-lucky cowboy in *Covered Wagon Days* used silver-tipped arrows; and a letter from Jewell, dated January 21, 1933, tells the writer to continue to use "the silver bullet and silver horseshoe gag." Besides, the idea of the silver bullet is universally known in folklore.

Fran Striker worked best under pressure. He was certainly under plenty of it now and he wrote all night every night. Trendle refused to take chances; he would not let *The Lone Ranger* go on the air until they had a number of scripts ahead, every one of which he, personally, checked through, word for word.

Meanwhile, incidentally, WXYZ had fallen behind in its payment of royalties. Would Mr. Jewell mind checking with the book-keeping department; Fran believed he was owed "around $200" (quite a bill at $4 a script!) and he had some "large obligations to meet at the end of the month." The WXYZ book-keeping department said the amount was nearer $126 and, through Jewell, promised to mail a check immediately. And did.

Two problems remained to be solved before the new show would be ready for its first broadcast: a musical theme, and the actors.

In music, Jewell was lucky. His chief sound man, Bert Djerkiss, came from a musical family and had a flair for finding passages in classical music to fit the emotional or dramatic content of radio dramas. He had catalogued a hundred such passages on Columbia, Brunswick and Victor Red Seal records. Jewell put Djerkiss to work listening to these records to find something appropriate for *The Lone Ranger.*

Howard Pierce said that he and "Jimmy ... sat down and we had to have some kind of music that went with a gallop. And we tried this one and the other one; and they didn't work. And all at once we had the *William Tell Overture* and that was it."

Jewell says it was finally a toss-up between the *William Tell Overture* and *Charge of the Light Brigade.* He liked the fanfare at the beginning of the "Storm" movement in the *Overture.*

As Jewell said: "The ever increasing cadence and roar of the brewing storm became a gallop whether or not Rossini intended it that way. I could almost hear those thundering hoof beats. So then and there, the music for the opening was set."

Now! Who was the first to play the famous role? According to Trendle, *Collier's Magazine, The Saturday Evening Post* and a very positive Ted Robertson who claims to have been there, the first to play the role was an actor named Jack Deeds.

On the other hand, the photograph of a handsome, dapper, slender, white-haired gentleman appeared only a few years ago in a Los Angeles newspaper, and under it was a caption that read: "The 'Lone Ranger' Heads $25-Million Company." He was identified as "The original Lone Ranger of 1930 radio fame—6-ft., 4-in. Texan, Lee Trent." He was boss of "the world's largest supplier of vitamin pills."

This Deeds-Trent mystery will be dealt with later.

James Jewell who, of course, directed the first show, is positive that the first Lone Ranger was an actor who had performed with the Jessie Bonstelle Company, George Stenius. Inclined to agree with Jewell is Ray Schaub who was the engineer at the control board on the first show. As sure as Jewell himself is George Stenius. Indeed, Stenius declares that so far as he remembers, Ted Robertson was not even there, and he declares he never heard of Jack Deeds until his name appeared in a magazine article.

But as to establishing the right original man Stenius feels the search is pointless: "This has all the historical significance of trying to discover who was the first scuba diver in Puerto Rico."

He is less emotional about the beginnings of *The Lone Ranger* than anyone we have talked with about them, except H. Allen Campbell. But this is how Stenius recalls it.

He had been on the road with *Elizabeth The Queen,* a stage play. He had returned to his home in Detroit when Jessie Bonstelle was in her last days. He received a call from WXYZ to audition for a program. He saw James Jewell whom he had known very slightly around the theater.

Jewell took him into one of the studios, gave him a script to read, and went into the control booth to listen. He told him to stand closer to the microphone and lower his voice. Stenius read again. When it was over, Jewell said, "I have other people to hear. I'll let you know."

Two days later, Stenius was called back to the studio.

"The part is yours," said Jewell.

A week later, they went into rehearsal for the first broadcast. It was now January 20, 1933. In the first scene, the Ranger was called upon to whistle. He was supposed to be calling his horse *Silver.*

Stenius tried to whistle. The kind of whistle you'd use to whistle a melody was not forceful enough for an action drama of this type. On the other hand, a whistle that was too shrill blasted the microphone.

"We'll take care of it later," said Jewell. And they went on.

The opening copy of the first script read: "The Jewell Players present a fiery horse with the speed of light, a cloud of dust and a hearty (LAUGHS)." The word in parentheses was a stage direction for the announcer. The announcer was Harold True. Jewell and Ray Schaub had been getting levels on the *William Tell Overture* when, only minutes before air time, True walked in and picked up his copy of the script.

"What the hell is this?" he exploded. " 'Laughs'!"

"It means you laugh," said Jewell, looking at the clock.

"Laughs! Holy Christ! Am I supposed to be an actor?"

Horizontal wrinkles appeared over the outer ends of Jewell's eyebrows.

"Come here," he said, leading the way out of the studio into the back hallway toward the freight elevator.

"What the hell am I supposed to be laughing at?" True demanded.

"It's Striker's idea of the *Lone Ranger.* We've got to change it. Trendle wants him serious." Jewell looked at his watch, the ferocious expression remaining on his forehead.

"Well, let's give it a cowboy yell," True suggested.

"Try one."

"Yi, yi!" yelled True.

Jewell fixed his eyes fiercely on True. "No good, Harold. It lets that whole opening down. You've got to top the words 'a fiery horse with the speed of light, a cloud of dust and a hearty' whatever. You've got to top it."

"Yi, yi, yi, Yippee!"

Jewell shook his head. "Sounds like a guy on a spree. Try calling the horse."

"Hi, yi, Silver," yelled Harold without much enthusiasm. Time was running out.

"That's a little better," said Jewell. "Do the whole thing. Run into it. Give it guts."

"The Jewell Players present a fiery horse with the speed of light, a cloud of dust, and a hearty HI YO—O, SILVER—R! The Lone Ranger!"

"That's it! That's it!" Jewell cried.

They raced back into the studio seconds before the red light went on. Jewell didn't even have time to get back into the director's booth, which was fortunate, because as True read the opening and the overture played, he remembered the whistle. On cue, Jewell brushed Stenius aside, stepped into position before the mike and gave a whistle.

Jewell recalled: "The whistle was deleted when I left WXYZ because I took it with me. I was the whistle."

For the first few weeks there was no listing of *The Lone Ranger* in the newspapers. There is evidence that the show had not yet been given its title.

On the 15th of February, 1933, Jewell told Striker in a letter that the reaction they were now getting was quite favorable and that in his opinion *The Lone Ranger* would click just as *Warner Lester* had. He warned Striker again in this letter that he was not making the character mysterious enough—a suggestion, he said, that came from "the company."

Jewell also promised Striker that if the two series were sold—and they were "trying their damnedest"—they would be glad to pay the $6 royalty he asked for. But now at least and at last, the show was on the air.

About the origins of the show, Howard Pierce felt: "The sum and substance of the whole thing was that *The Lone Ranger* was George Trendle's idea."

George Stenius said: "I would certainly trust Jimmy Jewell's recollection because it was his baby, in my opinion, from start to finish."

Perhaps the conclusions nearest the truth came from Jan Striker (Fran's widow): "There was always some difference of opinion as to how *The Lone Ranger* originated. He (Trendle) claimed it was all his idea and Fran claimed it grew from his scripts of *Covered Wagon Days*. We were not in a position to argue. In those days a job meant a great deal to us."

10

Things Were Looking Up—and Down

On January 31, 1933, eleven days after *The Lone Ranger* had its unheralded birth, the Michigan Radio Network first went on the air. Its key station was WXYZ. The newspapers ignored the event although they did list, without explanation, a "Dedicatory Program." The program ran for an hour and a half.

Campbell considered the cost of telephone lines which connected the six outstate stations with WXYZ, too high. He called Irving Bell, Chief Engineer.

"Irv," said Campbell, "who's our contact over at Michigan Bell?"

"Bill Waters."

"Know him pretty well, do ya?"

"Well, he's a pretty well-known man down there."

"Can you arrange for him to come in and see me?"

Bell set up an appointment. Campbell disarmed the phone company representative at once with a warm greeting and the strong white smile.

"We're hookin' up with several stations out through the state," said Campbell. "And it's costin' us a hell of a lot o' money. And I'm wondering if we might not do bettuh to talk to you and Michigan Bell than to A.T.& T. Hell, we're not a national network. We're local."

"I've gone over the situation with Irv Bell. You're paying for Class A lines," said Waters.

"I don't think we need Class A lines," said Campbell.

Bill Waters smiled. "You're not getting them, Mr. Campbell."

"What are we getting?"

"You're getting Class C lines."

"Well what the hell! You're chargin' us for A lines."

"Well, it isn't just like that, Mr. Campbell. We're giving you connections that are a little better."

"The Michigan network is payin' for over 1300 miles of cable."

"It depends a whole lot upon the connections, and you do have good connections."

"But Class C lines. We ought to be able to work out some kind of a deal."

"How do you work out the pay on this thing?" asked Waters.

"You bill us. We pay you. And we collect from the stations. These are thin times, Mr. Waters, and I'm tryin' to make money for WXYZ. There's nothin' against this."

"We'll work out something, Mr. Campbell."

And they did. WXYZ wound up making a profit on the line charges.

Another problem developed. Some advertisers were interested in the Michigan network but they did not have distribution of their products throughout the state. They wanted it but felt the additional cost of six extra stations too much of a gamble considering the times. Campbell tried to think of some way to remove the gamble. He knew that down in Cincinnati, Station WLW had a merchandising department, and he began to think of ways to adapt such a department to the needs in Michigan. He was in the market for a man, and like a miracle, the right man came to his attention through a telephone call from an acquaintance asking him to see a Harold Christian.

"Harold has done a remarkable job promoting Dr. Adams' Toothpaste," said the acquaintance. Campbell knew it; he'd heard of Harold Christian. The voice on the phone continued, "He's talked with me about merchandising and marketing and he has some ideas I think you could use."

Christian was not a big man. His voice seemed to persevere through obstructions in his nose. But Campbell found him quick, alert.

"I've been wonderin' how a Depression could kill a toothpaste," said Campbell to break the ice. "Do people stop brushin' their teeth?"

"No," said Christian. "They brush their teeth with salt and soda. Saves money."

The skin at the ends of Christian's mouth erupted into tight little wrinkles that extended down into his chin when he laughed.

"I understand radio business is not a whole lot better than any other right now," Christian continued. "In fact, I've heard WXYZ has some deals where you accept groceries and coal in payment for air time."

Campbell grinned noncommitally.

"I don't know where I got it," Christian went on, "but I have the impression somebody told me that Mr. Trendle accepts the coal and peddles it to somebody else."

Harold Christie

Christian laughed. Campbell drummed one finger lightly on his desk top. Christian became serious.

"I believe I can help this station, Mr. Campbell. You and Mr. Trendle have just formed a Michigan network with affiliates in all the larger cities in the state."

"That's right," said Campbell, biding his time.

"I have an intimate acquaintance with food and drug brokers, wholesalers and retailers throughout the territory." There was optimism and vitality behind Christian's little voice.

"What did you have in mind in the way of help?" Campbell asked.

"Are you familiar with the merchandising help Station WLW down in Cincinnati has been offering its clients?"

"I know somethin' about it," Campbell said modestly.

"Well, Mr. Campbell, I propose that WXYZ set up a merchandising department to promote the sale of products advertising on WXYZ and the Michigan network. I would start where WLW is now and go beyond—far beyond. I would work out special promotions tied in with the stores with mass displays, banners, shelf stickers—all of these things—and we'd get 'em up ourselves. We might even have demonstrators. I believe I could do one hell of a job for you and your advertisers."

Harold Christian was hired. He started with a staff of four men. The new department developed rapidly and went far beyond even his expectations. Things were looking up.

And they were looking down.

The Great Depression was sinking into its deepest dip. Construction of private homes and apartment houses had dropped 95%. A thousand homes a day were being lost by their owners through foreclosures. There were fewer suicides but more people were going insane.

Kunsky-Trendle financial resources were under a heavy strain. The size of their staff had increased substantially. When a sponser purchased Detroit only, a second program had to be fed to the network. This meant two programs on the air at the same time. Jewell's dramatic department was expanding. Four sound men were now on staff. They were not being paid much, and actors were getting only $5 a show. But it all added up. The station was supporting 65 musicians. Advertisers to pay for all this were few and far between. The station—meaning John Kunsky and George Trendle—was losing $4,000 a week.

H. Allen Campbell went to Trendle's office.

"We just can't exist this way."

"I know that," Trendle agreed. "I'm glad you recognize it."

"I have some ideas about it."

"That's good. What are they?"

"I have a payroll here," said Campbell, holding up the papers. "I have no damn business doin' this but, I mean—I had to do somethin'."

"Go ahead."

"So," said Campbell, looking at the papers, "since I'm goin' to talk payroll, I'm goin' to start first with you. I propose that you don't need it and we take you off the payroll."

Young Campbell looked up at his boss, his strong white teeth showing in a bright grin. Trendle's smile was miniscule with a wry quality. His china-blue eyes were steady.

"Go ahead," he said. "You're next."

"I just can't live," said Campbell. "I don't have the money to live on. So, if you expect me to stay here I'm goin to have to have $200 a week. I can't cut any further than that."

They went over the payroll and Campbell made the cuts. Ray Schaub had been earning $60 a week; he was now cut to $30. The cuts went all the way down the line to the mail boy, Jimmy Riddell. His pay was reduced from $12.50 a week to $10.

On February 14, 1933—St. Valentine's Day—the Governor of Michigan declared a "bank holiday." Banks were staggering under the nervous withdrawal of their depositors and the "holiday" was to give them a "breathing spell." In the preceding two years, 773 national banks and 3604 state banks across the country had failed. People began to think their savings would be more secure in a tin box, or buried in the back yard.

When the banks in Detroit closed, H. Allen Campbell paid a visit to the owners of the Fort Shelby Hotel. He offered the hotel $78,000 worth of air time in exchange for rooms and meals. The proposition was accepted and a number of WXYZ employees went there to live—especially new talent as it came to town and was hired. They received $10. a week plus room and board.

George Trendle was of the opinion that the Fort Shelby never took advantage of the air advertising.

Jimmy Riddell, however, felt sure they did. "Sure they did. Their advertising was for Salutaris. It was a Scotch soda—Salutaris Scotch Soda. I don't know what the hotel had to do with that but everybody was swapping their trade deals around. First of all, you were paid in city scrip. And secondly you traded. People didn't have money but they did need advertising and you would trade."

Things were so bad, Ted Robertson remembers, that on Friday, pay day, you would get a pink slip that said, "The Kunsky-Trendle Broadcasting Company owes you $12.50."

Robertson remembered, however, that the company never completely defaulted. "We went a long time with that and they finally paid off—fifty cents on the dollar. We never did get the full amount."

James Riddell agrees with Trendle and Campbell that dropping the Columbia network had been a costly mistake and gave a clue to who used the due bill at the Fort Shelby.

"It was the biggest mistake we ever made. The next mistake we made—whatever the network programming was on a rival station, we would program the same at the same time. Ruth Etting? *We* would put on a girl singer. Tony Wons? *We'd* put on a Tony Wons. Guy Lombardo was the Number One orchestra. We'd put on Carl Rupp against Guy Lombardo. We just programmed whatever they did. That's where we brought all that talent into town. That's where they stayed; at the Fort Shelby."

Trendle and Campbell made a swing through the state to check up on the reception of the Michigan Radio Network. The station managers were all well pleased. Driving home, Trendle was thinking of the $4,000 he and Kunsky were losing each week. He thought of the huge staff of talent they were supporting out at the station. Sales were down and he wondered just how Campbell was feeling about all this, and whether he might lose him.

"H.A.C.," he said. He never used Campbell's first name. "I have a great respect for your opinion."

Campbell started to reach for a cigarette but thought better of it. Trendle did not drink or smoke and the smoke might bother him in the car. It was winter and the windows were closed.

"You could get another job," said Trendle, "if you think things don't look so good."

Campbell waited.

"Maybe Kunsky and I would be better to drop this whole thing."

"Is that what you want my opinion on?"

"What do you think, Campbell?"

"In my opinion," Campbell said firmly, "we all have a lot goin' for us. I think you and Mr. Kunsky should ride it through."

So far, for Kunsky and Trendle, there was no profit in radio; but Campbell was right; it showed a great potential. Broadcasters were discovering the magic pull of the new medium and WXYZ was no exception.

Warner Lester, Manhunter was still the most popular drama on the station. There were some objections among the staff that the character of Michael Axford was too exaggerated—too Irish. It was decided that he would be dropped from the show. So Jewell had Striker, in Buffalo, write a story during which someone would try to shoot Warner Lester and hit Mike Axford instead.

Immediately after the broadcast the telephone began to ring. The next day, hundreds of letters poured in; listeners were inquiring about "poor Mike Axford." Bouquets of flowers were sent to the station to be forwarded to the hospital. James Jewell milked the situation for all it was worth. He had five or six bulletins broadcast every day as to Mike's condition. So popular did Axford prove to be that he "recovered" from his wounds and was restored to the series.

Anne Campbell had a similar experience. On one broadcast she asked listeners to send in jig-saw puzzles for some charitable purpose. The next day she had more than a thousand puzzles.

Meanwhile another major step was taken toward perfecting *The Lone Ranger.*

"There's too much talk and not enough action," Trendle declared. "The narrator talks too much."

"Somebody has to explain to the audience what's going on in the Ranger's mind," said Jewell. "He can't keep talking to his horse."

"Too much narration slows it down," Trendle insisted.

Howard Pierce suggested, "What about another character?"

"That might work," Trendle agreed.

"How about a girl?"

"No," said Trendle definitely. "That's out. No romance."

"Well maybe it could be an Indian," said Pierce.

"That's right, Mr. Trendle," Jewell agreed. "We had a letter from a boy in Chicago—wants to know why we don't have Indians."

"He might even rescue the Ranger if he ever got into a tight spot," said Pierce.

"This second character could scout out the evil and report it to the Ranger," Jewell thought aloud.

Also, at this time, the Ranger kept changing his identity. He would mix with other characters as an ordinary man gathering information and evidence. Then he would disappear, quickly don his disguise, and reappear to right wrongs as the mysterious Ranger.

"The Indian could be standing by to help him with his quick changes," said the old theater man, Howard Pierce.

"An Indian character might be all right provided he doesn't take anything away from the Ranger," Trendle conceded. "Always remember—the Ranger is the hero."

Jewell remembered going to a boys' camp on Mullet Lake, south of Cheboygan. Near the camp, on Indian River, was a place called Cross Village where lived a large settlement of Saginaw and Chippewa Indians. Jewell recalled it had been a custom to invite a Chief Thundercloud to Camp Kee-Mo-Sah-Bee every summer to interest the boys.

"We could call the Indian Chief Thundercloud," he said.

"Can't use that," Pierce objected. "There's already a Chief Thundercloud in the movies."

"Well—whatever," said Jewell.

"Just don't make your Indian too important," Trendle warned.

In his next letter to Fran Striker, Jewell suggested introducing an Indian who always stood ready at the Ranger's command to help him make his changes. Fran went back to his *Covered Wagon Days*. In that series he had a half-breed named Gobo. But the name didn't sound Indian enough. Fran got out an atlas and looked at maps of the southwest. He found a Tonto Basin in Arizona. In his next script—the tenth he had written—he introduced Tonto and told Jewell in a letter to change the name if he didn't like it.

Jewell was not too happy with it. Chief Thundercloud had been fond of "firewater" and when he had drunk too much the leaders of Camp Kee-Mo-Sah-Bee would send for some of his tribe to fetch him home. The Indians had called him "a tonto" when he was drunk and obstreperous, meaning, "wild, savage." An Indian who behaved badly was called a tonto. Hardly appropriate, Jewell thought, for the Lone Ranger's companion. But he was over-ruled; everybody else liked the name Tonto.

Tonto's expression came from a camp in northern Michigan.

Chosen to play the role was an actor of first-class experience in the theater—John Todd. He was bald, fiftiesh and a little paunchy; but Todd was a vigorous man and he had exactly the right voice for Tonto. He played the role in radio for 21 years, and the expression most identified with the character after "ugh" and "mm" was "kee-mo-sah-bee," meaning "good scout"—an expression which had been the name of a summer camp for boys in northern Michigan.

Much more was going on at WXYZ than the dramas.

Howard Pierce was putting out a 6-page paper publicizing the Michigan Radio Network and the personalities who appeared on it as well as the people who ran it.

On the front page of the first edition—undated, but it had to be early in 1933—was a picture of James Jewell. Under the picture, in headline style, was the caption: "Playwright for WXYZ." And the copy went on to identify Jewell as the man who "creates, writes, directs and plays in the famous *Manhunters* and *Lone Ranger* dramas."

The second edition front-paged a photograph of the *Lone Ranger*, and the one who posed for that picture was not the actor playing it, but Brace Beemer who looked the part.

Herbert Green, announcer, was described in the *Michigan Radio News* as "an ad lib expert for remotes." Remotes were programs that originated outside the studios, and there were many of them, wherever music was played by dance bands. There was also a daily program called *Lutheran Organ* which originated in a church.

George Trendle was very conscientious about "public service" and soon claimed that his station was "the official mouthpiece" of such organizations as the Masonic Temple Association, the Knights of Columbus, the Detroit Council of Churches, the Detroit Federation of Women's Clubs and many others. He himself was a Detroit City Fire Commissioner.

Because of his interest in civic affairs, WXYZ, in 1932, was the first station in Detroit, and probably one of the first in the country, to carry editorials. These were 15-minute airings of editorial viewpoint on a weekly program called *Town Talk*. *Town Talk* spoke out fearlessly for and against various aspects of Detroit's political and governmental life. Supposedly the scripts were written by someone on the staff. Actually, they were written by a *Detroit Free Press* reporter, Cliff Prevost, who didn't want the paper to know it. Every word of every quarter-hour commentary was okayed by George W. Trendle.

Harold True was the voice of *Town Talk* thus becoming, as it were, the official voice of the station. And he was doing the featured evening newscast, *Day in Review*.

"He speaks true, forceful and fast English," said the blurb in Pierce's *Michigan Radio Network News*.

One day, the mail boy, Jimmy Riddell, was sorting the incoming mail. One letter had been written by someone at the station; it had been misdirected, and returned to

John Todd—Tonto in "The Lone Ranger"

the station. But the writer had neglected to identify himself over the return address on the envelope; so Jimmy opened the letter.

It was a blast against the station, Brace Beemer and everybody right up to George W. Trendle. It was signed, simply "Del." Word got around and the letter went to Trendle. Had it been written by Del Sharbutt, the announcer, or by the orchestra leader, Del Delbridge?

It was believed to have been written by Del Sharbutt, and he was fired. Sharbutt went on to the network and won national fame as the voice of Campbell Soups.

"M-m, good."

John Kunsky was enjoying life and people. He had built a beautiful home and the Kunskys did a great deal of entertaining. Every Sunday night they had a party.

Kunsky would drop in at the Madison Theater Building occasionally, although he did not seem to be of much use there. He was admired by all who knew him. Some felt he was every bit as sharp as George Trendle and sometimes even more willing to take a gamble.

He liked to take people out for a drink. A pretty girl named Tommy Thompson worked in the office and Kunsky used the mail boy as his "beard." Jimmy Riddell was now going on 21.

According to Pierce Riddell was "A cute-lookin' guy—blond—and all the girls were after him."

So Jimmy would take Tommy Thompson to a nearby bar to have a drink with John Kunsky. There was nothing scandalous about it; Kunsky was devoted to his wife and family; but he just liked people, and he enjoyed having a drink with a pretty girl.

His "beard," Jimmy Riddell, often attended the Sunday night "gathering of the clan" in Kunsky's home.

Kunsky's partner was all business, and still sizzling on his front burner was *The Lone Ranger.*

"The Ranger always uses perfect English, no accent. Don't ever cast aspersions at any race or religious group. Be fair. Make him more serious. Don't let your scripts get far-fetched; keep 'em logical. Remember, the Ranger never shoots to kill. He is a sober-minded man with a righteous purpose. Make the kids look up to him. Make him their idol. And keep things moving. Give me more action."

When he eventually discovered who was actually writing the scripts, he picked up the phone and called Buffalo.

"Striker?"

"Yes."

"This is George Trendle."

"Who? Oh! Mr. Trendle! Yes, sir."

"I think it would be a good thing if you would drop over here within the next few days and let's sit down and talk about this thing."

"Okay, Mr. Trendle."

As he hung up, Fran Striker stood by the phone wondering why the call had come from Trendle instead of James Jewell. Maybe, he thought, he would now get more than $4 a script. Maybe now he could get $6.

11

The Mystery of Jack Deeds

It is a moving experience to talk with the widow and children of Fran Striker. Seldom has the memory of a husband and father been so wrapped in love and admiration—and protection. When they talk of him there is no trace of morbid obsession or shallow, dutiful sentiment. Jan Striker, his widow, and two of his children, Robert and Janet, are alight with happiness as they recall their lives with him. There are smiles and laughter, and a shielding tenderness as though he had been a fragile, sensitive being who had been taken advantage of.

Jan is a gentle lady. Her voice is quiet. Her smile is still that of a little girl. It is trusting and genuine and never far away. But she is firm.

Remembering her first days with Striker, Jan said: "We knew each other from the time we were five and we couldn't stand each other. He was on the track team in high school. I didn't like him. I thought he was a roughneck and a bum. And he thought I was too ladylike. We weren't married 'til we were twenty-six."

Fran went to the University of Buffalo but had to leave college before he finished and he later developed a complex about that. He went to New York and worked for a company that staged plays for ladies' clubs, fraternal organizations and other amateur groups. It had nothing to do with the professional theater.

He returned to Buffalo, and got a job announcing on Station WEBR. He also sold time, produced programs and sometimes ran the transmitter. Anyone who knew Fran Striker later would have difficulty thinking of him as an announcer. His voice was tight, constrained, seemed to come from his throat through a felt-lined vise. But in 1927, radio was new and except for the networks, everybody did everything. Fran did many "remotes" in tacky little joints that served food and had music. Jan began to go out with him to keep him company.

The Strikers were married in 1929 and Fran left WEBR to go to WTAM in Cleveland as continuity editor. He announced there, too, and may have written a little advertising copy. There, he also wrote his first full-length radio drama—a biography of Stephen Foster.

Herbert Howell who owned WEBR went to Cleveland and persuaded Fran to return to Buffalo as manager of the station. Fran organized a band and a company of actors. And now he did more writing—five-minute skits, half-hour mysteries, love stories, westerns and dramatizations of the classics.

According to Jan he wrote "All kinds of things. Just any old thing to fill up time—quarter hours, half hours. He would be finishing a script while the first part of it was being broadcast."

In Cleveland, Fran had earned $75 a week; now he was earning a little more; but still not enough. As manager of WEBR, he received a mimeographed copy of a Seth Parker script from Phillips Lord. Lord was syndicating the Seth Parker series and asking the impossible price of $25 a script. Fran thought, "If I could sell just ten scripts in syndication, I would make $150."

According to Jan, "Fran began to write things and mail them out. Howell told him he would either have to work exclusively for WEBR or quit. So he quit."

The Strikers were now living in a rented flat. Jan would stay up nights and work with him, typing. But when their first child, Robert, was born, she would sleep at night. His writing did not bother her; she was used to it. But out of consideration for his neighbors he would write one night at home and the next night at his mother's house nearby.

Mommie Striker was jealous of Jan. She brooded over her son like a huge German goose. Fran had a sister, but he was Mommie's strudel. She called him "Laddie Boy" and kept coffee brewing for him night and day. He drank quarts of it and smoked a lot, too. He was not a trained typist and his attack on a machine was almost savage. He pounded. His typing seemed faster than it was because it was so noisy. He inherited great vigor from his German mother.

Radio stations began to buy his scripts. Among his credits were NBC stations on the West Coast, the Yankee Network in New England and some fifty independent stations across the country, including WXYZ in Detroit. He was now turning out fifty thousands words a year. *The Lone Ranger* was on WXYZ three times a week alternating with *Warner Lester, Manhunter.*

Of *The Lone Ranger* Fran wrote later, "The complex sound plot made it a difficult show to produce. Few of the stations that regularly bought any other scripts would handle it, and none could use more than a single script a week."

Fran also wrote, significantly: "When I saw that my weekly output of three *Lone Ranger* stories was bringing in little more than the twelve dollars paid by WXYZ, I went to Detroit in the hope of doing something about it. It was my first visit."

Nothing about having been sent for.

At any rate, the first meeting between George Trendle and Fran Striker was mutually desired.

"I liked him and I guess he liked me," Fran commented later.

Campbell is about the only man alive at the time who admits having had nothing to do with the creation of *The Lone Ranger.*

"I sat on the sidelines not caring who the hell was the originator. So we finally begin to churn this thing together and whether it was Jewell or whoever that was doin' the writing of these shows we finally felt that we needed some help on the show that we believed was getting closer and closer. Striker was called in from Buffalo and Trendle and Jewell got together with him. I had nothin' to do with it.

"So now we have a meeting in the Maccabees Building after this has been experimented with and Striker has written some shows. Others have written some shows and Trendle is all the time feeding these ideas of what he's trying to get into the show.

"Finally, we get something we think is comin' close to what he's been trying to drive home. So! I get invited into the conference in the studio because of my commercial aspect of the thing. And we have Harold True, Brace Beemer, Jewell, Striker, Pierce, Trendle and myself."

Up to the time of this meeting the new show had had no title. Now, at this meeting, several weeks after the series had made its debut, it got its name. Campbell remembers and all agree that it was Harold True who first pronounced at a meeting in his crisp, vigorous voice, "The Lo-o-one R-a-a-aynger!" That became the title.

During his first visit to Detroit, Fran Striker saw a production of the show and was excited. Jewell's four man sound team surpassed in his mind anything heard on the networks.

"You make CBS sound silly," he told Jewell. "I've heard them use airplanes for automobiles and automobiles for airplanes. But here at WXYZ—."

Before he returned to Buffalo, he sat with Trendle. Striker was out-going, warm, friendly—but in business matters, extremely backward. No one ever needed a manager to speak for him more than he. It was a struggle for him to bring himself to speak of money. Almost apologetically, he hoped aloud that now perhaps—now that his scripts were getting closer to what Mr. Trendle and Mr. Jewell wanted—and because he couldn't sell them anywhere else, to other stations, because they were so hard to produce—perhaps they could raise his royalty to six dollars.

"You don't understand," said Trendle. "Kunsky and I are losing three to four thousand dollars a week. An increase in royalty is out of the question at this time."

Striker's heart sank; he could think of nothing to say. Trendle went on, "But I think we have a good show here; I think people are going to listen to it. Campbell is a good salesman and I think sooner or later he's going to sell it."

Striker's big heart did a little push-up.

"Now if and when it goes commercial," Trendle assured him, "we can afford to pay more for scripts. Meanwhile, if you ride along with me on the existing basis—. How much are we paying you now?"

"Four dollars a script."

"How would you feel about giving up your syndication and moving to Detroit?"

Fran was not sure.

"I'll mail you a contract," said Trendle. "Meanwhile, will you agree to continue on the present basis?"

Trendle waited for him to think it over. He was sure the writer was not really thinking; he was only feeling disappointed. Trendle was confident.

"Okay," said Striker at last. Nearly fifteen million people were unemployed. Six million were on welfare. Even at four dollars a script Striker felt a security he had not known for a long time.

Franklin D. Roosevelt had been elected President. Six days after he took office he broadcast the first of his historic Fireside Chats. The results convinced the new President that the most potent means of communication with the electorate and for promoting his New Deal was a radio broadcast. It was not uncommon for fifty thousand letters to be on their way to the White House on the morning after a Presidential broadcast. The day after the first Fireside Chat a few solvent banks here and there re-opened. Radio's power over the public continued to prove itself.

"Harold," Campbell said to Christian, "we've got to go to Chicago. Broadcast Hash is unhappy."

Broadcast Corned Beef Hash was a product of the Illinois Meat Company. General Manager of that company was Fred Martin, and he was not happy with his show on WXYZ. He was sponsoring a modest little country music stanza with three performers, *The Three Texans.*

"I just can't get what I want in Detroit," he complained.

"Your problem is you've got the wrong show," said Campbell. "You're riding the wrong horse."

"It's the horse you sold me," Martin muttered.

"I sold you *The Three Texans* because at that time it was the best program we had to offer," Campbell explained affably. "Now we can do bettuh. We've got a couple o' things goin' at WXYZ that you couldn't buy on any other station in the country. I think you'll be interested in hearing what Harold Christian has to offer."

Christian explained the merchandising department. He had definite, specific ideas of how they could increase distribution and sale of Broadcast Corned Beef Hash. Martin listened with interest. When Christian had finished, he thought for a minute. Then he looked at Campbell who was quietly watching him.

"You said you had a couple o' things," Martin said at last. "What's the other?"

"Another horse," said Campbell, smiling. "A great white horse named Silver. It's a new dramatic show George Trendle has worked out called *The Lone Ranger.*"

"Oh you have to hear this show!" Christian enthused. "It's a beaut; it's a beaut!"

"But I don't need a half-hour show," Martin declared. "I only want fifteen minutes."

"We'll give you fifteen minutes," Campbell agreed. "And I'll give it to you on a two weeks trial—free. You don't pay a dolluh until you decide to buy the show. That's how sure I am of this program."

"I'll let you know," said Martin finally.

Back in Detroit while waiting for Martin's call, Campbell did a fast job of selling George Trendle on the idea of trying a fifteen-minute version of *The Lone Ranger.*

"You can try it for a couple of weeks," Trendle agreed, "if you really think that'll sell it."

Trendle called Jewell to find out how cheaply they could produce three quarter-hour versions of *The Lone Ranger* per week. Jewell figured they could do them for $25 a week. Trendle then told Striker to re-adapt some scripts for the cut version and have them ready.

Martin called.

"Your merchandising department is one hell of an idea. Now, WXYZ doesn't have to give the Illinois Meat Company a damn thing. We'll take that show you were talking about on a two weeks' trial. But we'll pay for it."

So for two weeks, *The Lone Ranger* went on as a fifteen-minute show. Martin came over to Detroit to listen to it; it was being broadcast only in Detroit. Martin didn't think much of it. He dropped it.

"It's just as well," said Trendle. "It's no good in fifteen minutes. We'll keep it a half hour or nothing."

According to Harold Christian, Martin lived to regret his dropping of the Ranger: "Martin bought another program and I started merchandising Broadcast Corned Beef Hash, and if he told me once, he told me a dozen times that the biggest mistake he made in his life was not keeping *The Lone Ranger*."

George Stenius, who was playing the Lone Ranger, could also write. Three of his one-act plays were produced by a little theater off-shoot of the Bonstelle company. Jewell went to see them.

"George, you are absolutely crazy!" Jewell exclaimed afterward. "How can you waste your time acting in radio when you can write so well!"

"Well I don't think of myself as spending the rest of my life in radio," smiled Stenius.

"An actor who depends on acting for a living these days," Jewell declared, "is an idiot because there is no living to be made."

"I do want to get back to New York eventually and try my hand at writing," Stenius admitted.

"Take my advice," said Jewell. "Leave here and do it."

Stenius took the advice. He quit. He changed his name to Seaton, wrote pulp fiction and a couple of plays, went to Hollywood to write screenplays for Metro-Goldwyn-Mayer and branched out into directing. George Seaton (Stenius) wrote and directed *Apartment for Peggy* and *The Big Lift*. He co-produced *Bridge at Toko-Ri* and won Motion Picture Academy Awards for his screen play *Miracle on 34th Street* which he also directed, and for *Country Girl*. In 1971 he directed *Airport* which also won an Academy Award and was a tremendous box office success. He was President of the Motion Picture Academy of Arts and Sciences from 1955 to 1958.

Meanwhile a second actor went into the role of the Lone Ranger. His identity is the last, smallest and most baffling piece to fit into the jigsaw puzzle of the program's history. It took four years to put it in place. Ted Robertson, whose memory of those days was the youngest and freshest, said his name was Jack Deeds.

But somewhere along the line of Lone Rangers was a Lee Trent. And this is the only place in the puzzle here he can be fit in. A letter addressed to him seeking information was not answered. No one connected with the program in those days recognized his picture; no one remembered a Lee Trent. A letter was mailed to the company he was said to have headed. After three years, a reply was received from the President of the Con-Stan Industries of Canada.

"Yes," he wrote, "I was a voice of the Lone Ranger in the beginning, *under a different name*" (Emphasis mine).

The letter was signed by Lee Trent.

The only conclusion that can be reached is that Lee Trent and Jack Deeds are one and the same. A letter seeking confirmation was never answered.

According to Ted Robertson, Deeds was "a dapper little man with slick, straight hair which he parted near the middle." He had been acting in the dramas more or less right along, but when the Big Chance came, he "fell apart." The ferocious frown and the perfectionism of James Jewell robbed him of his confidence. He was "scared to death." The harder Jewell tried to get the Lone Ranger out of Deeds, the less Deeds was able to give.

The day after his first performance, Jewell had a visit from Howard Pierce.

"Look, Jim," Pierce said. "The feller you had on last night is just not going to do."

"I know," Jewell agreed. "There's a boy over at Wayne University—he's got a gorgeous voice. He auditioned for me the other day."

"Well, you better get him in," said Pierce.

"Has no experience whatsoever," Jewell declared. "But I'll work with him. I'll let him play a couple of parts. As soon as he's ready, I'll put him in."

"Better be soon," said Pierce.

It was sooner than Jewell planned. Deeds took to preparing alcoholically for his ordeal of rehearsals with the ominous Mr. Jewell. The rehearsal time came when he arrived too well fortified. He was, as Ted Robertson recalls, "absolutely stoned."

Jewell fired him on the spot and that night played the Lone Ranger himself. Although Trendle had insisted that the Ranger speak perfect English with no trace of an accent, one witness recalls that Jim Jewell started out with a sort of Texas drawl which slipped all the way to Alabama during the show. It was Jewell's first and last performance of the famous hero.

The "gorgeous voice" from Wayne took over the next show, ready or not. The boy's name was Earl Graser and as Jewell had promised, he "had the timbre of voice, the masculinity, the virility."

Unfortunately, Earl Graser was only five-feet-six-and-a-half; but no one had ever foreseen the day of personal appearances. And for radio, Graser was a great improvement. A listener's flesh tingled when he yelled, "Hi yo-oh, Sillver-r. Awa-a-ay!"

The true test came on the 16th of May, 1933, nearly four months after the show's first broadcast.

12

It Was A Very Good Year

It is difficult today when talking to the gentle, mellowed Ted Robertson to recall how highly charged he was by contrast when he was younger and worked at WXYZ. Besides his main duties as sound man, he, like almost everyone else, did a little announcing on the side. Concentration was required and occasionally some fast work among announcers at WXYZ, especially when two programs were on the air at the same time—one to the Michigan Network, the other to Detroit only.

One day Ted Robertson realized as he signed off a local newscast in one studio that *The Mountaineers* were ending a Michigan network program in the studio across the hall—with no announcer to sign them off. So the moment he had the newscaster safely off the air, he raced to the rescue. Dynamic in emergency, Ted charged into the door of the studio across the hall. Half opened, it stuck on a piece of torn carpet. Ted ran into the door and knocked himself out. Pete Angel, leader of the Mountaineers, managed to sign off the show. It was some time before Ted regained consciousness.

Pete Angel and his small group of musicians were mountaineers on some programs, Hawaiians on others, Italian serenaders on others.

Drama was by no means all. There were the Dole Brothers and the Frohne Sisters; the Melody Maid and Three Little Maids; the Spartan Singers, the Three Texans, and Jack and Earl. Studio A housed orchestral programs conducted by Frank Gahagan, Del Delbridge, Leonard Seel, Eddie South, Rex Battle and Seymour Simon whose arranger, Jerry Marks, composed "Santa Claus Is Coming to Town."

A kid named Amos Jacobs sang the theme on a series of police dramas. And there was a new boy in the sound department.

Ernie Winstanley was born in England and smuggled into the States by an aunt and uncle through Canada. His father returned to England and left him there with his mother and brother. His mother died the month after *The Lone Ranger* went on the air. Ernie lived with his brother for a while; then he rented a little room not far north of the Maccabees Building.

In 1932 he won a statewide essay contest in grammar school. Winners from all parts of Michigan were brought to Detroit to receive prizes and be interviewed on WWJ. The program bore the ponderous title *Secretary Hawkins' Fair and Supper Club,* and was produced by a newspaper woman named Jane Schermerhorn.

Wynn Wright, the dramatic director of WWJ, happened to be looking for a boy to play a role in a series he was producing. The remains of an English accent caught his ear and he asked Ernie to read for him. For a year, Ernie played a character called "Scrub Tory."

Then he saw in a newspaper that James Jewell was auditioning people at WXYZ. Ernie auditioned and became a regular with the Jewell Players whenever there was a part for a boy. He also worked in the sound department, mimeographed scripts for *Children's Theater of the Air* and became the show's master of ceremonies. He also appeared every afternoon on *Kiddies Karnival* at the Colonial Department Store as "Little Sambo."

As soon as classes were over at Cass Tech High School he went to the studio and worked until eleven o'clock at night. This left little time for home work and he barely scraped through, but he was happy. James Jewell was like a father to him—a stern, exacting father who taught him to take his job seriously and to be responsible. He loved the camaraderie of the studio; actors, announcers and engineers were his family.

It was still not the end of the Depression. President Roosevelt, who had promised to balance the budget, tossed it into the trash can instead. The government

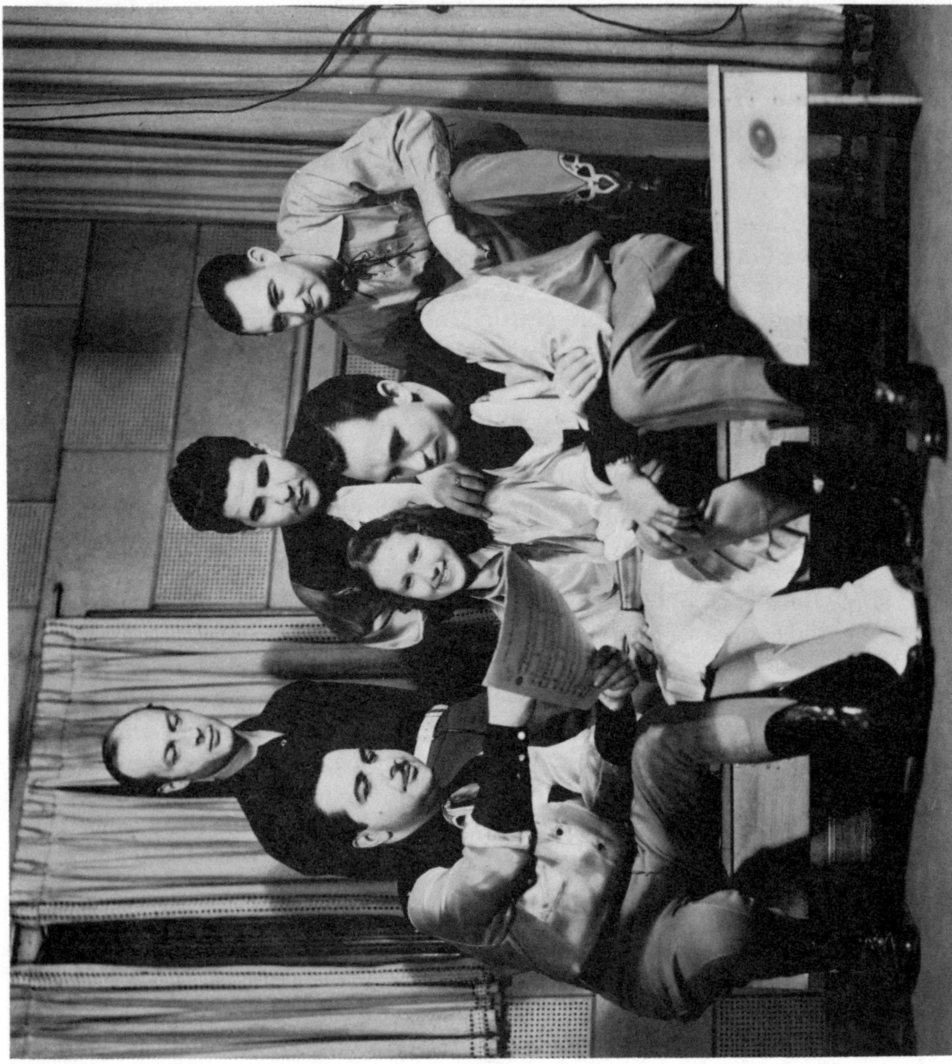

Linda Ware with Moutain Pete and his Mountaineers

broke out in an alphabetical rash: FERO, CWA, PWA, NRA, NIRA, CCC, et cetera, et cetera, et cetera—each combination a new department designed to prime the pump.

One day H. Allen Campbell's telephone rang. It was John Fetzer.

"Al, this meeting tomorrow—is it important?"

"I wouldn't call one unless I thought it was important."

Who was John Fetzer? According to Campbell, "John Fetzer had a Kalamazoo station of 500 watts, daytime only. This was like a cow pasture. And John Fetzer was just losin' his shirt and the reason he was querying like that—he didn't have the expense money to drive down to Detroit. This guy really had a patched pair of pants. But look at him today."

John Fetzer owns the Detroit Tigers.

The meeting was important. It was a pep meeting to introduce a new sponsor who had purchased WXYZ and the Michigan Radio Network to introduce a new product. It was manufactured by Miles Laboratories, a campaign to promote it in California had flopped, and the company was thinking of discontinuing it. The product was Alka Seltzer. The sale was a triumph for Campbell and the new merchandising department under Harold Christian.

Remembering the campaign, Christian said: "When Alka Seltzer hit the market, we placed over 1700 of their dispensers in drug stores plus 2000 window and counter displays."

The program and the merchandising campaign were so successful that the president of Miles Laboratores could not believe the sales figures he was getting. He "gave his staff hell" for making mistakes; according to the figures, Alka Seltzer was being sold in Michigan by the *freight car* load. When he was convinced the figures were correct he complained that Miles Laboratories just could not produce Alka Seltzer fast enough to meet the demand and suggested the company cut down on promotion and advertising. Christian persuaded him to enlarge production facilities and continue the campaign. Miles had to put on a double shift until the plant could be enlarged. Success on WXYZ in Detroit and the Michigan Network persuaded Miles Laboratories to proceed with Alka Seltzer and marked the true beginning of broadcast advertising by that company.

On *The Lone Ranger* program of May 16, 1933, at the end of the story, the Ranger rode in on Silver—via cocoanut shells pounded on manly chests—and said, "Boys and girls, I have in my possession ten thousand Lone Ranger guns. The first ten thousand boys and girls who write in for a Lone Ranger gun will receive one absolutely free!"

Two nights later, the Lone Ranger had to announce that all the guns had been given away. At the end of the week, he begged his listeners to stop writing letters; there were no more guns. In six days, WXYZ received more than 24,500 letters addressed to the Lone Ranger. And, according to Ted Robertson, they then knew that they had listeners.

But the program had still not been sold. Even so, 1933 stacks up as one of

WXYZ's greatest years—a year of creation, struggle—and triumph. An unexpected victory was achieved at an event sponsored by the Detroit Department of Recreation.

The Fifth Annual Children's Circus took place in front of the Casino on Belle Isle, one of Detroit's best known public parks. The event was in the planning stage for months. John J. Considine, supervisor of mens' and boys' activities for the Department of Recreation, was General Chairman of the affair. He contacted James Jewell. Jewell stored his marionettes at the Recreation Department headquarters; he had often done shows for the Department in the past. Considine suggested he put on an intermission act for the circus.

"Oh," Jewell objected, "I haven't done a marionette show for too long. I don't have the people to help me."

Considine was disappointed; but then he said, "I listen to *The Lone Ranger*. That's a great show. The kids love it. Is there any chance of getting the Lone Ranger?"

Jewell promised to work something out. Thus, he and Howard Pierce planned the first appearance of radio's most famous hero. They decided not to use Earl Graser because he was too short. Instead, Pierce took Brace Beemer to a tailor and had a cowboy costume made to fit him. He rented a cowboy hat from George Kane, a little costumer around the corner from the Madison Theater Building.

At last the day arrived, July 30, 1933. Twelve hundred juvenile performers had rehearsed for months and were now on Belle Isle made up as clowns and wooden soldiers and dolls and animals.

Stories differ as to where Jewell had obtained a horse. Pierce says they hired him from a farmer. Jewell remembers renting "what they call a steady horse," a bareback rider's horse, from a small one-ring circus found on the outskirts of the city.

"He was white but he wasn't such a hot looking horse," Pierce recalls.

"I was there," Ted Robertson says, laughing. "This was a bi-ig horse. The darnedest old plow horse I've ever seen."

But the 1200 juvenile performers and the 10,000 people who had gathered to watch them had no eyes for the horse. All they saw was the Lone Ranger. His appearance was intended only to fill a lull between the two main parts of the circus, but when he went galloping onto the field yelling "Hi-yo-oh Si-ilver-r!" something electric happened.

As reported the next day on the front page of *The Detroit Free Press:* "The elephants took themselves apart and became boys again." Clowns wiped their grease paint on their sleeves. Wooden soldiers suddenly lost the stiffness in their legs and went sprinting after the radio star as he rode off the field, 1200 kids and 10,000 spectators behind him. The intermission lull turned into the feature and ended the circus.

Not once did the reporter mention *The Lone Ranger* on WXYZ.

"He was followed," he wrote, "by boys who had read much about lone rangers of the west")"Read"—not heard).

"Horse and rider," he said, comprised a "speciality stunt contributed to the circus from an outside organization." (No mention of radio or WXYZ.)

"The horse," he noted, "ended by dropping to his knees in a bow to the spectators." (This seems to confirm Jewell's recollection of a circus horse.)

The newspaperman got in a final dig at the rival medium in the last sentence of his article: "Despite the rout that made the final acts impossible, the first part of the circus was a grand success."

The grand success as far as WXYZ was concerned had been the intermission. The crowd had practically ripped off Brace Beemer's costume in its eager recognition.

The show was still not sold, but George W. Trendle mailed a contract to Buffalo. By its terms, Fran Striker would be paid a weekly salary of $100 for writing whatever scripts the broadcasting company required, with the stipulation that the company become sole owner of whatever he wrote.

Fran thought long and hard. Men with college degrees were lugging hods of cement in building what is now called the Hoover Dam. They were digging ditches in a big irrigation project in the Tennessee Valley. They were working as labor on a great Triborough Bridge to connect New York City with Long Island. Men with college degrees were laboring on highways, digging sewers and collecting garbage.

He had not even finished college; he and Jan had a son, Bob; and they were expecting another baby. Security seemed more desirable than ownership of a radio series. Fran signed the contract.

But he continued to live in Buffalo.

In 1933 another juicy, profitable plum fell into Trendle's hands. The Kunsky Theaters which had been sold to the Paramount-Publix Corporation were losing money. In fact, Paramount-Publix had declared itself bankrupt. S.A. Lynch, receiver, offered Trendle a five-year contract to take over the management of the Michigan, United Artists, State, Fisher, Riviera, Eastown and Ramona Theaters. He signed an agreement in October. In November, Detroit celebrated his return to theater management with a big "Welcome Week." The theaters were now called United-Detroit Theaters and Trendle was President of the company.

John Kunsky was in no way involved. Howard Pierce, on the other hand, was delighted to abandon radio and go back to the theaters. Trendle also moved Alice Gorham out of news and publicity for WXYZ into his new theater operation as head of publicity and advertising.

Trendle restored his old policy of clean entertainment for the family. The National Board of Review, even the Detroit Police Censors, might okay a film. Still, Trendle would not permit it to be shown in any of the United-Detroit Theaters until he had screened it himself and it passed his own code. Every smutty line or suggestive piece of "business" in any visiting stage presentation was pencilled out for its Detroit engagement.

He carried this same policy into his radio operation. Cigarette advertising was turned down. There were other restrictions.

Replacing Alice Gorham with WXYZ was a feisty little newspaper man named

Felix Holt. He was short and skinny, had sharp blue eyes and a big nose, almost never removed his hat, and hopped around like a drop of cold water on a hot skillet. Felix soon became a dominating personality around the studios and if he was not the closest one there to George Trendle he managed to create that impression.

Another personality of ultimate importance was added to the roster in 1933. Charles Livingstone had started acting with the Jessie Bonstelle company in 1928. When the economy crashed in the fall of 1929 he had been touring with a company of *After Dark* which had been taken over from Bonstelle by the New York producer William A. Brady. *After Dark* closed in Boston. There was so little work in the theater that Livingstone returned to Detroit, his home. He looked up Adams Rice who was preparing his tent show for the summer tour. He hired Livingstone as his business manager.

They set up camp at Utica, Michigan. Two or three weeks later, actors began to arrive for rehearsals. Among them was a "chubby, cute little person" with whom Chuck Livingstone immediately fell in love.

Harriet had come from Ohio to attend the Bonstelle School of the Theater. At the end of the school session, she was offered an opportunity to get some experience on the summer tour with Adams Rice's "Detroit Players Tent Show." Part of her experience was falling in love with the dark-haired, black-eyed business manager.

At the end of the summer, Harriet found a job with William Hodge in a touring company of *The Old Rascal* and Chuck wound up with a stock company in Wheeling, West Virginia. On the 9th of November, 1930, Harriet interrupted a jump from Wilmington, Delaware, to Cleveland, to marry Chuck in Wheeling. Her tour ended in Cleveland and she returned to Wheeling where she played leads opposite her new husband.

The couple spent a second summer with the tent show, followed by another frustrating winter season in New York and Hollywood. They returned again to Detroit and Chuck went to the Madison Theater Building to ask for a job as announcer at WXYZ. Trendle sent him out to the studios to see Brace Beemer. Beemer was now studio manager. Livingstone did an audition a la Floyd Gibbons. He did not get the job.

But the Livinstones had known James Jewell slightly both at the Bonstelle Playhouse and with the summer tent show and Jewell called Chuck occasionally to play parts on *Warner Lester* and *The Lone Ranger,* and when a new series was added, *Thrills of the Secret Service*, he gave Chuck the leading role of Bob Barclay—a name that would be revived fifteen years later in another spy series. The girl who played opposite Chuck in *Thrills of the Secret Service* was a student at the University of Michigan—Martha Scott.

For six weeks in November and December, 1933, the talent of WXYZ was on display in personal appearances at the magnificent Fisher Theater. These appearances were arranged, written and produced by James Jewell. Strangely, although the Fisher was one of the finest of the United-Detroit Theaters, neither

Charles D. Livingstone

Trendle nor Howard Pierce could recollect what appears to have been a unique event.

A glittering ad in the Saturday newspapers, November 18, announced: "ON STAGE! POPULAR WXYZ RADIO STARS—WXYZ PLAYERS (THE JEWELL PLAYERS). STARS OF RADIO PROGRAMS: THE MANHUNTERS, THE LONE RANGER, IN POOR AUBREY—ADAPTED FROM GEORGE KELLY'S STAGE SUCCESS, THE SHOW OFF. ALSO, THE THREE FROHNE SISTERS. WXYZ RADIO STARS."

The movie was a Marx Brothers comedy. After reviewing that, the *Free Press* critic, Len Shaw, wrote of the appearance of The Jewell Players, "Elinor Flynn, Ruth Dean Rickaby, Marjorie Manning and Cledge Roberts kept the audience laughing heartily."

The following week, on stage at the Fisher: "THE GREAT RADIO HIT, MANHUNTERS, PRESENTED BY WXYZ PLAYERS—WARNER LESTER, MICHAEL AXFORD AND ALL THE REST, IN PERSON."

Beginning Friday, December 1, sharing the bill with the movie *Dancing Lady*, starring Joan Crawford and Clark Gable, "ON THE STAGE THE LONE RANGER WITH HIS HORSE SILVER RIGHT ON THE FISHER STAGE—TONTO, THE INDIAN, AND ALL THE REST OF THE ACTUAL CAST THAT HAS MADE THIS WESTERN RADIO PLAY THE MOST POPULAR OF ALL PROGRAMS—IN PERSON!"

For this, Jewell rented the same horse that had appeared on Belle Isle. But Brace Beemer did not ride. Posing as the Ranger was Harold Bud Olenkamp. Pat Dennis, radio columnist for one of the papers, wrote that he had seen a personal appearance of *The Lone Ranger* during which "the heroic masked rider had to be boosted to the saddle by co-workers."

One of the actors recalls an even more embarrassing experience. Malcolm McCoy had come to Detroit "on a hunch," had auditioned for Jewell, and played the title role in the *Dr. Fang* series. He had also voiced Trendle's pet program, the weekly editorial, *Town Talk*; nobody suited him; almost everyone seems to have had a crack at it at one time or another.

Malcolm McCoy had a kind of moon face, took short steps, seemed always to be leaning forward, his fingers always together—one hand picking at imagined hangnails on the other.

It fell to his lot during the Lone Ranger sketch at the Fisher Theater to hoist a small calf from one level on the stage to another. He would grab the little beast around the four legs and lift it.

Of this particular episode, McCoy had vivid memories. "And this matinee performance—we did four shows a day—the calf—up and did it! So when I was supposed to pick it up, I thought —N-no-o! I'm not! What the hell am I going to do? So I walked over to the calf and—I certainly wasn't going to put my arms around that calf's rear end. So! I took—*gingerly*—I took the front legs and set them up on the higher platform and then got around to the back legs and lifted them up—by the ankles. Off stage they just howled."

This is The Lone Ranger cast making a personal appearance at the Fisher Theater for a week in November, 1933—not as The Lone Ranger but as the Jewell Players. The man in the mask never played the famous role but made the appearance Earl Graser was too short. "Tonto" here was a "phony," too. Todd was middle-aged and bald, so he would not do for personal appearances. The mean looking villain below the "Ranger" is Charles Livingstone, later director of the dramas. The happy couple he glares at are Beatrice Lieble and Malcolm McCoy. Man at the right is Herschell Mayall, once prominent in movies. Woman in doorway was Ruth Dean Rickaby, once married to Robert Strange, later to the brother of George Sanders. (John Conway?)

So did the audience. Nevertheless, the engagement of *The Lone Ranger* at the Fisher Theater was a great success. To get an audience the next week, the Fisher ads read: "FISHER ON STAGE. THEY GAVE YOU A GREAT SHOW LAST WEEK. THE WXYZ PLAYERS WHEN THEY PRESENTED THE LONE RANGER. NOW—SEE THEM THIS WEEK IN THE FIRST STAGE PRESENTATION OF ANOTHER OF THEIR WIDELY FOLLOWED AND FAMOUS DETECTIVE PLAYS, INSPECTOR MARTIN."

The movie was *Tillie and Gus* starring Alison Skipworth, W.C. Fields and Baby LeRoy.

The following week seems to have been a kind of throwaway although the actors remember it best: "THE FISHER PLAYERS IN THE DRUMS OF OUDE." For it, Malcolm McCoy had his hair dyed blonde and he, Charles Livingstone and John Todd wore kilts. And of all people to appear in this sketch as an aging East Indian native was the very Irish Jim Irwin. To his credit, nobody recognized Michael Axford.

The six weeks ended with a great Christmas show: "WXYZ RADIO REVIEW ON THE STAGE WITH DEL DELBRIDGE AND HIS RADIO ORCHESTRA, SPARTEN SINGERS, JACK AND EARL, THE SUNNY SINGERS, RUSSELL NEFF, THE THREE TEXANS, THE FROHNE SISTERS, THE MOUNTAINEERS—WITH GUEST STARS."

McCoy said about the hurlyburly of these weeks: "We were doin' all the radio shows besides, and tearing back and forth in make-up from the theater to the Maccabees. It was a madhouse!"

The actors were paid $50 a week for their stage appearances. It was a six weeks bonanza for them, and WXYZ added prestige to its growing reputation for showmanship.

Still, *The Lone Ranger* had no sponsor.

13

Silver Was A Plough Horse

Immediately after the sensational surprise of the Lone Ranger's personal appearance on Belle Isle, Campbell held a meeting with his salesmen. They reviewed the "circus" triumph, and Campbell gave each salesman samples of mail received during the pop-gun giveaway earlier.

"Now," he told them. "These are the results; so put your thinkin' caps on and go to work. You are at liberty to go out and talk about *The Lone Ranger*."

The first to get a potential client under his "thinking cap" was two-fisted, hard-hitting Garde Wallace. The Gordon Baking Company had the most beautiful horses in Detroit hauling its delivery wagons. Jim Gordon was proud of those horses and Wallace knew it. Furthermore, Wallace had already sold spot announcements to the Gordon Baking Company.

Garde visited Gordon's advertising manager, Everet Wilshire, and pushed hard on the horse angle, knowing Wilshire would confer with Gordon. The Gordon Baking Company produced a bread called Silvercup. Wallace enthused about this coincidental tie-in with the Ranger's great horse, Silver. It was a natural.

Late one afternoon in the early fall Garde Wallace strode into Campbell's office and crowed, "I've sold *The Lone Ranger.*" He lit a cigarette, dropped the match in a tray, sat down, and added, "It's all set. But—I do have a little problem."

Campbell waited, his shrewd little blues watching the salesman through frameless spectacles.

"He wants to buy this show in Detroit—just as it is."

"No problem," said Campbell

"But," Wallace went on, "he also wants to produce our scripts in Chicago and New York."

Headquarters of the Gordon Baking Company was in Detroit. In the beginning, Jim Gordon had been a small baker who sold his own bread to stores himself. Now the bread was sold in Chicago and New York and the company maintained branches in those cities. In fact, Everet Wilshire had come to Detroit from Chicago where he had driven a truck. Later, he was to become president of the company. There was a brief silence in Campbell's office.

"Garde," said Campbell at last, pressing a cigarette into the tray on his desk. "I know exactly what this man wants to do and I will not give you this kind of permission to talk to him any further."

"He wants to buy," Wallace insisted.

"If you want to make an appointment with Wilshire for both of us to go out and talk to him, we will do this."

"I'll call him right now," said Wallace jumping up. Garde looked at his watch as Campbell pushed his phone toward him. Wallace talked with Wilshire briefly and hung up.

"He'll see us tomorrow morning. This thing is hot!"

"Here's what you're goin' to have to do," Campbell said coolly. He had a way of emphasizing words without raising his voice, by separating them slightly, tapping one finger gently on the top of his desk with each word. "You're goin' to keep - your - mouth - shut. I'm the guy who's goin' to talk to them and I have some ideas about it and this will not hurt your commissions or anythin'—but—I'm not goin' to permit what he's tryin' to work."

Wallace's story was confirmed by Wilshire. He told Campbell he'd buy the show just as it was in Detroit, and he would pay to have the same scripts produced by stations in New York and Chicago.

"Mr. Wilshire," Campbell said calmly, "it isn't quite that simple. This is our show. We own it. We're offerin' to sell it to ya. You might be buyin' somethin' that gives you a terrific - image - in - this - market."

Wilshire pushed his desk chair back a little, put one ankle over one knee and leaned back, waiting.

"Now!" Campbell went on easily. "We have been experimentin' with this show since last January. We've experimented with it to the point that we've finally found out the pull on the thing."

Wallace shifted in his chair. He was aggressive even in repose.

"Now!" Campbell continued. "Do you suppose that I'm goin' to let *you* have a script that you take over to Chicago to WGN and let them come up with a lot o' screwballs in this show? Same at WOR in New York? No."

Wilshire raised his hands in a little shrug and slapped his thighs as though to say, "that's that!" Wallace took a deep draw on his cigarette and put it out. Campbell remained calm as he went on.

"I'll tell ya what I'll do, though. Ah'll letcha know how fair I'll be with ya. Ah'll give you a script. I'll do better than that. I'll go to Chicago and I'll give WGN a script—the same script that we have right here in Detroit. And I'll let you have WGN produce this show-as-they-see-it."

Wilshire's ankle slid off his knee and he sat up straight, resting his arms on his desk, watching Campbell.

"They have all the sound effects—everything," Campbell continued. "Let them produce it."

Wilshire looked at Wallace. Wallace was watching Campbell with raised eyebrows.

"This will not go out over the air. But if you will agree to have your own Chicago personnel and their youngsters and the families that they know gather in the studio, we'll let WGN produce this show and let them listen to their show-but-immediately followin' *their* show—I want to prove a point to you-*we* pipe *our* show into the *same* studio. The same audience will hear both shows and by their response we'll make our minds up. And this isn't goin' to cost ya a dime."

Wilshire rubbed his lower lip with a big, stubby finger.

"I'll go for that," he said finally.

"Garde can work out the time and everythin' with you," Campbell concluded, standing up. His tan face broke into a smile. The three men gravitated toward the door of Wilshire's office.

"To further assure you," Campbell concluded. "If you think you can produce this show in Chicago, if you think you have a drag with WGN that you can produce this show in Chicago cheaper than I can feed her over telephone lines, I'll assure you that you're wrong. I will do this! You give me the price that WGN quotes you for producing the show in Chicago and I'll match it."

A time was set within days. A script was mailed to WGN. To play the Ranger, WGN imported a one-time movie star who was now doing a lot of radio acting—Francis X. Bushman.

The afternoon of the audition arrived. Campbell went to Chicago and found the studio packed with young fry and families and friends of the Gordon Baking Company. They were not in the studio with the actors and sound effects. They were in a separate studio where they could not watch the performance any more than they would be able to see the show piped in from Detroit.

Garde Wallace who had accompanied Campbell to Chicago was holding himself down but Campbell could sense his high-pitched excitement. In fact, Campbell himself felt the juices running as he waited to hear what Chicago would do with *The Lone Ranger.*

Halfway through the Chicago-Bushman-WGN version, Campbell said softly in Wallace's ear, "They can't hold a candlestick beside what we have."

But Wallace still looked anxious.

At the end of the audition there was a nice round of applause. Then H. Allen Campbell made a little speech.

"Now we're going to give you the same show you just heard, but—as we bin doin' it in Detroit, with our own cast and sound effects. We want you to decide which show is bettuh."

Two minutes after the end of the WGN audition, the identical script, beginning with the same William Tell Overture, came into the studio over telephone lines into a loud speaker.

Campbell knew the superiority of his production: "You could almost see the kids notice the difference immediately. I mean, even the sound effects were better. So, immediately after this show that we fed them in there—holy smokes! I mean, the whole damn audience just went crazy!"

"You win, kid," Wilshire admitted. "You win. No question about it. Now how do you want to do this? How do we get into New York?"

"The same way we got into Chicago," said Campbell, his strong white teeth showing in a grin.

The Lone Ranger had finally picked up a sponsor and a new national network was about to be born. It was November, 1933.

Soon after Silver and Silvercup Bread became associated Garde Wallace ran into Jim Gordon himself. Gordon expressed enthusiasm for *The Lone Ranger* and Wallace told him it was directed by James Jewell.

"Jewell. Jewell," Gordon repeated, trying to remember.

"His family had a little grocery store on Van Dyke and Gratiot, he tells me, and they used to sell your bread."

"Oh yes. I remember. I delivered it to them."

Gordon expressed a wish to watch a broadcast and although this was strictly against Trendle's rules, Wallace took him and Wilshire to the Maccabees Building one evening. It was the only time James Jewell talked with Jim Gordon. The conversation worked around to horses.

"Kids really love Silver," Jewell remarked.

"Gee," Gordon said. "I wish we had a horse—a real Silver we could show the kids."

Jewell put an ad in a newspaper: "Wanted: A white horse. Must be gentle. To be ridden by children."

He received a phone call from a Captain Cary in New Haven, Michigan, twenty-odd miles north of Detroit.

Jewell recounted the experience: "Mary Martha and I were going to be married that week and the first thing I ever asked her to do was to go with me while I bought a horse!"

They drove out to Captain Cary's farm in Jim's new Hudson Super Six. There seemed to be nobody around. They walked out to the barn. There were stalls for, perhaps, twenty horses, but not a horse in sight. They called but nobody answered. They walked back to the barn door. Way down across a big field, Jewell spotted a farmer, ploughing. He and Mary Martha walked across the field and waited at the end of a furrow.

"Hi!" the farmer called, stopping the horse. It was a white horse.

"Hello," said Jim. "My name is Jewell."

"Oh yeah. You're the man I called about the horse."

"Yes, that's right."

"Well," said Captain Cary. "This is the horse."

"*This* is—the *horse*?" Jewell exploded. Horizontal wrinkles appeared above his eyebrows. "A *plough horse?*"

"Oh now wait a minute," said Captain Cary. "My daughter is in school and she can't exercise the horse any more. And my leg won't led me ride. So this is the way I exercise the horse."

Jewell looked from Captain Cary to the horse, the ferocious frown still wrinkling his forehead.

"I don't want to buy a *plough* horse," he insisted.

"Please don't call it that," begged Captain Cary.

He unhitched the horse and made a hand signal. The horse romped around in a complete circle and returned to Captain Cary.

"This is a circus-bred horse," he said. "This is no plough horse."

Without a word, using only hand signals, Captain Cary put the horse through its tricks. Finally, the horse sat back on its haunches and pawed the air.

"This is the most amazing thing I have ever seen in my life!" Jewell exclaimed. "What is the horse's name?"

"White Gold," said Captain Cary.

"How much do you want for him?"

"I want two hundred and fifty dollars for her," said Cary. "She's a mare."

"Well," said Jewell, using it for a bargaining point. "I was looking for either a gelding or a stallion. Er—did you say two hundred and fifty dollars?"

"Can't do any horse trading," said the farmer.

"Can't come down—just a little?"

"Not one dollar. There's not a horse in the world like White Gold—a *real* horse, that is."

Jewell took a wallet from his breast pocket and started counting his money.

"You notice I said a *real* horse," said Captain Cary. "Did you ever hear that *Lone Ranger* program?"

Jewell felt his face turn warm.

"I've heard of it," he managed.

"Well, watch this," said the farmer. And he yelled, "Hi-yo, Silver!"

White Gold reared up-and whinnied.

"I have to have this horse!" Jewell cried. He counted out two hundred and fifty dollars.

"You have to take her right now," Cary told him, pocketing the money.

"What do you mean—right now?"

"Immediately," insisted Captain Cary "Now."

"But—I have to get a trailer."

"Well then, the deal's off," said the farmer reaching into his pocket for the money. "My daughter will be home in a little while and if she sees this horse go from here I'll never live it down."

So Captain Cary put a rope around White Gold's neck, and with his new Hudson Super Six in lower gear, Jewell and his bride-to-be led the horse all the way from New Haven, Michigan, to a farm at Utica Junction owned by Jewell's uncle.

Two days later, Mary Martha and James Jewell were married and moved into a new home in Grosse Pointe Park. The day they moved into the house, a big truck arrived from the Department of Parks and Recreation, and neighbors watched big doll-like figures carried into the garage. The marionette theater was also taken in; it took two men. The very next day, a van drove up and delivered a horse. The neighbors began to circulate a petition.

But Jewell kept the horse in his garage only one night. The Boy Scouts had asked him to arrange a personal appearance of the Lone Ranger at the State Fair Grounds, and tomorrow was the day. That day, Harold Bud Olencamp made another appearance as the Ranger, and his great horse Silver was a white mare Jim Jewell had found pulling a plough.

The contrast between the continuing Great Depression and the rising new broadcasting industry was an interesting phenomenon of the early 1930s. Apples were still being sold on street corners, thousands of families continued to lose their homes through foreclosures, and white collar men were blistering their hands levelling mountains, building dams and planting trees for the Civil Works Administration.

Radio was a grace note of optimism in the general dirge. The penthouse of the Maccabees Building was a happy refuge. The pay was small, but it was pay. And hearts of talented people beat faster as their abilities were used. Between 6 A.M. and midnight musicians, singers, announcers, sound men, writers, arrangers, actors and specialists of different kinds rode fourteen floors up and down on six elevators in joyful contrast to passengers bound for dentist chairs and doctors' offices that dominated the rest of the building.

One of the least egotistical among those who climbed the stairs to the 15th floor penthouse was Earl Graser. He was settling nicely into the role of the Lone Ranger. He took direction well and was liked by all the players.

Graser was 24, still studying at Wayne University. He had mild blue eyes, fair skin and was inclined to be chubby. Born in Canada, he had never been west of Michigan, couldn't ride a horse, and shot a gun only once in his life. When he was 18, he ushered in Kunksy's Michigan Theater and announced selections played on the organ. Once, he had even appeared on the Michigan stage as an Alpine shepherd. He had been called upon to drive six sheep along a ramp in a theatrical thunderstorm while the orchestra played part of the *William Tell Overture.*

As an actor, he had limitations. One of these was brought to the attention of Fran Striker by James Jewell on one of the author's frequent visits to Detroit.

"Fran," said the director, "do me a favor. Please don't write the word 'probably' in any dialogue you give the Ranger. Earl can't say it. It comes out 'prob'ly', and you know Trendle wants the Ranger to speak perfect English. If Earl gets out 'probably' okay, he invariably fluffs somewhere in the next few words. It throws him every time."

Jewell had surrounded himself with quite a company. There was a fiery little Italian, Jack Petruzzi. There were a heavy character woman, rather dowdy, with big brown eyes named Gwen DeLany (pronounced L-A-H and not L-A-Y, if you please); another older women, Bertha Forman, who brought an English accent with her and a gentle sense of humor; and Fred Reto, a small man with a distinctive gravel voice. There were Jewell's sister, Lee Allman; Ruth Dean Rickaby who owned a slobbering Boston bull and drank too much; Beatrice Leiblee, a very handsome woman with a heart-shaped face and very dark green eyes, straight hair and a friendly smile; and Marjorie Richman who mixed her own perfume and wore too much of it.

The Lone Ranger was now being carried over WGN in Chicago. Harry Sale, advertising counsel for Gordon Baking Company in Chicago, had a gripe.

"It's ridiculous," he said. "As great as Silver is, why should two men ride him?"

Up to this time both the Ranger and Tonto rode Silver together.

"Tonto should have a horse of his own," Sale suggested.

WXYZ was agreeable, meaning ultimately George W. Trendle.

"But don't name him," said Sale. "We'll run a contest."

The name Scout was the result of the contest. Thereafter, when the listener heard the Ranger urge on Silver, he would usually hear John Todd as Tonto follow with, "Git 'em up, Scout."

Just as Fran Striker, in a letter from Buffalo, had suggested a Lone Ranger Safety Club early in 1933, James Jewell claims he was the first to have an idea which would later be credited to another.

As he recounted it: "The first comic strip that was ever written resembling *The Lone Ranger* was presented to a man by the name of Neebe. Howard Pierce and I went to see Mr. Neebe who was then with Campbell-Ewald, and Mr. Neebe owned the rights to the Tarzan comic strip.

"And the name of the comic strip which we presented to him was called *The Runaway Stagecoach,* and I remember the sequence we had drawn up showing the Lone Ranger galloping ahead and finally, in the last second, the crucial second, stopping the runaway horses as they were going down this mountainside. This predated *The Lone Ranger* cartoon strip, I would say, five or six years. This was in 1933."

Thelma Watson emerged from George Trendle's office, admiring the young man with her smiling cinnamon eyes.

"Mr. Trendle will see you now."

The young man stood up, decided to leave his top coat on the bench where he had been waiting, and stepped into Mr. Trendle's office. Trendle, tall and immaculate, stood up and extended his hand.

"Hello, Fred." He indicated a chair and both sat down.

Trendle had known Fred Flowerday's father. The house where he had lived was right behind the Flowerday greenhouse. Fred's grandfather had started the florist shop in 1882 and had later become head of the horticultural department of the City of Detroit.

"How is everything?" Trendle asked, pleased with young Flowerday's clean-cut appearance. "I was sorry to learn about your father."

Fred's father had died leaving a widow and three growing sons. Trendle had learned of the father's death through Harold Christian. Christian, also a family friend, had been in the habit of checking with Mrs. Flowerday to see that the boys had everything they needed in the way of clothing.

"Well, Mr. Trendle," Fred said, smiling. "We have given up the flower business. This depression—not very good for the flower business. So—we're out of it and I need a job."

Trendle lifted the phone. "Miss Watson, get me Beemer."

Trendle liked the poise of young Flowerday. He did not fluster. He seemed quietly confident and cheerful despite hard times.

"Brace," Trendle said into the phone. "I'm sending a fellow out to see you. His name is Fred Flowerday. See if you can find him a job."

Beemer's first question when Flowerday saw him was, "Do you know any jokes?"

Beemer was trying to get some material for a morning program. Flowerday couldn't think of any jokes.

"Jim Jewell will see you." said Beemer.

Jewell wanted to know if he did any acting.

"I never have," Flowerday admitted.

"Well," said Jewell, "I need another sound man. How old are you?"

"Eighteen," said Flowerday.

"One of my boys is going to switch to singing," said Jewell. He was speaking of Bert Djerkis. Now working in the sound effects were Bert's brother, Shields, Ted Robertson, Ernie Winstanley and a blonde kid named Larry Kelley.

"I'll put you on the payroll now," Jewell offered. "You watch the boys work for a week or so and I'll start you in the first of the year."

Charles Livingstone's first crack at directing radio drama was as Jewell's assistant on *Dr. Fang*. It was Livingstone who introduced exact timing to the WXYZ dramatic department.

Livingstone remembered it with some interest: "Jim (Jewell) would just watch the clock and you'd be reading your lines and all of a sudden a hand would come across your mouth. They'd cut the studio dead and bring up some music and he'd say, 'Take three pages out and start here.' Then the music would fade out, they'd open the mike again and you'd start reading where he pointed on your script.

"If we were running short, in order to lengthen the show—we'd be playing a scene in *the Lone Ranger* and all of a sudden a character not in the script comes walkin' in. He'd say 'Howdy, pardner. What you fellers doin' 'round here?' And this would be Jim! Jim would come in and ad lib and we'd ad lib until he figured we were probably back on time; then he'd point to the script and we'd pick it up again where we'd left off."

When Livingstone started timing *Dr. Fang* he would mark the time at the beginning and end of each scene. He found that even this did not work because actors varied their pace between rehearsals and performance. So in what was called the "dress" or timing rehearsal, Livingstone marked down every fifteen seconds on his script. Then, during performance, if an actor got carried away with his own performance, he'd get a signal to hurry up; or if he was too fast, he'd get a signal to slow down.

"From that time on," says Livingstone, "we never ran over or under."

Exact timing was vital in broadcasting, especially when tied in "live" with a network—either receiving or "feeding." There were then no minute commercial announcements between programs. One program would end exactly thirty seconds before the beginning of the next one, whether it be local or network—and it might be either. A half-hour program had to be exactly twenty-nine minutes and thirty seconds; a quarter-hour show was actually thirty seconds short of fifteen minutes.

Livingstone's system of exact timing was not new in the broadcasting industry but it was a definite contribution to the operation at WXYZ.

Yes, 1933 was a very good year, a momentous year. It marked the beginning of the Michigan Radio Network, the beginning and first sale of *The Lone Ranger,* and the beginning of profit. In a dark, impoverished world, radio was an oasis of sunshine and as far as George W. Trendle was concerned, he was in the middle of it. He had reason to enjoy a merry Christmas and to rub his hands together in gleeful anticipation of the year ahead.

14
Birth of the Mutual Network

Seventeen days after the new year rolled in, *The Lone Ranger* added Station WOR in New York to WXYZ in Detroit and WGN in Chicago. Within a few weeks, Campbell, who was adding business sophistication to his natural magnetism, sat in the office of Tucker Wayne in Atlanta, Georgia. Wayne represented American Bakeries, makers of Merita Bread which was sold through seven states in the south. Campbell was trying to sell *The Lone Ranger.* His competition in the sale was a popular transcribed series.

"Where do you come up with these prices?" Wayne wanted to know. "I can buy this other show for a third of what you have quoted me."

"Buy it if you like it," said Campbell, relaxed and confident. "I mean—I'm givin' you something that shows results—in New York, in Detroit, in Chicago—. I don't know what this other show's goin' to do."

"I know you've got a good show," Wayne admitted.

"I don't care what price this other one will quotcha. The price I'm quotin' is our price take or leave it."

Campbell was grinning and Wayne laughed.

"You're sure a tough character," he said.

The problem was how to get *The Lone Ranger* into those seven southern states. The answer was by electrical transcription. The E.T. was comparatively new, but Campbell knew that if others could sell programs that way, so could WXYZ.

Contracts with Tucker Wayne were signed on Friday. The first recorded shows were to be in the mail three days later. James Jewell remembers it well and observes how characteristic it was of "the downtown office" to issue a command with no thought of what was involved in obeying it "out at the station."

Jewell complained a little: "Allen Campbell sold the Ranger to seven stations in the south without telling me about it. I was informed on Friday that we were to send three half-hour transcribed shows out of WXYZ to these stations that following Monday, in addition to all the other work we were doing.

"Here's what had to happen. Al Hindle who was head of RCA in Chicago was one of my friends. Arrangements were made with Hindle to open a line between Detroit and Chicago and we would pipe in those three half-hour shows on Saturday and Sunday.

"So the troupe of actors—and God love 'em, God bless 'em—just said, 'All right, Jim, it's going to take us two days to do this and we won't have time to eat or sleep and let's get to work.'

"So this is literally what happened. We piped those first three shows over from Detroit and they were recorded by RCA in Chicago. And we met the deadline for Monday. The records were mailed out on time. And nobody ever said thank you."

Next, Campbell called a baking company in Kansas City. It, too, wanted *The Lone Ranger*. It, too, hesitated over the price.

"Look," said Campbell. "What do you want to do—have your competitor buy this show? I mean—they'd grab it!"

Kansas City fell into the bag.

Campbell's memory was ebullient: "We made money off of our telephone lines. We got double what the lines cost us. This was another bit of profit over and above our profit over actors and staff. So these were terrific sales. And all this stuff was a part of WXYZ at that time; it went into WXYZ's till." (*The Lone Ranger* was not incorporated until 1935.)

Campbell was now out of Detroit much of the time and needed an assistant in the home office. Harold Christian was too busy with the merchandising department.

He remembered the activities they were engaged in: "We developed sales contests, premium offers that were used by manufacturers for the first time before starting their national campaigns. We worked out special promotions tied in with the super markets with mass displays, banners and shelf stickers that we made up ourselves."

In 1934 Christian was invited to join the Detroit Association of Grocery Manufacturers, giving him—and WXYZ—more of an "in" than they had had before. Christian now had five men working under him.

The assistant Campbell found was a plump, nervous, stocky little man from Toledo named Junior Sutton. Junior was as wound up as Felix Holt, never stood still, had a habit of teetering from one foot to the other. He had been hurt in an accident; his jaw had been wired together, and not too well. Junior had a lot of good ideas, some of them wild, but he worked right under Campbell who expected him to develop commercial ideas and take care of many details. One of his assets was a knowledge of music.

Trendle was as busy as his Sales Manager. Campbell closed no deal for *The Lone Ranger* without his being present—in Chicago, in New York, in Atlanta, in Kansas City and in Boston where they signed up the Yankee Network.

So Trendle needed help, too, especially on the theater end. He hired a man from Hollywood, Earl Hudson, who had to his credit some movie scripts and a production or two. The Depression had knocked him out as it had so many, and he was glad to start with the United-Detroit Theaters for $75 a week.

Out at the station, Harriet Livingstone, actress, was doing a daily *Woman's Hour*.

Commenting on it Harriet revealed the ad lib characteristics of many WXYZ activities: "Every day, Anne Campbell rhymed a recipe. Had she left out the main ingredients I would not have known. But I read the little rhymed recipes and people could send in for them. I didn't even know how to bake a blotter."

Trendle also hired a new Chief Engineer. He was not satisfied that the station was received everywhere as it should be. The new Chief Engineer, Lynn Smeby, a rotund young man with spectacles, tore down two towers on top of the Maccabees Building and erected a single-tower Truscon transmitter. He had it covered from base to tip with red neon lights. The tower was taller than the building itself and many tenants of the building were worried because in a high wind they could see the tower sway.

At times there could be a real worry. Al Schaub recounted one: "One time I know they took precaution on Woodward Avenue, blocked off traffic until the wind died down. It was a three-leg tower which was the strongest type, and they put a counterpoise underneath the tower for the ground. However, the Maccabees Building was not a solid steel building. It was half steel and half stone block and we never had a very good signal output there."

With the success of *The Lone Ranger,* a serious rift developed between James Jewell and George W. Trendle. It reached a climax at a showdown meeting in February, 1934.

"You were not honest with me," Trendle declared. "You told me you were writing those first Lone Ranger scripts and you weren't."

"I am sorry, Mr. Trendle. I did write the first Lone Ranger scripts."

"They were being written by Fran Striker in Buffalo," Trendle insisted.

"Not the first ones, Mr. Trendle. I wrote those and you said they were no good. That's when I contacted Fran Striker."

Trendle pushed at a paper on his desk.

"What is it you want?" he demanded.

"I feel that now *The Lone Ranger* is beginning to earn money, I should have a certain share of it."

"What, according to your thinking, might be a certain share?"

"Certainly more than I'm getting."

"Are you not now the highest paid member of my staff out at the station?"

"That's what you tell me."

Trendle looked noncommittal. He shoved the paper toward Jewell.

"I want you to sign this assignment of authorship."

"I don't think I can, Mr. Trendle, but I'll talk to my brother."

Ralph Jewell, the musician, was now an advanced student of law. He read the release, shaking his head.

"*Lulu and Leander, Teeing Off,* the Ranger—everything. It's not reasonable."

Lulu and Leander was one of the few comedy bits ever allowed on WXYZ. It was a daily dialogue between two characters. Jewell wrote it, and Trendle liked it. (Lulu was played by Lillian Randolph who later did many network shows from Chicago.)

"No," said Ralph Jewell to his brother at last. "Don't sign this thing until he gives you some sort of assurance that you'll get something for it."

Trendle breathed a little harder but did not otherwise betray his impatience when Jewell relayed his brother's advice.

"Jewell, you know you have a good thing here. You are the highest paid man on the staff out there. Have I ever denied you anything? You have everything you want—all the sound men you ask for. I never tell you to cut down on the number of actors you want on a show. You think it over. You have a good set-up here. And you know me well enough to know I'm not going to let you go hungry. You think it over. I'll send Ray Meurer out to see you."

Raymond Meurer, boy singer in the Kunsky Theaters, banjo player in a WXYZ orchestra, was now practicing law and Trendle was paying him a retainer fee to negotiate with unions, take care of contracts, and protect him legally.

By the time Meurer had arrived in the Maccabees Building, Jewell had decided that he did indeed have a good set-up for such depressing times. He signed over his rights to his writings.

Nothing seemed changed on the surface but behind the china blue eyes of George Trendle, Jewell had recognized a cold opponent, of immovable hardness, and the friendship was severed. From that time on, Jewell did very little on his own initiative.

Campbell's big project in the Spring of 1934 was to land the Tiger baseball games. The team was owned by Frank Navin. Tiger broadcasts were all tied up with WWJ and Ty Tyson. But Campbell was determined to get baseball on WXYZ and the Michigan Radio Network. He knew he had to have "a real, topflight guy" to compete with Tyson. After some weeks, his thoughts zeroed in on a former idol, Tiger outfielder and hitter, Harry Heilman.

His search for Heilman took him to the sheriff's office. Heilman was working as an under sheriff. Campbell took him to a bar and ordered drinks.

"I suppose you're wonderin' why I looked you up," said Campbell.

Heilman's thin face wore a tight-lipped smile.

"I bin thinkin' about getting WXYZ into baseball," Campbell continued. "Have you ever thought you might do some broadcastin'?"

"No, I never thought about it," Heilman admitted.

"The News is fightin' me," said Campbell. "But I think I can sell the idea to Frank Navin."

Heilman leaned into his beer noncommittally.

"How does it strike ya?"

"Sounds good to me, Mr. Campbell."

"I'll get back to you," Campbell promised.

Frank Navin resisted the idea but Campbell was persuasive.

"What the hell. You're not goin' so damn hot. You got your trainin' expenses down in Florida. We'll give you $25,000."

Navin gave in. WWJ would continue its broadcasts with Ty Tyson, but WXYZ would also broadcast the games. Campbell signed Harry Heilman for fifty dollars a week!

On May 22, 1934, Fran Striker sold, assigned and transferred all his Lone Ranger scripts, Warner Lester Scripts, *Thrills of the Secret Service*, and anything and everything he ever wrote in the future, as well as any profit, benefit or advantage that might arise from printing, publishing, vending and broadcasting his scripts, now or later—to the Kunsky-Trendle Broadcasting Corporation. The agreement, buried in 258 words of legal mumbo-jumbo, translated into simple terms, meant that Trendle now owned the creative firepower of Fran Striker, lock, stock, barrel and cartridge supply.

Campbell went to Chicago and sat down with Bert MacFarlane who was General Manager of the *Chicago Tribune*. The *Chicago Tribune* owned Station WGN.

"You're 50,000 watts here in Chicago," said Campbell. "And you don't have a network."

"We don't want a network."

"WOR is 50,000 watts in New York and they don't have a network. WLW in Cincinnati is 50,000 watts—going on 500,000 watts. They take whatever they want from the networks. I mean—they just kick the networks around. They want to be independent."

"You're independent in Detroit," said MacFarlane. He was not quite sure what

Harry Heilman

Campbell was driving at. "We're all independent. As far as we're concerned here in Chicago, we like it that way."

"I have an idea where we can still be independent and exchange programs and save some money," Campbell ventured. "For example, we do a musical show with Eduard Werner. He conducts the orchestra at the Michigan Theater. Now. If we're spendin' a thousand dollars for an Ed Werner show and all we do is just feed it to Detroit, why the hell don't we feed it to you in Chicago?"

"Well—"

"We could save a hell of a lot o' money feeding programs to each other without bein' kicked around by a network. And each of us could offer a market for the others' sponsors—like we're usin' you for Silvercup."

"It sounds good, Allen, but Detroit—Christ! You're only 5,000 watts. What the hell have you got that you could feed us?"

Campbell's tan face broke out in a big, white smile.

"We can feed you *The Lone Ranger*."

MacFarlane remembered the competition over *The Lone Ranger*. He pretended to wince, good-naturedly.

"Well, what else do you have, besides Ed Werner?"

"Bert, all I can say to you is this. We will make more good shows available to you than you will ever make available to us."

Campbell laughed.

"You guys don't produce as good shows as we do. I've listened to 'em. Why do you think I took a chance on the Ranger? You should be able to get bettuh talent in Chicago and in New York than we get in Detroit, but hell!—we go out and steal your own damn talent. Ah hired Hogan from Chicago. There's a number of announcers that we've done that with."

Campbell went to New York and talked with Alfred J. McCosker, once a flamboyant press agent, now top man at WOR. Campbell also talked with John Clark of WLW in Cincinnati. This station had been added to the Lone Ranger outlets for Silvercup.

Trendle was well aware of every step his Sales Manager was taking. Finally the day arrived when Bert MacFarlane from Chicago, John Clark from Cincinnati and Trendle and Campbell from Detroit met with Alfred McCosker in New York. All agreed that a combination of their four stations might work to mutual advantage in a loose kind of combine. None wanted to expand to a full network with power to give the run-around.

"But wait a minute," said McCosker. "We would like to go a little further than these four stations, I mean!"

"We have the nucleus of these telephone lines," Clark pointed out. "It's nothing to go down to Washington and have a Washington outlet."

"And it wouldn't be anything," said Chicago's MacFarlane, "to extend a little bit to St. Louis. St. Louis is a good market."

"St. Louis might be okay," said New York's McCosker. "But WOR would just as soon have Philadelphia out of it."

"How about Pittsburgh?"

"Pittsburgh's okay."

"We ought to have Pittsburgh."

Chicago, New York and Cincinnati looked at Detroit.

"I'd have no objections to Pittsburgh," Trendle conceded.

"Allen?" McCosker asked.

"All right," said Campbell. "We'll put in Pittsburgh. We'll put in St. Louis. And we'll put in Washington. And that'll be seven stations and that's all it'll ever be."

Clark nodded, noncommitally; MacFarlan studied his fingernails; McCosker looked at his cigar.

But it seemed to be agreed. A contract was signed on the 19th of September, 1934. The seven stations were hooked together with Class A telephone lines, arranged by H. Allen Campbell; and because the expenses were shared mutually, it was called the Mutual Network.

Just after Thanksgiving, Fran and Jan Striker, and their boys Bob and Don, moved to Detroit into a house Fran had found for them. He had stayed in Detroit several weeks before they moved. As soon as they were settled, Fran took them for a ride downtown. He had a little surprise for them.

They drove down Jefferson from the eastern edge of Detroit and turned north on Woodward Avenue. Fran pointed out a trick store on the left and a burlesque theater on the right. They drove past the old city hall and came to a department store; the windows were lighted and the Christmas displays were already installed.

Jan's heart did a little jump. The windows were filled with a Lone Ranger display. Dummies portrayed the masked hero, the Indian, the white horse, the little black horse. The Strikers sat looking at it for five minutes. Then Fran drove north to Grand Circus Park, drove around the park pointing out the several theaters Trendle managed and the Madison which he and Kunsky still owned and where Fran's office was. Then he headed back south on Woodward Avenue and stopped again, across the street from Hudson's.

They sat gazing at the display. Jan was speechless. She loved this man and she was proud of him. Seeing the Lone Ranger display in one of the largest department stores in the country made them both feel that all of Fran's hard work might pay off at last.

In his mind, Fran belittled the agreements he had signed. Who would write *The Lone Ranger* if he didn't? Besides, if they didn't like it here, they could leave—and his talent would go with them. Fran looked at the display once more, laughed, leaned over and kissed his wife.

Jan looked down at Donnie, asleep on her lap, his round face a duplicate of her own. And she looked at Bob, sitting wide-eyed between them on the front seat.

"Dad's a pretty big man," she said.

It was a happy night for the Strikers.

15
The Girl On the Couch

Roma Davis looked up from her typewriter as the blonde walked into the reception room. Her brown eyes started back to her work, stopped halfway, shot back at the girl who stepped up to the little metal fence.

"I'd like to have an audition," she said.

"What do you do?" Roma asked. Roma was short, well put together and the whites of her eyes usually showed tell-tale signs of a late, wakeful and exciting night before.

"I want to act," said the blonde. Her voice was arresting. It was low, cultured and positive.

No flighty blonde, this one, thought Roma.

"Why don't you sit down?" Roma suggested. "I'll tell Mr. Jewell you're here. What's your name?"

"Vitez," the girl told her. "Joan Vitez. Mr. Jewell doesn't know me."

"He may see you before he goes into rehearsal," said Roma, swinging through a little gate and stepping down a corridor. In a moment she returned with James Jewell.

"Come with me, please," said Jewell.

The blonde followed him into his office. There was just enough room in it for a desk, two chairs and a short couch. Jewell indicated the couch. She faced a window, Jewell swung his desk chair around and stared at her. She was wearing a cheap dress, but it fit and was well-chosen.

"Vitez, is it?" he asked.

"Right."

"Vitez, Vitez—what's Vitez?"

"Hungarian."

"Were you born in Hungary? You speak beautifully."

"My parents were born in Hungary. I was born right here in Delray, but I'm Hungarian still. You don't change that."

"What's your experience?"

"None - professionally," the girl admitted. "Just school. You know. But people constantly say to me that I have an interesting voice and why don't I go into radio."

"You have a good voice all right," Jewell admitted. "If you can do anything with it."

A tall, well-built man with wavy blonde hair appeared at the door. Jewell invited him in and introduced him. It was Brace Beemer. He sat in the second chair. There was something very personal in the way he looked at the girl, but it was not offensive.

"What's your family?" he asked.

"Mother, father and brothers," Joan told him. "We have a little Hungarian restaurant on Second near Warren in the basement of the Corinth Hotel.

"She wants to act," Jewell explained.

An idea passed between the two men as they looked at each other; Beemer's face lit up.

"Mills!" he said with a laugh in his throat.

"So it's decided?" Jewell asked, turning serious.

"What do you care?" said Beemer. "You have enough to do."

Beemer looked back at the girl. Jewell looked a little glum. The Mills Baking Company had been on the air for a year and a half with a little show called *The Happy Home Village*. Jewell had played a cartoon character on it called Cy Jones, and Mills had used a cartoon of the character in its newspaper advertising. Charlie Hicks, salesman on the account, had indicated the sponsor was in a mood for a change of program. Jewell and Beemer had discussed it before and they had had an idea.

"Roma," he ordered. "Tom Dougall is up in the studio; he's on the Ranger tonight. Will you get him down here?"

That Jewell and Beemer sent for Tom Dougall rather than discuss their idea with Fran Striker proved the gap existing between the executive office downtown and the studios in the Maccabees Building.

Tom Dougall was a thin, physically hard, blonde Irishman in his thirties. He was introduced to Joan Vitez.

"We-ell!" he exclaimed, revealing his instant impression. He took her hand and sat at the end of the couch viewing her with an arrogant smile. Dougall wore his clothes well. In fact, he modelled. He had on a grey flannel pinstripe. He looked urbane, sophisticated—even a little blasé.

"You tell me you want to write," said Jewell.

"Yes."

"You wrote the anniversary show for the University of Michigan?" Beemer asked.

"Collaborated," Dougall corrected.

"Jewell also tells me you wrote special material for network shows in New York," said Beemer.

"Yes," said Dougall.

"I want you to write a series for this girl," Jewell told him. Dougall's flirtatious manner changed to intellectual inspection. The three men talked around her. They listened to her low, vibrant voice and her perfect diction. There was no nonsense about her. She was a girl who had taken family responsibility, and it showed.

"It's got to be something very simple," Jewell suggested.

"Simple?" Dougall exclaimed incredulously.

"Down to earth," Jewell added.

"Down to earth, maybe, but hardly simple," Dougall commented, turning on his arrogant smile again.

"I don't think I have the right voice for a charwoman," Joan agreed.

"Definitely not," cooed Dougall.

"I mean something easy," said Jewell, seriously. "Don't dream up an actress's challenge because this girl is no actress—yet."

"I can get it on the air in a couple of weeks if you get something ready," Beemer advised them. "A quarter hour in the morning, five days a week."

"Okay." Dougall slapped his thighs and stood up. "Simple and down to earth."

"Keep that beautiful voice in mind," Beemer advised.

"Ye-es," Dougall agreed, giving her another look. "We'll see what we can do."

Tom Dougall had begun working on WXYZ in 1934, as an actor. He had played in *The Drunkard* in Detroit with Harriet Livingstone and before that had gone to the University of Michigan with Chuck Livingstone. He had also played a leading role in the show on which he had collaborated at college, and had toured east with it, playing at the Metropolitan Opera House in New York.

Dougall was proud of his experience: "I am the only actor of the Trendle group who can say he sang at the Met."

He and Livingstone had acted together at the university and when *Warner Lester, Manhunter* had toured the Butterfield Theater circuit after appearances at the Fisher Theater and Jewell had had to leave the unit because of his work load at the station, Livingstone had taken over the role of Warner Lester, and Tom Dougall had been hired to play the role vacated by Livingstone. When the tour had ended, Dougall had begun to play juveniles in the Jewell radio dramas at WXYZ.

Something of New York had rubbed off on Tom Dougall. He had enjoyed intimate acquaintance with such celebrities as Tallulah Bankhead and Noel Coward and was not hesitant when drunk—which he often was—about supplying bawdy details of his experiences in Manhattan.

He now went to work on a series for Joan Vitez.

In 1934, almost a year after the Lone Ranger's personal appearance at Belle Isle and many months after the shows at the Fisher, the Woodward Grand Theater advertised a personal appearance of "Silver, Star of the Lone Ranger Broadcast." According to Raymond Meurer, the station knew nothing of this until the ad appeared in the newspaper. Meurer, who had given up banjo and singing and was devoting full time to legal practice, immediately filed an action claiming that this was *not* Silver of *The Lone Ranger* broadcasts and had no right to claim any connection. He was able to establish to the court's satisfaction that this was simply a horse that had been rented for a fee, and that a horse could not be transmitted through the air, and that such a "personal" appearance would perpetrate a fraud upon the public. He won an injunction and there was no appearance of the rented horse.

Meurer made an appointment with Trendle.

"Oi propose that we incorporate *The Lone Ranger.*"

"I don't think that's necessary," Trendle said. "You got your injunction."

"This toime I was lucky," Meurer declared, smiling. "I claimed infringement, but infringement in law implies some kind of patent or trademark or copyright. If we incorporate this program under its title we would be in a much safer position at court. Take this case of Correll and Gosden. They were successful in blocking the use of the name Amos and Andy on work overalls, because they were incorporated."

Trendle rubbed one side of his nose with a very clean finger but he continued to look skeptical.

"How much would it cost?"

"Consider it an investment in security." said Meurer. "Putting *The Lone Ranger* on the air is coh-ahsting you a lot of money. The protection of setting up a firm identified with the title is, in my opinion, important. We could stop things loike that horse situation from happening."

"Make The Lone Ranger a separate company? Apart from the station?"

"Separate from the station. We might even have an advantage in the event of law suits where the liability might be borne strictly by the corporation."

"Rather than the station?"

"Rather than WXYZ. We might even get into secondary property rights, such as patenting the expression 'Hi-yo Silver'."

"Well," said Trendle finally. "It's something to think about."

Meurer's memory was clear on convincing Trendle: "I actually had quite a job selling Mr. Trendle on the idea of incorporating. We always laughed about it because I advanced the coh-ast—and billed him later. We incorporated under the laws of Michigan and we were able to file our articles of incorporation with the United States Patent Office in the realm of trademarks and we had some very interesting trademark cases later on where the secondary property right, 'Hi-yo, Silver,' came into play."

The Lone Ranger was incorporated in January, 1935, "to keep it apart from the station and to protect the station in case of infringement suits," as Trendle put it.

Pierce and Kunsky, of course, were part of the separate corporation. Both were still WXYZ stockholders, but they were seen less and less around the station. There is said to be a feeling among Kunsky heirs that John was pushed aside.

Making a separate corporation of *The Lone Ranger* widened the gap between Trendle and James Jewell who had felt a promise was broken when he was not made part of the corporation. Harold True even consulted a lawyer about suing Trendle, at least to be given credit for what he believed to have been his contributions to the program's creation. But the lawyer refused to take the case because, he claimed, True was working for Trendle and anything he thought of belonged to the man he was working for. True, however, never recovered from bitterness.

One man who devoted his life to *The Lone Ranger* was not heard to complain about being left out of the corporation. Fran Striker still comforted himself with the feeling of security for himself and his family and with the confidence that if he became dissatisfied and left the station he would take his talent with him.

Within two weeks of her first interview at WXYZ, Joan Vitez was summoned back to his office by Jewell.

"I have a show for you," he said.

It was to be called *Ann Worth, Housewife*, and it would be on the air five mornings a week. She was to play a concert pianist who gave up her career for marriage. Her husband would be played by Malcolm McCoy and there was a

Ann Worth—Housewife. Joan Vitez and Harriett Russell Livingstone. Photo courtesy Gordon Conner.

Cast of Ann Worth, Housewife.
Standing left to right: Ted Robertson, Ruth Dean Rickaby, Malcolm McCoy, John Todd, Marjorie Richman, Harry Golder, Tom Dougall, Beatrice Lieblee, ?, Fred Reto, ?, Lee Allman and Jack Petruzzi
Seated: John Vitez and Gillie Shea.

running part for a maid which was good news for Gwen DeLany.

"Now, Joan," said Jewell. "We're putting you on the air. That's what you wanted—to act on the radio. Now we're going to ask you to do something for us. Will you be willing to do this for thirteen weeks—gratis—until we find a sponsor?"

Joan had performed in two Lone Rangers without being paid; Jewell had called them her audition. Joan was green; she was naive; but she was no fool.

"Look," she said. Her voice was low but it had a ring. "I know there's a Depression—and *how* I know it! I know you can't pay me a great deal but if it's going on the air, it's worth something to somebody and therefore I would like to be paid."

"How about $18 a week?"

"That's all right ," said Joan, "until you find a sponsor."

Another writer was hired downtown to work with Junior Sutton. She was a full-bodied teutonic blonde named Marie Meyer. Her first job was to write commerical announcements for the theaters. They were advertising on all stations in the area. They were also producing promotional programs when special movies were opening. Scenes from the picture were dramatized and performed by Jewell's actors on the radio.

But the big job every week for Marie and Junior Sutton was preparing a script for the Michigan Theater program. This was a big variety show that went on the air Sunday noons. Eduard Werner, musical director of the Michigan, conducted an augmented orchestra at the station, and the numbers were based on the Michigan stage show.

The new show opened at the theater on Friday; Junior and Marie wrote the radio program on Saturday; and it went into rehearsal Sunday morning for the broadcast. It was a tight schedule for preparation. Werner was unable to get the list of musical numbers to the Madison Theater Building before late Friday afternoon.

Marie Meyer remembered the haste in which all things were produced: "I can remember Junior Sutton walking back and forth as he dictated this program. I didn't take shorthand and he was in a rush."

Sutton was a good man for this show. He had played drums with Seymour Simon's orchestra. And he was not only a musician; he also had had experience on the engineering end at WSPD in Toledo. He knew radio. Marie Meyer called him "a very talented man."

James Jewell's forehead wore its ferocious frown.

"Where's Ruth?" There had been a break between rehearsal and broadcast time was now coming up. Ted Robertson banged into the studio, breathless.

"Jim, she just can't make it. Ruth."

"What do you mean she can't make it?"

"She's drunk as a loon!"

"Where?"

"On the 14th floor. She can hardly stand up. She'll never get up those stairs."

Jewell looked at the clock. Fifteen minutes.

"Fred Flowerday! Ernie! Go with Ted! Get her down to the lobby, take her around back to the freight elevator, and get her up here! Goddam it—*Move!*.

Jewell knew from experience that Ruth Dean Rickaby could act—drunk or sober, if they could just get her into the studio. The freight elevator at the back of the building ran up to the penthouse.

"Jeez," Ruth muttered when the sound boys got to her. "I don't know what hit me."

The boys got her down to the lobby and supported her through a back hall to the freight elevator. Ernie Winstanley ran ahead and found a wheel chair stored there. They dumped the actress into the chair. Jewell was waiting for them when the freight elevator reached the back hall of the penthouse.

"Goddam it, Ruth!" Jewell exclaimed, more in exasperation than in anger.

"Don' worry," said Ruth.

She tried to stand up; couldn't; they wheeled her into the studio. Jewell shoved her script into her hand.

"We've got just three minutes. Let me hear you read."

To everyone's amazement, plump, black-eyed Ruth Dean Rickaby read perfectly.

"Don't try to stand up," said Jewell. "We'll lower the mike."

All through the halfhour show the other actors crouched low on their feet to adjust to the lowered microphone. Between lines, they watched Ruth Dean Rickaby anxiously.

Flowerday was lavish in his praise: "She gave a great performance; she was a great actress. I was always very naive about it, but Ruth always smelled very heavily of perfume. I understand she drank it to disguise the booze."

Such was the camaraderie of the players that they had a great affection for her and helped her through. Even Jewell was kind and understanding. He never stopped using her.

Trendle always spent the month of March in Florida. While he was there in 1935, *Ann Worth, Housewife* went on the air. Tom Dougall wrote extremely well from an actor's point-of-view. His dialogue was good; his characters were clearly realized. Ann Worth was soap opera but it was good soap opera and within a couple of weeks, the Mills Baking Company picked up the sponsorship and Joan Vitez was raised to $35 a week. Marie Meyer wrote the commercials.

Trendle was not enthusiastic. On the morning of his return from Florida he summoned Striker to his office.

"What's this thing I just heard? This Ann Worth thing."

"I don't know anything about it, Mr. Trendle."

"Well, who's writing it?"

"One of the actors."

Trendle's face was twisted a little as though he had just bit into some dry charcoal. Fran was enjoying his reaction.

"Well, it's not much of a show," said Trendle. He called Campbell.

"Campbell, what's this soap opera we've got on in the morning? Mills Baking Company."

"Mills is Charlie Hicks' account," Campbell explained. "They wanted to make a change. I haven't been too close to it."

"It's not much of a show," Trendle muttered. He called Jewell.

"Where'd this soap opera come from we've got on in the morning?"

"Charlie Hicks called Brace Beemer about the Mills company wanting a new show—and this is what they bought."

"I don't think much of it," said Trendle.

But *Ann Worth, Housewife* was incorporated by Meurer and remained on the air for three years.

Two Detroiters emerged from an elevator on the ground floor of 1440 Broadway, New York, and walked toward Times Square.

"You've gone beyond us," said Trendle. "I know I can't keep you."

"Why do you say that?" asked Campbell.

"Mutual wants you."

They found a table in the Astor Grill and ordered lunch.

"It is true, isn't it?" Trendle demanded. "Mutual wants you?"

Campbell had been offered the presidency of the Mutual Broadcasting System.

"I can't compete with that money," Trendle observed.

"Oh, we might make some kind of a deal." Campbell's strong white teeth showed in a smile.

"Obviously, you have something in mind."

"Frankly, Mr. Trendle, I'd like to own part of the station."

Trendle raised his eyebrows. The waiter put their food before them and they began to eat. Trendle kept his eyes on his food, thinking. Campbell ate his mutton chop silently, with a confident eye on Trendle.

"What if I cut you in for ten per cent?" Trendle finally asked him.

"I'll take it," said Campbell.

"I'll talk to Howard," Trendle promised.

Pierce sold ten per cent of his stock to Campbell, and Trendle named Campbell, officially, General Manager of the Kunsky-Trendle Broadcasting Company. Junior Sutton took the title of Sales Manager although Campbell continued to be the station's commercial mainspring.

Ed McCosker became head of the Mutual Broadcasting System. Despite the agreement in the beginning that the network would be restricted to seven stations, the others were so delighted with what they were doing that they started to add stations, and more and more stations.

"It's getting too big," Trendle complained. "This was not our idea."

Campbell agreed. They also agreed now that dropping CBS had been a big mistake. But they were dubious about continuing with Mutual. Campbell talked in Chicago with Niles Trammel, Vice President of the National Broadcasting

Company. It was Trammel who had discovered Amos and Andy. Campbell admired him for his reputation as a great time salesman, knew he was on the way up, and predicted he would one day be top man at the National Broadcasting Company. Men of this type were glamorous to Campbell. He had no great interest in "talent" but he was very conscious of the status of executives.

On September 29, 1935 WXYZ joined the Blue Network of the National Broadcasting Company and became the only station in the world to be on four networks at one time. It continued to feed *The Lone Ranger* to Mutual; it remained the key station of the Michigan Radio Network; it broadcast a big musical variety show called *The Band Box* to The Canadian Broadcasting Commission; and it received programs from NBC-Blue.

That same September, Harold True hired a new announcer who was to become one of the station's hardiest personalities. He was a chubby young man from Ohio whose head seemed to be wider at the chins than at the top which was covered with well groomed thick black hair. Despite his weight, he had a healthy clean-cut appearance with a smooth, pink complexion and rather deep-set green eyes. He was a jolly fellow named John Slagle.

"You had a phone call from Dick Osgood," said Jan Striker when Fran came home.

"Albany?" Fran asked.

"No. Cleveland," said Jan with a quiet little giggle.

"What are they doing in Cleveland?"

"They've been fired," said Jan, laughing.

"They sure get around," Fran noted. "New York, Boston, Hudson Falls, Albany, Cleveland —. What are they doing in Cleveland?"

"They'll have to explain it to you. I couldn't understand what they were talking about. They'll be here tomorrow."

"Oh?"

"I invited them to stop off on their way to Chicago."

The Strikers had first met the Osgoods on a blind date in Syracuse the year before. Osgood had used Fran's *Warner Lester, Manhunter* series on the Yankee Network. A native of Boston,he had acted for twelve years in the New York theater, and on the networks in *True Story Hour, True Detective Story, Silver Flute, Harbor Lights, Empire Builders, Gilbert and Sullivan* series and others until the Depression drove him home to Boston where he sold a radio series he wrote, *Royal Romance*, to Station WNAC. He was hired to direct the series, and organize a dramatic department. Like Jewell in Detroit, Osgood had produced a half-hour drama every night except Sunday on the Yankee Network.

There, he had met and married Nancy Howe, certainly one of the first female disc jockeys in radio whose morning and noon-hour record programs literally supported WNAC through the Depression.

Osgood was fired for some reason or another and went to a summer stock company in Brookfield, Mass., where a play he had written, *No-Account,* was performed. Then he became Program Director of WGLC, a 50-watter in Hudson Falls, New York. His wife followed him and there they had purchased and played in Fran Striker's *Hank and Honey* series. When WGLC moved to Albany and became WABY, one engineer and the Osgoods moved with it.

Then they joined a fantastic project in Cleveland which was expected to wipe out space radio. Programs were sent into homes over wires. Collaborating in this epic experiment were Ohio Bell Telephone, Philco Radio, the Warner Brothers and a seventh heaven unheard-of organization known as North American Power. Special transcriptions were made for the project by Andre Kostelanetz, Fats Waller, Gertrude Neissen, Waring's Pennsylvanians, St.Bartholomew Choir, an opera company and many others. Before the plans went awry and the mammoth dream dwindled into Muzak, the Osgoods were fired.

As they drove into Detroit and passed a big stadium on their left there rose a mighty roar. The city when they drove through the center of it exploded into hysterical, jumping, yelling people; cars blew their horns and paper floated down from tall buildings. Laughing people waved at the Osgoods.

"What a reception!" Nancy exclaimed.

They didn't learn until the Strikers told them that the moment they had driven past the stadium the Tigers had won the World Series.

What's in Chicago?" Fran asked that evening.

"Who knows?" said Dick.

Nancy shrugged, turning brown eyes from Fran to Jan.

She was smiling, but Jan thought she detected a trace of bitterness. The Strikers had not escaped from insecurity too long before not to be sympathetic. The warm heart of Fran Striker took over.

"Before you go to Chicago," he said, "I want you to meet Allen Campbell and Brace Beemer."

The timing was right. George Trendle had been complaining about ad libbing announcers. He wanted prepared copy, written introductions to musical numbers. Words thus put into announcers' mouths might not be clever but what they said would at least be neat, not sloppy.

Allen Campbell and Brace Beemer were both aware of Trendle's thinking and on October 13, 1935,they hired Dick Osgood to write sustaining copy for the announcers of WXYZ at $65 a week.

The championship fight in New York that year between Germany's Max Schmeling and Detroit's Joe Louis had drawn the first million-dollar gate since 1927. Now, after six years of sweet, soft-flowing music by Wayne King and Eddy Duchin to sooth jagged nerves, jazz was brought to life again by Benny Goodman. The worst seemed to be over; the Depression seemed to have turned around; better times were on the way.

16
Hornet's Nest

Trendle was pleased. WXYZ was sounding better all the time, and now with NBC-Blue they had Jack Benny, Helen Hayes, Ben Bernie, Believe-It-Or-Not-Ripley, Ozzie Nelson, Walter Damrosch, Seth Parker, Lowell Thomas and Walter Winchell.

Fred Flowerday had promoted the sound of horses' hooves from actors slapping their chests to the use of cocoanut shells in a long trough. He had arranged the trough in different sections with broom corn in one section, gravel in another and sand in the third so that by changing positions at the trough, sound men could move their horses through sage brush, over rocky terrain or along a sandy trail. Flowerday and Ernie Winstanley attained the peak of perfection by discarding the cocoanut shells and substituting bathroom plungers.

Grosset & Dunlap and the Whitman Company had been licensed to publish Lone Ranger novels written by Fran Striker. And the Lone Ranger Safety Club, just formed, promised to be a great success. They had given out 475,574 badges in six weeks.

And now something new was cooking. Trendle sent for Striker. Fran's walk was made up of short steps to which was sometimes added an eager little bounce. As he stepped into Trendle's office his grinning response to Thelma Watson's greeting faded and his lower jaw sagged a little. The whites of his eyes looked yellow. He was tired. He had been writing all night.

On Trendle's desk was a copy of a book called *The Adventures of Jimmy Dale.*

"I had Felix Holt read it and tell me what he thought of it," said Trendle holding it out. "He likes it."

The Adventures of Jimmy Dale was about a young millionaire who was the very image of respectability by day but who took on another identity at night as a sort of modern Robin Hood.

"Read this," Trendle said to Striker. "We can do something like this with our new series. *The Lone Ranger* appeals to the kids. Now I want to put something on the air to interest young people who are about to vote. I want to do something to show young men how crooked office holders can be, and what they have to do to stop it—that they have to get out and vote and see what's going on in the world, watch things so we can elect candidates to office who will *be* something. See?"

"Kind of expose crime—and set things right—like the Lone Ranger," said Fran.

"But modern. Work something out and let me see it."

Fran started out; then turned back.

"Oh!" he said. "You know that plot you gave me for *Thrills of the Secret Service?"*

Trendle's face had a mischievous look on it.

"I heard it last night," he said. "The one where your hero got trapped under the capitol in Washington and no way out."

"How *do* I get him out?" Fran asked.

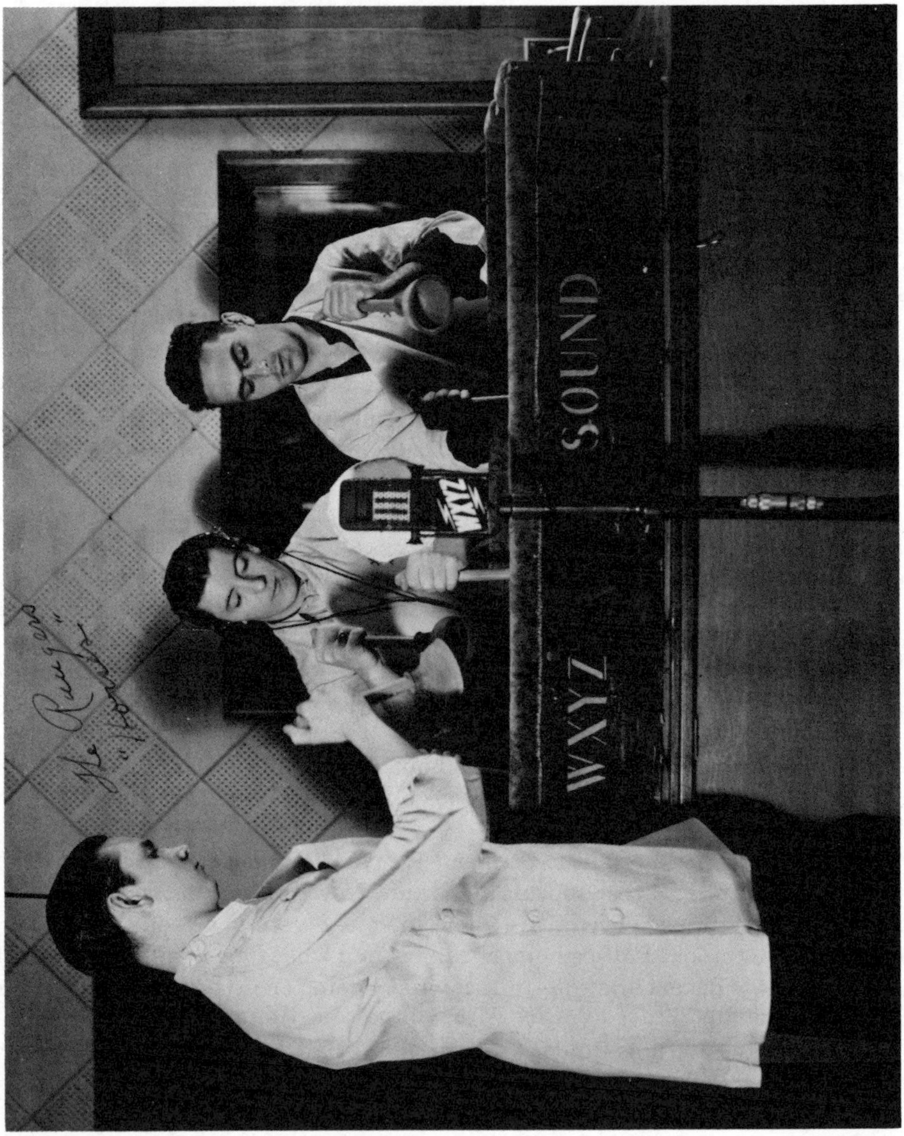

Half of the sound crew

"You don't," said Trendle. "That's the end of the series."

"You son-of-a-gun." Fran laughed, but he had liked that series.

"I want to see what you've got on this new thing in a couple of days."

Dick Osgood was writing continuity for twenty-two singers: Cyril Wezmael, Virginia Hooper, Russ Neff, Salvatore Cucciara, the Escorts, the Girl Friends, the Sunny Singers and others; and there were piano programs, organ programs and orchestra programs. He needed help and obtained permission to hire an assistant for $35 a week. He hired Al Hodge. They had worked together in the wired radio experiment in Cleveland. It was coming to an end and several of those involved were drifting to Detroit—Charles Wood, Bill Kennedy, Bill Crago. Others were going back into the theater. Jimmy Backus would try Hollywood.

Al Hodge was over six feet tall, lean, hard and handsome, although when he smiled he showed poor teeth and seemed self-conscious about it. His complexion was florid, his eyes varied from grey through blue to bright green; his hair was brown and straight and thick. His father had worked with Buffalo Bill. Al had been born in Ohio.

He now went to work writing sustaining copy for programs that went to the Michigan Network. Since he had had experience putting record programs together in the wired radio experiment, he now made up similar programs for the Michigan stations. In fact, he had the same records to work with, for those which had been specially produced for the Cleveland project had now been made up in a library which WXYZ leased. Al also wrote *The Bandbox Review* which WXYZ broadcast to Canada.

George Trendle was telling Campbell and Ray Meurer about the new series.

"You got a name for it?" Campbell asked.

"No."

"Well, you tell me this fellow goes out and stings the crooked politicians and the gangsters. Why don't you call him the Bee or the Hornet?" Campbell suggested.

"Not enough action in a bee," Trendle complained. "Hornet might be all right. I like that. *The Hornet.*"

"I'm afraid, Mr. Trendle—that won't do," said Meurer —troubled oil on peaceful waters as usual.

"Oh?"

"I think there's been a Hornet. It might be legally dangerous when it comes to copyrights."

"Well, couldn't you call him some *kind* of a hornet and be different?" Campbell asked, amused.

"We moite do that if we gave him some descriptive title," Meurer conceded.

"Well—The Buzzing Hornet, Green Hornet, Red Hornet, Hornet with a Sting—" Campbell threw in casually. The problem did not seem vital to him. Neither did Meurer.

"Green Hornet. *The Green Hornet,*" Trendle repeated. "That's not bad. *The*

Green Hornet.

So that was it. James Jewell began auditioning actors for the leads in the new series. The news even spread to New York and Chicago: WXYZ was looking for a leading man. "That's the Lone Ranger station, you know."

Trendle was as fussy about the new series as he had been about *The Lone Ranger*. Striker wrote script after script. The night-time identity of the hero of *Adventures of Jimmy Dale* was Grey Seal. His mark was a grey seal. Striker, Trendle and Jewell had decided to call their hero Britt Reid. He would be descended from the Reid family of which the Lone Ranger was a member. And for his night-time activities he would be known as the Green Hornet.

Ted Robertson's memory of researching for one aspect of the Green Hornet was infallible: "I remember when we first created the sound of the Green Hornet, Mr. Trendle, somewhere in his past had been lying in a hotel bed and he claimed he had heard this sound. And so we had to develop this sound—Fred Flowerday and Ernie Winstanley and I. And Jimmy Fletcher had joined the sound department. The four of us started to try to create the sound. We scrounged sirens; we did everything we could with sirens. But it was not what Mr. Trendle remembered.

"We went to an alarm corporation in Detroit and they brought out all kinds of buzzers and all kinds of things and that was not what Mr. Trendle remembered.

"Actually, we didn't know what we were looking for. He said it was like a giant bee."

Fred Flowerday remembered his troubles: "I even brought a jar of bees from my cottage to record them. I think for ten years I was working on that buzz."

Robertson added: "It ended up—we finally—on a record turntable there is an arm that comes out and that has the pick-up head with the needle that you drop on the record. Fred and I got a piece of wood with two raised ends and we put a violin string between the two ends and drew it taut. And in the middle of that violin string, we put a little piece of wood that had a hole in it. We threaded the violin string through the hole. Then we put the pick-up head into that and we had a buzzer attached to this thing and we vibrated it and it—we could increase the volume and we made records."

Jewell gave it a final touch, auditioned it for the boss, and Trendle accepted it tentatively until something closer to what he had in mind could be worked out.

The Tonto of the new series was to be a Japanese named Kato. He would be Britt Reid's trusted valet and driver of a mystery car.

Cullen Landis, once a leading juvenile in the movies, was now directing commercial films for Jam Handy. He got James Jewell on the phone.

"You've got a couple of Chinese actors working for you over there," he reminded.

"That's right," said Jewell, "if you can call them actors. I use them on *Dr. Fang.*"

"Will you give me their names and let me use them. I need a couple o' Chinese."

"I'll trade you two Chinese for one Jap," said Jewell. "You have a Jap?"

"It's a deal," said Landis. "We use a Jap who runs a restaurant down the street."

So one day, a neat little Japanese wearing rimless spectacles, appeared in

Jewell's office. His name was Tokataro Hayashi. Jewell renamed him Toyo for convenience and sometimes he was called Raymond Hayashi, and Raymond Toyo.

It was decided to make Britt Reid a newspaper publisher. This would put him close to politics and crime during the day and give him leads for his night-time attacks in the name of justice. Two other running characters were created. The publisher was played by Malcolm McCoy; Reid's secretary, Miss Case, meant a job for Jewell's sister, Lee Allman.

Jewell asked Trendle where they'd put the new series.

"It'll alternate with *The Lone Ranger*," said Trendle.

"You're moving *Warner Lester* and *Dr. Fang*?"

"We're killing them, said Trendle. "Even if the Hornet goes once a week, Striker's got enough to do."

Trendle told Striker to begin a wind-up series in which the two heroes, Dr. Fang and Warner Lester, would meet in mortal combat and kill each other off. This would bring the heroes of two popular series together and serve as a listener build-up for *The Green Hornet*.

"You're going to get a lot of complaints," Jewell predicted. "People love Michael Axford."

It was decided to save the character of Axford and popular Jim Irwin who played him; Axford was carried over into *The Green Hornet* as a member of the staff of the newspaper published by Britt Reid.

The sound department met a new problem: it needed a telephone. The telephone company would not give them an instrument; so they invented one.

Fred Flowerday: 'We devised this thing where you put the coin in and it slides down a little slot, hits a bell, and falls into an inverted cow bell."

Flowerday treasures this invention in his office to this day. Everything was ready for the new series—approved scripts, supporting players, sound effects, *Flight of the Bumble Bee* for the theme—everything except—a leading man. They had still found no Britt Reid.

"Why don't you recommend Al Hodge?" Nancy Osgood suggested to her husband. They had all acted together on the wired radio experiment in Cleveland.

"I don't know whether he's good enough for this," said Dick.

"It wouldn't hurt to suggest him." Nancy, not on staff, was doing a daily *Women in the News,* acting in some of the dramas, modelling, and working in industrial and slide films at Jam Handy's. She was also coming to the conclusion that marrying Osgood and leaving her star position at WNAC in Boston had been a bad mistake.

Al Hodge was suggested and auditioned. Jewell made a recording of his voice along with all the others and sent them over the line to the fireplace in the board room at the Madison Theater Building. George Trendle chose Al Hodge.

While *Dr. Fang* and *Warner Lester* were killing each other off, H. Allen Campbell called on his old friend, Harry Hyman, President of United Shirt Shops, and sold him on sponsorship of *The Green Hornet.*

One of the announcers, Fielden Farrington—black hair, black eyes and with a permanent expression of wry amusement—was chosen to be the announcer. On

Al Hodge—The original Green Hornet 1936-43

Gillie Shea, Al Hodge, Ted Johnstone, Herschell Mayall, Gwynne Fermin, Lee Allman, Mary Barrett, Raymond Toyo, Chuck Webster, Fred Fry, Jim Fletcher, Ernie Winstanley, Dewey Cole, Fred Flowerday.

January 31, 1936, three years after the debut of *The Lone Ranger, The Green Hornet* was broadcast for the first time. Al Hodge was "scared absolutely to death." He had been close enough to the preparation to know the degree of perfection expected of him. Further, Mr. Jewell's acting company was a tight little family with only a token welcome for newcomers. And James Jewell himself with his ferocious concentration and familiar scowl could be a veritable tyrant. When the first broadcast was over, Hodge went to his rented room in Webster Hall behind the Maccabees Building and broke out in an all-over rash of hives!

17

Which Way Did He go?

1936 at WXYZ was the year of The Big Shuffle. Actually, it began on pay day before Christmas, 1935. Many of the envelopes were thicker than usual. It was customary at Christmas time to find a sheet of such instructions enclosed with the pay as: "don't spit on the floor, "men wear neckties and jackets," etc. etc. But twenty-two envelopes were even thicker than the others.

"Oh boy!" exclaimed one of the singers tearing at her envelope. "A bonus!"

It was not a bonus; it was her "notice." She was fired. In all, twenty-two performers, mostly singers, were "given the axe." Among those fired was Nancy Osgood.

Brace Beemer was not fired but he was moved downtown. Trendle thought he might have a brain; he seemed to possess more than a voice; so he was moved to the Madison Theater Building to work into management. Trendle was not too fond of Junior Sutton, didn't like the way he greased his hair.

To replace Brace as Studio Manager, Tom Ashwell was hired. Tall, handsome and in his thirties, Ashwell came from the west coast. He was a victim of the Depression; his one suit was shabby and his trousers were patched. He tried and he did have ideas but none were accepted. He was soon fired, drifted down to Florida, was eventually found guilty of rape and murder, and was executed.

Following Ashwell as Studio Manager was a man named Ricker who had worked with Major Bowes at the Capitol Theater in New York. He also tried hard, held endless conferences, and picked everybody's brains for ideas. Ricker was short and fiery with black eyes and wavy black hair, well greased.

Fred Flowerday was naive enough to believe the dramatic department gave Ricker a nervous breakdown. Actually, it was alchoholism that destroyed his association wtih WXYZ.

Alcohol was also cause of the downfall of Charles Gibbons, newscaster. He made it to the weather report one night then got stuck in a groove: "Gonna be cool. Cool. Oh yeah! It's gonna be cool, cool, cool—" whereupon he gave up trying and let his head sink to the table. That was his last broadcast for Trendle.

Some new people were hired. Chief Announcer now was Carl Gensel, older than the others, with gray hair and gray face. He was liked, and he was capable. Carl auditioned another vicitm of the "wired radio experiment" when it folded in Cleveland. The applicant who stood before a microphone waiting for Carl's signal to begin was a short young man with a large head. His face was distinguished by one feature: huge, mellow, dark brown eyes set wide apart.

On the floor below, his friend Al Hodge who had arranged the audition was in Russ Neff's office to hear it. Neff was now the studio manager. Harold True dropped in. The three men heard Carl Gensel Say, "Give us your name, a little of your background."

In a deep, cultured voice the applicant began:

"My name is Charles Washington Wood—"

Downstairs, Harold True exploded.

"Washington! For Chris' sake."

"It's a conspiracy," said Neff. They laughed.

"Did you put him up to it?" True demanded, turning to Hodge.

"No. I didn't tell him" Hodge laughed, "that must be his name."

"George Washington Trendle will love that," Neff exclaimed, his snappy little eyes laughing merrily.

"Goddam it, get Felix Holt in here. This is a story for Radio Jake," yelled True.

"He isn't hired yet," Neff warned.

"Sounds good, though," Hodge pointed out loyally.

Wood was now reading a commercial.

"Too goddam heavy," said True.

"For the Ranger maybe," Neff agreed. "But for those classic records at night or—the Detroit Conservatory of Music?"

"Yes." Hodge agreed, seeming to nod from the waist up.

Charles Wood was hired. His middle name was not mentioned again.

The station also hired a new musical director, Benny Kyte, rather short, thick waisted, easygoing with huge blue eyes. He had been with WJR. One of the first programs he created came out of a Ricker conference. It was called *Blue Chiffon.* Every measure of music had to be arranged by Hank Finney; it was shimmery as Shep Fields' "champagne" music would be later, and Lawrence Welk's style after that. Osgood wrote special copy for the program.

But with the firing of twenty-two performers, the pressure of supplying sustaining copy was considerably lessened. This, plus years of insecurity in the New York theater, put Osgood in a chronic state of fear and when a phone call came from H. Allen Campbell he felt a hot flash of apprehension.

"Dick," said Campbell. "We're goin' to discontinue your department."

Osgood's heart turned to lead.

"But we're not taking you off the payroll," Campbell continued. "We have something in mind and I'll get back to you."

Days passed. There was no call. Osgood became more and more worried. His wife was back on the air with *Women in the News* and she played occasionally on *Ann Worth, Housewife* and *The Green Hornet*. But she hated WXYZ; she hated Detroit; and most of all she hated her husband.

After several days of stewing in insecurity and domestic discord, a wave of native New England pride overcame Osgood's fear and with some impatience he called Campbell's office.

"You tell Mr. Campbell there's no point in my sitting around in the Maccabees Building. I'll be at home if he wants me."

Instead, he spent the next few days watching trials in a downtown courthouse.

Al Hodge continued to write copy besides playing *The Green Hornet* three times a week. He laughs heartily today as he recalls two James "Jewellisms" which he had difficulty adjusting to.

Hodge: "Jim had a theory that at the end of each scene, every sentence had to be finished either in a higher register or with more volume, guts. It had to be tagged some way. And he would tell you how to do it. For instance—" Hodge breaks out in another hearty laugh as he recalls the gimmick—"you have a scene with Kato and — uh' you finally found out who the crook was and you would say to Kato—"—Hodge's voice becomes louder and higher as he builds to the end of the sentence—"you say to Kato, 'Look through these newspaper reports, Kato. I have discovered that without a doubt the man who is *behind this whole racket* IS A GUY NAMED *SO—AND—SO!!!*' "

Hodge laughs again.

"You could never say it naturally. 'That gets 'em,' Jewell would say, 'They're going to hang on there because now you're excited.' No matter whether it was the crook or something you ate for dinner, you had to feel it this way."

What was the other "Jewellism?"

"Well—you're doing a show. I'm not talking about an audition or a rehearsal. You're on the air in the middle of a broadcast and he would come in and take two pages of your script right out from in front of your nose because the show is running long. You have to be a little flexible to adapt yourself to this kind of an operation."

James Jewell: "You remember what a bastard I was. Even Howard Pierce would complain about my treatment of people. But—all that work! I was using the druggists downstairs; I was using a beautician; I was using a mortician; and I got 'em all together and I had to pound into them in three hours rehearsal nineteen years of experience. And I would make people bleed until they did give me what I wanted—even my own sister."

Tokataro Hayashi, or Raymond Toyo, was a constant source of amusement. The fastidious little man was big on pool. He taught other actors and the sound men how to play; he could even beat the musician, Chink Daughterty, considered the studio champion at the time.

Raymond had first worked in Detroit for a prominent automotive executive who had given him the financial backing for his Japanese restaurant. By the time he was sent to Jewell he was prosperous. He was having so much fun playing Kato on the radio that he forgot to collect his pay checks. They did not amount to much, of course, but they did pile up. Finally Jewell explained that he was being paid for his acting. His triangular little face was without expression or interest but Jewell forced him to take the checks.

Then, without warning, he just didn't show up. Inquiries at his restaurant revealed that he had gone to Seattle to greet some girls from Japan. One he married; the others he brought back with him to be waitresses in his restaurant.

Another actor managed to fake Kato until Raymond's return. Jewell explained firmly to the little man that actors had a responsibility to be present when required. To make amends, Raymond invited the entire cast to his restaurant and served them a Japanese dinner of many courses—with saki.

Soon after his return to Detroit, Raymond looked up through his rimless spectacles at the handsome Hodge and said, "Missuh Hodge—you want maid?"

Hodge looked down at him and laughed.

"Is Japanese custom. Wife work." said Toyo.

"I live in a two-room apartment," Hodge laughed. "I need a maid like I need a hole in the head."

"She have to do something," Toyo insisted. "Please. She work for you 'til I get her job. You pay nothing. She have to work."

Hodge: "She had been an upper-floor maid in a very wealthy family in San Francisco and used to arrange flower decorations and so forth. Most of her time was spent flicking little dust spots off the furniture—totally of no value as a maid. But she worked for a while."

Toyo was an enterprising little fellow. He wanted Ted Robertson to go into business with him. There was a garage around the corner from the Maccabees Building and Raymond proposed that they take it over, rent out space and wash cars. Robertson was not interested; he was now Jewell's assistant.

Dick Osgood was eventually summoned to the Madison Theater Building. He would now work downtown in the executive offices—"a special assignment in commercial continuity." At the same time, the mail boy, Jimmy Riddell, was being promoted to "traffic." Traffic was in charge of The Log. If radio ever had a Bible of its own, it was that inviolable document known as—The Log.

The Log told to the second when every sound was to be broadcast. The studio was to treat it as the law of Moses. No announcer, director, engineer, newscaster, sportscaster, singer, pianist or orchestra leader dared stray one second from its dictates; and sometimes the dictates could be tricky and subtle. One program might go two ways—to WXYZ and to the Michigan network. But a sponsor might buy the program in Detroit only. This meant that while one announcer would read the advertiser's commercial to Detroit, another announcer in another studio would read at the same time a "public service" announcement of equal length to the state. There were countless variations on this theme, especially at WXYZ. It was said that any announcer who could "cut it" at WXYZ in Detroit, could work successfully anywhere.

Mimeographed copies of each day's Log were distributed to all so that each might know exactly when he was expected to be on the air, for how many minutes and seconds and to what destination: WXYZ only, the Michigan Radio Network, CBC, NBC-Blue, or MBS (Mutual Broadcasting System)—or a combination of these.

The original copy of The Log, known as the Salmon Log, went to the announcers. They must initial every item on the Log as its dictates were executed. The initials attested to the fact that what was designated to "go" at a given time—went—and to the right destination. If there was any variance whatever, that variance was to be noted exactly as it happened on The Log. Master Control had a duplicate copy which was counter-initialed by engineers so there could be no mistake.

Jimmy Riddell approached Traffic and The Log with a characteristically light and happy heart. He was now twenty-four years old and married.

If one were limited to a single word with which to describe Jim Riddell, that word would have to be "sunny." He was bright, zestful, warm. His smile was wide, white and toothy. His clear Scotch-blue eyes were alert. His complexion was clear—rosy in wintertime, tan in summer from the golf course. His attitude was open, healthy; there seemed to be nothing hidden about him. He would drum on the top of his desk with his pencil and whistle softly as he studied the intricate problems of Traffic.

Working for him were two women. Bernadine Ryan was new—a pretty Irish girl with bright eyes, a tight little smile and red hair. The other was a dumpy woman in her forties with luminous brown eyes and a dominating manner—Irene Donnelly. In the same, square, ample office, without a desk but working at a deal table, feeling very insecure and out of place, was Dick Osgood.

He found the atmosphere very different in the executive offices. No one chased anybody through an office firing a prop pistol, as often happened in the studios. Laughter was differently inspired and less boisterous. Footsteps were less urgent, their sounds muffled in carpeting. Even the clatter of typewriters seemed subdued. The words "Mrs." and "Miss" were spoken more often. People concentrated on their work over longer periods of time. There was less conversation.

Out at the studio work was fun; downtown it was important. Talent's attitude toward the Madison crowd was one of general suspicion, even dislike, mixed with fear. On the other hand, downtown executives and salesmen felt that studio personnel was a little unreal and certainly less important; its job was simply to obey and deliver. Downtown was where the orders were given and decisions made.

Ted Robertson (studio): "There was a terrible conflict. We never understood each other and nobody ever took the pains to try and get us together and say, 'you know, we're all seeking a common goal; we're a company. *We've* got to show a profit so that *you* can earn your salary.' Nobody ever did that. Trendle was a very remote man—a very cold, odd individual. Ray Meurer also. And nobody ever appreciated the terrible hours and days that we put in up at the studio. And—and—we had fun."

James G. Riddell (downtown): "It's really two worlds, you know: the business world and the talent and production end. Of course, we had a physical gap as well, which you don't have in most places. We always thought it was good. If you get too close to people and they're not working out, it hurts you to have to let them go, whereas if you are a little aloof, you can do it. Sounds like a cold approach but it's a business approach."

"Commercial continuity," Osgood's new responsibility, meant all advertising copy. It went to the studios in a looseleaf book. Two books were made up every day in obedience to The Log, one containing all commercial copy up to 6 P.M., the other all copy after 6 P.M. to sign-off at 1 A.M. Perhaps Crowley-Milner, department store, wanted to advertise a special in ladies' hose on Tuesday. Square Deal Miller might order announcements for engagement rings on Wednesday and coffee percolators on Thursday. It was up to Irene Donnelly to make up the books each day. Her responsibility was to see that the right copy for each account was in place as per contractual agreement.

Osgood's job was to check the copy in the book before it was sent to the studio and see that it did not violate the policies of George W. Trendle. Trendle was as fussy about the advertising heard on his radio station as he was about what was seen on the screens and stages of the United-Detroit Theaters. His first thought always was of the listeners. He refused advertising dentists and usually turned down makers of patent medicine. He insisted that advertisers prove the honesty of their claims. He reserved the right to censor any copy.

Osgood was ordered to see that no announcement ran over one hundred words. Spots were not sold by the "minute"; one hundred words was maximum and Osgood was ordered to edit any additional words. He was told to watch for exaggeration and for anything of questionable or offensive character; nothing flamboyant was allowed in commercial copy.

No man was allowed to buy time to make a speech on WXYZ unless George W. Trendle could read the speech beforehand and put his mark of approval on it. Every piece of copy that went on the air must bear on it a scrawled and almost indecipherable "T."

Junior Sutton and Marie Meyer handled one really big commercial account: Conn's Clothes, for men. Junior and Marie concocted one of the best-remembered advertising slogans of the period: "Wear 'em, Tear 'em, Compare 'em, Don't spare 'em." They wrote ten-minute skits, then five-minute skits for this apparel company. They were used not only on WXYZ but by two of the most popular morning men ever in Detroit radio, Ralph Bingé and Joe Gentile on Station CKLW in Windsor, Ontario, across the river from Detroit.

For Conn's Clothes, WXYZ may have broadcast the first singing commercial.

James G. Riddell: "There was a Vere Wirwille and a guy by the name of Ray Gonzales and somebody else; there were three of 'em. And they used to play private parties. They used to do everything. And Vere—for many years—now it comes to me—did the Conn Clothes commercials, singing commercial. 'Conn's Clothes, Conn's Clothes. They're neat and nifty.' "

Could this have been radio's first singing commercial? Or was it "MacDonald for Coal"? Riddell says that Conn's is the first one he remembers.

Conn's was on five days a week, skits and commercials. Then spot announcements in dialogue for which actors were hired. Eventually, these were recorded and used on all stations in the city. Junior Sutton produced most of the recordings in the Maccabees Building. The others were done at CKLW by Binge and Gentile.

Jack Pickering, Detroit newspaper man, recalled that Binge and Gentile once did a skit burlesquing *The Lone Ranger* and used the *William Tell Overture.*

According to Pickering, WXYZ was outraged: "Meurer threatened suit on grounds of copyright infringement for playing the William Tell Overture. Ted Campeau, the manager, was out of town at the time and Dick Jones, commercial manager then, received the call and reminded Meurer the overture had been in public domain for years and years."

Commenting on other activities of the time, Marie Meyer said: "Another thing we did was half-hour Sunday shows for the openings of big movies and they were picked up by all stations—WJR, WWJ—and I don't know whether CKLW and WJBK were included or not. But the idea was if you had it on all stations everybody had to hear it. We did one on *Blue Hawaii.* We did one on *Hitler's Children.* Oh, I'd say there were about a dozen of them."

Once one of these half-hour sales came up at the last minute. Junior and Marie rushed out a thirty-minute adaptation of a new movie and Jewell collected his actors and sound men waiting for the script. Marie knew it could not go on the air without Trendle's initial. It was a holiday and Trendle was not in the office. Marie, who did not drive, grabbed a taxi and raced miles out to Trendle's house in Grosse Point to get his initial for the broadcast. It had all the drama of rescuing the program (Pearl White) from a buzz saw (Trendle). The script with the big "T" scrawled on its front page was rushed across town to the Maccabees Building by cab in the very nick of broadcast time.

One Friday afternoon in early June, 1936, three men were in the office of H. Allen Campbell. Campbell, wearing a tailored blue flannel suit, sat relaxed at his desk. Junior Sutton in dark grey stripes moved restlessly with little steps backward and forward. Sitting across from the desk, slouched down, his knees crossed in baggy tweed, was Irv Stoll.

Irv Stoll was a salesman—inanimate, enigmatic, unexcited and unexcitable. He was over six feet tall and thin. He wore rimless spectacles. His face was thin and colorless. His smile was smug; what he had, he had, and it was up to them to deliver.

"Floyd Tracy will be up here on Monday for an audition," he said.

"But this is Friday!" Junior complained. "Benny Kyte's no problem; I can have the orchestra and quartet and Cyril Wezmael. But what's this feature thing you're talking about?"

"That's up to you, Junior," Stoll smiled easily.

Campbell sat watching the two. Sutton's restlessness bothered him. So did Stoll's inertia. But he liked the account.

Floyd Tracy was Advertising Manager for Hickok Oil in Toledo. They had Hi-Speed Gas Stations throughout Ohio, Indiana and Michigan. They had advertised on WJR for years—a quarter-hour series called *Jimmy Allen*; and during the summers, Benny Kyte's music. Now they were in a mood to follow Benny to WXYZ, five nights a week.

"Floyd wants some kind of a story feature—industry—or—I don't know."

"Dramatic?" Junior demanded, turning from the window.

"Not necessarily," said Stoll.

"WXYZ and the network?" Campbell asked again.

"The Michigan Network plus stations in Ohio, maybe Indiana and probably additional stations in the Upper Peninsula."

Campbell began to think of the profits from the line charges plus money for time and program.

"Hell," he said. "This might even pay for the orchestra."

"But *Monday*!" Sutton complained, teetering from foot to foot. "Here it is Friday. Werner just sent in his numbers and we've got the Michigan Theater Hour to write. And we've got this special hour to do for Universal Pictures. Does it have to be Monday?"

"Tracy is coming up to Detroit to hear it on Monday," said Stoll, quietly, firmly, still with that smug smile.

"Well, Jesus, Irv," Sutton exclaimed seriously. "Can't you let us know about these things?"

"Not'll I know about 'em, Junior."

"What about Osgood?" Campbell suggested. "He's a writuh."

Stoll pulled a leaflet from his coat pocket.

"Tracy gave me this. Put out by a paper manufacturer. Might give you an idea of what he's got in mind."

Unruffled Stoll pulled a circular from his pocket and proceeded to light a cigarette. Sutton took the leaflet and stared like an owl through his thick lenses at Stoll. Why hadn't he offered this in the first place?

"I thought you might have a better idea," said Stoll as though reading Junior's thoughts, blowing smoke casually at the ceiling.

Sutton left the office. He stood in the corridor outside, glancing through the folder. It gave a brief history of the paper industry. Sutton proceeded along the corridor and passed through the door at its end into Traffic. At the left just inside the door sat Osgood at the deal table.

"Dick," said Sutton. "We've got a very important audition Monday morning. Benny Kyte and the orchestra, male quartet, soloist—and a story feature, five or six minutes. The Story of Paper. Here. This'll help you. You write it."

Osgood's face was sour as he took the leaflet. His weekend was shot, but he didn't complain out loud.

There wasn't enough material in the leaflet for more than a minute. Osgood spent Saturday in the public library. The more he read about paper, the more interested he became. He took notes.

Sunday he studied his notes, marked the facts he considered most interesting, and wrote his history of paper around them. He re-wrote the story. He timed it for six minutes. It was two-and-a-half minutes too long; eight-and-a-half minutes. He made cuts. He timed it again. He fussed over it, reinstating one fact he had eliminated, and cutting another he decided was less interesting. The story had to be re-written and timed again. It timed all right but there were two or three places that were not easy to

The acting and sound staff at WXYZ-Radio, 1935. 3rd from right, top row, Earle Grasor (The Lone Ranger). 7th from left, middle row, John Hodiak (later movie star). 10th from left, same row, the original Kate.

read. He re-wrote those bits into simpler language and shorter sentences. Then he typed the whole thing over.

Monday morning, he handed the script and the leaflet to Junior Sutton. He was not invited to listen to the audition with the others but sat at his table working. He could feel Importance in the corridor outside the office as the Possible Sponsor was ushered into the board room with the fireplace.

After the audition, that door opened. Important Voices emerged and made a Grand Exit down the creaking carpetway on Their way to lunch. It was late afternoon before Junior Sutton came into Traffic.

"How'd we do?" asked Jim Riddell.

"They bought it," said Junior. "The story did it. You're going to have to write five a week, Dick. Line up your subjects a month in advance. Send a list to Floyd Tracy in Toledo for his okay, and another one to Mr. Trendle for his initial."

"When do we start?" Osgood asked.

"Monday, June 20," said Sutton. "6:15 P.M."

"How do we list it on the The Log?" Riddell asked.

"*The Factfinder*."

"Who writes the commercial copy?" Irene Donnelly wanted to know.

"Floyd Tracy," said Sutton. "He'll mail it to us."

"Who's going to do the Factfinder on the air?" Osgood asked, concealing his hope.

"Russ Neff," said Sutton.

Osgood was disappointed but comforted himself with the realization that he had not been lost in the Big Shuffle; he was safe. But more important, *The Factfinder* in June, 1936, began a congenial relationship that lasted more than 15 years. It proved very profitable for WXYZ; and the Hickok Oil Company which did no other advertising increased the number of Hi-Speed Gas Stations from 228 to 1800.

18
You Win a Few; You Lose a Few.

There was excitement at the studio. Jack Benny was going to do one of his broadcasts from Detroit. He was on NBC's Blue Network then which meant the Detroit broadcast would originate from WXYZ. No sound men were coming with him, so a script for the Detroit show was sent in advance and Ted Robertson and Fred Flowerday went over it listing sound requirements. One was to be the sound of a wheezy old automobile.

Flowerday remembered it vividly: "Ted and I got this big wash tub and an old electric fan. We removed two of the blades of the fan and rewired the thing into the tub and that was a live prop on Jack Benny's show. I'm not saying that the sound was not improved upon later but this was the first time he had used the Maxwell.

"We didn't have the advantage of a tape recorder. We did not have the advantage of a wire recorder. They had not been invented. Some of our sounds were on records done by the Jennett Company or Major Standard Sound Effects. But most of our sounds were live sounds which we had to invent ourselves."

One standard prop was an empty cigar box with one end partly removed. Each actor had his own with his name on it. To give the effect of talking from the far side of a closed door, the actor would speak his lines into the cigar box.

The Benny broadcast was on Sunday night. Sunday morning, Robertson and Flowerday showed up for rehearsals with their paraphernalia at Masonic Auditorium, a huge concert hall where the broadcast would originate before a live audience. Don Wilson, Benny's announcer, was already there.

"Can either of you do any acting," he asked.

"I can't," Flowerday admitted.

"I've done a little," said Robertson.

"Let me hear you read this bit."

Robertson read it for him.

"Well, we'll see."

Detroit actors appeared throughout the day hoping to pick up a little work. By afternoon, one of them had been given Robertson's bit. But Ted had the last laugh. The actor's bit was not funny. But the sound of the old car invented by Ted and Fred Flowerday was a sensation. It won the biggest howl in the show. It was even applauded. It broke up Jack Benny.

The WXYZ sound department, now made up of six men—Robertson, Flowerday, Ernie Winstanley, Dewey Cole, Jimmy Fletcher and a newcomer named Fred Fry—was often considered the best in the country. The National Broadcasting Company tried to lure it to New York in toto, but for varying reasons the six men decided to stay in Detroit.

Ted Robertson had been sitting in the reception room of the executive offices in the Madison Building for two hours. He was ready to explode. He didn't care why he might have been sent for; he knew only that it was rehearsal time at the studio and he wasn't there. Finally, he jumped to his feet and stepped over to Ad Werner at the switchboard.

"This is nonsense!" he exclaimed. "You tell Mr. Beemer that I was here for our appointment at one o'clock. It is now three. I am going back to my work at the studio. If he wants to see me tell him he can bloody well come out to the studio."

And he walked out of the office. A message was waiting for him at the studio. "Call Brace Beemer." He ignored it. In the middle of the rehearsal he was summoned to the telephone. It was Beemer and Beemer was angry.

"I told you I wanted to see you."

"Forget it, Charlie," said Robertson. "I did my best. I went down there and that's a long trip and I could best use my time up here in the studio."

"I *won't* forget it," said Beemer.

And he didn't.

H. Allen Campbell was on the west coast trying to sell *The Lone Ranger* on the Don Lee Network. Gingham Bread was interested but was insisting that for the west coast the name of the Ranger's horse be changed from Silver to Gingham. "After all," they argued, "you named it for Silvercup Bread in the middle west." Campbell explained that this was not true; Silver had been named before the show was sold to the Gordon Baking Company and could not be changed. Well, they said, they would let him know.

As he rode back to his hotel, Campbell reviewed the possibility of selling *The Lone Ranger* to a single sponsor coast to coast. It would save a lot of trouble to have one national sponsor. He had begun to work on General Mills. Meanwhile, at the end of the second year, Gordon Baking Company had dropped the show; it had been picked up by General Baking (Bond Bread) with headquarters in New York; and now it seemed likely that Bond would not renew for another year. Merita still carried the show in the south. And still another sponsor carried it over the Yankee Network in New England.

These important matters were on Campbell's mind when he checked his messages at the hotel and was handed a telegram which read: "Beemer has offered his resignation and I have accepted. Don't worry. See you when you return to Detroit."

It was signed, "Trendle."

As soon as he was back, Campbell reported to Trendle. They discussed developments on the west coast and laughed at the idea of changing Silver's name to Gingham.

"What happened with Beemer?" Campbell finally asked.

"Oh he came in here and gave me a big speech about all the things that you do that are wrong," Trendle said with a wry smile.

Campbell laughed, "What might they be?"

"I didn't pay any attention to it," said Trendle. "All the time he was talking, I was thinking, 'This is great! Here's this guy in here trying to undercut the man who hired him, while *he's* out on the road selling property that's making us a fortune.' So I fired him."

Beemer landed right side up. He had made a good friend of Ed Jeffries, Mayor of Detroit, and through him met Sam Gilbert. Gilbert was president of the tobacco company that made R.G. Dun Cigars. Gilbert liked Beemer and he admired Beemer's flair for writing commercial copy.

At one of their frequent Gin Rummy games, Beemer broke the news that he had "left" WXYZ.

"Open your own agency and I'll give you the R.G. Dun account," Gilbert suggested.

And that's what happened.

Fran Striker was now getting some help from Tom Dougall and a stolid, brown-eyed, soft-spoken young man named Gibson Fox. But Fran was still mass-producing six half-hour radio dramas a week—Rangers and Hornets.

He had a "sort of understanding" that when and if the Ranger expanded, he would be paid more—so much per station, as in his own syndication days. But he was offered a new contract with an impressive raise. He couldn't resist that raise; so he signed the contract. The ink was hardly dry when the news broke that Gingham Bread had bought *The Lone Ranger*, Silver and all. On January 18, 1937, the program for the first time went coast to coast; and Fran Striker had signed away a fortune.

"Earl Graser is here to see you." It was Campbell's secretary.

"Tell him to wait."

Campbell swivelled in his chair and sat looking out a window. It was contract renewal time for Earl Graser and Trendle's spy system had reported that the actor who was playing the Ranger wanted more money. It was said he would probably ask for twice as much as he was getting. Trendle had somehow received the information that Graser was talking about hiring a manager. He had asked Campbell to "take care of it."

Campbell turned back from the window, pushed down a couple of levers on his inter-office communication box and got WXYZ's new auditor, Adolph Stoffer.

"How much are we paying Earl Graser?"

There was a wait while Stoffer looked it up.

"A hundred a week."

They disconnected. Campbell decided to give the actor time to work up a little sweat, let him wait a while. He finally left his desk and walked leisurely along the carpeted hallway toward the reception room, but stopped at Sutton's office.

"What do you hear from Bill Stockwell?" he asked.

Stockwell's agency handled the advertising for the Automobile Club of Michigan.

"They're happy," Sutton reported.

Marie Meyer gave Campbell a smile and drawled with a kind of happy whine, "They like 'em real well."

She referred to five-minute scripts she was now writing for the Auto Club. She was also writing five-minute scripts for General Motors. Campbell beamed pleasantly, killing time.

"Anybody hear Osgood's story on *The Factfinder* last night?" he asked.

"It was on radium," said Marie. "I thought it was interesting."

"How much time does he spend on those scripts?"

"He spends three hours every morning in the library gathering material," said Sutton. "And it takes him a couple of hours in the afternoon to write the story."

"Lot o' time for a six-minute story."

Campbell sauntered down the hall and smiled at Earl Graser.

"Come in," said Campbell, turning his back and leading the way to his office. Earl followed, carrying his coat. Campbell indicated a chair, walked around his desk, slow and relaxed, and sat down.

"How's the driving out to Farmington this time of year?" he asked pleasantly. "Pretty wet?"

"I hardly notice, Mr. Campbell. I'm pretty sleepy by the time I drive out there at night. Grand River isn't bad this time of day."

"I hear you got picked up for speeding," Campbell said, smiling.

"Thirty-eight miles an hour," Graser told him, wondering how he knew, and shifting his top coat in his lap.

Campbell observed that Graser seemed to be tightening up a little. He took his time lighting a cigarette. Finally, he said, "I hear you want to talk a new contract."

"It's about that time, Mr. Campbell." Graser looked uncomfortable.

"There's a rumor goin' round," Campbell smiled, pinning the actor down with his shrewd blue eyes, "that you want higher pay, that you might even hire a manager."

Graser shifted his coat again. "I don't see how anyone could have heard that."

"You're doin' pretty well with us," Campbell pointed out, keeping the conversational lead. "Don't you feel you can talk with us?"

Graser was too surprised to speak. Campbell continued.

"Let's see, you were gettin' seven-fifty—seven dolluhs and fifty cents—when you started on *The Lone Ranger*. Per show. Now you're gettin' a hundred a week. That's about four times as much. How much do you think you should have now?"

He waited. Graser cleared his throat. "The show seems to be doing very well. It's going coast to coast now, and we broadcast each show three times. That's twice as much."

"A third again as much," Campbell corrected. "Six thirty to the east, seven thirty to Chicago and the Central Time zone; ten thirty to the west coast. You were doin' it twice before we added the west coast."

"The show is doing very well," Graser repeated.

"We're doin' damn good," Campbell agreed. "So are you. *The Lone Ranger* is putt'n' you through law school. It's bought you a house in Farmington. You have a great voice, Earl; you are fine for the Ranger. But you know you're not an actor. You're right for the Ranger but what else could you play?"

Campbell's words came out quietly but slightly clipped and staccato, and one finger added emphasis to key words by tapping gently on his desk as he spoke them.

"I know my limitations, Mr. Campbell; and I have no great ambition to be an actor. But I *am* right for the Ranger and I should be paid all it's worth."

Earl shook his top coat a little and settled it back in his lap. Campbell leaned back in his chair easily. His air was that of confiding in a friend as he said, "You know, I don't get into the program end o' things very much. My interest in the Ranger is only to keep it salable. We've been on for some time now and I'm interested in keepin' it commercial. In fact, I'm workin now on a new sponsor—a national advertiser, a big sponsor; and I been thinkin'—maybe we ought to freshen up the Ranger a bit. Nobody ever shot the Ranger. It might add interest to have him shot. We could have him in the hospital for two or three shows. We might even have him die and his nephew could take his place—the kid, Dan Reid. Hell! That's what we're after—is the kid audience."

Earl Graser sat frozen. His pink and white complexion had suddenly faded. He had nothing to say. Campbell waited calmly, watching. Then he said, "But there's no great hurry. We're not goin' off half cocked. I'm willin' to let things continyuh as they are for another thirteen weeks if you'll sign a contract for $125 a week."

Earl Graser signed the contract.

Dan Reid, the boy character, had to be dropped temporarily from *The Lone Ranger*. Ernie Winstanley's voice changed. In a short time, he plunged into romance with red-eyed Roma Davis. Off mike, his voice was soft and engaging and he could turn on a great deal of charm. He continued to "work sound" and play occasional roles. But he graduated from the *Children's Theater of the Air* and went into a 'teen show called *Junior Matinee*. Among others on *Junior Matinee* was a handsome boy with brown hair and brown eyes from an Italian family, another graduate of the *Children's Theater,* Johnny Desmond. *Junior Matinee* was sponsored by Bunty Candy Bars and was on the air just ahead of the evening news, five days a week, for about a year.

Another kid who had dropped out of sight came back again. He was a nervous Russian-Ukranian boy from Hamtramck. He had appeared in some high school plays but under the driving direction of James Jewell he went into a freeze and was unable to perform. His name was John Hodiak.

Joan Vitez was having a similar problem. One morning was unusually rough. Ruth Dean Rickaby had appeared for Ann Worth with perfume on her breath. She wore the big, floppy brown hat and peeked up at people from under its wide brim, thinking no one would know. One wavering hand held her script, the other gripped the cloth cyclorama, which hung from a circular rod behind the actors to improve the sound. Hanging onto the "cyke" was the only way Ruth Dean Rickaby could stand. She was drunk again.

Jewell, as usual, was kind and understanding. He spent his anger instead on Joan Vitez. He came into the studio from the director's booth, stopped the rehearsal and stared at her.

"What you sound like!" he exclaimed.

"Well tell me what you want, Jimmy. Just tell me what you want."

"To begin with, I want some acting." He was ice cold, sarcastic. He pointed to the director's booth. "If you could just go in there and hear yourself. You're *reading!* Get those lines off the paper into your head—and your heart, if you have one!"

Rehearsal was resumed, Ruth swaying against Joan as she clung to the cyclorama. During the brief break before broadcast, Joan escaped to the girls' room. Bea Leiblee followed and found her in tears.

"Now, now," said Bea, putting her arm around Joan.

"He does the one thing that absolutely ruins me," said Joan, wiping her tears. "He puts me down."

"You're taking it for Ruth," Bea explained.

"He puts me down all the time. He takes away my pride."

"Joan, just remember, he's teaching you how to act. Come on, honey, it's air time."

"I'm good enough to go on the air," Joan said. "But I'm not good enough for Jewell."

Late that winter, 1937, Malcolm McCoy and Beatrice Leiblee were married. They had known each other a number of years, had worked together in stock companies before Detroit, and it seemed like a good idea to get married.

Bea had a friend who took it rather hard. Everybody called her Wacky. Wacky wore tailored suits and a stark hair-do. But for the wedding, she curled her hair and dolled up in a frilly dress and high heels and the gang hardly recognized her.

Two weeks after the wedding, to celebrate an anniversary of *Ann Worth, Housewife,* Frank Dougall, one of Tom's two brothers, gave the cast and crew a party. It was a disaster. Everybody was tired and everybody got drunk and everybody said things they should not have said. Gillie Shea, to his credit, quietly backed up against a wall, gradually sank to the floor, and went fast asleep.

Tom Dougall, on the other hand, became more and more shrill, arrogant and insulting. Sometimes it's wise to call a spade a rake, but at this mad March party, spades were spades—in spades. And the casualty was the two-weeks marriage of the McCoys. A few nights later, Malcolm McCoy returned to his apartment to find Bea's farewell message scrawled across a mirror in lipstick. She had left and returned to her old apartment, preferring friendship to marriage. McCoy was knocked out. He finally recovered with the help of alcohol and studio camaraderie, but it took a year. Bea went on as though they had never been married.

In August, Russ Neff was going on vacation. The question was: who would do *The Factfinder* during his absence? Neff asked Harold True; True called Allen Campbell; and Campbell took it to Junior Sutton. Sutton stood at his office window, teetering from one foot to the other, chewing at his wired jaw. He thought of calling Jewell. Maybe one of the actors—then Osgood popped into his mind. Osgood would be a new voice on WXYZ. He had been an actor. Besides, why toss the decision back to the Maccabees Building? It could be made right downtown. So it was arranged; Osgood would read his own scripts on the air for two weeks.

Everybody was surprised. He sounded good. Floyd Tracy called from Toledo. He preferred Osgood to Neff. It was decided to keep Osgood on the air and make Neff the studio manager. Osgood moved back to the Maccabees Building, feeling as though he had been returned from exile and restored to society. For the first time, he began to feel that he really belonged. In his home, however, he belonged less and less.

"Al Hodge to see Mr. Campbell. I have an appointment."

The star of *The Green Hornet* looked tan and handsome in a light, tailored suit. He broke into a smile and shook hands with Junior Sutton as Sutton walked by. Irene Donnelly had heard his voice and found an excuse to leave the traffic department for her first look at the handsome Hodge. She floated heavily past him, smiling up at him broadly.

"My daughters adore you," she said.

Al acknowledged the compliment, grinning and nodding from the waist up.

Campbell would see him. He walked in. Campbell was standing, showing his strong white teeth, holding out his hand. Hodge shook it and sat down.

"How was your vacation?"

"Fine, Mr. Campbell. I drove to California."

"You must have spent most of your time on the road."

"Just about."

"What can I do for you?"

"Well, Mr. Campbell—." Hodge laughed. "I took the northern route and I happened to hear *The Green Hornet.* From Canada."

Hodge watched Campbell for a reaction. There wasn't much. Campbell was on guard.

"Yeah. They carry it up theyah."

"Well, it's the doggonedest thing." Hodge broke out laughing again. "It's sponsored up there."

He was still smiling, but there was a glint in his eyes which today were cold grey.

"I'm glad it's making some money somewhere," smiled Campbell, cool and easy.

"Yeah," Hodge laughed, getting a little harder. "A tire and rubber company. It threw me a little."

"Why was that?"

"Have you looked at my contract lately?" Hodge laughed but the steel grey eyes watched Campbell steadily.

"No," said Campbell, smiling. This was a game.

"I think it says in my contract that I get paid a little more when the show is sponsored."

"That's a fine point," said Campbell. "The show isn't sponsored in Detroit—not since United Shirts dropped it. You'll be getting some more I expect when Mars Candy takes the show on Mutual."

"But it is already sponsored in Canada."

"I knew it was sponsored up theyah," Campbell admitted. "I forgot your contract. We're goin' to have to change that when we write up the next one. We carry you fellers a long time on a sustainin' basis and I figure we have a right to make somethin' back on our investment when we sell you."

"Well—." Al laughed but his eyes stayed sharp and hard.

"I'll see what I can do to raise the ante a little," Campbell promised.

Al Hodge walked from the Madison Building around Grand Circus Park to the Tuller Hotel. In the lobby, he called a room number.

"Gwen," he said. "It's Al Hodge."

"Come up."

He told her what had happened in Campbell's office. He laughed about it but obviously was still angry. Together Hodge and Gwen DeLany composed a letter to Paul Dulzell, Executive Secretary of the Actors Equity Association, asking for information about a new broadcasters' union. It was called the American Federation of Radio Artists—AFRA.

On the 26th of September, 1937, they received a letter from the Executive Secretary of AFRA, George Heller. He suggested a meeting with Ray Jones who represented the new union in Chicago.

On Sunday evening, October 10, ten nervous actors from WXYZ met in a hotel room with Ray Jones who had come to Detroit from Chicago. They were one person short of the number required to get a charter. A second meeting was called for the end of the month. That meeting was held in Webster Hall, behind the Maccabees Building, and twenty-nine people were present. It was agreed that representatives would contact other people at other stations. The first signed applications were accepted at a meeting on the 7th of November. The initiation fee was $10. One dollar was accepted as part payment. At that meeting, the fledgling Detroit Local of AFRA collected $47.

In 1937, "Dynamic Detroit" was leading the nation out of the doldrums. The Roosevelt regime in Washington had primed the pump, and the automotive industry was pumping hard. There was an air of unbeatable vitality in Detroit, and WXYZ was definitely part of it.

In the fall of that year a group of dynamic Detroiters came together at a cocktail party. Mayor Jeffries was there. Leta Wales Beemer was there with Brace. Joan Vitez arrived with Harry Heilman. Heilman was much older than she, but he appealed to her above all other men at WXYZ. She liked his tight little crooked smile. He was a gentleman. A drink with him at the Alcove and watching him play the horses at the bookie joint behind the cigar store next door lifted Joan out of her Jewell miseries. So occasionally Harry and Joan had a date, and now they were at the cocktail party.

In the shuffle, Heilman disappeared and Brace Beemer stood before Joan. As always he looked tall and elegant. Leta was drawn into the group behind them.

"Let me buy you a drink," said Brace. They gravitated toward the bar.

"I hear you're teed off because they fired you," said Joan. "I don't blame you. Do you know why they fired you?"

"No, and they don't either," Beemer chuckled. The two moved away from the bar and stood in a corner. There was something overpowering about Brace Beemer. It could have been his size; it might have been his ego; it probably was a combination of both.

"Joan," he said. "Do you realize what a jerk you are?"

She thought, this is the man who hired me. What's he trying to say? She looked up at him steadily, without an answer.

"Do you realize you're getting only fifty per cent of what they're paying for you up there?" he asked.

A vertical crease appeared above Joan's beautiful nose.

"I'm telling you," Brace insisted. "Did you ever know this? Did you realize that you're getting only half of what the sponsors are paying for you?"

Joan's laugh was a kind of hoot. "Is that so? Who gets the rest?"

"You never heard of the clip, did you?"

"You mean Mills is paying $150 for me and I'm only getting $75?"

"I don't know what you're gettin', honey," said Brace with a wide, boyish grin.

"But I do know you're gettin' half of what Mills is payin'. That's the clip."

"But Brace—who gets the rest?"

"Who cares? The thing is—*you* don't get it."

The next morning, Tom Dougall directed Ann Worth as he now often did. After the broadcast, Joan Vitez found Jewell in his office and told him she was leaving the show.

"What's *this* all about?" Jewell wanted to know.

"Jimmy," said Joan, without temper, but firmly, "for an actor to be good—you've got to have your ego. Without your goddam ego, you're nothing! You know? I've got to go in there with confidence that I'm me and I can do this and this is right. And then somebody puttin' you down! I was green. I was absolutely new green."

"Sit down, Joan. We'll talk."

"No, Jimmy. I'm going to New York."

Five minutes after Joan had left his office, Jewell went to the bank on the street floor of the building. It was pay day. Joan Vitez was drawing out her savings. He got in line behind her.

"Look, Joan, come back upstairs and we'll talk; we'll straighten it out."

"I've got it all straight, Jimmy."

"New York is *full* of actors. Competition is terrible."

"I'm not as green now as I was, Jimmy. I've made my plans. I will give you one more week, to find a replacement, that's only fair, but that's it and thank you very much."

Joan took her savings and walked away. She played Ann Worth one more week, with a girl named Lenore Collins watching her, and listening to her, and learning to imitate her. Lenore was a buyer from Hudson's Department Store, pretty and with a low voice.

Ruth Rickaby gathered some of the cast together for a little farewell. Joan told them about "the clip." They had all suspected; but none of them knew.

"Look," said Malcolm McCoy, picking at a hangnail and looking at Joan. There was something impish in his laugh. "Do us a favor. Go to Mills Baking and tell 'em what's going on."

"That would be great," said Petruzzi.

"What have you got to lose?" Ruth asked.

"You're leaving anyway," Petruzzi pointed out.

"All right," said Joan. "Why not?"

True to her word, Joan Vitez looked up a man whose name, she remembers, was Russ. He was in the advertising department at Mills Baking. She had met him before in connection with Ann Worth. She told him now she was leaving the show. He tried to persuade her to stay.

"No," she insisted. "I've kind of passed the point of no return here. I've already made the decision and that's the way it is. But the other actors wanted me to come to you and ask if you knew about the clip?"

He didn't know what she meant. She explained it. As Joan Vitez walked away from the Mills Baking Company she erased the whole affair from her mind. She

Johnny Slagle, Benny Kyte and studio orcehstra 1935-45

never went back to WXYZ. Lenore Collins took over the role of *Ann Worth, Housewife* but, for whatever reason, Mills Baking cancelled two months later and the show went off the air.

An afternoon in December, 1937. The award lay on top of Trendle's desk. He and John Kunsky stood looking down at it. They had been running the station seven years and eight months. John Kunsky looked up from the citation, sentimentally overcome.

"Ain't - that - somethin'!"

His eyes were moist; he was very proud. This was one of his increasingly rare visits to the station. He had changed his last name to King. He had always disliked his last name but promised his mother he would not change it while she lived; and he had kept his promise. The company that owned the award-winning WXYZ was now known as the King-Trendle Broadcasing Corporation.

The award was Variety's 1937 Citation for Showmanship in Program Origination. It was presented each year to the one radio station in America chosen as the best in producing programs. In Trendle's mind his best programs ranked in this order: *The Lone Ranger, The Green Hornet, The Factfinder, Hollywood Impressions* (the half-hour program promoting movies), *Children's Theater of the Air, Junior Matinee* and a morning program for home-makers called *Happiness House.*

Trendle shrugged off the award, made light of it, but it was obvious that he was proud of it. Seven years before, he had known he had a long way to go before WXYZ would have his own personal sound. Today, he thought, the station did sound good. Less sentimental than little John Kunsky-King, he nevertheless held out his hand with a rare display of warmth, smiling.

"Merry Christmas, John."

"Merry Christmas, George."

Trendle left for his home in Grosse Point. King rounded up everyone he could find in the downtown offices and took them all over to Cliff Bell's for a drink to celebrate.

19
Nothing Stands Still.

When *Ann Worth, Housewife* went off the air, Tom Dougall was moved downtown and given a desk in the same office with Fran Striker. It was an awkward arrangment for both. Two men could hardly have been less alike. The lean, hard bachelor Dougall with his sophistication and intellectuality was a complicated personality. Fran Striker was a warm, simple person, somewhat naive, and dedicated to his work and family. The two had nothing in common, except that they both wrote Hornets and Rangers. Each respected the other and there was no friction; but there was no real congeniality.

Dougall had inherited an income just large enough to make him a little lazy. His attitude toward the Trendle dramas was half serious, half amused. He said one of his

Rangers was based on Romeo and Juliet.

Instead of mining ideas in Shakespeare, Striker spent hours watching western movies. Actors sometimes recognized dialogue in scenes by Striker as something they had heard in a theater. Because he felt he had created the Ranger, Tonto, Britt Reid, Axford and Miss Case, Fran never got used to other people writing them. It was not jealousy. It was more as though strangers were walking through his yard.

Once he had licked the limitations of the Lone Ranger format imposed by Trendle's rules and policy, Fran reduced the mass production of plots and characters to a formula so clearly and exactly worked out that he could take ten cards, each designating a type of character, a problem the character could be stuck with, the manner in which the Ranger learned of it, how he and Tonto became involved in it, how they licked it, and how he disappeared. After shuffling the deck and placing ten cards face up, the rest was almost automatic.

Striker worked at night—alone. Dougall played at night with the studio crowd.

Early in 1938, the first Lone Ranger movie serial was released by Republic Pictures. It did a lot of business and won the Exhibitor Award as "Best Serial of the Year." Net profit for King-Trendle: $60,000. An agreement had almost been reached for the Lone Ranger newspaper cartoon strip. For the first time, Ranger shows were recorded for stations not tied with the live network broadcasts. In little time, eight stations contracted for the transcriptions.

Although at the end of each Lone Ranger broadcast, it was announced that the programs originated in the studios of WXYZ in Detroit, the show became so popular nationally that people overlooked its point of origin. For example, Al Jolson's last stage musical, about a radio cowboy who was afraid of guns and horses, had its premiere in Detroit. The show might have fulfilled its intention to be funny if its authors, director and producer had visited WXYZ and watched the Lone Ranger broadcast—actors jumping on and off invisible horses, the sound effect of a gun shot in one studio and an actor dying of it in another, six men making galloping horses with bathroom plungers, and so on. But they were apparently unaware that the source of their inspiration was only four miles north of the theater. The show flopped.

When Horace Heidt played Detroit with his band, he offered a prize to the one in the audience who could give the Hi-yo, Silver call and sound most like the original. Earl Graser was in the audience with friends who persuaded him to give his famous yell. He didn't win.

The Green Hornet picked up the Detroit Creamery over WXYZ, the Michigan Network, WSPD in Toledo and two or three neighboring stations. It was about to go sustaining to the Mutual Network and to stations on the west coast. A movie company was showing some interest in it.

In February, 1938, *The Factfinder* was replaced by a transcribed quarter-hour dramatic series called *Black Flame of the Amazon*. Floyd Tracy had been visited by a Chicagoan who sold Hi-Speed not only on using the series but on financing the production of the transcriptions in California.

"We'll sell a lot of sponsors so you will get your money back and then some," the man promised.

It was a bad deal. The Chicagoans never sold another sponsor and the whole thing fell through. At the end of May, *The Black Flame of the Amazon* burned out and Hi-Speed resumed with *The Factfinder.*

While the black flame was flickering and the factfinder was put to rest, a plum fell into the lap of Dick Osgood. Ramon Novarro, the movie actor, was booked to appear at the Michigan Theater. He had many hits to his credit including *Prisoner of Zenda, Scaramouche, The Student Prince, Mata Hari, Cat and the Fiddle* and *Ben Hur;* but Metro-Goldwyn-Mayer had dropped him and he had made a couple of duds for Republic Pictures before plunging into a personal appearance tour. Trendle had a feeling he was a has-been and would need some special promotion to bring in business. He summoned Earl Hudson and Alice Gorham into conference. Out of this meeting came a series of five little five-minute programs put on WXYZ every noon for a week before the star's arrival. The series was called *Hollywood Highlights* and although some movie news and gossip was made part of it, its main purpose was to invite Novarro's fans to mail in questions they would ask him if they had a chance to meet him. Thursday would be the deadline, and Friday noon 20 winners would be announced. They would meet Novarro at the Michigan and ask him their questions in person on a special half-hour broadcast early Friday evening.

It was decided that to make the promotion sound important it should not be done by a regular announcer. Trendle did not want one of his actors to be used. Earl Hudson suggested Harold True. Alice Gorham suggested Dick Osgood.

On Monday, March 6, 1938, the Factfinder became for one week a Hollywood gossip. Letters poured in. Novarro arrived in Detroit on Thursday. He had lunch with Osgood and picked the questions he would answer. Twenty people were called and appeared with Novarro on the special broadcast from the musicians' room under the stage of the Michigan Theater.

The promotion helped Novarro to a pretty good week at the huge Michigan Theater; and Trendle decided to keep *Hollywood Highlights* on the air, but move it to 5 P.M. There was at that time only one show like it and that was Bide Dudley at WCAU in Philadelphia. Osgood did Hollywood news four nights, and on Fridays a similar show promoting radio, called *Listen, Detroit*! on which he interviewed various members of the staff—announcers, musicians and actors. Finally, the scope was broadened to include *all* entertainment—legitimate theater, night clubs and concerts, as well as movies and radio; and the program was given a new title, *Show World.*

Entertainment news came over the wires. Items were picked up in trade papers and newspapers from New York and Hollywood. Osgood kept a file. He wrote features about people who were playing in Detroit. He began to interview celebrities regularly. Alice Gorham invited him to attend screenings with the three newspaper critics of pictures booked into United-Detroit Theaters; he covered night club openings with Herschel Hart of *The Detroit News*; and went to opening nights at the Cass and Schubert-Lafayette Theaters. And he reviewed what he saw. It was one of the first programs of its kind in the country; and it survived for 32 years.

The unionization of broadcasters progressed slowly in Detroit. Questionaires distributed to all major radio stations about wages and working conditions were ignored. There was a reluctance about joining a union; performers were egotistical, individual—and afraid. The insecurity of the Depression was still fresh in mind.

In January, 1938, a meeting was held for the first election of officers and Al Hodge was unanimously elected to serve as the first President of the new Detroit Local of AFRA. He missed the first two Board Meetings, and many others. Gwen DeLany was the first Executive Secretary. She served without pay and received less and less radio work. The union did pay some of her expenses. It was still without a contract.

Other stations were paying no more for their actors than WXYZ. Ted Robertson recalls acting on a program for WWJ. Rehearsals began at 8 o'clock in the morning. There was some time off in the afternoon. The broadcast was in the evening, before a live audience. The actors had to wear tuxedos and evening dress. The pay was $2.50.

Even so, there were many who were eager to act in radio. One of the most persistent was John Hodiak. He had now graduated from high school and was working at the Chevrolet plant, but he came to see Jewell again. Jewell didn't give him any work, but granted his request that he might be allowed to watch rehearsals.

One night, Fred Reto who was recognized from coast to coast by his gravel voice, suffered a gall bladder attack 15 minutes before air time. As luck would have it, he had a very small part that night. Jewell turned from his stricken actor to John Hodiak who was watching in a little booth outside the studio.

"*You!*" Jewell yelled through the window, pointing at Hodiak. He beckoned to him to come into the studio, shoved Reto's script into the hand of the startled Hodiak and said, "You are now one of the Lone Ranger actors."

Still, Hodiak was seldom used, and he stopped coming to the studio.

There was no air-conditioning at WXYZ. One steaming hot Sunday in July, fifteen musicians, eight actors, four sound men and James Jewell were crowded into a stifling studio. They were sweating through the preparation of a special half-hour broadcast to promote a new movie. Junior Sutton and Marie Meyer who had prepared the script were in the studio control room with the engineer. The program would originate at WXYZ and be carried also by WJR and WWJ. Musicians' shirts were wet; actresses' hair clung to damp foreheads; and perspiration dripped from the conductor's chin.

"Ernie," said Jewell. "Did you get the tubs?"

"Yes, sir. They're in the sound studio."

"Jim, Ted. Help Ernie get those tubs in here."

Three sound men lugged into the studio two big wash tubs. Jewell ordered several 25-pound cakes of ice delivered to the studio and had them put in the tubs. He had five electric fans placed behind them so that cooled air would blow toward performers and sound men. Jim Jewell had introduced air-conditioning to WXYZ.

Three men were in Trendle's office: Harold Christian, small, eager and nasal; H. Allen Campbell, calm and astute in tailored brown woolen Cheviot; and Trendle himself, his linen shirt as fresh as though he had just dressed after a shower.

"What's the story on this Cream o' Nuts Soup?" Trendle wanted to know.

"The product was not up to expectations," Christian explained. "It wasn't what was represented to us."

"We cancelled the program," said Campbell.

"It's the first failure we've had in the merchandising department," Christian pointed out.

"Well—who *is* this Cream o' Nuts Soup?" Trendle asked.

"An independent company, Mr. Trendle," said Christian.

"Did they have any spots?"

"No," Campbell told him. "Just the program."

The whole situation left Christian amused. "Oh, the stores had an awful time. They had to take most of the stuff back after they sold it. The product was just no good. The company that made it just never got off the ground. They're going out of business."

"It's not good for the station to have that kind of stuff on," Trendle remarked as they broke up the meeting. "Oh! Campbell. Wait. I want you to take care of Jewell. He's pestering me for more money."

James Jewell was now producing *Children's Theater of the Air* at the Regent Theater every Sunday, *Kiddie Karnival* at the Colonial Department Store every afternoon, special shows promoting movies, *The Green Hornet* and *The Lone Ranger*. He felt justified in asking for more money. Furthermore, he was convinced that he was indispensable. So, for that matter, was everybody else at the studio.

It became impossible for him to establish contact with Trendle and he found himself in Campbell's office. He tossed a telegram on Campbell's desk. It was an offer from the west coast for a lot more money.

Unhurriedly, Campbell read the telegram. He was convinced it was phony. Finally, his shrewd blue eyes looked up from the telegram and narrowed a little as he looked at Jewell.

"You considerin' this offer?"

"What can I do, Mr. Campbell?"

"Hell! Grab it! We can never pay you that kind of money. If you can get this, take it."

The two men stared at each other in silence. Campbell believed he was calling a bluff. Jewell was stunned; he simply could not believe Campbell's reaction.

"I'll think about it, Mr. Campbell," he said finally.

Jewell believed afterwards that Trendle killed the offer and had him blackballed on the west coast. Meanwhile the situation at WXYZ had become impossible for him; he couldn't bear what was happening to his pride. On May 17, 1938, he addressed a note to George Trendle in which he stated that when his contract expired on the first of June, he would not renew. He wrote that he had another job and he offered to train anyone Mr. Trendle would pick to be his successor.

Trendle's reply was brief, almost curt, expressing surprise and wishing him luck. Plans were made at the studio for a big farewell party. A banquet auditorium was engaged at Webster Hall. It was sold out to the last table in the balcony. The sound men took up a collection and Fred Flowerday went to a jeweler who specialized in trophies and ordered a set of miniature gold plungers to be made and inscribed to Jewell. When the time came to pick up the gift, Flowerday had spent the money and the sound department had to take up a second collection. The actors had a plaque made up inscribed to Jewell with all the actors' names engraved on it with a farewell message.

It was a big and sentimental evening. No one was present from the executive offices, but Jewell received a handsome watch bearing the inscription: "To James Jewell, a token of friendship and appreciation." And it was signed with the names of George W. Trendle, John H. King, H. Allen Campbell and Howard Pierce.

The following week, Station WWJ announced that James Jewell had joined its staff. He took some of the kids with him from the *Children's Theater of the Air*. His stay at WWJ was comparatively brief, however, and he moved on to Chicago.

The people with whom he had worked remembered Jewell with mixed emotions.

Fred Flowerday felt: "Jim was a hard guy to get along with and a hard guy to please but I think he knew what he was trying to get all the time; and he was in complete command of himself. And I think sometimes people interpreted his demand for the way he wanted things to be done as being a tyrant—other than just being more demanding. But Jim, I think, did a lot for the technique of broadcasting. He was the originator of a lot o' things."

Ted Robertson had reason to have even more mixed feelings: "It's been a long, long time. I worked for Mr. Jewell and Mr. Jewell fired me regularly three times a week. And we had terrible scenes because we were both tempestuous.

"But I have to say about Jimmy Jewell; he was a genius. He was an absolutely dedicated man. He had no sense of humor whatsoever, and he used to drive us to distraction because he made demands upon us. He was an absolute perfectionist. I want to say I have met a lot of people but I have never known a genius like Jewell.

"He worked so hard!"

Charles D. Livingstone remembered him with a somewhat heavier hand: "Jim was a sort of tyrant in the way that he handled people. If you did a good job for him and you were for Jim Jewell all the way, everything was smooth. But as with any director, if you crossed him in anyway there was hell to pay."

Howard Pierce was somewhat condescending: "Jewell was a nice boy. I don't know anything about his being a genius, but Jimmy always did his work. He did a good job."

The *Kiddie Karnival* did not long survive the departure of James Jewell. Russell Neff and Tom Dougall took over the writing, directing and production of *Children's Theater of the Air* which was moved from the Regent Theater downtown to the Broadway-Capital, now the Grand Circus Theater.

"We bin thinkin' of somebody to take over the dramatic shows." Campbell dropped a match into the ash tray on his desk.

"Yes, sir," said Livingstone, obviously in suspense.

"We bin thinkin' of Osgood. He did some directing in Boston. But he's doin' *The Factfinder* and he's got this movie show." Campbell leaned back easily in his chair and eyed Livingstone with a perceptive and steady gaze. "Felix Holt your friend?"

"Yes," said Livingstone. "Felix got us our boy, our adopted son, Russ. He and his wife, Marjie. Last January."

"Felix thinks you're the guy for this thing and Mr. Trendle is inclined to go along with him. So we're goin' to give you Jewell's job. You're getting $75?"

"Yes, sir," Livingstone said hopefully.

"We'll continyuh to pay you $75 a week until the end of the year. And then we'll pay you up to a hundred."

So Charles D. Livingstone became dramatic director of WXYZ. Tom Dougall was made his assistant. If Livingstone was apprehensive about moving into the vacuum created by Jewell's departure, he showed no sign of it. Actors and sound crew co-operated with him enthusiastically. Generally, the actors were less afraid of him and he brought back to WXYZ a few who had been unable to perform because of their fear of Jewell. Among these was John Hodiak who developed into one of the best villains on the station.

Almost immediately, Trendle got on Livingstone's back about *The Green Hornet*. "I still don't like that hornet sound. I want you to go to work on that. It still doesn't have the right kind of a sting to it."

So the sound crew went back to work on that. Among the immediate improvements made by Livingstone was the extension of his timing technique to all dramatic programs. No longer did a character not in the script prolong a program that was running short by interrupting a scene to ad lib; programs no longer ran short. No more did actors undergo the harrowing experience of having pages snatched from their hands during broadcasts.

Listeners had complained that the musical segways were too loud, that they had to turn down the volume on their sets when the music came on, and then couldn't hear the actors and had to turn it up again. Livingstone took this complaint to the engineers. They explained that music sounds louder in decibels to the human ear than it really is. So it became the practice to set the actors' level at zero and the music level at minus four. Livingstone then made it a habit to warn the studio control engineer when a musical interlude was coming so he could change the level of output. There were no more complaints.

Livingstone worked out another directorial invention that was very effective. About two minutes before a gun shot or an explosion, the level of the voices was gradually lowered, like dimming the lights, so that when the shot or explosion occurred, it sounded much louder than what the listener had been hearing, creating surprise and reality.

Chuck Livingstone also planned the remodelling of the studio lay-out for more convenience and workability.

Two announcers of importance were hired by Russell Neff in 1938. One, Jack McCarthy, had majored in Speech and Drama at college, and headed to the West

They'll Dramatize 'Paths of Glory' on Air Tonight

A radio playlet will be heard over station WXYZ at 10:15 p. m. today when the WXYZ Players, under the direction of James Jewell, will broadcast episodes from the stirring war story, "Paths of Glory," which starts tomorrow in The Detroit Times. "Paths of Glory," by Humphrey Cobb, is the novel which France banned! It strips the tinsel from the glamour of war. Rehearsing for the broadcast are (left to right): MALCOLM McCOY, KLOCK RYDER, JAMES JEWELL, RICHARD E. OSGOOD, FRED RETO and CHARLES LIVINGSTONE.

Coast as soon as possible: "I got out of University of Michigan in June of '33 and took off for Hollywood immediately. I mean, I didn't want to keep them waiting."

The movie studios were obviously not ready for Jack but he was offered the role of Lorenzo in a stage production of *The Merchant of Venice*, starring Lucille LaVerne. Its run was short and a combination of the Depression and a withering indifference to his presence on the part of the movie industry sent McCarthy back to Michigan. He was accepted into a training program for executives at the J.L. Hudson Department Store in Detroit, and got a job as announcer at WMBC, a small station at Grand Boulevard and Woodward Avenue. For three months, he worked at Hudson's during the day and at WMBC from six to midnight. Then he dropped the Hudson project and stuck to radio.

One of WMBC's popular programs was a 60-minute daily feature called *The Happy Hour*. Amos Jacobs, who grew up on the program, worked the bars with his bride. He had a Semitic face with dark brown eyes and a large hooked nose, and he told Jewish jokes.

He and Jack McCarthy were fooling around the studio one hot day in their shirt sleeves. Something Jacobs was wearing around his neck popped out from under his shirt. McCarthy, a devout Roman Catholic, saw it and was shocked.

"What's *that!* he exclaimed incredulously. "That looks like a *scapular!*"

Jacobs gave him an equally indignant look and said, "It *is* a scapular!"

"What the hell are *you* doing with *this*?" Jack demanded, grabbing the scapular in his hand as though he might jerk it off Jacob's neck.

"Who has a better right?" Jacobs said, defiant and offended.

"A *Jew*?" cried McCarthy.

"A *Catholic*!" yelled Jacobs.

After two years at WMBC, McCarthy heard that WXYZ was looking for a new announcer. He went to Russell Neff, auditioned and was hired.

McCarthy was a tall man whose height was mostly in his legs.He had strong hands with long fingers. His face was caught somewhere between pretty and handsome. His hair, parted on the side, was black, straight, short and thick. And his Irish eyes changed from grey to green according to his health at the moment. The main thing in radio, of course, was the voice, and his was right up to the Trendle standard—resonant, masculine, and when relaxed and not pushing a product, ingratiating.

The other announcer was Douglas Edwards. Short and round with brown eyes and brown hair, this gentle southerner was already determined to make the networks as a newscaster. He started in radio when he was 15 at Station WHET at Troy, Alabama, then moved to WSB in Atlanta. From the beginning Edwards wanted to do news and in a deliberate move to prepare himself, while announcing at WSB, he did some moonlighting on the *Atlanta Journal*. For a year, he put a Sunday page together.

In September, 1938, a friend phoned him from Louisville to tell him the Lone Ranger station in Detroit, WXYZ, was in the market for a new announcer. Immediately, Edwards cut an audition record and sent it off to Russ Neff with a letter giving his background. Neff offered him fifteen dollars more a week than he was getting in Atlanta, and two weeks later, Douglas Edwards showed up at the Maccabees Building in Detroit.

As he remembered the move, Edwards said: "I went to Detroit to get exactly what I got there. Actually, it was a network experience. You couldn't find a better training ground anywhere in the country. I listened to Harold True's delivery. Very few people could deliver a more crisp copy than Harold True. Harry Golder was a superb workman. Harry liked the beer; liked it in the morning and, I don't know, I guess he probably drank a little Scotch, too, here and again; but he *was*, I think, one of the better men at our business. I learned a lot from Fielden Farrington."

One of Edwards' first assignments at WXYZ was to go north to Alpena, Michigan, with Traffic Judge John D. Watts to do a special remote broadcast of Watts' *Sandlotters* program, devoted to sports. Doug, from the south, had never seen ice skates in his life. This was winter, bitter cold, and he had to stand on the ice for 45 minutes describing what he saw—"color," they called it.

Edwards suffered somewhat: "It was a little chattery, boy, because I was cold. So then I just went on a regular routine. Carl Gensel was the Chief Announcer. Some of the other boys then were Fielden, Golder, Jack McCarthy, Joe Hurd, Charles Wood, Rollie Thomas, Don Hawkins, Jack Lellman. And I'll never forget those salmon-colored Logs!"

Slowly and surely, Douglas Edwards steered his course toward the news. WXYZ had initiated something new in straight news reporting. Two men had been hired to re-write the material that rolled off the news wires. One was a quizzical, capable little man named Henry Alexander. The other was Rollie Hoover. Presiding over this news department was the other man who had hired them, that bouncy little bundle of exposed nerve ends, Felix Holt.

Douglas Edwards haunted the newsroom, watching the preparation of copy. He talked with Felix Holt whom he admired tremendously. Like everybody else, too, he believed that Felix was the studio's best pipe line to George W. Trendle.

"*What—the—hell!*"

The used car dealer lost no time getting the WXYZ salesman on the phone.

"Who the hell you got readin' my spots?" he complained. "I just heard some guy say my salesroom is on *Leevenwah*. Who the hell ever heard of *Lee*venwah? Where's that?"

The salesman called H. Allen Campbell and Campbell called Russ Neff.

"Get a hold of the guy who just read Jack's Used Car spot and tell him how to pronounce Livernois and if he can't say it the way we say it in Detroit get rid of him."

Neff pounded up the stairs to Studio E. Harry Golder was there. He couldn't have been the one. Joe Hurd was there. Nope. It must have been Charles Wood.

"Yes," said Wood in his deep, cultured voice. Golder had a wide grin on his face.

"Listen," said Neff. "Livernois is pronounced Liver-*noy* Not Leevenwah."

"Oh?" Wood's upward inflection gave him an air of superiority.

"The sponsor called up and Campbell's on the war path."

"Right," Wood agreed. His attitude was that of the United Nations Building toward an outhouse in Kansas.

Neff exchanged a sly grin with Golder and left the booth.

"Crazy place," mumbled Charles Washington Wood. "Beaubien is Boh-*bean*, Sault Saint Marie is the *Soo*, and now Livernois is Liver*noy*."

Joe Hurd laughed.

"Yeah, Charlie, and *Dertrwah* is pronounced De*troit.*"

20
Tale of the Phantom Broadcast

In the Spring of 1938 the King-Trendle Broadcasting Company and the United-Detroit Theaters abandoned the Madison Theater Building and moved across the Grand Circus Park into what is now called the Mutual Building (then the Stroh Building). Trendle and the United-Detroit Theaters occupied the 16th floor; the executive offices of the radio station took up the 17th floor.

The Lone Ranger was taking more and more of his time. A second movie serial was being planned. The radio program was now sold in foreign countries by transcription. Lone Ranger novels written by Striker were very successful; and by September, the Lone Ranger cartoon, also written by Striker, was appearing in the *New York Journal-American.* Lone Ranger merchandise was a big thing on the market. Trendle allowed nothing to be sold that might be harmful; there would be no Lone Ranger knives or bows and arrows nor anything that might lead to a fire. He was the controlling power behind all these activities—broadcasts, serials, books, cartoons, toys—and the Lone Ranger Safety Club which now had well over a million members.

Sponsorship of the Ranger continued to be a complex problem. Merita still had it in the south. Sponsors were coming and going in different parts of the country. Scripts had to be written and produced in such a way that different commercials could be inserted in different markets. It was complicated.

Trendle found it more and more difficult to wear three hats: WXYZ Radio, the United-Detroit Theaters; and the incorporated dramas. All three divisions were prospering. The Ranger alone was expected to net better than half a million dollars this year. His policy of clean family entertainment had restored the theaters to popularity. And as an indication that the Depression was over and the station was in the black, he noticed that Campbell and the salesmen had adopted the latest symbol—the camel hair coat.

Howard Pierce retired in 1938. He had built a home in Fort Lauderdale. He still received ten per cent of the mounting profits and held an executive title. He was happy to get out of radio, if not the theaters, his special love.

"I loved the theater," he said. "I was only interested in the theater. I wasn't particularly interested in broadcasting."

Trendle was constantly trying out new shows. He set one in the time of the American Revolution. Its main characters were twin brothers. It was a turkey and soon was beheaded. He had a number of flops. One that almost made it appeared from nowhere toward the end of 1938. It was called *Ned Jordan, Secret Agent.*

Instead of riding a great white horse or driving a fabulous automobile, Ned Jordan travelled by railroad for which he supposedly worked. Instead of a Tonto or a Kato, Ned Jordan had a girl friend and an FBI agent. In this series, the FBI would bring a problem to Jordan, and Ned would solve the mystery and in the final scene turn over the villains to Proctor, the FBI agent. The tag line of every story came as Proctor seized the villain and said, with gusto, "Uncle Sam wants you!" These were words printed on enlistment posters on which Uncle Sam was pictured pointing at the passerby. Hitler and Mussolini were becoming more and more ominous in Europe and appeals to American patriotism were increasing although, so far as the public knew, the United States at this time had no idea of going to war.

There was the usual rash of auditions. A plump young inexperienced blonde, named Shirley Squires, won the role of the sweetheart. Dick Osgood landed the role of Proctor. And chosen for the title role of Ned Jordan was the new annnouncer, Jack McCarthy.

Fred Flowerday finally produced a sound that satisfied George Trendle for the fantastic car driven by the Green Hornet. Beginning with a recording of a Dusenberg, he equalized and filtered the sound until all that was left was a kind of hum.

Mr. Trendle also had a thing about doors. He wanted the door of Britt Reid's office in the Hornet to sound exactly like the door to his own office. Flowerday studied the sound, got the help of a carpenter, spent hours and a lot of money, and built a heavy prop door against the wall of the sound studio. It was used only for Britt Reid's office, and Trendle was satisfied.

But he still was unhappy about the sound of the Hornet which served as the program's signature. Fred was still experimenting.

Meanwhile, H. Allen Campbell had a problem.

Campbell: "We had trouble gettin' *The Green Hornet* goin'. We put it on the network sustaining. From that, we sold it to General Mills. They sponsored it for a year. We couldn't keep the damn thing sold. We had one piece of bad luck after another with our sponsors of the Hornet."

Campbell had no such problem with *The Lone Ranger;* as soon as one bread dropped it, another would pick it up. He was still working on one national sponsor. Meanwhile, Ned Jordan landed Mars Candy Bars. WXYZ now had half-hour dramas going coast to coast Monday through Friday: *The Lone Ranger,* Mondays, Wednesdays and Fridays; *The Green Hornet* Tuesday nights; and *Ned Jordon, Secret Agent,* Thursday nights.

Then the writers received a black-on-pink office memo from Trendle ordering them to come up with another action show like *The Lone Ranger*, but with a dog for a hero.

"Go-o-oh, Mogo-o-oh, GO-O-OH!"

Tom Dougall.

Tom Dougall could hear it in his mind—another Hi-yo, Silver. With his instant intellectuality, a whole phase of his life flashed into his mind as he read the memo from George Trendle.

Dougall's father, now dead, had been a builder at one time in the Canadian Pacific territory. He had been an admirer of the poems of Robert Service and possessed a collection of books by the Canadian writer. He gave his handsome son, Tom, five dollars every time he memorized a poem by Service, and would rouse him from his bed in the middle of the night to recite for his guests.

It was natural that Tom Dougall should set the drama for Trendle in the Yukon. He had already solved the problem of the dog. It couldn't be a dog like Lassie because this, Trendle said, must be an action story. It had to be a working dog. And a working dog in the wild days of the Yukon could be nothing but a Husky.

The dog's master, of course, must be a fine, upstanding member of the Northwest Mounted Police, established to keep peace in 300,000 square miles of wilderness, in 1873, 31 years before the word "Royal" was added to the group's title. And who would be the heroic policeman's Tonto? Who, his Kato? Who else but a French-Canadian guide.

Tom thought of a plot and wrote his story. Trendle had suggestions. Tom rewrote his story. Trendle criticized. Tom wrote it over again. And a few more times.

"I don't like Mogo," said Trendle. "You've got to find a better name. Mogo sounds like a mutt. This dog is a hero."

"He's a king," Dougall agreed. "King of the Huskies. His name is King."

"That's better," said Trendle.

So the yell became, "O-on, King! O-o-on, Husky!"

On January 3, 1939, *Challenge of the Yukon* first went on the air as a fifteen-minute series, broadcast in Detroit only. Big, heavy Jay Michael with moustache and frizzy hair landed the role of Sergeant Preston, switching from villains on the Rangers and Hornets which was quite a triumph all things considered. Harriet Livingstone's father, Frank Russell, was the French-Canadian. And sound man Dewey Cole barked, whined and made other appropriate dog sounds as King, the Husky. There was eventually a to-do with AFRA about that. The union felt that impersonating a dog was acting and therefore earned the performer an actor's fee; and the union won the argument.

The show was sold, not by Campbell, but by George W. Trendle, himself, and he was rubbing his hands together with a broad grin on his face when he sent for Campbell to tell him.

"Quaker Oats," he said. "They bought it."

"Maybe I should put you on my sales staff," Campbell smiled, sitting down.

"They were kind o' wishy-washy about the whole thing," Trendle grimaced. "Finally, at the end of our first meeting in Chicago, the feller said to me, 'Mr. Trendle, we have two programs under consideration, but we'll add your *Challenge of the Yukon*'."

"I remember," said Campbell. "You said they were goin' to have a child psychologist or some such evaluate the three shows."

"From Western College."

"What were the other two shows?"

"I've forgotten. One of 'em was *Wild Bill Hickok.*"

"That was two weeks ago."

"Well, I went back yesterday and we had this meeting in their office and this professor feller was there. So after about a half an hour's discussion, the guy from Quaker Oats says to him, 'Professor, which of these three program do *you* think we ought to take?'

"The professor said, 'There isn't any choice,' he says, 'there's only one program you should take,' he says, 'that's *Challenge of the Yukon.'*

"So I'm sitting there and he says, 'Why?'

"'Well, the psychologist, he says, '*Wild Bill Hickok* is a sort of—oh, doesn't mean a hell of a lot. And the kids get over six years old and they can see through him and they're disgusted with it. And this other one might last 'em until they're twelve years old. But,' he says, ' Sergeant Preston will be good from six to sixty'—that was exactly his expression; 'the people from six to sixty will enjoy that program and that's the program I think you ought to buy'."

Campbell laughed. "I can think of a couple of accounts I should let you handle."

"Anyhow, they bought the Challenge. I'm not sufficiently educated in that kind of thing to know whether the psychological effect was this or that or the other. I just went in there with what I thought was a good show. I may make a half hour show of it later. Oh! Tell Charlie Hicks I've got a secretary for him. Name's Elizabeth Ashton. She's been working for me in the theaters. Bright girl. A booking agency offered her more money. I asked her if she'd like to move upstairs in radio where we could pay her more money—and she would."

"I'll tell Charlie."

"Elizabeth Ashton. They call her Bessie—Bess—something like that."

"I'll tell him. How about lunch?"

"Can't. I have a meeting of the Fire Commission—which reminds me. Why can't you land Nate Shapero?"

Nate Shapero was on the Fire Commission with Trendle. He owned a chain of drug stores. Campbell laughed.

"What do you ask me why can't *I* get him? Why don't *you* get him? *You* know him better'n I do. That's one of the couple of accounts I was thinkin' of."

Jim Irwin was dead. The ex-minstrel man who had played Michael Axford in *Warner Lester, Manhunter,* and in *The Green Hornet* suffered a stroke in January, 1938, and died in July. He was 72. Fans who loved him jammed the Catholic cathedral where his funeral was held, and overflowed into the street.

The role of Michael Axford went to Gillie Shea.

Before his death, Jim Irwin agonized through a tragi-comic experience peculiar to broadcasting. Most veterans of the industry on the talent side have favorite yarns about air fluffs and on-the-air disasters. The one involving Jim Irwin is rememberd in New York by Al Hodge, in California by Ted Robertson, and in Detroit by Fred

Jim Irwin (the original "Axford" in "The Green Hornet"). Photo courtesy Fred Flowerday.

Flowerday.

As an example of how taped conversations can be used authentically as dialogue in a history, three people widely separated geographically are re-united after 35 years on the printed page as though they had been gathered together in a single room to recount in their own words, better than anyone else could do, the tale of the phantom broadcast.

Al Hodge in New York: "We used to do a repeat show of *The Green Hornet* to WSPD in Toledo with a thirty-second break between the two broadcasts. One night the repeat show was cancelled and we didn't tell anybody."

Ted Robertson in Hollywood: "I was in that little booth affair with Larry LaDue, the engineer. I sat there with my turntables. And Larry was off to my right. And Chuck directed from in the studio because he usually played a part also. And this was *The Green Hornet!*

"What happened was, the second broadcast was cancelled because the President was on, or some political reason. Chuck knew about it and he thought this was a great chance to pull a gag. So he told Larry LaDue, the engineer, about it. He told Al Hodge about it. He told the announcer about it, Fielden Farrington. And Rollie Thomas in the announcers' booth knew about it.

"Well! We finished the first broadcast. For thirty seconds we fanned the door to air out the studio. Then the light came on for the repeat and I was given the cue to bring up the theme, which I did. I brought the theme up. And then I was given the cue to fade it down, which I did, and the cue went to Fielden Farrington, the announcer. It began to tickle him—the joke. And he started to talk. And he broke up. Completely. He couldn't say a word.

"So I brought the theme up until he got himself—what I thought to be—controlled. And I brought it down again and pointed to him. And again he broke up.

"Now! Rollie Thomas was in the announcers' booth and I ran in and said, 'Something is wrong with Farrington. Here is the copy. For God's sake will you please read it?'

"And Rollie, who knew all about the gag and was a real cool cat, got in and he read it all the way down to the bottom where he was supposed to say, 'The Gree-ee-een Hornet!' Instead, he said, 'The Lo-oh-ohne Ray-aynger!'

"Well! Poor old Herschel Mayall and Lenore—Jewell's sister who was pregnant at the time, and Freddy Reto—were standing at the back and they could see their whole world going to pieces.

"Well now, I sat—I didn't know *what* was going on and LaDue was — uh — you know, you couldn't get to Larry. He would sit there with a grin on his face no matter what happened.

"Now! Hodge came in and started—uh—three pages too late and stumbled over a chair getting to the microphone. Now Chuck was standing in there and he got hysterical but in a funny way. He hung his head. I thought he was sobbing,—because he was a very emotional man. And I thought, my God! He's gone out of his skull."

Fred Flowerday: "Chuck Livingstone was in on the gag and Ted Robertson wasn't. Chuck was doing his best to disrupt everything. And they grabbed people's scripts out of their hands—mixed up the pages. Jim Irwin—he was Axford."

Hodge: "I remember old Jim Irwin because we skipped a couple o' pages. And poor Jim! He had trouble holding his upper plate in and he could not find the place and he kept saying, 'Yash, Mis' Britt. Yash, Britt'—and he's thumbing his pages and they're going on the floor, and he's sayin', 'Yash, Britt, begorra I'll find it.' And he's going on—could not find it."

Flowerday: "Jim Irwin nearly had apoplexy. Ruth Rickaby's about to faint."

Robertson: "Lenore was standing with tears streaming down her face."

Hodge: "Ted Robertson came out and hit Livingstone."

Flowerday: "Robertson was in the control room running records. He went in and grabbed Livingstone, put him out in the hall there, and slapped his face to snap him out of this."

Robertson: "So I finally got Chuck out into the hall and I said, 'It's all right. It's all right.'

"And he said, 'It's only a joke.'

"And I said, 'You're out of your mind!' And I hit him across the face thinking I was bringing him back—out of hysterics, you know. And then he ran into the studio screaming, 'Stop! Stop! Robertson is killing me!' "

Hodge: "Somebody came in and put Jim Irwin back together again."

Robertson: "It was an odd situation. I tell ya, I was shaken. And I was afraid for Lenore. I thought she would lose that baby. You know, we were all—gosh, I think the best paid of us at the time probably made 35 dollars a week. And that meant our livelihood, and we could see it just going down the drain."

21
World Premiere

The Detroit local of the American Federation of Radio Artists still had no contract with a station, but in the Spring of 1939 held its first AFRA Ball. Everyone dressed for the evening and it was an attractive crowd. There was a live band for dancing and liquor flowed along with shop talk. Socially, the surprise of the evening was the arrival together of Malcolm McCoy and his ex-wife, the beautiful Bea Leiblee. Bea was glamorous in a long off-white evening gown. She wore red shoes and Mac had given her red camelias to wear in her dark hair. This was one of their first dates since the divorce.

The evening, however, sagged; it became more and more in need of a lift. That lift finally came from an unknown little guy with a hooked nose, brown eyes and Jewish jokes. The crowd from WXYZ dominated the union and cared little for outsiders. As the little comic began his monologue, those in charge of "entertainment"—of which there had been almost none—were outraged.

"What's *that*?" one demanded of the other.

"That's Amos Jacobs," said Gwen DeLany.

"Who's Amos Jacobs? Who let him in?" growled Malcolm McCoy. "Get him off!"

"Mac, you can't do that," Gwen insisted, her broad face firm, her big brown eyes snapping. "He's a member!"

"He's spoiling the party," another exclaimed.

But he wasn't. He had the crowd laughing. The drooping affair perked up with the little man's talent while members of the Entertainment Committee looked on uncertainly from a far corner.

Gwen DeLany, old enough to be his mother, had great faith in Jacobs' ability. It is said she bolstered his morale when it sagged, which was often.

Another year, another baseball season, and Harry Heilman was back on the air.

According to James G. Riddell, "Harry was so difficult. At the beginning of the year, he'd be great. As the year progressed, he'd sulk; he wouldn't even talk. He was tired! Don't forget, at the beginning he reconstructed a lot of games from ticker tape. At the end of the year, he would just go into a hospital—just stay there and rest a couple of weeks."

Harry Heilman, baseball great, had a gentle way of speaking. Tall, thin, with a happy face, he bent over and flicked his ash into the tray on Allen Campbell's desk and told him he was going to Jackson the next day to make a speech. Campbell's eyes narrowed behind his glasses.

"What do you get for that" he asked. "Anything?"

"Well, it's tough," Heilman admitted. "but I'm —uh—I like it."

"Harry Heilman," Campbell announced flatly, "you're not goin' to do that any more. I'm goin' to be your manager. You're not goin' to any meetings unless you get paid for it. Minimum—$50."

Campbell: "We never got any of that money. I didn't want any part of it. I just wanted him to be happy with us."

In 1939, the second Lone Ranger movie serial was released by Republic Pictures and, as the first had done, won the Exhibitors' Award as "Best Serial of the Year." Chief Thundercloud played Tonto in the movie serials and Lee Powell was the Ranger.

In August, Trendle found it impossible at last to wear three hats, and resigned as President of the United-Detroit Theaters. Earl Hudson replaced him and soon proved his caliber by landing the world premiere of an important Paramount Picture called *Disputed Passage.* Seven stars were imported from Hollywood to add to the hoop-la and Detroit approached a state of hysteria.

On October 20, 1939, the day following the event, a headline in the *Detroit Free Press* read: "50,000 JAM DETROIT STREETS FOR GALA MOVIE PREMIERE," and the *Detroit Times* printed pictures of Robert Preston, Dorothy Lamour, Susan Hayward, J. Carroll Naish and one identified only as "the master of ceremonies."

The master of ceremonies had walked out on the stage unannounced—"cold" as they say. Also, in this case, cold sweat! By what circuitous route those in charge of the premiere had finally lowered their signs to fix on WXYZ's Dick Osgood no one alive remembers. Not until the morning of the premiere was he called by Junior Sutton, and such was the emergency that Osgood was given no chance to demur.

Harvey Campbell, official greeter of the city, Vice President and Secretary of the Detroit Board of Commerce, originally scheduled to introduce the celebrities, had for some reason in the final hours backed out of the assignment, and it was then too late to get anyone of any importance.

Instead of going to the public library for his usual morning research for a Factfinder story (they were written two or three weeks in advance), Osgood, already feeling like something scraped off the hull of a wrecked ship, raced downtown to Capper & Capper, a men's store, explained his predicament, and tried on tuxedos. He had not worn one since his years in the theater. The store promised to have a suit delivered to the Michigan Theater by early evening.

Osgood hurried to the Maccabees Building, researched the stars he would introduce from material collected in his files, and prepared his five o'clock *Show World* broadcast, except for the entertainment news of the day which he would have to assemble later.

Early in the afternoon, as ordered, he reported to the Book-Cadillac Hotel to meet the stars and be briefed on how each wanted to be presented to the premiere audience. J. Carroll Naish, who played the villain in the movie, suggested that for a laugh he act tough after being introduced and Osgood pretend to be afraid of him. Dorothy Lamour, the star of the picture, was late flying in but Osgood worked out routines with her leading man, John Howard, with Robert Preston who was supposed to be romancing Miss Lamour at the time, with a juvenile from the cast, Joe Allen, and with two Paramount starlets, Olympe Bradna and Susan Hayward.

"You can get with Miss Lamour at the theater tonight," he was told. "But whatever you do, do *not* mention a sarong. Miss Lamour is sick of the sarong bit after all the *Road* pictures with Hope and Crosby. She wants to get away from that. So *lay off the sarong!*"

Osgood raced back to the Maccabees Building, collected show business items from the newswires, wrote the first two or three pages of *Show World,* including a report of the afternoon's experience, and prepared interviews with Olympe Bradna and John Howard who came to the studio to be interviewed on the 5 o'clock broadcast.

That over, he escorted his visitors to the elevator, grabbed his Factfinder script for the evening and ran up the stairs to rehearse with Benny Kyte and the orchestra, the quartet, the soloist and announcer John Slagle. The Hi-Speed show was now being broadcast over a regional network of eleven stations.

As soon as he was off the air at 6:30, he walked fast from the Maccabees Building to a garage around the corner where he kept his car. He drove several miles out Woodward Avenue to where he had lived with Nancy and her daughter, Mary-Jane Hays. The Osgoods had separated but in an old wardrobe trunk left behind in the basement, Dick had a dress shirt, black tie, studs, patent leather shoes and make-up.

Unable to find a key to the trunk, he smashed the lock, grabbed what he needed and hurried back to his car. As he drove downtown, feeling the pressure of time and a fast-developing case of opening night nerves, he went over and over and over his introductions for each of the stars, and what he was supposed to do on the stage with them. He had acted in stage plays for years, but this was different: he had no character to hide behind, and no script to lean upon.

Most of the fifty thousand people seemed to have collected around the Michigan Theater. To save carrying his shoes through the mob, he changed in his car in the parking lot. He was wearing a green suit. Patent leather shoes with a green suit!

He grabbed his shirt, his studs, his black tie and his make-up and jostled his way to the stage door which was on a stubby little dead-end alley off a side street. The street and the alley were jammed with excited fans. The stage door was locked. Osgood pounded on it.

"Which one is that?" cried some girls.

"Hey! Which one are you?" demanded some men.

"You can't come in here," came a voice from the other side of the door.

"Let me in, please," yelled Osgood through the door. "I'm the master of ceremonies."

"He's the master of ceremonies," said some men, disappointed.

"O-oh," moaned the girls, very let down.

"You'll have to let me in," Osgood yelled.

"Let him in!" cried boys and girls together.

"I won't unlock this door," shouted the doorman from inside. "But you can crawl in the window."

With that, the window beside the stage door was opened and Osgood climbed in to the cheers of the crowd. Inside, he inquired about his tuxedo.

"What tuxedo?" the doorman asked.

"Capper & Capper was supposed to have delivered a tuxedo." Osgood explained.

"I haven't seen no tuxedo."

"I can't go on in this green suit. Will you please check out front?"

Bob Corbin, manager of the Michigan Theater, was extremely busy. Osgood was told which dressing room to use and went up by elevator. He stripped to his waist and started putting on a little make-up. He heard the elevator door open and raced out into the hall. Robert Preston stepped off the elevator.

"Oh," said Osgood. "I was hoping you were my tuxedo."

"Go on that way," Preston suggested, grinning. "You look great."

Somebody found the tuxedo. It had been delivered at the front of the theater. To the dismay of the crowd at the stage door, the stars were also being delivered at the front of the Michigan in new 1940 models provided by the automobile manufacturers. Bagley Avenue, in front of the lobby, had been torn up and posts driven into the street to provide safety lanes. John Slagle was on the sidewalk announcing the stars' arrival over a loud speaker.

Dorothy Lamour had arrived. She and several others joined Preston in the dressing room next to Osgood's. Miss Lamour opened the dressing room window and called to the fans below.

"Come on up!"

There was a roar and fans began jumping up trying to get hold of the suspended stairs of the fire escape to drag them down.

"Are you nuts!" exclaimed Preston to Lamour, slamming shut the window.

As soon as he was dressed, Osgood went to the dressing room next door.

"You looked better the other way," grinned Preston.

"Miss Lamour," said Osgood. "I'm your master of ceremonies. How would you like to be introduced?"

The glamorous Dorothy Lamour tucked one hand under Osgood's arm and guided him back to his dressing room.

"Tell me," she asked. "What's the nicest store in Detroit?"

"Most everybody shops at Hudson's, I guess," said Osgood. "It's one of the best department stores in the country."

"But—where are the smartest *little* shops? The smartest in town?"

"Oh—I would say along Washington Boulevard."

"All right, Now I'll tell you what we'll do. You ask me what I've been doing since I got to Detroit and I'll say, 'I've been Christmas shopping on Washington Boulevard.' And you ask me if I found anything, and I'll say, 'I found just what I wanted for my mother.' and you say, 'Oh? What did you find for your mother?' And I'll—"

Lamour opened the large hand bag she was carrying—and pulled out a sarong.

It brought down the house. The huge audience roared again when J. Carroll Naish grabbed the lapel of Osgood's new tuxedo and Osgood acted scared. A newspaper photographer got a picture of it, and that was the one published in the *Times* the next day with the unidentified "master of ceremonies."

Also the next day, Earl Hudson wrote Osgood the following letter:

Dear Dick:

You have challenged my confidence in my ability to distinguish between professionals and amateurs.

Until this morning I was voicing compliments about the splendid performance you gave last evening as master of ceremonies which impressed me as outstanding for even a professional. To my amazement I am advised on excellent authority that your appearance on the Michigan stage last evening was the first of its kind you have ever made.

When an amateur at that sort of thing, and a very difficult task it is, so conducts himself that more than twenty years of observation of both professional and amateurs is upset and an amateur is rated without question as a professional, you may from here on draw your own conclusions as to how effective and impressive you were. And there is no slightest need for you to have any reservations or any modesty in drawing those conclusions to the fullest limit.

For United-Detroit Theaters Corporation and myself I want to congratulate you for your masterful handling of a fast moving succession of impromptu situations and dialogue and express sincere appreciation for your very important and extremely efficient contribution to the success of the occasion.

Cordially yours,
E.J. Hudson (signed)

Osgood received letters from others, including the man for whom he had substituted, Harvey Campbell, who wrote, in part: "—frankly, from this distance, you were the only element that was very much worth remembering in that entire evening."

And what was WXYZ's reaction to all this? There was a phone call from Junior Sutton.

"How'd it go?"

"Okay."

And that was all.

Ruth Dean Rickaby lost her best excuse for drinking when the stage and screen actor Robert Strange lost his wife. It is said that Ruth and Strange had been in love for years but that his marriage could never be disrupted. As soon as his wife's death left him legally and morally free, there was a tearful farewell party in Detroit for Ruth Dean Rickaby. She flew to California, married Strange, and lived happily until he died.

Other, and more important changes were announced in March, 1939. Trendle caught Russell Neff job-hunting, and fired him. A former WXYZ announcer had become an executive with the Knox Reeves advertising agency in Minneapolis and wrote Neff, suggesting that the two meet in Chicago for a talk. The letter was opened before it reached Neff. When he returned to Detroit he learned he was no longer wanted at WXYZ. It was mutual. Russell Neff became vice president of Knox Reeves and director of radio and television advertising. He had a long and prosperous career.

Harold True replaced him as Studio Manager of WXYZ. At the same time, Ted Robertson and Al Hodge were named assistants to Chuck Livingstone, and Fred Flowerday became chief sound man. Flowerday took the job seriously. He often remained in the sound studio until one o'clock in the morning rehearsing effects for the next day. It was also his job to find recorded music for the interludes between scenes in the dramas. He had no musical education; his selections were intuitive; he spent hours in a music store listening to classical records. When the *Cleveland Plain Dealer* contacted him for a story, the paper simply did not believe that a sound man could pick such appropriate music without a musical education and wound up getting a story from James Jewell in Chicago, who, of course, had been out of the picture for some time.

John Hodiak developed into an excellent voice and WXYZ put him under contract for $37.50 a week. Toward the end of one of the thirteen-week contract periods he was appearing on almost every show and felt he earned more money. The station told him not to be ridiculous. So he auditioned in Chicago and won a running part in *Love of Life,* but the part was not coming up for some weeks.

He returned to Detroit and said he was leaving. The station then offered him a new contract at $75 a week. *Love of Life* would pay him $175. However, he didn't have money enough to tide himself over. When time came to go back to Chicago, he didn't even have train fare. So the actors at WXYZ took up a collection and staked

Hodiak to his future. Tom Dougall staged a rip-roaring stag farewell party for him. The running part Hodiak played in *Love of Life* soon lost its breath and sat down, and John had lean times until he landed the title role in the Chicago-based network radio series *Li'l Abner.*

Brace Beemer also ran into hard times. He lost the R.G. Dun account and was rumored to have lost his savings in politics. He needed a job. The subject came up in the Stroh Building when Campbell, Sutton and Trendle were breaking up a conference. Bouncy little Sutton interrupted his headlong charge toward Trendle's door, turned back and bumped into Campbell who was walking behind him.

"I suppose you've heard about Brace Beemer," Sutton said jovially. "Harold True tells me Brace wants to come back to work for us."

"He'd make a better narrator on the Ranger," Trendle remarked, testing Campbell with a smile.

"Hell," said Campbell. "If the guy is available and we need a good voice on *The Lone Ranger*, let's bring him back."

"Do you mean that this man who came in here and cut your throat—?" Trendle demanded. "Do you mean you'd hire him back?"

"What the hell!" Campbell said, his white teeth showing in the tan face. "He didn't get my job, did he?"

Beemer was re-hired to narrate *The Lone Ranger.*

Bill Morgan left the downtown YMCA where he was living and walked along Adams to Woodward Avenue. This was his first morning in Detroit and this was the biggest city he had ever lived in. He was from Decatur, Illinois. Detroit, he thought this brisk February morning, was "just a marvelous city."

His step was brisk and the brown eyes imbedded in his strong features were very alert. He walked a little as though his back were stiff, and there was something about his shoulders that gave the impression his torso had stopped growing a little too soon.

He stood on the corner and looked up Woodward Avenue. He saw the WXYZ transmitter on top of the Maccabees Building, and took a street car. He had come to Detroit to attend the winter semester at Wayne University, having decided to teach literature. He had had some radio experience on a small station in Decatur, and thought he might make some money to help him through school. He got off the streetcar and went to the fourteenth floor of the Maccabees Building. After a wait, he got in to see Chuck Livingstone who liked him immediately.

"You're in luck," said Livingstone. "You sound a little like John Payne and we're doing a special half hour for the Michigan Theater—a dramatic version of his new picture, *Wings of the Navy.* Come back in two hours."

Morgan was thrilled. When he returned, actors had gathered in the organ studio. Sutton was bouncing around in the control room behind Livingstone, feeling his authority. Morgan had read about ten lines when Sutton said abruptly, "He won't do."

"Give him a chance," said Livingstone.

"I don't think he's right. He won't do."

Livingstone stopped the rehearsal, went into the studio, took Morgan by the arm, and led him into the back hall.

"Sorry, kid," he said. "They don't think you're right for it. But stick around, we'll get parts for you."

So Morgan stuck around: "He got a few for me. Not many. But enough for me to say truthfully that I was on *The Lone Ranger*. And I was on *Challenge of the Yukon* with Jay Michael."

But it petered out, and Morgan concentrated on his studies at Wayne. He would return to Wyxie.

One of the most unique announcers to work at WXYZ was hired in late 1939. There was an incongruity in the package labelled Bob Hite that gave him character. Physically, nature had drawn him with sharp, clean-cut lines. Of average height, his hair was dark and slick, his features were nicely outlined, his clothes fit, and he was neat. Mentally, on the other hand, he didn't seem to be quite with it and he occasionally made colossal blunders. But his voice and delivery, like his appearance were clean, added to which he projected such persuasion that he could read the telephone directory and make you want to buy the numbers.

Bob had been a cookie salesman when a friend (female) told the owners of Station WEXL in Royal Oak, Michigan, they should listen to him. Hear and admire him, they did for a week—no pay. Chief Announcer at WEXL was Kirk Knight and Bob helped him one night. After that, Kirk Knight (who later went to WWJ in Detroit) took Monday nights off and Bob Hite ran the station. He also did a newscast.

One Monday the engineer brought in a recording machine. This was new to radio in the 1930s. The engineer made a recording of one of Hite's newscasts; Hite listened to it and he thought it sounded pretty good. He played for another friend. This friend (male) knew Felix Holt and told him Hite was even good enough for WXYZ. Through Felix, Carl Gensel auditioned Hite and told him to go away for two months and practice; and he told him what to practice.

Remembering it Hite's recollection was vague: "I can't remember what it was, but I probably still have the same faults."

However, he went back after two months and Gensel hired him at $35 a week. He had been making $40 as a cookie salesman; but radio was more fun. Bob's naive quality soon made him a "fall guy" among the announcers. He tried to be on his guard after a while. Even this got him into trouble.

One Sunday afternoon, he was assigned to announce a program of classic music played by teachers from the Detroit Conservatory of Music.

Jack McCarthy remembered: "They always had a 'dress rehearsal' in those days because, of course, everything was live. Bob came back to the announcers' both and as usual there were six or eight of us sitting around there waiting for our next assignments. Bob said, 'How do you pronounce this word, fellers?' The word was f-u-g-u-e. Harry Golder told him how to pronounce it. Bob decided we were pulling his leg. He went on the air and announced it as a Few-gew."

Between 1932 and 1939, the volume of radio news almost doubled but station owners and advertisers were timid about expressing opinions. Walter Winchell got away with it, and WXYZ with *Town Talk* continued to express the strong political opinions of George W. Trendle a quarter of a century before editorializing by management became common practice. On the other hand, one of the most literate broadcasters in radio history, Alexander Woollcott, was cancelled by his sponsor in 1935 because of his "unkind" remarks about Adolf Hitler and Benito Mussolini.

With few exceptions, newspapers did not attain the level of advertising they had reached before the arrival of radio and the Depression. Advertisers turned increasingly to the personal, vocal appeal of the commercial announcer. Advertising on radio increased every year of the Depression until in 1939 it reached nearly $200,000,000. By 1940 there were 800 licensed radio stations in the United States.

The WXYZ newsroom had taken on a new writer named Tom A'Hearn—thin, droll, relaxed, tough and stubborn—a blue-eyed Irishman with thick curly hair and eyebrows like bushes. Black-eyed Rollie Hoover faded out of the picture.

While news was expanding, the quiz show was born. *Professor Quiz* started the trend in 1936. In November, 1939, WXYZ decided to follow the trend with its own *Children's Theater of the Air.* Dramatic skits were dropped and a question-and-answer exchange with the audience substituted. The name of the show was changed to *Radio Schoolhouse* and Dick Osgood, fitted out with black gown and mortar board and called "The Professor" became master of ceremonies. (The somber costume was soon discarded.) Between musical numbers, he fired questions at members of the audience; their answers were given into microphones carried up and down the aisles by youngsters; and at the end of the program, Osgood went out into the audience himself to fill until sign-off theme.

Tom Dougall was now producing the program. There were forty juvenile regulars on it. Each week, ten children had solo spots and the other thirty did choral harmony behind them, and arm movements, and sometimes dancing. The groups of ten rotated so that every youngster had a solo spot once a month. The children were auditioned very carefully. Once accepted, they stayed on the show until they were sixteen when they automatically "graduated." Graduation Sunday came to be an annual event that jammed the theater. It was not unusual for people to line up around the block to get in. The audience paid admission and stayed to see the movie. Fred Rivard, ex-vaudevillian who ran a dancing school, was choreographer; Eve Gladstone, staff pianist, trained the singers; and Benny Kyte and several musicians accompanied in the orchestra pit with special arrangements made by Harry Stafford. A week's work went into every broadcast of *Radio Schoolhouse.*

Lone Ranger rehearsals were called at 2:30 P.M. and the narration was so important that it was necessary for Brace Beemer to rehearse with the actors, sound men and music. His first week back, he arrived on time. Then he began to be late.

This was embarrassing to Ted Robertson, who, as Livingstone's assistant, directed some of the rehearsals. One day when Beemer was an hour late, Robertson took him into the back hall.

"Brace! Look! I know this is embarrassing to you because you were my boss before. And it's kind of difficult for me because I was a little feller under you. But now you are under me. That's the way the dice have fallen and let's make the most of it. I must ask you to be on time."

Beemer promised to be on time and for the next two rehearsals he was prompt. Then he began to slip again. Robertson called him into the director's booth. There was a glint in his eye.

"Look. You made a deal with me. Now keep the deal or I'm going to embarrass you."

Three shows later, Beemer arrived late again. Robertson opened up the "talk-back" and said for all to hear, "*Mister* Beemer. Rehearsal is 2:30. All the actors and the stars get here on time. I expect the narrator to be on time. If you are *not* here, you're going to get fired."

Ted Robertson was very strong for the union; he was active in its formation. Beemer, on the other hand, was very unenthusiastic about the American Federation of Radio Artists which was now in the process of negotiating its first contract with WXYZ.

Harry Golder cornered Robertson one day between rehearsals and said, "Hey! The word is—I understand you're pretty much on the left wing."

"What is this nonsense?" Robertson asked.

"Well, I get it from downtown," said Golder, "that you're considered to be pretty far left."

"Where did this come from?" Robertson demanded.

"From Trendle's office," said Golder.

The explosive Ted Robertson lost no time bearding the lion in his den.

"Mr. Trendle," he said, his face red with anger, "there is a disturbing rumor going on."

Trendle's smile was frigid.

"They say it came from this office," Robertson continued. "I have a good idea how such nonsense got here but if you have anything to ask me, or if anybody is saying anything about me, I want you to bring him down here and let me confront the man."

Trendle poked at the calendar on his desk. "I don't know what you're talking about and I don't think you do, either."

That night between the second and third broadcasts of *The Lone Ranger*, Robertson shot pool with some of the boys in the pool hall on the same block. When he went back for the 10:30 show to the west coast, he found Livingstone with tears in his eyes.

"What's the matter, Chuck?"

"I don't know how to say this to you, but — I have to let you go."

"Why?"

"Orders. From downtown."

"Well, Chuck, I know it's not your fault. It goes right back to Brace Beemer. He is not one of my favorite characters. Frankly, I think he's an arrogant bastard! That was my first impression of him and it hasn't changed. But-*you* know. That's the way the dice have fallen."

Robertson did his last broadcast for WXYZ that night. After it, Beemer stood before him in the sound studio.

"I understand you've been let go."

"That's right."

"I want you to know if there's anything I can do to help you, if you need money or anything, I'd be happy to help you."

"No thanks."

Ted went to Chicago and saw James Jewell and John Hodiak. Hodiak took him to meet the producers of *Li'l Abner,* and several other men he knew. Ted got a staff job with CBS in Chicago as a producer and eventually became Vice President of Compton Advertising, Inc., manager on the west coast of radio and TV programming.

The buzz sound symbolizing *The Green Hornet* still failed to satisfy Trendle, and the show had now been on the air for three years. It was Fred Flowerday who finally became the hero.

He found the sound in the office of a dentist. It was not a drill; it was the dentist's wife. She played a novel musical instrument known as the Theremin. The Theremin may have been the first electronic musical gadget. The "musician" stood in front of it and controlled pitch and volume by using her hands, moving one up and down and the other back and forth without touching the instrument.

Flowerday had the Theremin set up in a studio, persuaded the lady to play it, and Trendle listened.

"Now you've got it!" said Trendle, very pleased. "You have finally found the sound of *The Green Hornet.*"

22

Four on the Way to Wyxie

The Great Depression lasted ten years. In 1940, the poison which had paralyzed America was definitely being expelled. By August, the production of durable goods finally had regained the peak it had attained in 1929, and the gross national product was up to $100,000,000.

The Strikers felt secure at last. Not only was Fran in charge of all radio scripts for the Ranger, the Hornet, Ned Jordan and Sergeant Preston; the comic strip for which he supplied the story line was now printed in 120 newspapers; and he was writing Lone Ranger books for children published by the Whitman company in Wisconsin, as well as hard cover Ranger books for adults published by Grossett & Dunlap. Fran was now 36 years old, working 14 hours a day, and, according to

Deltasig, a fraternity magazine, earning $10,000 a year. He had bought a home in Grosse Pointe.

Under the direction of Raymond Meurer, all dramas had been incorporated; and so had *The Factfinder. The Factfinder* was the only program other than the dramas to be incorporated separately.

There was one sour note in WXYZ's prosperity of 1940; Bond Bread cancelled its sponsorship of *The Lone Ranger* in July. For ten months the program was without a sponsor in the major markets of the east and middle west. H. Allen Campbell was applying all his affable magnetism, persuasive charm and great sales ability to the capturing of General Mills.

Meanwhile, there were the usual moments of fun.

Malcolm McCoy was playing a banker in *The Lone Ranger.* His bank had been robbed and he swore revenge. As he read the punch line his fist was clenched, his eyes were slits, and his voice was gutsy. He capped the exciting emotional buildup with: "—and I will catch them and prosecute them and take everything they have—LOCK, BOCK AND STARRELL!!!!"

WXYZ was the first station in Detroit to sign a contract with the American Federation of Radio Artists. The contract went into effect February 1, 1940. At the very next Board meeting, Raymond Meurer appeared to plead for "flexibility and elasticity" in the interpretation of the contract, especially in four areas. He argued that WXYZ should be allowed a 10% discount when signing actors to "long term" contracts. He believed that juvenile actors should not have to be paid the full rate. He wanted actors to continue to double from the story proper to dramatized commercials without drawing extra pay. And he asked that all existing contracts with actors be continued at the pre-union rate.

He lost on all four counts. He appeared before the Board several times that year to ask for favors, and to try to save the station from paying special fees in line with the contract which he had negotiated.

Emily Holt, who was now National Executive Secretary of AFRA, advised the Local to stand by its contract with the station, without compromise. She pointed out that nationally AFRA had grown rapidly because "the need for organization among radio workers was urgent due to widespread unfair exploitation effected on radio workers by management groups."

In the first six months of the union's existence, $35,000 had been collected in initiation fees and dues "without resorting to force or coercion of any kind." In this half year, the American Federation of Radio Artists already had collected twice as many members as belonged to the stage performers' union, Actors Equity Association, which had been in existence for 27 years. Mrs. Holt pointed out, further, that it had taken the American Federation of Musicians nine years to effect its first contract; the Screen Actors Guild, four years; and the Actors Equity Association, five years. In less than one year, AFRA had negotiated contracts with NBC and CBS in New York, covering minimum wages, maximum rehearsal hours "and other vital points." The first contract in Detroit was signed two years and four months after the inception of the union in October, 1937.

In June, 1940, the Detroit Local was given a big lift by Bob Hope. Hope was appearing in person at the Fox Theater and would do his weekly network broadcast from WWJ. WWJ, owned by *The Detroit News*, had refused to negotiate with the union. Jay Michael, President of the Local, visited Bob Hope backstage at the Fox and asked him if he would wear an AFRA button into the WWJ studios.

Hope went even beyond that; he needed some local performers and he refused to work on his broadcast with any but AFRA members. WWJ immediately called the union. It was ready to negotiate. In December, AFRA and WWJ signed a contract.

WJR still refused to talk.

Harold True and Carl Gensel were listening to an audition.

"Sounds pretty good," Gensel commented.

"Who is he?"

He's been up at WOOD in Grand Rapids. Graduated from the University of Michigan." Gensel looked at the letter in his hand. "His name is Myron Wallace."

"Myron! For Chris' sake! We'll have to change that."

"But he does sound good."

"Hire him."

Harold True maintained it was he who changed Myron to Mike—Mike Wallace.

Mike Wallace met George Trendle only once. It was a memorable encounter. Mike had been announcing at WXYZ for two or three months when he arranged to have a few days off to get married. The day before he was leaving to drive to New York for his wedding, he fell heir to an accident.

1940 was an election year and Harold Ickes was making a speech that night at Olympia Stadium in Detroit. As it was broadcast over WXYZ, it was to be transcribed for a rebroadcast later over the Michigan Radio Network.

Technically transcriptions were somewhat in their early stages. As Wallace remembered: "We didn't have engineers on our transcriptions in those days. We put on our own big platters for replays."

Mike was assigned to the Michigan Network replay, cued up the transcription, made the opening announcement to the Michigan Network, started the turntable, the speech was on and he stepped out into the hall outside the announcers' booth to join the remnants of a wedding party. The curly headed little Italian actor, Jack Petruzzi, had been married that day.

Suddenly, in the back of his ears, Wallace heard the faulty transcription repeat: "Moe Annenberg is one of the most corrupt one of the most corrupt one of the most corrupt one of the most corrupt one of the most—"

It had repeated a dozen times before it really registered with Mike that he could jump into the booth where the transcription was being played. He hit the turntable and the transcription jumped ahead and continued without further catastrophe.

As soon as his heart stopped pounding and his face was cool again, Mike leaned back in the chair he had sat down on and thought calmly. This transcription went only to the outstate network; nobody in the downtown office would have heard it. He was safe. He drove off the next day to be married.

But Wallace was not to get off so easily: "I was on my honeymoon in Woods Hole, Massachusetts, and picked up a copy of *Time* Magazine and there was the whole story."

"Oh, my golly, I've lost my job," he said to his bride.

Sure enough, when he returned to Detroit, there was a black-on-pink note from Trendle: would he please drop in when he had a moment?

Mike lost no time getting to the Stroh Building. And he waited and he waited and he waited, watching Thelma Watson typing letters, and wondering where he could go for a job. Finally a buzzer sounded. Thelma rose tall from her typewriter and trod the deep pile to Trendle's door, closing it behind her. She emerged in a moment, the cinnamon of her eyes alight with humor, and said, "Mr. Trendle will see you now."

The only thing on top of Trendle's desk was a copy of *Time* Magazine. Wallace felt the heat in his face; he was sure this was the end of his career. Trendle held up the magazine.

"This story just about—the way it was? This story here in *Time*?"

"Yes, sir. That's the way."

Trendle put down the magazine, leaned back and studied the announcer with cold, appraising eyes. His linen gleamed white; he wore Masonic cuff links.

"You just got married, didn't you?"

"Yeahp," said Wallace.

"Need the job, don't you?"

"Yes, Mr. Trendle. I do."

Trendle closed the magazine and tossed it into the trash basket under his desk. "Okay. Go on back to work."

Bob Hite was not so lucky. The primary election of 1940 was a hot contest in Michigan. Two Democrats and eight Republicans were fighting for the Governorship. Attorney General Tom Read was favored by the experts to win. (He didn't. Governor Dickinson, 81, was re-elected.) Read was well organized and his campaign was so well financed that he could buy fifteen minutes every night on WXYZ and the Michigan Network. On Friday, September 6, a sizable ad in the papers called attention, as usual, to his broadcast that night at 8 o'clock.

Another "political talk" was listed in the regular newspaper program schedule for one hour earlier, at 7 p.m. The name of the politician who would talk at seven was not announced; he had no ad; and not even Bob Hite can remember today which of Read's seven rivals it was; but the manager of his campaign considered that seven o'clock speech important enough to hire auditoriums throughout the state and pack them with voters to hear it. Like Read's it was also to be sent over the network.

The speeches of both candidates were delivered to the station on big 16-inch electrical transcriptions with live copy for announcers' introductions and sign-offs. All had been okayed by Trendle. It was Bob Hite's fate to be assigned to both broadcasts. He carefully checked both transcriptions for the number of turns each required before each speech began. He looked over the live copy. He put both transcriptions back in the file, and the copy back in the book.

"Come on down and have dinner with Kathy and me," Mike Wallace suggested.

It was six o'clock. The three went to the Southland Restaurant on the ground floor of the Maccabees Building. At quarter of seven, Bob excused himself, took the elevator to the 14th floor and hurried upstairs to the announcers' booth. Fielden Farrington was there. Hite cued up the speech.

"Will you start the transcription for me?"

"I'll do that for you," said Farrington.

At 6:59:30, *The Lone Ranger* was signed off. Fielden Farrington made the local station break. At 7 o'clock on the nose, all the right plugs were in and all the right buttons were pushed, and Bob Hite, in his most commercial style, introduced the speaker to Detroit and the state of Michigan. Farrington flipped the switch and the transcribed speech began. It was at least two minutes before the phone began to ring.

Bob Hite remembered the predicament well: "The Party had filled auditoriums all over the state with people to hear their god. I introduced their god all right; but played his opponent's speech. As soon as the phone calls came in, I got the right transcription out and put it on the other turntable and started it and faded out the wrong speech and faded in the right speech. That's the best I could do. Probably threw the timing of the program off; I don't remember; I was a little up in the air."

Not only *up* in it, but *out* in it; he was fired. Bob Hite had a thin time trying to free lance in Chicago until an agent found him a job with KWK in St. Louis.

One remembers Fielden Farrington as relaxed and business-like. He was a first-class announcer, but his eyes were a brown combination of humor and boredom. He eventually went to CBS in New York and announced several of the most popular daytime soap operas. Finally, the boredom won out and he abandoned the broadcasting industry to enjoy success as a novelist.

There were many comings and goings at WXYZ in the year 1941, but under Trendle's fussy perfectionism and the relaxed dynamism of Allen Campbell, the station gained steadily in prestige and profits. Several personalities who were to figure prominently in the history of the station were in the region of its wave length though not yet on it. Fate, however, was moving them in the right direction.

Bill Morgan was living in a garret and working at three part-time jobs to earn his way through Wayne University. One of these was as waiter at Martin's Feed Bag, a little horseshoe shaped lunch counter near the "campus." One night, a fellow student who was head of the University Radio Guild, Ralph Nottingham, brought his girl into the Feed Bag and tipped Bill that Station WCAR in Pontiac had asked the Guild to send out some candidates for part-time announcing.

The next morning, on his way to Pontiac on a Greyhound bus, Bill met an acquaintance, Ed Lesic, son of a Polish Orthodox priest. They found WCAR on the 6th floor of the Riker Building in Pontiac. Both auditioned and both were hired. Bill Morgan was at least in radio. He was also falling in love with the Woman's

Director of WCAR, Jane Knight.

Another on his way was Charlie Kocher. From the time he was 16, Charlie Kocher dreamed of being Chief Engineer of a major radio station like WXYZ. For ten years, he repeatedly applied to Lynn Smeby for a job without getting one. Meanwhile, he worked a bit at WEXL in Royal Oak, and at WJBK's old transmitter in Highland Park. It would be a long time before he got a call from Wyxie.

Eventually, Lynn Smeby was not at the station for Kocher to visit with. Smeby ran counter to Trendle. He had improved the reception of WXYZ with his tower on the Maccabees Building, even if the red neon lights with which he trimmed the transmitter did get covered with ice in the wintertime and break off and fall to the roof and somtimes to the street below.

Trendle wanted a still better transmitter, and he wanted more power. He wanted 5000 watts. To increase his power, he would have to move his transmitter outside city limits, and since Smeby was not going along with his wishes, he called in an outside consulting firm. Smeby, meantime, had married Harold True's daughter.

Trendle called True one day and said, "Harold, do you mind if I fire Lynn?"

"Christ, no!" exclaimed True. "You'd be doing him a favor."

So Smeby was fired. Within a week he had a job with I.T.T. to build a 50,000-watt station in San Juan, Puerto Rico, after which he became a successful consultant in Washington.

Vern Alston, a short man with black hair and black eyes was raised from the ranks to replace Lynn Smeby, and a new 5,000-watt transmitter was built on a lot at Joy and Greenfield about seven miles from downtown on the western edge of Detroit. On February 21, 1941, at 7:30 P.M., there was a "BIG RADIO CELEBRATION FEATURING WXYZ'S 5000 WATTS NEW HIGH POWER TRANSMITTER. LONE RANGER, GREEN HORNET, NED JORDAN *ALL ON ONE PROGRAM.* $300 IN CASH PRIZES. 10,000 LONE RANGER LUCKY TOKENS FREE WHILE THEY LAST" etc. etc. etc.

Chuck Livingstone's plans for remodelling the drama studio were carried through. A director's booth with windows on all sides was built between the sound studio and the actors' studio. Both the engineer and the man who ran music and sound effects records worked with the director in the booth. Beside the booth, a short, narrow passageway connected the two studios. Also, at this time, the sound effects studio was wired to serve as an echo chamber; but when it was used for this, everybody in it had to be very quiet. It was crude, but it worked.

Everybody—sound men in their studio, actors in the other studio, off-mike people in the passageway, and engineer and record man in the booth—took their cues from the director. When Chuck's young son was asked what his father did, the boy said, "He points at people."

Sound effects improved constantly. For the door of the Green Hornet's super-car, a real, heavy automobile door was mounted firmly and permanently in the sound studio.

Early in 1941, Malcolm McCoy decided to try his luck in Chicago. Jim Jewell, Ted Robertson and John Hodiak were all there; and so were two other former

Edythe Fern Melrose

Detroiters, Earl Ebi and Wynn Wright from WWJ. McCoy did fairly well, but when Livingstone invited him back, he was happy to return to WXYZ.

One morning in 1940, a Grosse Pointe Chrysler dealer and his top salesman appeared at the reception desk of Station WJLB in Detroit. They asked to see the manager.

"I'll bring her right out," said the receptionist. "She's expecting you."

"She!" exclaimed the dealer. "I asked for the manager."

"She is the manager," the receptionist told him.

Edythe Fern Melrose emerged, hand outstretched, smiling disarmingly.

"You're Charlie Record," she said, beaming at the dealer. "I've heard about you. Come in. Come int' the office.

She led the way brightly. The two men exchanged wide-eyed looks as they followed her. She gestured to chairs and sat down happily behind her desk. WJLB's sales manager had told her they had sold the account and didn't know what to do with it. She had told him to have the dealer come in.

"Now!" she enthused. "Tell me what it is you want to sell."

"We want to sell Chrysler cars."

The femininity of Edythe Fern Melrose was misleading. Her face was round, her eyes were blue, her hair was blonde, and her smile was wide. Her laugh was rather like a high-pitched chime. But underneath the butterfly was a spider that spun a web of steel and many a big fat account got happily caught in it—happily, because there the simile ends. Having caught an advertiser in the web of her cheery charm and persuasion, Edythe neither poisoned nor devoured him. On the contrary, she invariably increased his business.

"Tell me, Mr. Record," she wanted to know, the ends of her lips hiding behind dimples in her round cheeks as she smiled joyfully at the dealer. "Why are Chryslers better than other cars?"

The dealer stared at the feminine phenomenon before him.

"Why don't you come out and see?" he managed to say.

"Fine!" Edythe chirped brightly. "I'll be out this afternoon."

Edythe went out to Grosse Point. Charlie Record drove her around the block in a New Yorker. Then Edythe insisted that she be allowed to drive it herself. She impressed the man with her business-like silence as she manouvered the car.

"This is the greatest thing that ever happened," she said. "Have you ever done anything about trying to sell this car to women?"

"What do you mean?"

"You've got the fluid drive," she enthused. "A woman is not a monkey when she stops for a traffic light and can't shift gears promptly."

The dealer's mouth dropped open.

"You've got a turn signal," Edythe continued. "Women have always been laughed at 'bout dryin' their nail polish because they stick their hand out the window. A woman is a lady in this car."

Charlie Record shut his mouth and turned to look at Edythe.

"Look at your interior," Edythe commanded. "You've got broadcloth in here. This is what women want in their homes."

They were back at the dealership and got out of the car.

"You've got the greatest thing a woman ever saw!" Edythe exclaimed, patting the New Yorker affectionately with her gloved hand.

"Write it in a piece of copy," said Charlie Record.

She did. The next day she read it on her program. He called her on the phone.

"Would you put that on again tomorrow? The Chrysler executives are going to listen."

She read the same copy again the next day. Charlie Record called her again.

"Could you do that on another station?"

"I could not. I am under contract to WJLB as the station manager."

"For how long?"

She told him several months.

"All right," he said. "We'll wait."

Early in 1941, Edythe Fern Melrose submitted an idea to WXYZ through Junior Sutton. As part of the idea, she suggested building an actual dream house—a "house of charm"—and broadcasting a woman's program from it. She proposed that all products be tested in this actual house before being accepted for advertising. After six weeks, her presentation was returned to her with a brief note from Sutton. WXYZ was not interested.

She continued as manager of WJLB and she worked on an idea for Chrysler. When she had it crystalized, she recorded a program and took it to them. They listened and bought it.

But they did not want WJLB. She made an appointment with H. Allen Campbell through Harold Christian. One day in September, 1941, the two supercharged commercial personalities met head on. It had to be quite a meeting.

Campbell was wearing a perfectly tailored suit of expensive woolen the color of straight Scotch. His shoes were polished dark brown brogues; his tie was dark burgundy; and his white-on-white shirt seemed tailored, too. The French cuffs showed just enough below his coat sleeve to reveal gold cuff links. Behind his usual relaxed, affable manner, the perceptive little blue eyes were sizing up Edythe Fern Melrose.

Edythe, in a beautifully tailored blue suit feminized by a blouse with a frilly neck and frilly cuffs, was smiling sweetly, mouth-ends disappearing into rounded cheeks. Her manner was forthright and friendly. From all he had heard of her ability, Campbell had expected someone a little more angular and rocky.

"What can I do for you?" he asked, settling down behind his desk.

"Nothing," said Edythe through the wide smile. "I have something for *you*."

Blue eyes met blue eyes steadily; smiles held. Campbell would not ask the obvious question. He waited.

"I've got a contract. For *you*!" Edythe announced. The high chime of her laugh filled the office. "Five days a week, fifteen minutes a day."

Campbell's strong teeth gleamed white through the natural tan. "What is this going to cost us?"

"It will cost WXYZ nothing," said Melrose. "It will make the station money. It's an account for you. How would you like to have Chrysler on WXYZ?"

"I can't think of any objection off hand," Campbell laughed.

"I am bringing the contract to you," Melrose chimed, her manner positive and assuring. "And I come with it."

"What do we pay you?" smiled Campbell.

"Nothing," Edythe beamed. "My contract is with Chrysler. Chrysler is paying me."

Campbell had no more objections. It was a deal.

Melrose remembered the details precisely: "I was getting $100 a week from Chrysler and in 1941 we were still throbbing from our latest depression. We were mobilizing for war and advertising didn't know what in the world was going to happen to it.

"But it was a good deal for me and that's how I came to WXYZ."

WXYZ added prestige and balance to its programming that year with a memorable musical program sponsored by the Stroh Brewing Company. Gus Haenschen was imported from New York once a week to conduct a large orchestra made up entirely of strings, played mostly by symphony musicians. Thomas L. Thomas and Margaret Daum also flew in each week from Manhattan to sing on the program which was broadcast from Orchestra Hall, the home at that time of the Detroit Symphony Orchestra.

This program was worked out in detail in the downtown offices in conference with Campbell and Sutton. It was a program of dignity and beauty.

By an odd coincidence, when Chrysler and Stroh were both on WXYZ, a young man destined to be an important character in the station's history, was working for one and on the bowling team of the other.

Fred Wolf started working for Chrysler in 1934. He was a time keeper, then a production clerk; then he went into time study. In 1941 as the nation prepared for the possibility of war, and the Chrysler Corporation won the job of building the General Sherman tank, Fred Wolfe was the first man drafted into the estimating department.

Fred was also an excellent bowler. He rose through some smaller leagues and started bowling with what was to be a championship team at the Stroh Brewing Company. He had no thought whatever of going into radio and nobody in his life would have dreamed of connecting the two.

It was not unusual for employees at WXYZ to receive their mail already opened. In May, 1940, Nancy and Dick Osgood were divorced. That spring, Osgood received an offer from Sherman K. Ellis & Company—an ad agency in Chicago, and the letter had been opened. Osgood turned down the offer and sent a copy of his reply to Trendle, attaching to it an inter-office memo on which he typed: "Since you have already read the letter addressed to me from Chicago and so already know of the offer, you may want your information to be complete. Attached is a copy of my reply."

By return delivery, Osgood's communication was returned to him. Scrawled across the face of it was the word "Thanks." And under it, the letter "T."

Trendle's curiosity was not always so well rewarded. There came to WXYZ an out-doorsy one-time newspaper columnist, Lee Smits. Lee was a good writer, a colorful personality and a capable news man. He came to WXYZ under the sponsorship of the Michigan Consolidated Gas Company to do an evening newscast.

Soon after his arrival, Lee Smits wrote a curt note to the executive offices declaring that opening other people's mail was against the law and that the next time any letter addressed to him was opened before he received it, he would report the matter to the postal authorities. His letters were not opened after that.

Douglas Edwards, ambitious, well-organized, and with his eye to the future, had applied himself diligently. He had been impressed with the news savvy of Felix Holt and he had studied Hank Alexander's procedures in the news room. He now felt the time had come to lift a foot to the next rung. He had Larry LaDue cut a transcription of one of his newscasts and sent it off to CBS in New York. The record was returned from CBS with a letter from the then Director of Operations, Gerald Moseby: "While your recording is good, we must limit our choice to the best."

Plainly, Mr. Moseby did not consider Doug Edwards the best. So Edwards did the next best thing. WSB in Atlanta, the station he had left when he joined WXYZ, had changed hands. The new owners knew him and offered what he was getting at WXYZ to return to Atlanta and three newscasts a day. He would prepare them himself. Douglas Edwards returned to Atlanta.

In 1941 there was an important expansion of service in Harold Christian's merchandising department.

Christian: "One of our clients asked if we could help him find a few food demonstrators whom we could work into super markets. We trained four women and that was the start of our demonstration service for grocery, drug and department stores.

"This new department developed so fast that within three years we had 80 women demonstrators on payroll. They had special uniforms and wore buttons that showed they were affiliated with WXYZ.

"It was not unusual for a demonstrator to sell three hundred cases of a product in a three-day sale in a single market. We gave the client a report on each store—as to the number of people served, the samples used, and the comments of customers.

"We made this service available not only to WXYZ clients but also to our many broker friends to promote their products even though at the time they might not be using WXYZ radio. The brokers were so grateful for this service that when future advertising contracts came up, they invariably recommended WXYZ. In fact, in

Lee Smits

many cases, they insisted that their principals use WXYZ.

"Our relations with the food brokers were second to none and this relationship produced many new clients and kept our clients on for longer contracts."

Nate Shapero had developed into a kind of joke between Campbell and Trendle. Shapero was still a member of the Detroit Fire Commission of which Trendle was made President in March, 1940.

Nate Shapero had been a pharmacist in the United States Navy. He had purchased a drug store in Detroit and multiplied it into a chain of drug stores throughout the country. At the time, he owned two chains in Detroit—the Shapero Drug Stores and the Cunningham Drug Stores.

Campbell was amused by one of Shapero's antics: "Nate Shapero would ride on the train with me goin' to New York. He used to have his secretary keep in touch with my secretary to find out when I was goin' to New York and plan his trips so's he could talk to me the whole evening about radio—and never buy anything."

One evening, Campbell and his secretary were working late. He was leaving the next day for New York to keep an appointment with "Mix" Dancer of the Blackett, Sample & Hummert agency (later Dancer-Fitzgerald-Sample, Inc.). A deal for *The Lone Ranger* was getting warm. At six o'clock, Campbell finished his dictation and was on his way out the door when his secretary stopped him.

"It's Larry Michelson," she smiled apologetically, holding her phone. Campbell hesitated. He had promised his wife he would be home for dinner and he was late. But he turned and went back to his desk.

"Allen," said the agency man. "Could you have dinner with Nate Shapero and myself at the Book tonight?"

"Hell no," said Campbell laughing and without emphasis.

"Allen, it's important!"

"I'm on my way home to dinner," Campbell insisted, "and I promised to be there at quarter to seven."

"Allen, Nate is ready to made a deal."

"I've missed dinner with my wife too many times, and particularly after ridin' the goddam rockers with Nate Shapero for two years I can't afford to miss this one tonight."

Campbell laughed pleasantly. Michelson was serious.

"This is going to be different," he promised. "Please. This has gotten down—as a *result* of your two years of riding—now we're getting someplace."

Campbell gave in. After two years of "ridin' the rockers," it was at a table in the Book Cadillac Hotel that night when Nate Shapero agreed to advertise on radio for the first time. He wanted a newscast. The clouds of World War II were gathering and Campbell, Michelson and Shapero agreed that the newscast should be introduced by the sound of an airplane and the program should be called *Cunningham's News Ace.*

Cunningham's News Ace "zoomed" into the homes of WXYZ listeners several times a day, eventually spread to other stations, and remained on radio and television for a quarter century.

Jack McCarthy recollected: "I can remember doing a *Cunningham's News Ace* with Harry Golder. Harry was doing the news and I was doing the commercials. We were sitting across the table from each other and the microphone was suspended between us.

"And this commercial was for Dr. Grabou's Pipes. These pipes required no break-in period; their slogan was 'pre-smoked'—'Pre-smoked Pipes.' And the order was to punch this copy and of course, McCarthy was giving his all all the way through, probably a full minute commercial.

"And the concluding sentence then was, as I punched it out: 'So tomorrow, stop in at Cunningham's and get one of these Dr. Grabow's Pre-smiked Popes! I mean Pre-Smiked Popes! PRE—SMIKED POPES!!!

"*Three times*. I could not get it right. Golder just folded his arms and put his head down and there was silence for about fifteen seconds."

23
Jeanie, and a Dead God

On January 1, 1941, broadcasters sent Jeanie with the Light Brown Hair into battle with the prevailing writers of popular music. Radio had refused to sign a new agreement with the American Society of Composers, Authors and Publishers. Stations had been paying fees for use of popular songs from movies, Broadway shows and Tin Pan Alley; and they felt the fees were too high.

The American Society of Composers, Authors and Publishers had been organized in 1914. Seeds for its creation had been planted in 1897 when a copyright law was passed ruling that music could not be performed for profit in public without the consent of whoever owned the copyright. The law was not put to a test until Victor Herbert took it to court seventeen years later.

In 1914, Victor Herbert was conducting one of his lesser musicals in a New York theater. He went to dine at Shenley's Restaurant, and there he heard an orchestra play music from the show. He thought it unjust that, while people paid permission to hear his music in the theater and he received royalties for it, musicians in a restaurant played it without paying a fee, and without even asking his permission. He took his complaint to a lawyer. The case went all the way to the Supreme Court.

Supreme Court Justice Oliver Wendell Holmes upheld the copyright law of 1897 and ruled that a composer's music could not be publicly performed for profit without the composer's permission. He further stated that in granting that permission a composer was within his rights to attach any conditions he desired—such as the payment of a fee.

Having won his case, Herbert, with the help of John Philip Sousa and George M. Cohan, organized the American Society of Composers, Authors and Publishers to enforce the copyright law, thus protecting its members. Further, ASCAP would make obedience to the law easy by serving as a central clearing house where performing rights could be obtained and fees established. Fees were collected wherever music was played commercially—in theaters, dance halls, night clubs, restaurants, hotels, concert halls—and radio. These fees were distributed among ASCAP members according to the volume of their output and the prestige of their compositions.

Part of Franz Werner's work as music librarian of WXYZ was keeping a record of every number performed on the air by pianist, singer, organist or orchestra. Later, it was made easier to pay a "blanket" or annual fee based on the power of the station and a calculation of the amount of music played in a year's time.

In 1940, as time approached for renewal of their agreement with ASCAP, the broadcasters of the nation decided to boycott the society's music. A quick survey soon showed that no popular music would remain to be played. Fifty thousand creators of the most popular music belonged to ASCAP, directly or by association. Stations would be limited to folksongs, country music and music in public domain. So, in 1940, radio set up its own music publishing business—BMI, Broadcast Music, Inc.

The boycott of virtually all popular music in America went into effect January 1, 1941. Almost nothing from BMI achieved any degree of popularity during the first years of its existence. Music of English and European origin could not be used because foreign composers belonged to societies which had reciprocal agreements with ASCAP. For some reason, however, music from South America was available and it was during the 1941 boycott that tangoes, boleros, cariocas and rhumbas from south of the border enjoyed a prestigious amount of exposure. In the States, the best source of music in public domain had been composed by Stephen Foster, and "Jeanie With the Light Brown Hair" was played so often for so many months that it became a joke.

While radio's search for usable music was like panning for gold in Lake Erie, ASCAP had its own bitter disappointments. Listeners were surprisingly undisturbed by the disappearance of popular music from their sets; the expected roar of indignation failed to materialize. And the staunch support the society had expected from performers who lived off the output of ASCAP's members—the Gershwins, Cole Porters, Hammersteins, Berlins and Sammy Cahns—was not forthcoming, either. Singers, big bands and star instrumentalists struggled along with low grade stuff that was available just to keep working.

After several months, the U.S. Government stepped in to end "the silly nonsense." Very little pressure was required. ASCAP settled for about half of what it had collected before.

The public expressed almost no reaction when the flood of favorite music returned to its speakers, but the trade was jubilant. However, bitter feelings festered between ASCAP and the broadcasters until they surfaced again seventeen years later when rock and roll emerged from BMI and ASCAP cried loud and clear, "Payola!"

The force of H. Allen Campbell's personality was of such quality that he invariably made his points effortlessly and without stress. He commanded attention with his relaxed mental self-assurance. His speech possessed almost no exclamation points. The success of his salesmanship was never to push anyone into something they didn't want. His policy, conscious or not—as with Harry Hyman of United Shirts—was to seek the client's good.

He was convinced that *The Lone Ranger* was a good buy for General Mills and he finally succeeded in passing along his conviction to Mix Dancer of Blackett, Sample & Hummert. Dancer finally sent young John Sample to Detroit to talk a deal. Campbell took him to Al Green's popular restaurant on the edge of Grosse Pointe for dinner. The talk went round and round as Campbell worked his guest gently toward the deal in dead center. The next day Johnny Sample met again with Campbell and they talked with George Trendle. The sale was set. But there was a catch.

"There is one thing you cannot do," said Sample. "We won't want you to do *anything* to take Gingham Bread off the air."

Campbell's white smile was set but his foxy blue eyes narrowed as he looked steadily at John Sample. What he had hoped to accomplish by landing General Mills was to simplify the whole set-up by having a single sponsor in all major markets coast to coast.

"You have to realize," said Sample. "General Mills makes Gold Medal Flour. And Gingham buys it. So does Merita. They are General Mills customers. We would *like* to have the whole country—south and west coast as well as the rest of it—we'd like that—but we do *not* want you to do *anything* to take our customers off *The Lone Ranger*.

Campbell leaned forward and pressed out his cigarette.

"There's no problem there," he said affably, smiling.

The deal was set. General Mills would sponsor *The Lone Ranger* as of May, 1941.

Campbell kept his word. He did nothing to dump Gingham. On the other hand, he did nothing positive to keep them happy. He simply stopped his friendly, personal contact. Gingham withered away; finally cancelled.

As Campbell remembered the affair: "General Mills was very upset about it, but finally we got the whole ball o' wax—with the exception—they never did get the south. But Merita Bread and General Mills sponsored the whole country."

Fran Striker now had an assistant writer named Sheldon Stark. Stark began writing at nine o'clock in the morning and quit promptly at five. He wrote Green Hornets mostly and was an excellent writer.

One month before General Mills assumed sponsorship, the program suffered a shocking blow. Earl Graser, 32 years old, fell asleep at the wheel of his car, rammed into the tail end of a parked truck, and at 5 o'clock on the morning of April 9, 1941, was dead.

Felix Holt was roused from his sleep by the news media. He awakened George Trendle.

"The wire services and newspapers want a statement."

"Oh—tell 'em we'll have another Ranger and a better one tomorrow," said Trendle carelessly.

"Mr. Trendle!" said, tense, excited little Felix Holt. "You can't make a statement like that. This man was a god."

The program was now heard by transcription in Australia, New Zealand, Hawaii and Canada, as well as throughout the United States.

"Well—you make up a statement, Felix," Trendle said. "You handle it."

The New York Times was published that morning with the headline: "DEATH LIFTED THE LONE RANGER'S MASK AT DAYBREAK TODAY."

It was the first time the world learned who had played the role. For the first time, the world learned that Earl Graser was a graduate of Wayne University, liked to garden and play badminton, was heard on 75 independent stations and 80 stations of the Mutual Broadcasting System, had strained his upper register and ruined his singing voice singing with a pit orchestra in Detroit, and had planned to live in Connecticut when *The Lone Ranger* ended.

Time Magazine said, "All over the United States, hundreds of thousands of children to whom the Ranger had sent toy lariats, six-shooters, ten-gallon hats and bristling wild west mustaches, whom he had commissioned Rangers and pledged to good conduct and fair play, mourned the most adored character ever to be created on the U.S. air."

On the morning of the 10th, Trendle called a meeting—Campbell, Livingstone, Holt, Sutton and Striker.

"It's too bad and I'm sorry it happened," Trendle said coldly, "but the show goes on. The question is who plays the Ranger and how do we make the change?"

Different voices were discussed.

"We could get a duplicate of Graser's voice if we had the time," said Trendle. "We could go to New York or Chicago. But there isn't time; the show goes on again tomorrow night."

"What about moving Al Hodge or Jack McCarthy up into the Ranger role?" Sutton suggested, teetering in a corner, chewing his jaw, looking lively and earnest.

"Hodge and McCarthy are too well identified with what they are doing," said Trendle. Hodge, of course, was playing the Green Hornet and McCarthy was Ned Jordan. Someone suggested Jay Michael. Trendle felt he had played too many villains on the Ranger. They discussed various announcers.

"Mike Wallace."

"Can he act?"

"Voice is right but I don't believe he can act."

"Brace Beemer," said Chuck Livingstone as though thinking out loud. His face was terribly serious. He had been closer to Earl Graser than the others. He was shocked, grieving. He continued: "He's a perfect change-over.

Holt, who had been perched on the front of a chair, his thin legs crossed, forearms folded on his lap, hands dangling, jumped up.

"Brace does all your personal appearances," he reminded Trendle.

All eyes turned to Trendle. The office was silent.

"Who?" Striker asked, looking from one to the other. Fran had been writing all night and he was tired and when he was tired he couldn't hear well.

"Brace Beemer," said Holt.

"We'd have to break him in easy," Trendle warned, looking at Striker.

"He can get shot at the beginning of the next script," Striker suggested. "He can be wounded—unable to talk."

"You'll have to write new scripts," said Trendle. "Keep him out of all of 'em for a week or two. If he has to talk have him talk in a kind of whisper. You can have him moan—just to keep him in the story. But you'll have to write your action around him."

So that's what they did.

Livingstone remembered the episode: "I think in Trendle's story he says six but I think there were only three scripts. We used the recording of Graser saying "Hi-yo, Silver—awa-a-ay'—for quite some time until Brace could do it. But Brace cut the word 'awa-a-ay' because he couldn't make it sound right. So he cut it."

Tonto nursed the Ranger back to health in a cave, and when at last he could talk again, it was the voice of Brace Beemer.

The New York Times printed a beautiful editorial on April 10, 1941:

> Earl H. Graser was killed in an automobile wreck Tuesday morning, but the rumor that the Lone Ranger is dead is unfounded. It was a man who died . . . he didn't take the Lone Ranger with him. The Lone Ranger doesn't die . . . his trusty steed waits to carry him on his errands across the face of the wondrous west where the air is crystal and virtue never lacks for its reward. Listen! There is the beating of the hoofs as, in the nick of time, he swings into action. Ride, Tonto! Ride, Lone Ranger! Hi-yo, Silver!

But for some the famous yell was never quite the same again.

Station WJLB which Edythe Melrose had managed before moving to WXYZ developed into quite a source of supply for the announcers' booth in the Maccabees Building.

Originally, WMBC—where Jack McCarthy had tussled with Amos Jacobs over a scapular—the station had been purchased by John L. Booth, who changed the call letters to his own initials and moved it from a basement at Grand Boulevard and Woodward Avenue to the Eaton Tower in downtown Detroit. It was then only a 250-watt station, but Booth hired a big staff orchestra conducted by Herb Mertz—and everything was live and big.

Booth had an ear for excellent announcers. First, there was Bill Morgan who had found himself unlucky in inter-office feuds at WCAR in Pontiac and managed to escape to a job at WJLB and take Jane Knight with him. She became WJLB's Traffic Manager. Besides Morgan, there was a fellow who resembled Tyrone Power, named Larry McCann; an eager, blue-eyed, clean-cut youngster named Hal Neal; B. Robbins Downey who later joined the faculty of Michigan State University; and a tall, solidly built blonde named Fred Foy.

Morgan and Foy shared a problem. It was a sense of humor which in any other activity would have been treasured as a rare congeniality. In radio it was a hazard; the two found it impossible not to laugh at the wrong time—that is, when they were

Jewell Players at the Fisher Theater

on the air.

Bill Morgan: "We could not work together without breaking up. They had a two-man news format there, and they put Fred Foy and me on that thing, and just sitting across the desk from each other, we'd still break up. They tried putting us in different studios and finally they had to put us on different air shifts as the only way not to fire one or the other of us."

All the announcers wanted to get to WXYZ; WXYZ paid more because it had an AFRA contract. Fred Foy was the first of the WJLB announcers to make it. Jack McCarthy hired him. Carl Gensel had moved to WCAU in Philadelphia; McCarthy was now WXYZ's Chief Announcer.

In 1941, Chicago was an important center of radio production. *Li'l Abner* with John Hodiak originated there. Several daytime "soap operas"—series aimed at lady listeners and usually sponsored by soaps, toothpastes and detergents—were created in Chicago by a prolific genius named Irna Phillips. A former teacher of English and dramatics in a normal school, she had played bits in radio dramas during a summer vacation when someone suggested that she try to write. She was now dictating some 2,000,000 words a year and earning $3,000 a week (to Fran Striker's $200!). Among her creations were *Today's Children* which was broadcast twice a day and ran for five years; *Road to Life* about doctors; *Woman in White* about nurses; and *Guiding Light* about ministers.

Ted Robertson, now at CBS in Chicago, heard that Irna Phillips was looking for an announcer for one of her shows. She wanted a new voice. Ted knew the producer.

"Look," he said. "I think I know an announcer in Detroit who could cut it."

"Will he come over for an audition?"

"I'll call him."

Robertson was talking about Mike Wallace. "I had become friendly with Mike Wallace. I thought he sounded good. I used to drop him suggestions, or I would even criticize him—heavily sometimes when I thought he was lousy. And we became personal friends."

Wallace went to Chicago, auditioned for Irna Phillips in competition with several top announcers, and got the job.

Wallace remembered the episode: "I left WXYZ in September of 1941 because I had a chance to go to Chicago and that seemed to be the Big Time. I felt later on that I should probably have stayed in Detroit a little longer, rather than going into Chicago insufficiently prepared. I would probably have done better and been happier."

With Mike Wallace, Douglas Edwards and Fielden Farrington gone, word got around that WXYZ needed announcers, and the word reached Bob Hite in St. Louis. He wrote a letter to Jack McCarthy, was offered a job and drove home to Detroit. Bob took over the announcing on *The Green Hornet* when Wallace left for Chicago.

Harold True, the man who had fired him originally, was also leaving WXYZ.

He had personal reasons: "I wanted to get out o' there. I mean, I knew I'd never get anywhere with Campbell there 'cause Campbell hated my guts and it was mutual. So I wanted to get out o' there and I—I knew the only goddamn quick way was to get into politics, 'cause that was strictly against the rules. So I put my name up for a Councilman of Detroit and I goddam near got elected! George Edwards beat me by 6,000 votes."

True moved over to WWJ at the end of 1941 and continued his popularity as a newscaster until he retired. At WXYZ, Al Hodge replaced him on *Town Talk*. And Earl Moore who had been assistant auditor was sent out to the Maccabees Building to pass on downtown orders as Studio Manager. In this late 1941 shakeup, Adolph Stoffer who had been auditor was replaced by Joe Kassel. Charlie Hicks who had been advertising and promotion manager was also lost in the shuffle and replaced by a big, warm Dutchman from the east, Lambert Beeuwkes. To work as his assistant, Trendle moved in a nephew from the theaters with which he still maintained a close relationship. The nephew by Trendle's first marriage was Bill Hendricks, a little man with quiet arrogance and an air of being constantly amused.

Elizabeth Ashton worked on as Beeuwkes' secretary for a while, then was moved to work with Raymond Meurer who now occupied a small office next to Campbell. Ashton was not too content with Meurer.

Her reasons were several: "His main task was, you know, handling the licenses for the products—the toys—it seemed everything on earth. I know one as—I made up a list—but that's a long time ago. I couldn't possibly—. And it wasn't only *The Lone Ranger. The Green Hornet*, too. And all those licenses. And also he was the attorney for all those corporations—yah, yah, yah. And I know I didn't like that end and—I didn't scream exactly, but I told Mr. Trendle in no uncertain terms and he said—just rest along for a while—so— —"

The real shocker of the 1941 shakeup was the firing of Junior Sutton.

Witness One: "He was fired for two reasons. First, he had run dry of commercial ideas. Second, almost his only account left was Conn's Clothes and there was a suspicion that some kind of hanky-panky was going on between him and the sponsor."

Witness Two: "Well, Junior was a big boozer, too, you know—*Big* boozer. And he was a conniver. I wouldn't be surprised if Junior was doing some conniving with Conn's Clothes. He was sloppy and dirty and—."

Witness Three: "Well, it had an awful lot to do with alcohol. Not only that. I was with him when he made contracts with certain buyers and he'd make promises that we couldn't keep and Campbell was left holding the bag."

Witness Four: "Trendle didn't like the way he put grease in his hair."

It was a tragic day for Junior Sutton. He had a nervous breakdown from which he never fully recovered. He finally wound up in a restaurant, we're told, washing dishes.

Campbell did not go far afield to find a replacement. He found his man in the Traffic Department—the sunny, optimistic, self-confident Jimmy Riddell, one-time mail boy. Campbell was now Treasurer and General Manager and James G. Riddell was named his assistant and Sales Manager.

A man was brought in from out-state to take Riddell's place in Traffic. He proved unsatisfactory and the dumpy little woman with the Theda Bara eyes took over—Irene Donnelly.

The lonely devotion of Gwen DeLany was lost to the Detroit Local of AFRA that same year. Her work had dwindled to nothing and she decided to move to the west coast. The Minutes of a General Membership meeting held on September 18, 1941, read: "President Dougall read a letter from our Executive Secretary, Miss Gwen DeLany, tendering her resignation from that office to take effect October 31st, 1941. At the conclusion of President Dougall's reading of the letter the deep silence that settled over the membership, told more than words can speak, of the sincere regret and realization of the loss which Miss DeLany's resignation means to Detroit AFRA."

Gwen DeLany found no great degree of happiness in California. Jobs were scarce; she no longer was young; and she was not in the best of health. Word reached Detroit finally that she was dead.

Edythe Fern Melrose had been with WXYZ two months when she received a call to visit Trendle. She says she went to the Stroh Building with "fear and trembling," but it's hard to imagine; Edythe Fern Melrose was not the type of person to be frightened; and she was collected enough to take her House of Charm presentation which had been turned down.

Whatever foreboding she might have felt disappeared the moment she saw Trendle. He was tall; he was clean; he was charming. He invited her to sit down. His baby-blue eyes were alive with personal appraisal. He smiled.

"You are a great salesman," said Trendle. His voice lacked depth and resonance to go with his height and figure and position. It seemed to need moisture; it was a dry voice.

"I was a very *poor* salesman," beamed Edythe, sitting straight in her chair, her feet together.

"Why do you say that?" There was something Teutonic and personal in Trendle's manner toward the ladies.

"I brought an idea to your station," said Edythe. "And you didn't buy it."

"What kind of an idea?"

The high chime of Edythe's laugh sounded as she rose from her chair. She spread out her House of Charm presentation on the top of his desk which, as usual, was clean and like his mind, uncluttered. He examined the sketches as Edythe explained her idea.

"Who turned this down?" he wanted to know.

"I don't know, Mr. Trendle, but the book was returned to me by Junior Sutton."

"Well he was so dumb I fired him," Trendle told her.

"He was not astute," she agreed, the ends of her lips lost in her round cheeks.

Whatever ogre men sometimes felt in the make-up of George Trendle rarely showed around women.

"You go right on with your idea," he said. "I'm for you. And I'll back you."

Edythe chimed again and held out her hand. He took it and she knew it was a man looking at her. He handed her the presentation and she left his presence.

Melrose was encouraged: "To me, his backing meant only that I could continue my shows on that station because when we went into World War Two, advertising took a terrific drop. We didn't know where we were going. Chrysler cancelled their contract because they couldn't make any more cars."

Toward noon on Sunday, December 7, 1941, Fred Rivard was on stage at the Broadway-Capitol Theater warming up the audience for Radio Schoolhouse. Dick Osgood stood at the control booth in the wings, listening to the station as the program was fed down to the theater. Just before the kids went on the air, the bulletin came over: three American battleships destroyed by the Japanese at Pearl Harbor, sixteen other warships sunk, planes bombed, 3,000 casualties.

Congress made it official the next day. The United States was at war.

24
Wartime at Wyxie.

Raymond Toyo Hayashi sat down opposite Dick Osgood at a table in the Southland Restaurant.

"Mistuh Ossgoo', what I do?"

"What's the matter?"

"They want send me to Japan. I never been Japan."

"Who wants to send you to Japan?"

"United States Government."

It was lunch time. Waitresses in white uniforms served with calm efficiency. "Ma" Sweeney who owned the restaurant bustled about near the kitchen. From across the room came the bright laughter of John Slagel and Marjorie Domini; they were working up to a marriage. Marjorie taught at the Detroit Conservatory of Music next door.

"Why does the United States Government want to send you to Japan?"

"This war! I don' know!"

Osgood dipped the fried butterfly shrimp into the delicious sauce Raymond had taught Ma Sweeney how to make. It was a delicacy not to be found anywhere else in Detroit since Raymond had closed his restaurant. Hayashi's brown eyes peered hopefully at Osgood through rimless spectacles.

"You can do something?"

Osgood was now broadcasting a series called *March of Victory* for the Hi-Speed Gas Stations. After five years, three months and 883 stories ranging in subject from "Abrasives" to "Zinc," the Factfinder series had been dropped. Floyd Tracy of Hickok Oil felt something tying in with the war effort would be more timely. *March of Victory* was devoted mainly to the contributions of science and industry to the armed services. Every script—and there were five a week—had to be cleared through six departments in Washington. Possibly the little Japanese who played Kato believed that because of this, Osgood might have some "in" with the government. But it was not so and Osgood could do nothing to aid Raymond Toyo Hayaski. He disappeared, presumably to a concentration camp in the west. No one at WXYZ ever saw Raymond again.

On the air, Kato changed miraculously overnight from Japanese to Filipino. The actor who fell heir to the role was Rollon Parker. Parker was in his mid-thirties, about five-feet-nine, strong, solid and slender. He was as talented as any actor who ever worked in Detroit radio or TV, and modest to a fault. He developed a passion for a diminutive, black-eyed girl of French-Canadian extraction, named Leonide. Rollie and "Leo" double-dated with Dick Osgood and one of the actresses, Gwynne Fermin, and often wound up in Osgood's apartment for drinks and close harmony. The Osgoods were divorced; Nancy had gone to WRC in Washington and would later become feature writer on a newspaper in St. Petersburg.

Rollie and Leo married, divorced, re-married and divorced again. The process turned Rollie's modesty into a first-class inferiority complex and he gradually became the slave of vodka and cheap whiskey

World War II temporarily interrupted a new development in the broadcasting industry. Science had been experimenting with television and by 1940 was sending through the air live images from political conventions. Only scientists witnessed the miracle, and they had to build their own receiving sets because none was on the market. The general public knew what was going on only by catching inconspicuous little pieces buried by more important matters in the newspapers.

The Radio Corporation of America and the National Broadcasting Company conducted extensive experimentation, and commercial application of the marvel was in sight when plants producing the necessary equipment were diverted to the making of war material.

Something else was happening that would be vital to the story of WXYZ. Several years before, the Communications Act of 1934 had created the FCC, the Federal Communications Commission, to regulate, among other things, radio broadcasting. The FCC conducted a three-year study of what some felt was the ominous power of the networks and in May, 1941, the Chairman of the Commission, James L. Fly, issued a batch of new rulings. No longer need radio stations sign away their independence for five years to join a network; no longer would a station on a network be restrained from accepting a program from a rival network; no longer could a network control more than one station in one area.

Particularly hard-hit by Fly's edict was the National Broadcasting Company. It owned or controlled more than one station in more than 38 different areas! Also, it operated, not one, but two networks—the Red and the Blue. Blue was the step-child. Shows like *Jack Benny* or *Fibber McGee and Molly* were tried out on the Blue Network and as soon as they became popular and captured a sponsor were transferred to the Red Network. There was a suspicion that NBC operated the Blue Network for no other reason than to keep anyone else from having it in competition. The FCC ruled that this monopoly must end. Assistant Attorney General Thurman

Arnold also took up the cudgel. An investigation resulted in the U.S. Senate.

Suddenly there was a mighty juggling of executives at NBC. Sidney Strotz, Vice President in charge of the programs, was moved to California and put in charge of all west-coast operations. And now, for the first time, the Blue Network had its own separate sales department, its own manager, its own program director, and its own sales promotion and publicity staff.

A plan was said to be in the works to make separate corporations of the Red and the Blue networks, Red to remain NBC, Blue to become the baby of the Radio Corporation of America. RCA also owned NBC and it was obvious to insiders there was still hope all might continue as one big, happy family. But in January, 1942, it was clear that neither NBC nor RCA would hold the Blue Network. The Blue chain of 101 stations was put up for grabs. The FCC eased up on its pressure for an immediate divorce in order to prevent a forced sale. In the next year-and-a-half, there were more than a score of bidders for the network. WXYZ was on Blue, and one of them was George W. Trendle.

Meanwhile a less spectacular and more immediate problem at WXYZ was what to do with Edythe Fern Melrose. Chrysler had cancelled; it had no cars to sell; it was turning out war material. James G. Riddell, WXYZ's new Sales Manager, asked Miss Melrose to drop downtown for a talk with him. The two went to Campbell's office. All three were cheerful, positive, optimistic people; there was no sense of disaster in the air; it was simply a matter of working out a problem. Miss Melrose had already made a start; as soon as Chrysler cancelled, she had gone out and sold a few accounts herself.

"Here's the billing we had on her show this week," said Riddell, placing a statement on Campbell's desk. The amount was $500. Peanuts today. Good billing in 1942. Campbell studied it a moment; then he looked up at her shrewdly.

"Would you like to continyuh with us on a participating basis?"

Edythe was just as shrewd as he was, although it didn't show in her round face and innocent smile. She was ready for him.

"I'll sell your announcements at the regular price," she said. And she added, *plus* 25%. I will bill you for the 25%."

"Sounds like a good deal," Riddell commented thoughtfully after a moment.

"Do you want a contract—some kind of guarantee?" Campbell asked pleasantly.

"Mr. Campbell, we don't need a contract," she said positively. "And only if I didn't know what I can do would I want a guarantee."

Riddell retrieved the billing statement from Campbell's desk.

"However," Melrose continued, "I don't want to work in the studios. I want to establish the idea of a house—a House of Charm. But while I'm planning it and until I have it, I want to continue in my own studio in the New Center Building. I want to remain anonymous."

"I don't see any objection to that," Campbell agreed after deliberating.

"Harold Christian will be a great help to you," Riddell promised.

"Oh, I know he will be!" Melrose agreed enthusiastically. "But I will also sell my own accounts."

The soprano chime of her laugh filled the room as she stood up briskly and crossed the carpet to Campbell, her hand extended. He was standing and took her hand, his teeth gleeming. Edythe Fern Melrose remained with WXYZ for 25 years without ever having a written agreement of any kind.

Douglas Edwards finally made it to New York from Atlanta. He worked hard at WSB. He did something that had not been done there before, something he had learned at WXYZ: he re-wrote the news after it came off the wire.

"They found it reasonably impressive," he recalls. He won top rating in Atlanta and WSB was a 50 kilowatt station. He figured he was as ready as he ever would be when his old friend from WXYZ, Fielden Farrington, phoned him from CBS, New York, and told him an announcer's job was open. With cool calculation the gentle but ambitious Doug Edwards went over the CBS news staff in his mind. Robert Trout and John Daly had both been promoted into news from the announcing staff. If they could do it, he could do it.

In December, 1941, Edwards started as an announcer at CBS. He let it be known at once that he was aiming at news and he wasn't long getting a crack at it. Paul White, then CBS Director of News, liked what he heard. Doug was told to stay close to John Daly and watch him work for five days and then try to do the same thing.

"Looking back," Edwards remembered, "I realize with what tender, loving care John Daly handled me. He let me know every trick in his book. He was doing a program called *The World Today*. This was an overseas pickup show where you had to be ready in case the signal failed—as it did in those days. No such thing as tape then. And we didn't use recordings on the network-at *all*. Not one! So, he showed me how to get ready for it, to anticipate it, how to lead into it."

Edwards did his first fifteen-minute CBS newscast on a Saturday night. The following Monday, John Daly came to him and said, "You have nothing to worry about."

Daly was right. Doug Edwards went to Europe with Edward R. Murrow during the war; and later he became anchor man on the evening news at CBS television. Thirty years later, Doug Edwards was still a CBS staple.

"What well-known Detroit employer hires a full-time detective to check up on his employees' personal activities?"

This was an item on Dick Osgood's *Show World*. He had a call from Trendle.

"Last night you had something on your broadcast about a detective," he said.

"I believe so," Osgood admitted.

"Some employer hiring a detective to follow his employees around."

"That's right."

"Who were you talking about?"

"Oh, I don't want to go into it any more than I did on the air, Mr. Trendle."

"Well, where did you get it?"

"It would betray a confidence to tell you, Mr. Trendle."

"Well—I don't mind your putting stuff on to get listeners, but you better be sure of your facts."

"Yes, sir."

This fact had come from a source very close to the employer. The employer who hired the detective was George Trendle.

The Traffic Department of a radio station stood between sales and programming. Matters could become complicated. Sponsors might make certain demands. Through error, two advertisers might be sold a spot announcement at the same choice spot. A salesman who was not informed might sell a spot on a program where time was not available. It was easy for things to go haywire. And they did.

To Elizabeth Ashton the memory was vivid: "Oi, oi, oi. Oh yes, there'd be lots of hassles. There were so many! It wasn't too long after Pearl Harbor I got into all that stuff. Jimmy got Junior's job, Jimmy turned into Sales Manager and they gave the job, they brought in a man from Grand Rapids to take over for Jimmy in Traffic and that didn't last, I mean, he made a mess of it. I don't know—one or two men. It was a holy mess and then finally they gave it to Irene Donnally and it just got worse and worse and worse. She was the last gasp before I went in and it was—oh God. What a mess! Ghow! What a mess it was! And so finally—one noontime—I went in there and took it over."

Bess Ashton's conversational style was meandering, but her mind was right on target. She had authority and savvy. She ruled the Traffic Department with a mind and a will of steel.

Ernie Winstanley was probably the first Wyxie-ite to go to war. He donned a uniform in early 1942. Fred Foy was drafted in September. Harry Golder joined the Navy. When Tom Dougall went in, the word spread, and like magic an unplanned farewell party "happened" in the Alcove, a bar on the same block with the Maccabees Building which had been given its name in a contest won by Ernie Winstanley and Fred Flowerday—AL for Al Burtenshaw who owned it, and COVE because it was a little place like a cove. Coves were common in the Ranger.

At the impromptu party for Dougall, Pete Van de Van appeared with his accordian and there was music; and somewhere, someone managed to hastily take up a collection and Tom was presented with cash—or a watch—nobody remembers. It was a wonderful, sentimental, alcoholic brawl. Fred Rivard took over as producer of *Radio Schoolhouse.*

In the summer of 1941, Dick Osgood rented a cottage in Canada on Lake Erie within commuting distance, to get in shape for the army. He was older than the others. He filed a long and complicated application for Special Services, and never received a reply. He passed his physical and was designated 1-A; and one day after news came over the wires that anybody over 40 was too old, he received his induction notice. But he was 41, and the induction was cancelled.

However, he found a new love in Canada, Anne Piper, from East Tawas, Michigan; and they were married the following Valentine's Day.

Bill Morgan was drafted; then rejected. He went back to WJLB for his job. Another announcer had been hired in his place, but he got his job back, feeling so insecure he arranged an interview at WXYZ with Jack McCarthy. McCarthy liked Morgan and told him when there was a vacancy he'd call him. McCarthy kept his promise and Bill Morgan followed his friends Ed Lesic and Fred Foy to the announcers' booth in the penthouse of the Maccabees Building.

Left behind now at WJLB were lean, dark Larry McCann and eager young Hal Neal. Neal soon left to go into training with the U.S. Army Air Corps.

McCann learned in conversation with Fred Foy of the advantages of working for a station that had a contract with the American Federation of Radio Artists. He persuaded the WJLB announcers to join the union and a contract was negotiated with WJLB.

The National President of AFRA was now the opera singer Lawrence Tibbett, who appeared regularly on radio. Tibbett was coming to Detroit to sing on the *Ford Sunday Evening Hour.* This program originated on WJR which was extremely resistant to unionization. Dick Osgood, now President of the Detroit Local, and Clyde Nutten, the new Executive Secretary, thought they might solicit the influence of Tibbett as Jay Michael had won the support of Bob Hope.

Mr. Tibbett proved evasive. He referred Nutten to the advertising agency that hired him for the program. The agency, of course, wanted nothing to do with the union. Clyde Nutten was fundamentally an estate lawyer and proved completely ineffectual as a union leader; and Osgood was far from militant by nature; Mr. Tibbett came and went without helping the local.

Actually, only the hot conviction of Larry McCann could overcome the chill sprinklings of discouragement showered weakly upon him by Mr. Nutten in the introduction of AFRA to WJLB. Then, having established a unit there, McCann promptly managed an audition with Jack McCarthy and joined Wyxie.

Later in 1943, Hal Neal and the U.S. Army Corps found themselves mismated in some way and Neal returned to the announcing staff of WJLB. He heard his friends Morgan, Foy and McCann on WXYZ and contacted them. McCann urged him to audition for Jack McCarthy, which he did, and Hal Neal, too, abandoned WJLB for the more lucrative and prestigious surroundings of "the Lone Ranger station."

McCarthy hired another announcer that year who was destined to become a man of importance in the industry. Don Faust was tall, had chestnut eyes and hair, oddly sunken cheek bones and a face that narrowed down to a firm, determined chin. He was working on Station WOL, then the Mutual outlet in Washington, D.C. Frank Blair was production manager there at the time, and the news director was Walter Compton. Blair was to become an important newscaster on NBC; Compton would head a national advertising agency, of which Ted Robertson would be west coast representative; and Don Faust eventually earned the title Vice President and General Manager of Station Operations of General Electric Broadcasting Company, Inc. But in 1942, the immediate ambition of Don Faust was to be a WXYZ announcer.

"It had become common knowledge," Faust remembered, "that WXYZ was a good place to get to."

He wrote to Jack McCarthy; McCarthy invited him to Detroit for an interview; he went on "spec" and McCarthy hired him. Don announced *The Lady of Charm* (Edythe Fern Melrose) in her studio in the New Center Building, and he announced *Radio Schoolhouse* on Sundays, becoming quite a girls' idol.

There was also a wild man from New Mexico named John Stebbins. He roared into town in a Cadillac three days late, but talked his way out of the dog house anyway. In later years he was to be a station broker working out of Chicago.

There was also a "special announcer" named Bob Carter. Special announcers were given that designation because they rarely did booth work and usually had programs of their own. Osgood was a "special" announcer because of his *Show World* and *March of Victory* and because he was master of ceremonies on *Radio Schoolhouse.*

Bob Carter "starred" on a morning program. He was very difficult to get along with. He ignored the union, and fell six months behind in his dues. AFRA finally used pressure on the station which in turn put pressure on Bob Carter.

Bill Morgan remembered the affair well: "The rumor was that he was being paid $125 a week which was quite hefty at that time. Benny Kyte and the orchestra worked on his show, only they were calling Benny "Larry Page." We all took turns announcing it, but Bob Carter was the star. He was a strange guy, highly competitive. And he had a persecution complex. One day I beat him at checkers. He got up and he knocked the board over with the remaining checkers on it, and he says, 'Of all the tough breaks'."

A player makes his own "breaks" in checkers.

WXYZ drew another announcer from WEXL that year—a strong, tall man who wore thick-lensed glasses and had one of the heartiest laughs ever created. It had a well-lubricated sound and came from the pit of his stomach.

Bob Hite recollected the man: "Norman Lenhardt. He was at WEXL and he was goin' to be there forever! And during the war when announcers were hard to get, I talked him into comin' to XYZ."

Lenhardt was a little more self-effacing: "All the good guys had either gone to war or moved on to fill the vacancies in the big spots—New York, Chicago, Hollywood. So what's a poor broadcaster going to do? Well—among the answers they were forced to, were guys like one named Norm Lenhardt, the nondescript itinerant who walked in one day and approached Jack McCarthy and filled out an application. McCarthy found that his announcing staff was so depleted by the draft—or people leaving—that he was forced to hire—this Lenhardt."

Actually, Lenhardt was a better announcer than he would admit. True, his dialect had the middle west stamped on it, and Tom Dougall said his thoughts usually got lost in his vocal chords, but he emulated David Ross who had been the idol of his youth. Norm's voice was deep and rich and fortunately he lacked the poetic preciousness of Ross.

Jack McCarthy rounded up a staff of excellent announcers.

"At that time," McCarthy remembered, "I think we had 13 or 14 staff announcers because we were doing WXYZ programming; and we had several programs on the national networks. So an announcer had a lot of things to do—a very tight schedule."

Al Hodge was doing many things besides *Town Talk* and *The Green Hornet.* He wrote copy; he did record shows and live shows; and he worked as a producer. In the fall of 1942, Coca Cola purchased a half-hour pre-football season pep rally at the University of Michigan. Hodge went out to Ann Arbor to produce the broadcast and decided to stay all night. The next morning he received a telephone call from Jimmy Riddell.

"Al, do you know anything at all about football?"

"Well, ya. I like football."

"Would you mind doing the color on the U of M football game today with Don Wattrick?"

"No."

"Fine. Do the football game."

Al did the football games with Wattrick the rest of the season.

There was also the broadcast of the B-24's. It answers young people who "can't imagine what you'd do without a picture."

The Ford Motor Company began producing planes for World War II, and the Blue Network wanted a broadcast of the B-24's coming off the assembly lines.

Telephone lines were installed for the remote broadcast. Al Hodge was to produce it and Bob Hite was to be the announcer. Hite was to describe the wings coming off one line, the bodies off another, and all being put together "for victory!"

When Hodge, Hite and an engineer arrived at the Ford plant to do the broadcast, they found a great expanse of empty factory. No noise, no planes; no anything.

Hodge remembers what they did in this emergency: "I got with the public relations man and we set up some practice riveting tables where they taught young fellows to use the riveting guns. Well, the public relations guy and I made the noise. We riveted on cue. And we did a half-hour broadcast over the Blue Network telling all about those airplanes coming off the line—and there wasn't one in sight. We riveted! And we yelled!

"New York called after the broadcast. 'Boy! That was great! We sure are turning out the stuff at Ford! We can't lose with industry turning out the stuff like that! Congratulations! That was a beautiful broadcast! Beautiful!' "

In March, 1942, the Lone Ranger Victory Corps was organized. Hundreds of thousands of Lone Ranger fans were enrolled. They collected waste paper and rubber and metals. They gardened. They were instructed in air raid precautions. In connection with this public service, prominent citizens appeared on the Lone Ranger programs.

The Lone Ranger, himself, was in demand to make personal appearances to promote the sale of war bonds. This inevitably led to requests for interviews wherever he went. But George Trendle and Ray Meurer controlled everything.

Instructions were written out for would-be interviewers suggesting questions that might be answered. Any questions as to the private lives of Brace Beemer or John Todd were *verboten.* No backstage information was to be given out. Beemer could not appear in costume without the mask. He could not sign autographs with his own name. He was forbidden to smoke or drink if children were present; he could never drink when wearing the Ranger costume. He was forbidden to make speeches on his own. He had to be introduced as the Lone Ranger—never as Brace Beemer. He could not make any appearance other than what was arranged by The Lone Ranger, Inc.

One enterprising, relaxed newspaper man unwittingly broke every rule by taking an oblique approach and it turned into an embarrassing experience. Jack Pickering was writing a column in *The Detroit Times* called *Here and Now.* He was most interested in how Fran Striker worked. Striker was now turning out the equivalent in wordage of four books a year the size of the Bible. Pickering thought Striker was a good subject for a column and went to Fran's big house on University Place in Grosse Point to interview him.

He had already met Fran through Dick Osgood and knew him as "a very pleasant host," agreeable and friendly—"just a nice guy." Fran liked people and he was easy to talk with. Jack Pickering became more and more interested as he listened during the interview. In fact, he found Fran so engaging as a hard working writer that he wrote not one column about him, but three.

Pickering remembered this as virtually unprecedented: "Sometimes I'd devote a whole column to one subject. More often, there'd be a lot of subjects in one column. But running a series of three! I don't know as I could ever remember doing three on one subject."

Fran gave credit to Trendle and made it clear that *The Lone Ranger* was Trendle's property. But Pickering was not at all interested in Trendle; he was interested in Fran's writing, how he put his scripts together.

The first column was published in *The Detroit Times,* June 1, 1943. Better publicity for the program could not have been purchased. Pickering wrote of the fifty million listeners—"maybe more than that." He pointed out that the Lone Ranger books, which Fran wrote, were now being translated into and published in several foreign languages.

"An energetic young man, somewhere along in his 30s," wrote Pickering, "he isn't given to the temperamental quirks usually attributed to authors, but he has a few peculiarities and one of them is going to work in the middle of the night."

He wrote about Fran's wife, Jan, and their three sons, Bob, 12; Don, 10; and Francis, Jr. 5. He told how Fran spent time with his family, helped his boys with their hobbies, made his own fireworks, developed his own pictures, and made his own movies. At the end of the column was the promise: "Tomorrow: How he gets his plots."

The world never learned from Jack Pickering. The very evening the first column appeared, Jack C. Manning, Editor of *The Detroit Times,* received an irate telephone call. It was to the effect that Trendle and no one else was the creator of *The Lone Ranger*, that the column was full of misinformation, and that if the other columns were printed there might be a law suit.

Pickering was called that night and told to kill the other two columns about Fran Striker. Several columns later, however, he quoted in *Here and Now* a letter from Raymond Meurer to John Manning—pouring troubled oil on peaceful waters—referring to "the very nice article written by Mr. Pickering," cutting down Fran Striker with "the first script had to be written about fifteen times before it was acceptable to Mr. Trendle" and "Whenever Striker gets off the beam, Mr. Trendle checks him on it" and repeating "that the real creator of *The Lone Ranger* was George W. Trendle."

The censorial powers of George Trendle and his legal counsel were not always as easily successful as in the case of *The Detroit Times*. One "J. Berry" wrote a feature about the program for *The Saturday Evening Post*. It was so unbearably laudatory that one suspected it might have been written by Felix Holt or Trendle himself. However, it is said that the piece did not please Mr. Trendle. He had insisted that he have a chance to read it before it was published and returned it all marked up with changes he wanted made. The magazine's response was, in effect, "Mr. Trendle, we promised that you could see it before it was printed but we did not say you could change anything."

Dick Osgood managed a trifling coup with the *approval* of George Trendle. He was to interview on *Show World* a boy actor who was appearing in a stage play at the Cass Theater. According to the program notes, the boy's hero was the Lone Ranger. Like most out-of-towners, the kid had no idea that he was working only a few miles from his hero; and Osgood thought it would be great to have the kid meet his ideal on *Show World*.

"You'd better ask Trendle first," warned Felix Holt. "He'll never allow it."

Holt had tried to scare Osgood once before. Osgood had wanted to do a story on "Divorce" on *The Factfinder,* and Holt had said, "He'll never allow it. Trendle won't even let us mention divorce on a newscast."

Osgood had sent a black-on-pink request to do the story, and Trendle had returned the request with "Depends on the story," scrawled across it. Osgood had submitted his story to Trendle and it had come back with the endorsement: "O.K.—T."

So now, Osgood asked if he might arrange for the juvenile actor to meet the Lone Ranger on *Show World*.

"I don't mind," said Trendle on the phone, "if you do it properly. Don't ask Brace any questions. Just let him say hello to the boy. And he must be in costume, wearing the mask."

The little actor had to stand on a platform to reach the microphone. He had no idea what was coming. Osgood's final question was: "Do you have any heroes? Anybody you'd like to meet or be like?"

"The Lone Ranger," said the kid, as expected.

"Turn around and see who's behind you."

There stood Brace Beemer, in costume, wearing the mask. He was so tall, sturdy and right for the role that even Osgood had duck bumps. The little actor didn't

Brace Beemer as the Lone Ranger

believe what he saw until he heard the voice. Then he almost fell off the platform.

Another *Show World* guest was the prominent movie actor Walter Brennan, who visited Detroit on a tour promoting *The Cowboy and the Lady*, in which he was appearing with Gary Cooper. Brennan knew the Ranger program originated in Detroit and wanted to meet Beemer. After the 5 o'clock *Show World* broadcast, Osgood introduced his guest to Charles Livingstone who in turn presented Beemer and the cast. The actors and sound men were, of course, thrilled to meet the prominent movie actor, but Brennan was just as thrilled to meet the bodies that went with the famous voices.

As Osgood drove Brennan and Mrs. Brennan downtown to the Book-Cadillac Hotel, Walter Brennan kept repeating, "I would have loved to watch that broadcast."

"Why don't you?" Osgood asked.

"Oh, they've arranged this cocktail party for newspaper reporters," Brennan grumbled. "I'm suppose—I have to be there."

"Why?" Osgood demanded.

Up to this time, Osgood had not been asked to attend a promotional press party, although he was screening all movies with the newspaper critics and was broadcasting reviews. If he had realized then how important these affairs are in the promotion of a movie and how imperative that the celebrity be present, he would not have dreamed of saying, as he did, "Look! This is crazy! There's a cocktail party every night but how often do you get a chance to watch a broadcast of *The Lone Ranger?*"

"Mr. Osgood's right," said Mrs. Brennan. "You may not come this way again for a long time."

The Brennans debated and succumbed. Osgood drove them back to the studios. Brennan watched the broadcast from the director's booth at Livingstone's side and was fascinated to observe how voices, sounds and music were all put together in the booth from three different sources: actors' studio, sound studio, and from the turntables in the booth itself. It excited him to see how the excitement was created.

"If you ever do one with Judge Bean in it," he enthused at the end of the broadcast, "let me know. Judge Bean is one of my favorite characters. I have always wanted to play him. Just let me know and I'll come to Detroit and play it."

Fran Striker who admired Brennan as an actor did write Judge Bean into an episode, but the broadcast came at a bad time for Brennan. On the day the story was broadcast he was working before the cameras in another movie with Gary Cooper, playing a role for which he won one of his several Academy Awards. The movie was *The Westerner* and the part he was playing was—Judge Bean.

25
Maccabees Radio Wrap-Up

Twelve months after the divorce of the two NBC networks, the Blue was still unsold and was still being operated by The Radio Corporation of America. Mark

Woods had been named President of the step-sister chain and was doing a good job. Additional stations had joined the network to give the Blue a total of 146 affiliates. Its own separate sales staff had sold $11,300,000 worth of time and the Blue had moved into the profit column.

Blue added prestige to its independent programming by signing Serge Koussevitsky and the Boston Symphony Orchestra to a series of 46 broadcasts. Mark Woods had also built up a stable of fifteen news commentators including Baukhage, Drew Pearson, Westbrook Van Voorhis, Martin Agronsky, Don Gardiner, Raymond Gram Swing and Walter Winchell.

Always dramatic and sensational, Winchell caused some embarrassment. He berated listeners for re-electing government officials who had "blundered" into Pearl Harbor. There was quite a reaction and Mark Woods warned his commentators to use better judgment or run into a blue pencil.

Time Magazine, January 11, 1943, reported that seven Wall Street investment houses, three insurance companies, two manufacturers, a newspaper and a retired capitalist had all considered purchase of the network but for various reasons had backed off. Raymond Meurer told us Trendle had put in a bid "by about $4 million short."

The amount paid for it was $8 million and the buyer was a vigorous, open-faced man who had graduated from Yale in 1905 and parlayed an idea purchased from a Cleveland candy maker into a $22 million business. He was Edward J. Noble of Life Savers. Former chairman of the Civil Aeronautics Authority, Noble had quit the Roosevelt regime as Under-Secretary of Commerce in 1940 to support the campaign of Wendell Willkie. He was now 61 years old.

The radio deal was made in midsummer, 1943. It was the first time a major network had been sold. Noble's purchase included three owned-and-operated Blue Network stations: WJZ in New York, KGO in San Francsico and WENR in Chicago which was only partly owned and part-time. Noble's new enterprise was incorporated under the name American Broadcasting Company. This took a little doing. An American Broadcasting Company had been operating Station WOL in Washington; and an American Broadcasting Corporation owned Stations WLAP in Lexington, Kentucky and WBIR in Knoxville, Tennessee. There also was an American Network, Inc., set up to provide program service for FM stations. This last was apparently years ahead of its time and was said to be happy to take $10,000 for its claim to the title.

Other, more serious problems remained to be solved. The Federal Communications Commission had ruled that no one could own more than one station in a city. Before acquiring WJZ, Mr. Noble already owned Station WMCA in New York. He would now have to dump it. Also, his purchase of the Blue Network must be okayed by the FCC. And he would have to invest in his own facilities and equipment; Blue had continued to operate interchangeably with the Red Network in the NBC studios. Now, with the war on, facilities were next to impossible to come by. NBC agreed to let Noble operate in its facilities temporarily, but he had lost a valuable experimental tie-in with NBC and the Radio Corporation of America on television.

However, the step-sister had been adopted by a new daddy. Like Apple Annie, she had a long, hard glamorizing and grooming process ahead, but many powerful men wished her well including George Trendle, Allen Campbell and James Riddell of WXYZ.

Things of less national importance continued to happen in the Stroh and Maccabees buildings in Detroit. Marie Meyer had a tiff with Riddell over extra fees for commercials written by her, and left WXYZ to join Alice Gorham and Earl Hudson at United-Detroit Theaters.

The ever restless Bill Morgan fell under the spell of a romantic build-up of Texas by Harry Heilman and went there to work at Station WOAI in San Antonio.

There was also an egg episode. Actor Paul Hughes had bought a farm and sold eggs to his pals at Wyxie. He delivered a dozen one day to Bob Hite who left them on the desk in the announcers' booth while he rehearsed with *The Green Hornet*. Mischievous fellow workers took his eggs down to the Southland where Ma Sweeney hard-boiled them. The hard-boiled eggs were placed in the box and the box put where Hite had left it. He took the eggs home with him.

Everybody hung around the next day for a reaction. There was none. But the next week, when Hughes was scheduled to deliver eggs again, Hite said to his friends before leaving the booth for the Hornet rehearsal, "Oh, hey! Listen, fellows, when Paul brings in my eggs, if you don't mind, would you put 'em out in the hall? These radiators in here really get hot."

It was night and all was quiet in the Stroh Building. Everyone had left for the day except the two men who relaxed in Trendle's office. Even after a long day, Trendle's collar and cuffs were unwilted and he looked fresh and clean as he tilted back in his desk chair, his baby blue eyes fixed on Campbell. Campbell, just as well-pressed and polished, slouched easily in deep blue broadcloth tailoring, his white shirt peeking just enough below suiting to reveal expensive cuff links.

"Out Jefferson might be worth looking into," he was saying. He had received a black-on-pink from Trendle saying, "I want to get out of the Maccabees Building. Any ideas?" Both men now owned beautiful homes in Grosse Pointe and drove on Jefferson Avenue to and from work every day. There was not yet an expressway in Detroit.

"The old Book home has been taken over by a group of doctors," Campbell continued. "And the U.A. Double-yuh is building a place out there."

"That's no great recommendation," Trendle commented.

Campbell smiled. "And Theodore Buhl's place has been turned into some kind of a mental hospital."

"That's more like it," said Trendle. And he added, seriously, "I just think it might be a good time to pick up a piece of property. We don't have room enough in the Maccabees Building. We could use some new studios. And Fran Striker and his writers—we could use that space down here if we could put them somewhere else."

"Out Jefferson could be the right direction," Campbell repeated.

"Well, there's a sign up on Louis Mendelssohn's place, just this side of a big church. I didn't notice what street, I'll look into it," said Trendle.

He did; he liked what he saw; Campbell looked it over; and they bought the Mendelssohn mansion for $34,000. Vern Alston was put to work planning the move.

Fran Striker was now supporting his wife and three sons, his mother and father, an aunt and an uncle, and his mother-in-law; and he and Jan were expecting another baby. He had discovered in New York how much writers were being paid and knew he earned more than he was getting. Jan never stirred him up about this, but Mommy Striker, who spent considerable time in their household, kept prodding him.

"*The Lone Ranger* was *your* show, laddie boy. *The Lone Ranger* is nothing but your *Covered Wagon Days* with little changes. You're being cheated."

What bothered Fran most was that he had been left out of The Lone Ranger, Incorporated. Helping him turn out scripts now was a writer named Bob Green; and his secretary, Betty Joyce, was also a good enough writer to do *Challenge of the Yukon.* He still needed help and when he read in *Broadcast Magazine* about a Dan Beattie who was writing and producing a series called *Government Girl* over Mutual from Washington, he wrote a letter asking Beattie if he would like to work in Detroit.

Beattie was not interested until, because of the war, Mutual decided to replace *Government Girl* and other "soap operas" with newscasts. When Fran wrote again he asked Beattie if he might now be interested and might knock out a couple of samples if he sent him a couple of scripts illustrating the Trendle format.

Helen Beattie, Dan's wife, urged him to try his hand; so he wrote a sample Ranger and a sample Hornet, mailed them off to Striker, and forgot about them. The William Morris agency in New York hinted that a visit there might prove fruitful.

Beattie then received a telegram from George Trendle inviting him to be his guest in Michigan for a week to see how he liked Detroit.

"Why don't you go?" Helen Beattie suggested. "You can stop off in New York to see the Morris Agency on your way home."

Dan Beattie, a dapper, tight-voiced southerner, made the same discovery in one day that Campbell had observed a dozen years before: Trendle never hesitated to go over the heads of his executives. It was two o'clock one afternoon when Beattie was able to appear in the offices of WXYZ, Inc. He had had difficulity finding a hotel room. During the war, Detroit was loaded with transients brought to the city by its massive production of war materiel.

At the switchboard in the reception room, Ad Werner saw a thin man of medium height, with greenish eyes and remarkably long eyelashes.

"I'm Mr. Beattie from Washington," he said. "I am expecting to see Mr. Striker."

"Yes. Mr. Trendle has been trying to reach you. He didn't know where you were," Ad told him, plugging in a line.

Thelma Watson appeared and conducted the immaculate Dan Beattie to Trendle's office.

Felix Holt
Publicity and head of news 1935-1945.

"I liked your scripts," Trendle said. "I'd like you to stay a while. I'll set you up in an office back there for a week and see how you like things."

"I'm anxious to meet Fran Striker," said Beattie.

"Oh, he's no longer with us. He wanted $2,000 a week or some impossible amount. Just got too big for his britches. Felix Holt is in charge o' things now. I'll take you in to meet him."

They found Felix Holt perched like a crow on a low radiator which had been made into a sort of seat. His arms were locked around his bony little knees and he was wearing a big western hat. Beattie didn't know what to make of him.

"Felix is the editor," Trendle explained. "Your scripts will go to him. Now I'll call Livingstone. He'll meet you out at the station. You go out there and talk to him and meet the Ranger and everybody. Tell 'em you're one of the new writers."

"How many writers do you have?"

"Me, Betty Joyce, Bob Green," said Holt, "and maybe you."

"How many shows do you do a week?"

"Three Rangers, one Hornet, one Ned Jordan and two fifteen-minute Yukons."

Back in Trendle's office, Trendle took a form from a drawer. He asked Beattie to sign it.

"Well—uh—I have to go back to Washington, Mr. Trendle. I can't sign a contract today."

"You can go back," said Trendle. "I'll let you go back every two weeks if you want to—weekends. At your own expense, of course."

In the end, Beattie signed an agreement to write one half-hour script a week. He alternated on Hornets and Rangers. Having coffee with Betty Joyce and Bob Green, he learned what had happened to Fran Striker. There had been no blow-up. Fran had simply gone to Trendle and said he thought he should be earning more money. Trendle suggested he try free-lancing and told him they would continue to buy Rangers from him on a free-lance basis as Holt accepted them. Fran was hurt. But he had confidence.

Beattie met Fran for the first time a week after he had started work in what had been Fran's office. Fran came in with two scripts.

Beattie recollected the meeting: "I remember he had a big, heavy great winter coat-double-breasted, grey. He shook hands, remembered writing to me and said, 'Let's go and have coffee while Felix looks at my scripts'."

Over coffee with Betty Joyce and Striker, Beattie was impressed by Fran's warmth and friendliness.

"If you have any trouble getting into the format, give me a ring. Maybe I can help you," Fran offered.

They returned to the Stroh Building and Fran went in to see Felix. In two or three minutes, he stomped past Beattie's door, walking fast and heavy. Beatty looked up in time to see Fran's heavy face pushed forward. It was flushed. He was obviously angry.

Betty Joyce came into the office, fell into a chair, and lit a cigarette.

"I feel just terrible."

"What's the matter."

"Fran."

"He looked mad."

"Felix turned down his scripts."

"What!"

"Both of them."

Beattie was shocked. Fran Striker! The original writer!

"Why doesn't he go to see Mr. Trendle?" Beattie asked, after a moment.

Betty took a drag, her brown eyes very troubled.

"Trendle won't talk to him. Trendle won't even see him."

Before WXYZ Radio left the Maccabees Building, a black-on-pink was delivered to Dick Osgood from the public relations director, Lambert Beeuwkes. It read:

"Thought you should know. *Show World* is the first WXYZ program to get top rating in Detroit. Congratulations."

Celebrity interviews had a great deal to do with it. It was the first, and for a long time, the only such program in the city. Before leaving the Maccabees, *Show World* had brought to the station 76 stage and screen stars, 25 names from the night clubs, 50 celebrities in popular music, 9 sports stars, 9 personalities from the field of classical music, 3 ballet dancers, 6 circus performers, 45 broadcasters and 40 "miscellaneous" ranging from Barney Oldfield, Dante the Magician and Robert Edmond Jones, scenic designer, to various and sundry beauty and festival queens.

Dick Osgood was riding high. He was featured on the Hi-Speed show; he was heard nationally as Proctor, the FBI man on Ned Jordan; and Lix Cough Medicine, being introduced by Harold Christian's merchandising department, was sponsoring *Show World*. Osgood even appeared in concert with Andre Kostelanetz and the Detroit Symphony Orchestra reading excerpts from *Alice in Wonderland* to Deems Taylor's *Through the Looking Glass.*

There was one nagging burr under his saddle, however. Although Fred Rivard was doing an excellent job producing *Radio Schoolhouse*, he was a hambone and loved entertaining the audience during the warm-up period, and Osgood felt strongly that Rivard wanted Osgood's job as master of ceremonies.

At the end of 1943, the Hickok Oil Corporation dropped *March of Victory*. It had been on the air for three years and two months. Osgood, Benny Kyte and the laconic salesman Irv Stoll drove to Hi-Speed headquarters in Toledo for a conference with Floyd Tracy and Mr. Hickok. Purpose of the meeting was to agree on an idea for a new series.

Osgood told of having seen three words in a newspaper ad that stuck in his mind as a good program title: In bold relief.

"Why not do a series on names that stand out In Bold Relief? For instance, you go into a grocery store and you see Kellogg's Corn Flakes. Who is Kellogg? Who is

Campbell of Campbell's Soups, Eastman of Eastman Kodak, Wrigley of the chewing gum, Fannie Farmer? People are coming along all the time whose names stand out In Bold Relief in the news."

Hickok bought it and it was decided to give Osgood a free hand in choosing subjects. *In Bold Relief* replaced *March of Victory* on January 24, 1944, while the station was still in the Maccabees Building. For the first time, Floyd Tracy insisted that his contract with WXYZ include the service of Dick Osgood, who, like Edythe Fern Melrose, never had a contract with the station.

One of Floyd Tracy's ideas was a Hi-Speed automobile horn. It played the first six notes of the Hi-Speed musical theme. One horn, used on the program, was stored with other props in the drama studios. One night, an aging, dignified actor named Herschel Mayall, who had a habit of getting drowsy when not on mike, fell asleep, relaxed against the horn, and in the middle of hoof beats of the galloping Silver on the western plains of yesteryear, the contemporary sound of "Hi-Speed Is On the Air," familiar only in Michigan, Ohio and Indiana, blew its way across the nation.

The most regrettable loss in moving from the Maccabees Building—after The Alcove and Ma Sweeney's Southland Restaurant—was the "fluff chamber." The choicest mistakes made by actors during performances had been written and printed on the walls of the passageway connecting the actors' studio with the sound studio. No one thought to copy down the errors of the years and the fun was lost under a coat of paint. Today, the best are forgotten. Surviving only are such minor sillies as "they must have a gozen duns" and "caught like tats in a rap."

Flowerday had appreciated them: "That wall was covered. If you were ever feeling bad, why, you'd just go in there and you'd come out laughing."

Flowerday remembers one that was not an actor's fluff. It was a writer's mistake. Fran had apparently forgotten to edit a script, and Trendle had not caught it, either. The actors noticed it in rehearsal—a line that read, "I hear a *white* horse coming."

In radio's final year at the Maccabees Building, *The Lone Ranger* took first place among children's programs in the *Radio Daily* poll, first place in the same category in a national poll of radio editors of magazines and newspapers, and won an award from the National Safety Council for "distinguished service to safety." And its author who had played a major role in the creation of the program was having his scripts rejected by his one-time friend, Felix Holt.

Radio would be fun in the Mendelssohn mansion, but with a difference. There had been, in the excitement of creation at the Maccabees, a sense of happy release, of escape from the Great Depression. And even though now the world was at war, there was in radio an exhilerating sense of usefulness and contribution. The Maccabees days had been rich and full. Radio had been an adventure. Now it was big business; its pioneers had settled down into "establishment."

There had been great change from that Saturday night in October, 1925, eighteen-and-a-half years before, when George Harrison Phelps had embarked upon his hobby surrounded by "yards and yards of velvet" in his crude studio in a one-time stable. The twenty years to come would see still greater change, as exciting as anything that had happened before. The change began in the Mendelssohn mansion.

Part Two
Stray Thoughts on a Picket Line

26
Penthouse to Posh

ON THE EAST SIDE of Detroit toward Grosse Pointe is a section known as Indian Village. Its streets extending several long blocks inland from Jefferson Avenue and the Detroit River, were originally named Iroquois, Seminole and Seneca. It is said Indians once camped there. Later it became part of a farm owned by Abraham Cook. Then, between 1895 and 1910 the land was subdivided and nearly three hundred imposing homes were built there by wealthy lumber barons, automotive giants, merchants, writers, bankers and other distinguished Detroiters whose names have outlived them on Detroit buildings and street signs: Arthur Buhl, Lord Griswold, Warren Booth, Walter Jennings, Edward Liggett and Hiram Walker.

The first lot sold was at the corner of East Jefferson and Iroquois. On this site in 1894 was built the first home in Indian Village. It was purchased from its original owner, Dr. Theodore A. McGray, by a little man who had been born in Germany in 1854, Louis Mendelssohn.

Louis was brought to America when still an infant. He became an architect and a manufacturing draftsman. In 1901, he was made manager and treasurer of the Modern Match Corporation and in 1909 became treasurer and chairman of its board of directors. Just when he purchased what came to be known as the Mendelssohn mansion is uncertain. He died in 1935 at the age of 81. It was this Mendelssohn mansion that was purchased by Trendle and Campbell in 1943 to be the new home of radio station WXYZ.

The move from penthouse to posh was made on a weekend in February, 1944. Thus one Monday night, Silver found himself galloping on what had been Louis Mendelssohn's sun porch, and busy actors mounted and dismounted their imaginary horses in a drawing room (Studio D) dominated by a handsome fireplace with a hand-carved mantel. Osgood and the orchestra broadcast their Hi-Speed show from the large ballroom on the third floor called Studio A for attic. Franz Werner and the music library were installed on the same floor, and Franz had a handsome young woman assistant who spoke with a delightful Austrian accent and was related to the movie star Paul Lukas.

Show World celebrities were interviewed on the ground floor at the front of the building in what had been Mr. Mendelssohn's library, now called Studio B for books. The shelves, protected from dust by leaded glass, were now empty, but it was a long sumptuous room, and Trendle had a large desk placed at one end of it for meetings.

"He would come in every morning about ten o'clock," Jack McCarthy remembered, "and we had to have that downstairs office ready for him. The door had to be half open, the chairs placed just right and then you'd have to line up outside. There'd be four or five of us—whoever had to see him on something."

Across the hall from Studio B was master control for which Verne Alston had built a horse-shoe console; and between it and the drama studios was Studio E, the announcers' booth.

The ground floor by the back door had a couple of offices and the little record library; and above this on the second floor in what had been the servant's quarters were small offices for Osgood, Lee Smits and the Chief Engineer, and a larger room overlooking the back yard which served as the news room.

The Hammond organ was set up in a one-time bedroom on the second floor over master control. It was now called Studio C (for chamber), and across a carpeted hall from it were two large square bedrooms connected by a bath. They were now offices occupied by program director-studio manager, Jack McCarthy, and dramatic director, Charles D. Livingstone. The hall was wide enough for one end of it to accommodate a large desk, a typewriter stand, a chair or two, and Bernadine Ryan. Bernadine, with the light red hair, Irish grin and delicate vitality had been rescued from the traffic department downtown and was now Jack McCarthy's secretary.

The back yard was macadamized and turned into a parking lot; and the carriage house, or garage, Studio G, was the new home of the pipe organ. A kind of floating floor had been built over three turntables which had been used by the Mendelssohns to turn around their carriages and cars. WXYZ now had two staff organists, Marguerite Werner and Armin (or Jack) Franz. Jack played the Hammond in the bedroom; the pipe organ in the garage remained the special domain of Marguerite. There was also a piano in Studio G. Fred Rivard and Eve Gladstone held auditions and rehearsals there for *Radio Schoolhouse.*

A side door opened into the mansion from the driveway but Trendle ordered employees not to use it. They were to use the back door and the back door *only*. Even he used the back door. Visitors could enter by the front doors on Jefferson Avenue—and a grand entrance it was. Two receptionists occupied desks near the inner, all-glass door; overstuffed chairs and a couch stood against the wood panelling; and the room was two stories high.

Once an eager young announcer, after an audition, felt so stimulated by the experience and impressed by the elegant surroundings that on his way out he bowed briskly to the receptionist on each side and walked straight through the plate glass door—without opening it, miraculously without damage, except to the door.

A grand staircase with hand-carved balustrade rose at the back of the reception room and divided halfway to the second floor, the right hand stairs leading to the McCarthy-Livingstone suite and the Hammond organ studio, the left hand flight taking one to a hallway leading to offices now occupied by the writers over the sun porch.

Felix Holt's inclination to seize power had now been thoroughly encouraged in his new position as chief of the script department. One morning, as Betty Joyce, Dan Beattie, Bob Green and Charles Livingstone assembled in his office, they found him perched on the edge of his desk, knees crossed, thin wrists crossed, bony fingers pointing stiffly to the floor. He was tense, as usual. A big brown, snap-brim hat covered his bald head and made his face look smaller. His pale blue eyes and thin lips betrayed a defiance as he looked from one to another.

"Now!" he began when the five were together. "I s'pose you think I was wrong to turn down Fran Striker's scripts. I've turned down six of 'em because I didn't like 'em. Now! There's goin' to be some changes. From now on I want a synopsis of every story before you write it. Hand in a synopsis first."

"Oh my God," Bob Green muttered.

Holt's defiant little eyes turned from Green to Beattie.

"I've never written from a synopsis, Felix," said Beattie. "I don't think I can write that way."

"That's the way it's goin' to be," Felix insisted. "Now! This script you threw in here this mornin'. You've got the Ranger saving a ranch owner. I want you to change the character into a woman. If I'd had a synopsis first, I could have told you and saved you a re-write."

"A—woman?"

"Give the Ranger a little romance. Make the ranch owner a woman."

"But Felix," Livingstone said, not unkindly but with consternation. "My God! You can't do that. Trendle—"

"That's the way it's goin' to be," Holt said crisply.

"Doesn't this change the whole character of the Lone Ranger?" Bob Green asked.

"That's the way I want it," Felix chirped. He jumped from his desk. "That's all."

The meeting broke up.

Fran Striker was an unhappy man but he tried to hide it from his family. He was still picking up $100 a week for the Ranger newspaper cartoon, and he received approximately $200 for the Lone Ranger books as they were written and published.

But he found that having been the original writer of the radio series did not open doors. New York seemed to feel that the Trendle dramas were flukes and looked down on Fran as a hack. He freelanced from his home in Grosse Pointe, and did some writing for *Voice of America*, a government enterprise presided over at the time by the eminent writer Robert Sherwood. Its purpose was to broadcast American propaganda abroad. It was part of the Office of War Information which broadcast some 3900 programs a week in twenty-five different languages from New York to San Francisco. Fran's pay was minimal and a half-forgotten feeling of insecurity was beginning to gnaw at him. He now had three sons, and a fourth child was expected.

Actually, he was supporting twelve people. He had purchased a farm beautifully located on a high hill outside Arcade, New York—a short drive southeast of Buffalo. There he had installed his father, his mother, his Uncle Gus and Auntie Bess, hoping the two men would set up a chicken business for him.

"They didn't want any part of it," Jan remembered. "It was too comfortable. It was nice to get a check from Fran every week and buy groceries and enjoy everything. So—that's the way it was. And Fran didn't resent it."

He did resent the fact that George Trendle refused to see him. And he was filled with anger because of Holt's rejection of his scripts. But he did not take it out on his family. Jan sensed his misery but bottled up her own apprehension and kept her house and family cheerful and in good order, except when Mommy Striker steamed in for a visit and churned up emotions. Fran put aside his resentment and anger when he was with his boys.

Bob Striker remembered: "When I got interested in chemistry, instead of buying me a little commercial kit that you really couldn't do much with, he made an acquaintance with a salesman at one of the chemical supply houses in Detroit. He purchased actual laboratory equipment for me and chemicals that most kids wouldn't be allowed to get within fifteen feet of at that age. I must have been fourteen or fifteen years old.

"And I remember when dad came home with a couple bottles of acid—one was concentrated nitric acid and one was concentrated hydrochloric acid—and he took the caps off and showed me how the fumes came out, and warned me what you don't do with these.

"One time I found the formula for tear gas and uh—started to make that without my dad's assistance and uh—it worked very well, except—I over-heated the little decanter that was holding the carbon tetrachloride, as I remember, and it cracked and the gas got loose in the house. It was a pretty bad day."

Bob and his mother laughed happily at the recollection—the love in the laughter a tribute to Fran.

Bob: "When I'd made a mistake like that, I'd get a lecture. But it was always good-natured: did I know what I'd done wrong rather than—why did you do it?"

Christmas was lean for the Strikers in 1943. The traditional Striker's New Years party was a strain on the budget but one which, as usual, filled the Striker home with love and friendship. Now it was 1944 and Jan knew her husband was a worried man. The demand for his work was not what he had thought it would be. And Trendle refused to see him; wouldn't even talk with him on the phone.

Chuck Livingston and Dan Beattie were alone in the raised directór's booth which had been built at the end of the dramatic studio. Dewey Cole, who ran records in the booth, was clowning with the actors in the studio. The slow, plodding often-interrupted rehearsal during which actors tried to follow instructions and sound men interpolated their effects, had ended; and now there was a break before the narrator and commercial announcer would join them for the "dress" or timing rehearsal. Tonight, they were doing *The Green Hornet*.

"Did you read my story?" Beattie asked.

"Tomorrow night's Ranger? Yes. It's a good one."

"Well, Felix wants me to change the whole thing."

"It's too late for that, Dan," Chuck said in dismay. "I've already called the actors for it. He's not going to change the characters, is he?"

"He complains that I didn't stick to the synopsis. You know the meek little character?"

"Yes. Ted Johnstone's going to play him."

"I didn't intend for him to take over the story. It just happened in the two days between the time I wrote the synopsis, and sat down to write the story. That meek little character just took over."

"It made a good story of it," said Chuck. "What's Felix want you to do?"

"He wants me to go back and write it from the synopsis."

Later that week, Trendle summoned Chuck Livingstone and the writers to his office in the Stroh Building. Felix Holt was not present. Dan Beattie was the first to arrive. Trendle was affable, friendly.

"You're still trying to make a comedian out of the Lone Ranger," he smiled.

In the first script Beattie had written, Rollon Parker had played a farmer driving a load of chickens into town. There was an accident and the chickens were flying all over. Beattie had received a black-on-pink from Trendle: "See me about this."

They had argued.

"I think it has humor in it," Beattie insisted. "It's not forced gagging; and it doesn't involve the Lone Ranger and Tonto, although it does give them a chance to chuckle. I just think it's human.

"Well—maybe," Trendle had conceded.

So Beattie managed to inject humorous scenes occasionally.

On this morning of the meeting, Trendle said, "I don't mind the humor so much as long as the scenes don't involve the Ranger. But I didn't like that script where you had him getting mixed up with a woman rancher."

"The woman rancher was a man originally, Mr. Trendle."

"Why'd you change it to a woman?"

"Felix," said Beattie, simply.

"That's why you're here," said Trendle.

The others arrived. Trendle stood up when Betty Joyce arrived and gave her a lingering look of appraisal as she picked a chair and sat down. As he resumed his chair, white daylight reflected from his head which looked as though it had been polished. His face was pink below it and fresh. His desk was clean.

"Holt tells me you're falling behind in your scripts," he said.

"We're spending too much time writing synopses," Bob Green complained.

"We don't need synopses," said Trendle. "We need scripts. Now I want each of you to speak frankly. What do you think of Felix Holt?"

"You want to know what I think?" demanded Green without hesitation. "I don't think Felix Holt knows his ass from his elbow."

Trendle looked at Green without a change of expression. Then he looked at Betty Joyce. She said nothing, but he sensed antagonism. He turned his eyes to Dan Beattie.

"We wouldn't be behind, Mr. Trendle, if you used Fran Striker's scripts," said Beattie. "I can't understand how *The Lone Ranger* gained the popularity it did with Fran Striker's writing, and suddenly his scripts are turned down."

"What scripts?" Trendle asked.

"Fran Striker's scripts," said Betty Joyce with a touch of anger. "Six of them."

"We'll check into that," Trendle promised. He looked at Chuck Livingstone.

"Leave me out of it," Livingstone said, laughing a little. "I'm not one of the writers!"

"All right." Trendle turned again to Dan Beattie. "Now I'll ask—what do you think of Fran Striker?"

"I think he's great. He's been very nice to me as a newcomer, giving me tips on what to write and what not to write."

"Do you think you could get along?"

"I think he should be back here, frankly," said Beattie.

"Oh!" exclaimed Betty Joyce, nodding her head back and forth as though Fran's return would solve the situation.

"I can get along with 'most everybody," Beattie continued. "But I just can't seem to communicate with Felix."

"It's impossible to get along with him," said Bob Green.

"Fran knows what it's all about," Betty Joyce enthused. "He's one of us. He works *with* us. He understands the problems."

"I think Fran should be back here," Beattie said again.

"So do I," Chuck Livingstone said at last.

As the four sat watching him, George Trendle swung around in his chair and looked out a window—at nothing—thinking.

"Six scripts," he mused. He brought the tips of his fingers together as he gazed out the window. "Maybe he's learned. He just got too big for his britches."

He turned back to face them.

"Well," he said. "Okay."

Felix Holt left WXYZ to try his luck in New York. Not much luck. He settled down in Buck's County, Pennsylvania, and wrote a novel about his native south, *The Gabriel Horn*. It was an excellent book, and successful. It was adapted to the movies (not too well), the title changed to *The Kentuckian* and Burt Lancaster starred in it.

Fran Striker returned to WXYZ in charge of the script department. All six of his scripts Holt had rejected were accepted by Trendle and produced. And the Lone Ranger had had his last experience with romantic female ranch owners.

27
The Shuffle of War

Men of WXYZ continued to be drawn into uniform. Within hours after Fred Flowerday went into Service, Trendle sent for Dan Beattie.

"How would you like to be assistant production director?"

As usual, Trendle went over the head of one of his executives, although he may have asked Livingstone for a suggestion.

Livingstone approved of Beattie: "Dan Beattie was a back-up for me. I always hoped that Fred would be my back-up. When he went into the army, I just couldn't wait for him to come back for him to take over."

Beattie signed a separate contract to be Livingstone's assistant.

Larry McCann went into the army in 1944. Billy Hengstebeck who had graduated from Radio Schoolhouse into the sound department, joined the Navy. More of a problem was created when Al Hodge left for service; someone had to be found to play the Green Hornet. Trendle listened to a number of his actors but their voices, he thought, were all too familiar.

Don Faust, the announcer, asked if he could have a chance. Don was 25 and felt he had been born too late for radio. As early as 1943 he had read an article about television and thought maybe he could get into it from the beginning. Chicago stations were already experimenting and he had gone there a number of times to try to get into it, but when war was declared TV development ceased for the duration. So he decided to stick it out at WXYZ.

Livingstone, who was going to New York to audition actors, promised to listen to him when he returned. Don took a couple of old Green Hornet transcriptions which were stored in the basement of the mansion to a recording establishment. Faust outlined how he used them: "I copied down the lines and I'd go over and over them to try to really make them sound as much as possible like Al Hodge. Over and over and over I recorded it and tried to emulate the sound. I was really a small talent in a field of giants back there. When you don't have that much talent, you work a little harder."

Faust got the job; he was the second Green Hornet.

Another announcer came to the station from WEXL. Heavyset Porter Bourne was happy at WEXL. He played records and liked it, and he was chief announcer. He was not possessed of that driving ambition which had driven Foy to California, Mike Wallace and John Hodiak to Chicago, and Douglas Edwards and Fielden Farrington to New York. It was Bob Hite who stirred him and arranged an audition for him at WXYZ. McCarthy hired him. McCarthy joined the Marines.

McCarthy explained that move: "Al Hodge told me that the Marine Corps was looking for officers from radio and the newspaper world. Like a darn fool, I went down and applied for a commission."

Dick Osgood replaced him as "program director," moved into a square office by the back door and was given an assistant. He discovered it was not much of a job; important decisions were made downtown.

He remembered a curious event. "I recall a very strange interview with George Trendle. He called me into his office, was very friendly, and put Marguerite Werner's job on thin ice in case I cared to drown it. He also hinted that he would welcome any reports I might care to make at any time about H. Allen Campbell. I was never close enough to Campbell to make any comments and I couldn't figure out why Trendle would encourage someone in my position to spy on his general manager."

Harold Christian's merchandising department was reaching a peak.

Christian remembered the way it had achieved its position: "Once a year we developed a visual sales and advertising plan which we projected for advertising agencies and manufacturers from coast to coast. This gave us an entree to the key personnel of such concerns as Del Monte, General Foods, Standard Brands and many others.

"We developed point-of-sale advertising pieces for broker salesmen. We conducted seminars with different brokers to help their salesmen merchandise products that we advertised on our station. We accepted speaking engagements from service clubs, advertising groups and national conventions on merchandising techniques and how they could be tied in with broadcast advertising."

This all produced business for the now-prospering WXYZ.

Big, blonde Fred Foy came home from the service late in 1944 and returned to WXYZ's staff of announcers—but briefly. He had met a major in the army who was chief of CBS production on the west coast. The major made promises. So after a couple of months, Foy quit WXYZ and went west. But the promises proved empty and after six months, he returned to WXYZ.

Charlie Kocher, who had called on Lynn Smeby time after time hoping for a job at WXYZ, was now 30 years old and working for the FCC in Cleveland. His office was in the Federal Building there. Down the hall was the office of the Department of Agriculture. In that office, a doctor who had worked for the Department of Animal Husbandry for 40 years was retiring.

"God," thought Kocher. "Forty years sittin' in this office!"

What he had wanted ever since he was 16 was to be chief engineer of a radio station. He put an ad in a trade magazine. There was not much response. He went to an employment agency. In two days he had a call; an organization in Detroit was looking for a chief engineer. He was to call a Mr. Campbell. Kocher called immediately.

Campbell had apparently made up his mind before he even met Kocher, but he took him up to meet Trendle.

"Where you from?" Trendle wanted to know.

"Originally I came from Detroit," Kocher told him. "Lived here all my life until the war. Then I got into federal service and travelled around a bit—mostly around the Great Lakes. But Detroit is home."

Trendle turned to Campbell. "You want him?"

"Well," said Campbell, "we'll give it a try."

Kocher was not very reassured by their attitude: "Here I am giving up a government job with retirement—and he says, 'We'll give it a try'."

Kocher knew the history of WXYZ with Chief Engineers. It was not very good. Only Lynn Smeby had lasted longer than two years. Verne Alston was leaving WXYZ to go to new NBC studios at Burbank, California. Kocher figured if he lasted only a year, however, he'd still break even because WXYZ was offering him double the salary he had been getting.

Kocher was aware of the irony of the situation: "Here I had looked for a job at WXYZ for ten years—willing to start at anything—and they didn't hire me! Finally, when I went after the *top* job—they gave it to me! All that Horatio Alger sink-or-swim stuff is for the birds. You got to get started at the top!"

Charles Kocher became Chief Engineer of WXYZ in August, 1944 and was still with the station 30 years later.

Bob Hite's announcing was in great demand. He did a vast amount of slide film and commercial movie work for the Jam Handy organization. He transcribed commercial spot announcements. He also engaged in an unusual off-the-air activity with a Detroit free lance writer, Mort Neff. Neff had a recording company.

It was customary during World War II for the government to give E Awards—E for Excellence—to industrial companies that performed exceptionally well producing war material. A big show was made of presenting these awards; admirals and generals would be sent to Detroit to make laudatory speeches.

Mort Neff had the idea of recording these ceremonies. The companies then purchased the recordings for their files. Bob Hite was paid to do the announcements and give the "color" of the ceremonies on these recordings. He remembers doing an E Award program at the Chrysler tank arsenal where Fred Wolf, curly-headed bowler of the champion Stroh Brewery team, now had 1200 men working under him in sub-assembly.

Hite had also become one of the most important announcers at WXYZ. He was chosen to do the important *Day in Review* 6 P.M. newscast when Harold True left. He was commercial announcer on *The Lone Ranger* and narrator on *The Green Hornet.* He was as content at WXYZ as Porter Bourne had been at WEXL.

Then, in four terrible days in September, 1944, his whole world fell apart. As he tells it, one gains new insight into the simplicity and true modesty of the man. He told us the story of the four-day tragedy one July midnight in an announcers' booth at CBS television in New York. He spoke quietly and unemotionally, but the horror of what had happened 27 years before still lingered in his words.

Bob Hite: "One Sunday, we played tennis, my wife and I, at the estate of Henry Reichhold. He was a friend of Mort Neff.

"On Monday she didn't feel so good; she thought she might be getting a cold. On Tuesday she stayed in bed with a sore throat. On Wednesday, the doctor wanted to test her for polio.

"Thursday morning she went to the hospital in the ambulance and I stayed around all morning. At noon, the doctor told me to go home, there was nothing I could do. I might as well go to work. He did tell me she had polio.

"I was doing *Challenge of the Yukon*. When I got off the air, I had a message to call the doctor. He said, 'You better get to the hospital quickly. Your wife is not well at all. She's fading.'

"And—I got to the hospital. Ford Hospital. She wasn't there. He told me to go to Herman Kiefer Hospital. They didn't have an iron lung at Ford Hospital. Instead of bringing the iron lung from Kiefer to Ford, they tried to take *her* to the iron lung at Herman Kiefer. They had her half way in an iron lung when I got there—just half way in—and she was—she was dead.

" 'Course that's pretty hard to—27 years old.

"I came right back to work. I did *The Day in Review* the day after the funeral. I can remember I couldn't concentrate on the news. I was reading the news and tears were running down my face. Nobody could tell in my voice. But I read it sort of mechanically. I just thought what the hell else am I goin' to do?

"Well—my sister Virginia was a buyer for J.L. Hudson's—came to New York every two weeks to buy millinery for J.L. Hudson. And she said, 'Why don't you come down to New York with me?'

"My wife died in September. And in October I came to New York with Virginia; and while I was here of course I came to CBS to say hello to Doug Edwards and Fielden Farrington; and—Doug said, 'They're lookin' for an announcer. Why don't you go down?'

"So they auditioned me and offered me a job. They asked me when I could start and I said in three weeks. I knew I'd have to give WXYZ two weeks notice. So he gave me an application to fill out and he said, 'Mail it in when you get back to Detroit.'

"I told him I'd take the job, but I knew I had a week to back out because I wouldn't have to give notice to WXYZ for another week.

"So I went back there and—speakin' about decisions, decisions do have to be made. I had the application. I filled it all out. And when I dropped it into the mailbox—that's the point of decision. I can remember walkin' out of the WXYZ studios when I knew I had to do it—and finding a mailbox—and dropping that in, because that was a hell of a move for a midwestern farm boy—you know?—who had only been in New York once."

28
Peak Year

"People laugh about Horatio Alger. We could use a little Horatio Alger. Too many people sit around with the notion that the world owes them a living."

It was late 1944 and Arthur S. Hickok, President of the Hickok Oil Corporation, was standing before his seated guests. Benny Kyte and Dick Osgood had been driven to the Toledo headquarters of Hi-Speed Gas Stations by laconic Irv Stoll. It was their annual pilgrimage of good will, and time for contract renewal. *In Bold Relief* had run a year and the sponsor was in the mood for a change.

Floyd Tracy, Hickok's advertising manager, sat with the guests beaming up affectionately at his boss who was thin, bald, a little stooped, in his sixties, and spoke in a high-pitched excited voice. As usual, Stoll sat smiling and silent.

Tracy remembered Stoll as ineffectual: "Irv Stoll never came in with an idea in his life. He made a living off me for years because not only did I give him a WXYZ contract, I gave him WHK in Cleveland, gave him WSPD in Toledo. I'd say we want to go to Akron, and he'd go down and make a contract. Or Lima. I just told him what stations to get lined up. So he got his commission from all those stations as well as WXYZ, and he never had to do anything for it. We spent well over $100,000 a year on radio."

No idea for a new program would come from Irv Stoll. None was expected from Benny Kyte; he was there because Tracy and Hickok were fond of him. It was up to

Tracy and Osgood.

"Why when automobiles first came out," Hickok was saying, "I bought bumpers and I'd go out on the street and when a man parked his car, I'd talk to him. I'd say, 'You should have a bumper on your car.' Cars didn't come with bumpers then; they were something new. Sometimes, while a man was in a store or an office, I'd attach a bumper to the front of his car and when he'd come out, I'd show it to him and sell him on keeping it. Times have not changed. You still have to work for it."

The Detroiters listened respectfully. Tracy's round face resembled a smiling moon. Only in the presence of the old man he admired would he be silent for so long a period. Tracy liked to dominate.

"I didn't know anything about the gasoline business at all," Hickok continued. "I started with just a couple of offices in a building. That's when I met *this* feller."

"I sold him his first multigraph machine," said Tracy, beaming at the Detroiters.

"Then we went on WJR in the early days of radio. We had 228 service stations," Hickok told them.

"Now we have more than a thousand." said Tracy.

"Do you have in mind a series of success stories?" Osgood ventured at last.

"I think that's a good idea," said Hickok.

"I'm afraid you'd run out of material," Tracy warned.

"We're talking a 52-week contract," said Irv Stoll.

"I think it's a good idea," Hickok repeated.

"It's amazing how much material you can find in the public library," said Osgood. "We did 883 Factfinders, and the only subject that was really hard to find material on was—the taxicab."

"But we're talking a year's contract, five nights a week," Stoll argued.

"I think we could do it," Osgood insisted.

And they did, beginning January 29, 1945. The series was called *Eyes on Tomorrow* and it ran four years.

There was one hitch: Osgood was now too busy to research and write the stories. Material had to be found and written, interviews had to be arranged and prepared, and many hours had to be spent in screening rooms looking at new movies for *Show World* which was on fifteen minutes a day; he was also writing and producing a commercial series for a home remodeling business; and he had been bought by the United-Detroit Theaters to do 5 minutes of *Hollywood Headlines* plus a guide to the 14 theaters every morning. These programs with detail work as pro tem program director made it necessary to purchase stories for *Eyes on Tomorrow* from free lance writers. Forty-six writers turned out 1,047 success stories for the series, and some of the best came from inmates of the Michigan State Prison!

Eyes on Tomorrow was in the 6th month of its first year when Arthur S. Hickok died; but the business he had headed and the program he had wanted continued.

The sponsor was appreciated at the radio station. As Riddell remembered: "Hi-Speed was sort of the backbone of things for a long time. That, and United-Detroit Theaters were both good accounts."

The Mendelssohn mansion had not been occupied long by WXYZ before it was invaded by Decca Records. Five-minute versions of *The Lone Ranger* were recorded. The place was in turmoil. Benny Kyte, rotund, affable and big-eyed, augmented the studio orchestra and among the additional musicians was a trumpet player from the Detroit Symphony, Leonard Smith. He became one of the leading band masters of the country. His silvery trumpet opened each Decca Record Ranger with the first notes of the famous theme. The music was "piped" down from the third floor to the drama studio and blended with acting and sound by the producer. It was a fussy, complicated job.

Also, one sponsor of *The Green Hornet* demanded live music, and for a time Fred Flowerday, when he returned from military service, Benny Kyte, and an arranger named Elbert Green, were busy composing incidental music. Flowerday would go over the script with Green and play recorded music he would have chosen for the interludes. Green would then compose similar music in the same mood. Production was tricky. Benny Kyte had to take cues in the ball room from what was going on dramatically two floors below. The program was live; there could be no re-takes; it had to be right the first time.

There was another change. Strangely, after playing the role nearly a year, it was decided that Don Faust was not right for the lead, and another announcer—a tall, curly-haired, blue-eyed kid in his twenties named Bob Hall—became the third Green Hornet.

Don Faust moved to Chicago, did an all-night program on Station WIND for a year, worked two years for no pay on an experimental TV station that became WBBM-TV, and made his first television pay doing production work for the J. Walter Thompson advertising agency. He co-ordinated the first sponsored remote football games in the middle west, and steered his career away from talent into management before he became very successful.

Late in 1945 the original Green Hornet, Al Hodge, returned from military service, spent a short time in Detroit solving a marital problem and playing a sort of hide-and-seek game with George Trendle, then went to New York and stayed. He appeared regularly on *Gangbusters, Mr. District Attorney, This is Your FBI, Portia Faces Life, Romance of Helen Trent* and many others—some directed by Ernie Ricca who had gone to New York from Detroit's WWJ. One, *Mystery Theater*, was directed by Frank Telford who had taught at Wayne (State) University.

A common image of the old-time radio announcer has him looking very eager and serious, with one hand cupped over an ear so he could hear how he sounded. This picture was rarely if ever seen in life at WXYZ. One of the least eager and serious announcers was Johnny Slagle. Jolly John took life very lightly.

One night, he put *The Lone Ranger* on the air, left the studio and went his happy way. Dan Beattie was directing the show because Livingstone was on vacation. It was one of Dan's own scripts, featuring a chase that took the Ranger to the Mississippi River. The chase continued on boats and there was a boat wreck. Beattie used more sound effects than the other writers.

He was proud of this show. The actors had all given fine performances; the sound effects had worked out excitingly, and the show was over, except for the sign-off. Jolly John had returned to the studio and stood ready, script in hand. Dewey Cole faded down the music.

"Boy, that was a good show!" Beattie enthused. He threw the cue for the sign-off. Loud and clear, John Slagel went into the *opening*: "—a fiery horse with the speed of light—"

"Up with the music! Up with the music!" Beattie cried.

Dewey Cole, who panicked easily, came up with the commercial. And it stuck in a groove. Slagle never did sign off. The half hour ended with: "CheeriOH, CheeriOH, CheeriOH..."

Five seconds after the fiasco, there were phone calls from Livingstone—and Trendle.

One day some weeks later, Trendle turned his powder-blue Continental into the driveway for his daily visit just in time to see Jolly John Slagle enter the Mendelssohn mansion thru the forbidden side door. Johnny did not announce *The Lone Ranger* that night. Johnny was fired—immediately.

He went to ABC in New York—the eighth WXYZ announcer to make the big town: Del Sharbutt, Bob Hite, Slagle, Fielden Farrington, Charles Wood, Hugh Downs, Douglas Edwards, and another who became extremely successful, John Cannon.

Harry Golder returned from the Navy to announce *The Lone Ranger*; Fred Foy came back from California; the restless Bill Morgan returned from Texas. Also among the announcers now were Hal Neal, Norm Lenhardt, Porter Bourne, Don Hendricks, Bob Henry, the former mail clerk and sound man, Larry Kelley, and El Prough whose first words upon returning from service were: "We shouldn't have stopped with Hiroshima; we should have gone right on and bombed Moscow."

There was also a new actor in the role of Sergeant Preston. Paul Sutton was imported from Hollywood where he had played villains in many, many movies. He was easily six-feet-four, well built and fairly handsome. His eyes were small and he wore thick-lensed glasses, but he saw Allen Campbell's secretary all right. She was almost as tall as he, which he liked. He pursued her and they were married. *Challenge of the Yukon* moved into prime time on a network as a half-hour show and changed its name to *Sergeant Preston of the Yukon.*

Some believed that Sutton was brought in to dangle over the sometimes temperamental head of Brace Beemer, but this is not true. The actor who was brought in for that purpose was old and as short as Brace was tall—a versatile Hollywood dialectician named Billy Fletcher. He sounded more like Brace Beemer than Brace Beemer when he tried, and he was around making Beemer nervous for about a year.

Working in the WXYZ newsroom now were Henry Alexander, beatle-browed Tom A'Hearn, a girl named Barbara Fearnside, a very talented dwarf named Bill Krueger, Dick Frederick who would eventually open his own public relations office and have among his accounts the Michigan State Fair, and on weekends, Harvey Taylor, a stuttering wit and great ladies' man who was to become one of Detroit's best music and movie critics.

Seymour Kapentansky was now writing morning newscasts for, of all people, Harry Heilman. One day as he and Kapentansky entered the mansion, Heilman moaned, "Why is Don Wattrick trying to get my job?" He was afraid he'd lose the baseball broadcasts.

It is generally considered by those who reminisce that 1945 was WXYZ-Radio's peak year. On November 30 that year, the 2000th Lone Ranger story was broadcast. Fran Striker figured this represented the writing of some 8,000,000 words—most of them his own.

In December, a testimonial banquet was given by the Officers Club of the Detroit Fire Department, honoring Trendle for having been a member of the Detroit Fire Commission for 21 years.

Also in December the contract of Benny Kyte who had been musical director of the station for 10 years was not renewed. No one alive has an explanation. One suspects the cause may have been the uncommon sales ability of the ambitious musician who replaced him.

In 1945 Raymond Meurer enjoyed another court room triumph. The Lone Ranger Inc. was sued by an aging one-time western outlaw, Al Jennings. Now 82 Jennings still did occasional lectures against crime, was still a little bit in the public eye. He wanted $100,000 because in one of the Lone Ranger stories, he charged, he had been defamed.

"The Lone Ranger shot my gun out of my hand—and me an expert," he complained. "It made me look like a sissy."

He said his neighbors avoided him. Also, he declared, the story had made him look like a common burglar. Proudly, he pointed out the difference betweeen a common burglar who sneaked around in the dark and an outlaw which he had been. An outlaw, he testified, committed crimes bravely right out in the open daylight. He claimed that because of the broadcast even little girls called him "robber." His privacy, he maintained, had been invaded.

Ray Meurer argued that Jennings' life had been so publicized, even by himself, that he had no privacy left, and that defamation of character would hardly be possible with a character already so infamous.

The case was dismissed.

The new musical director of WXYZ was Phil Brestoff. He was short and round, and his brown eyes windowed a character which seemed oddly contradictory. There was at once something "spoiled" about him as though he were the son of an over-indulgent mother, and something vaguely treacherous. He was a Clevelander, had gone from high school to the violin section of Phil Spitalney's orchestra at the Stillman Theater. He moved to Detroit in 1929, played fiddle in almost every theater orchestra in town and finally came to rest under the baton of Eduard Werner. When

Cast of Sergeant Preston. To left of him: Fred Flowerday and Ernie Winstapley. To right of him, Dewey Cole, Fred Fry and Jim Hengstcheck. Left to right in back: Rollon Parker; Billy Hengstebeck,

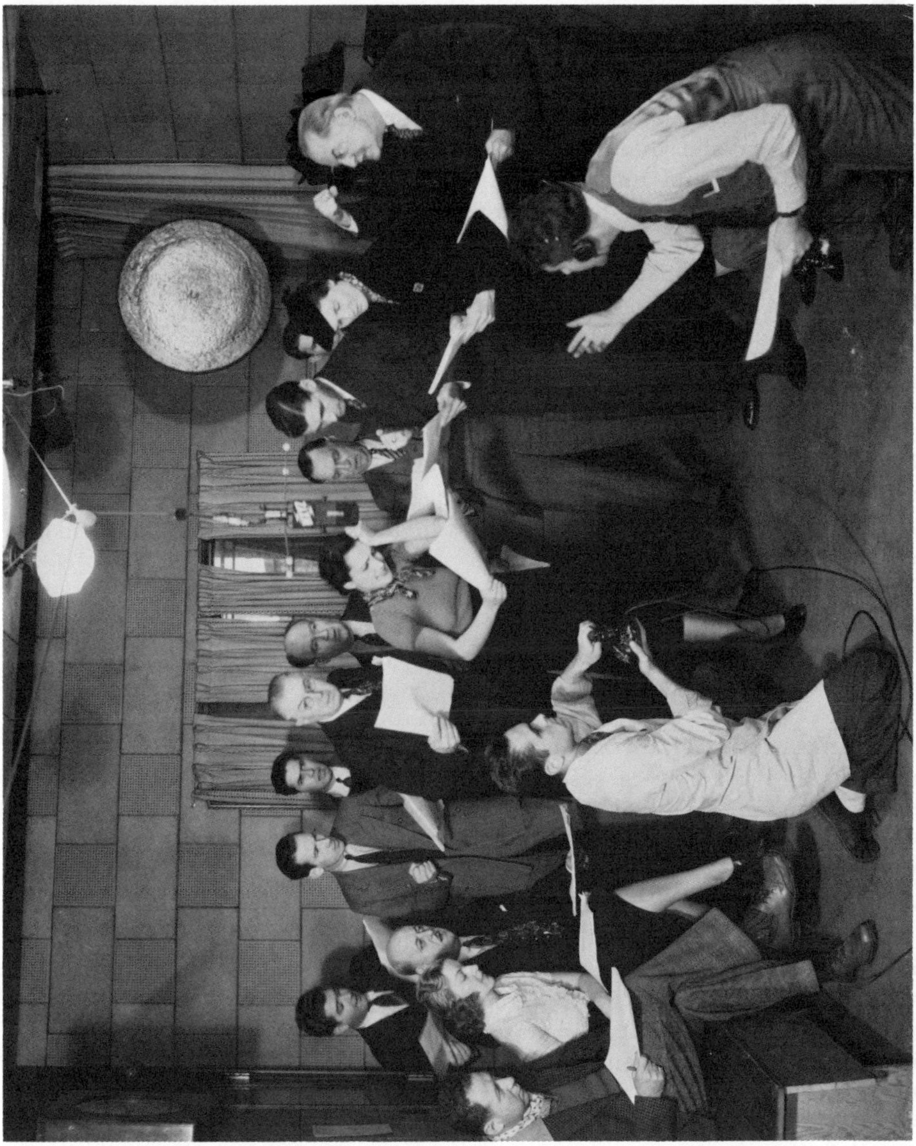

WXYZ actors 1935-1945

Werner left show business to practice law, Brestoff took over as musical director of the mighty Michigan Theater.

He had first appeared on WXYZ as part of a package which included his wife, Dee Parker. They had met at a dinner party, Phil had asked for a date. Dee had turned him down, so they were married.

Dee was small and solid. Her face was all curves, and attractive. She had big, soft brown eyes; and she sang like Dinah Shore. Phil had used her at the Michigan Theater, adored her, and nursed her ambition which was not negligible. One suspects he played an important part in pushing their way into WXYZ radio as a commercial package sponsored by a cigar. One also suspects he was smart enough to see the coming end of theater orchestras, to recognize the potential of radio for him and his wife, and to push their way into this lively new world.

He became musical director at WXYZ in January, 1946. The next few years were to be the best of his life.

It was also the retrospective view of George W. Trendle that the mid-40s were the best years of WXYZ. At any rate, it was the most profitable era of the Trendle regime. World War II had ended. Germany had surrendered on the 7th of May the year before. In August, atom bombs had been dropped on Hiroshima and Nagasaki, and Japan had given up. The Depression was ancient history—forgotten. The world was swinging—and WXYZ with it.

29
Big Deal

The end of World War II found Fred Wolf and Chrysler with 400 unfinished Sherman tanks waiting for turrets, but war's end did not cancel the contract with the government, and there were six months of turret work ahead. Chrysler was planning a future for Fred Wolf. The company considered him promising executive material and planned to move him up through various phases of the business in a grooming process.

Meanwhile, he had developed back trouble and was unable to bowl as well as he had; so he was made the Stroh team's manager. Besides being responsible for distribution of tickets, travel arrangements and other details, he also introduced players and gave scores over PA systems.

Norman White, then production manager of Station WJR in Detroit, was a great bowling fan and a special admirer of the Stroh team. He approached Wolf after one of the matches.

"Fred, did you ever think of doing a bowling show on the radio?"

"No. Why?"

"Well...Detroit's a bowling hot bed. A bowling show on radio should do well. We wouldn't want it on JR but you should go to WWJ or WXYZ. Do you know Jimmy Riddell?"

"I've met him but I don't know him."

"Well, look, come out to JR some night after bowling. Write a little copy and bring it over and read it and we'll record it and see how it sounds."

Meanwhile, Judge John D. Watts asked Fred to report activities of the Stroh team once a week on his *Sandlotters* program. No pay was involved; Fred did it to publicize the team.

At WJR, Norman White coached him. After they had recorded and analyzed his writing and vocal style on a dozen transcriptions, he said, "Get a guest. Write an interview. Write the whole show—opening and close—everything. But leave two open spots to indicate where commercials would go."

Fred went to one of the hottest bowlers of the time, Thurman Gibson.

"Thurman, you gotta help me. I need a guy to interview."

"Fred, microphones scare the hell out of me."

"Don't worry. I'll have it all written out."

Wolf wrote not only his questions but Thurman's answers. He spent over a week writing the quarter-hour show. He took his script and Gibson to WJR and Norman White transcribed the show.

"Go to places like Cunningham's Drug Stores," White advised. "Work out promotions. They're always interested in promotions. Who knows? Maybe Stroh's will buy it."

The vice president of Stroh's rival, the E. & B. Brewing Company, was a little Scotsman named Ken MacDonald, another bowling fan. He often attended the Stroh matches, and after Fred became manager of the team, the two became friendly.

"I'm thinkin' about doin' a bowling show on radio," Fred told him one night after a match.

"Might not be a bad idea," said MacDonald.

"I'm goin' down to see the Stroh people," said Fred.

"That's a good idea. They should buy it."

Fred remembered that Walter Zimmer, partner of the Zimmer-Keller advertising agency, had once bowled on the famous Stroh team. The Stroh Brewery Company was now a Zimmer-Keller account. Fred played his transcription to Zimmer.

"Hell, they gotta buy this!" said Zimmer.

But they didn't. Wolf called Ken MacDonald at E. & B.

"Well, why don't you come over and see *me.*"

"Are you kiddin'?"

"No. Come over. Let me hear what you've got."

Tape recording did not exist then. Wolfe borrowed a big, heavy transcription player and lugged it to MacDonald's office late one morning.

"First, we're going to lunch," said MacDonald.

He took Fred to the Butcher's Inn at Eastern Market.

"Okay," he said when they returned to the office. "What do you want to talk about?"

"*You* invited *me* over here," Fred reminded him.

"Oh. Yeah. Well. Let's hear it.

The show was about six minutes in.

"Shut it off," said MacDonald. "I don't have to hear any more. Let's put it on the air."

Wolf was flabbergasted.

"What station do you want it on?" MacDonald asked.

"I don't know. W—er—WXYZ, I guess."

MacDonald called a salesman at WXYZ and made an appointment for the next morning.

Wolf was overwhelmed: "This is pretty good. You walk into a radio station with a sponsor, it's not too hard to get on the air I'll tell ya."

Riddell was pleased and grinning when the salesman brought MacDonald and Wolf into his office. MacDonald introduced Wolf.

"Oh hell, I know Freddy," said Riddell, holding out his hand. He had heard Wolf on *Sandlotters* and had met him when both were guests at the Detroit Athletic Club. Riddell was also a bowler.

Wolf asked for time on Saturday night. He got it, to begin on the 15th of October, 1945. He had about a month to get ready.

MacDonald turned over the preparation of commercial copy to E. & B.'s advertising agency, Doner & Doner. The agency people didn't know Fred Wolf from a barrel of hay and they were up in arms. The whole thing had been sold over their heads; they felt part of the money being spent on a 26-week contract on radio should have been spent in print. They didn't want to write the commercials. But they did. Bill Morgan was assigned to the program as commercial announcer. The program was called *Talking Ten Pins.*

Wolf remembered that he got to go for 26 weeks, "and it was just a great experience."

Besides he got paid!

Jack Pickering: "I don't know if it was Fran (Striker) or somebody else that told me about visiting Brace Beemer. I'm pretty sure it was Fran who told me of visiting on Beemer's farm and here was Beemer complete with pistol belt, guns—and in the barn he saw a rat somewhere on a beam. Pulled out a gun, shot him dead center, and put the pistol away as though that was... Fran said he didn't know whether Beemer really could shoot like that. But, uh, you don't go shooting holes in your barn unless you're pretty sure you're going to hit what you're shooting at."

Leta Beemer: "Brace was as handy with a rope as he was with guns. We had a flock of muscovy ducks, those gorgeous great big white ones. They'd fly over to our lakes, then come back to feed. So, came hunting season, Brace was afraid somebody would shoot 'em; so as they came in, Brace would get this rope going, and he's just swirl it up in the air and get a duck by the neck, and bring it down, and put it in the barn. And he'd keep on doing that until he got 'em all in."

Ed Noble, owner of the American Broadcasting Company, created an Advisory Committee made up of men representing affiliated stations in various sections of the country. The ABC stations of Michigan, Ohio, Illinois and Indiana chose H. Allen

Campbell to be their representative, and the members of the national Committee made him Committee Chairman. He saw more of the leaders of the industry now than Trendle did. Trendle was not a good mixer; he didn't play golf; he didn't smoke; he didn't drink.

One noon, Campbell was having lunch in the English Grille beneath Radio City at Rockefeller Plaza with two network men.

"Allen, when are you going to join us here in New York?" Keith Higgins asked over their drinks.

Campbell laughed.

"That's what Sol Taishoff wants to know," he said. Sol Taishoff was editor and publisher of *Broadcasting,* a trade magazine. "He called me and said, 'When are you goin' to announce that you accept the presidency of ABC?'

"What did you tell him?"

"I said, 'Nevuh'."

John Norton, Jr., ABC vice president for station relations, drained his glass and set it down.

"You fellows are doing just a terrific job out in Detroit," he said. "You know, we're in a bad set-up at ABC. We only partially control a station in Chicago; we have stations in San Francisco and New York—and that's all. We'd love to own WXYZ. Would you think of selling?"

"Might," said Campbell, his eyes narrowing a little. "Whatta you got in mind?"

"Exactly who owns WXYZ—besides Trendle?" Higgins wanted to know.

"John Kunsky, Howard Pierce and myself," Campbell told him.

"Trendle never mentions Kunsky. Is that King? John King?"

"Yes."

"Never mentions him."

"Hell," said Campbell. "His interest is equal to Trendle's."

"Funny, he never mentions him."

"They are not on the best of terms," Campbell revealed.

"Why not?"

"Oh, Kunsky has some cock-eyed idea that Trendle wants to squeeze him out."

"What would you take for the station?" Norton asked.

"Why don't you make an offer?"

"We wouldn't know what to offer."

"How's two-and-a-half million for a starter?" Campbell asked.

Back in Detroit, he reported this conversation to George Trendle. Without a moment's hesitation, Trendle said, "Sell it!"

He looked at Campbell, thinking. "We'll keep the dramas," he said. "It's a deal—*if* we can sell the station and keep the shows—*and* dump Kunsky."

"You want to dump me, too?" Campbell asked, cutting through the Trendle ice with warmth and an unworried smile.

"No," Trendle said, still cold. "You'll be a partner in my new set-up. I'm just tired of the old connection. I'd be glad to get my money out of it."

"You want to talk to Kunsky?"

Trendle made a little grimace of distaste and waved one hand. "Oh, you talk to him, H.A.C. You're the salesman."

It is surprising that a man who had devoted so much time, effort and money to the building of a major radio station could be so ready to sell it.

Trendle explained: "Well, Kunsky had 40%, I had 40% and Howard Pierce had 10% and Campbell had 10%. Campbell could have died the next day and Kunsky would sit there. I'd have 40% and go into the courts and the damn thing would muddle around for years. I was tired of the whole connection. I just had the feeling that if I could push out of that thing and take these shows that I had on, and run my own business...."

Campbell called John King at his lodge in Wisconsin. The little man was pleased as a puppy to be called.

"Mr. King," Campbell said after a minute of small talk. "I have a deal whereby we could sell WXYZ for $2,500,000."

The puppy sounds ended. There was a silence.

"Why should I sell WXYZ?"

King sounded a little shocked, a little defensive.

"What good is it to you?" Campbell asked. "You have an office heeyuh but you're nevuh around. You're up north theyuh all the time."

"Allen, I love that station. I love being a part owner."

"Look at it this way," said Campbell. "The station didn't cost you more than a couple o' hundred thousand dollars. If you could come up with a million for that two hundred thousand wouldn't it be a pretty good deal?"

"I entertain a lot o' people up here at the lodge, and I'm proud of being part of *The Lone Ranger* and *The Green Hornet* and WXYZ."

The best that Campbell could report to Trendle was that King had promised to think it over.

"We can get more than two-and-a-half million," said Trendle. "We'll throw in Channel 5."

They had applied for this unassigned TV channel in Detroit and had a lawyer working on it in Washington. It was a big secret; few people even at WXYZ knew it.

Charles Kocher recalled that "the applicants included United-Detroit Theaters, the Detrola Corporation, WJR and one or two others."

Meanwhile, New York learned from Campbell that Trendle was interested in a deal. Campbell and Howard Pierce would be no problem. The thorn under the fingernail was John B. King.

At Christmastime in 1945, John King came to Detroit to spend the holidays with his daughter and her husband. He also called on old friends, as usual, in the executive offices of WXYZ. Campbell caught him on one of his visits and said, "Have you done any thinkin' about the set-up?"

"Oh yes, some. But not much," he said. And, he added, "Campbell, why are you so interested in this? Don't you like WXYZ?"

"Yes," said Campbell. "But, Mr. King, you're 70 years old; you got plenty of money. You don't need this money. I mean, I would think that you could do a hell of a lot more things with this than what you're doin' at the moment."

Then he changed the subject. Later during the holidays, King dropped in again. They were having a Christmas party in the offices.

"How's that deal comin' along that you're working on?" he asked Campbell.

"Well, I'm doin' pretty well with it," said Campbell. "I think I can get about three million for it, *plus* the cash that we have in the till. It's growin' all the time. It might even get bettuh."

"Well, let me tell you something," Kunsky grinned. "If you can sell this for 2 1/2 million dollars, I'll kiss your ass on the city hall steps."

American interest in television came sharply into focus in 1939 when a speech by President Roosevelt was televised at the New York World's Fair. Only a few hand-made receivers existed then, but those who saw were men who could envision a powerful new force in the world.

At war's end, RCA, CBS, the DuMont Laboratories and others resumed efforts to perfect process and equipment so that the scientific marvel of transmitting pictures through space (for that matter, through brick walls) might become practical and—hallelujah!—profitable.

George Trendle was not sure he wanted to go into TV, but he certainly did not want to be left out. So now, in 1946, he found himself in competition with several others who had applied for Channel 5 in Detroit.

During Charles Kocher's preparation for the hearing, the new chief engineer discovered that Trendle did not like being ignored.

Remembering this aspect Kocher noted: "Trendle sent me those pink notes, of course, like everyone else got, and I let 'em pile up and he finally wrote me *the* note which said, 'I think you're doing, generally, a pretty good job out there except for one thing—you don't answer my notes.'

"So I started giving his notes top priority."

One morning Kocher found a television construction permit in his mail. He lost no time contacting Trendle who was not surprised. He had already heard from his Washingon lawyer, William Hennessy, and had sent for Campbell.

"We got it!" he said. "We got Channel 5!"

But he had no intention of going into television right away, if ever. Equipment alone for a first-class station could run up to a million dollars, and it cost much more to operate a TV station than it did radio. It could cost a million dollars a year to run a station before it made a profit. Trendle recalled the struggle he had had to establish WXYZ radio. He was now 62 years old. But, as Campbell noted, "This gives us a better package to sell Noble."

Campbell and Trendle had lunch together one noon in the winter of '46 in the Terrace Room of the Statler Hotel. Campbell had just returned that morning from several days out of town. Among other places he had been to Flint, Michigan, but more importantly, he had been in New York.

Aside from the comfort and ease of the moment, the two men felt an excitement and well-being. Things were going well; the dramas were sold; the station had grossed $2,357,000 in 1945; and there was a big deal in the air.

"I think I'll have a glass of wine," said Trendle.

"Well, if you're goin' to get drunk, I'll have a Scotch on the rocks," Campbell kidded.

Trendle was eager to hear the results of Campbell's latest travels.

"I was right," Campbell announced as they raised their glasses. "Channel 5 has sweetened our pot."

"*How* sweet?" Trendle asked.

"Over three million," said Campbell. "And gettin' sweeter every day."

"Keep it stirring. Now what about Flint?"

"We can have WTAC," Campbell told him. "At our own price."

The two men ordered food.

"Now," said Trendle. "What's your idea on the production company?"

"I want to further *The Lone Ranger* between the two of us, and we can do more with *The Green Hornet*. Hell! It's a better show. Whenever I walk into the club, or anywhere in New York, they say, 'Here comes the Green Hornet.' They never say, 'Here's the Lone Ranger'."

Trendle did not share his enthusiasm.

"More than 80% of what I make goes to the government as it is. What's the sense of pushing for more. The Hornet is gravy for the government. Our meat is the Ranger."

Campbell laughed. "Hell! That's all right for *you*."

"H.A.C.," Trendle promised, "you have stayed with me and I'll give you a better part of the thing."

"How much bettuh?"

"You won't suffer. Pierce is no problem; he'll go along with anything I want. It's Kunsky we may have trouble with."

"I'll go to him again," said Campbell.

He went to Kunsky by telephone. He reported that the deal with Noble looked good. Then he said, "Mr. King, I know that Mr. Trendle cut you in for 40% interest in *The Lone Ranger*. He *gave* it to you; it didn't cost you a cent beyond your investment in the station. Now. Would you entertain an idea of sellin' your share of the Ranger? I mean, at a reasonable figure?"

"I just hate to give up being a part of *The Lone Ranger*," said the sentimental Mr. King.

Campbell reported back to Trendle.

"We'll print up cards for him to show his friends," Trendle suggested. "We'll give him an honorary title. But how much does he want in cash?"

"I think he'll take a hundred thousand."

"I can do better than that."

Trendle then called Kunsky and said, "If I can get $3,700,000 for the station, that would give you another hundred thousand in your share of this stock. Would you then turn *The Lone Ranger* back to me?"

King agreed.

Campbell was dumbfounded: "So Trendle got it for *nothin'*! Now we're goin' to turn around and sell *The Lone Ranger* for 3 1/2 million."

That did not come along until later.

"No, but you could see what the hell cooked." Campbell laughed at the recollection.

Part of the deal finally consummated with the American Broadcasting Company was that station WOOD in Grand Rapids would be included in the sale, as well as Channel 5. Trendle and Campbell rented production space in the Mendelssohn mansion for next to nothing.

"And," Campbell exulted, "we weren't confined to just feedin' our programs to ABC. We could feed 'em any place."

The sale was completed on the 24th of April, 1946. Trendle assembled his studio staff in the mansion and broke the news. He assured all concerned that their jobs were safe; New York, he said, had purchased the station because it was doing extremely well and ABC would want to keep the staff that had made it so successful. That meeting came closer to being sentimental than any other involving Trendle.

He also called together his executive staff in the Stroh Building.

Bessie Ashton remembered with a certain cynicism: "Ah, yes. That was a, that was, marvelous. You know, he broke, and he said, at the time, there would be no change, in personnel. Oh yeah, yuh, yuh. Good heavens."

Howard Pierce: "We sold.... It was for $3,650,000. We didn't get the $50,000, but we got the rest of it."

"What happened to the $50,000?"

"God knows. I don't."

After an association of more than 30 years, John Kunsky (King) was no longer a part of George Trendle's activities. He was given cards by the new Trendle-Campbell Enterprises, Inc. indicating that he was still part of *The Lone Ranger*, which, of course, he wasn't. His family, reportedly, is still bitter about it.

The first question in New York after the sale was, who will run WXYZ? Campbell had made it clear that he would no longer be affiliated with the station; he was going with Trendle. The logical person from Detroit's point-of-view would be James G. Riddell. Or possibly Harold Christian. Maybe even Raymond Meurer. Actually, Riddell had been responsible, under Trendle, during the absences of Campbell. But aside from casual contact with John Norton of Station Relations, nobody at New York headquarters of ABC knew Riddell. Nobody knew much about Christian. Nobody knew Meurer.

Detroit broadcasters, especially executives, were no less interested than New York in watching to see who would replace Campbell at WXYZ. One trade publication hinted that Merritt Schoenfield of the Detroit ABC office was "expected to take over in a supervisory capacity." There was also a report that one "Swanee" Hackman would be brought in from Minneapolis to run the station. There was even a rumor that Ray Meurer was playing a few cards in New York.

Jim Riddell received a telephone call. It was the one-time sales manager of WXYZ, now assistant general manager of Station WJR, Owen Uridge. He and G.A.

Richards, owner of WJR, would like to take him to lunch.

"Don't call me," laughed Riddell. "I'll call you."

Shortly after the sale, Campbell, still a member of the Board of Governors, returned from a meeting of representatives of ABC-affiliated stations which had been held on Ed Noble's private island off the coast of Georgia. He brought Noble's financial man, C. Nicholas Priaulx, with him. Campbell introduced Priaulx to Riddell.

"Anything to the rumor that Swanee Hackman's coming in from Minneapolis to run the station?" Riddell demanded at once.

"Oh, I'm sure nothing is decided yet," said Priaulx, giving the local boy a big network smile.

"You're *still* the sales manager," Campbell assured him.

"Nothing doing," said Riddell, his face flushed. "I'm not going to stay on under those conditions. I'd do all the work and he'd get all the credit."

Campbell moved on to show Priaulx a small office next to Riddell's where Priaulx would work in Detroit, going over the station's books with the WXYZ auditor.

Riddell called Owen Uridge.

"How about lunch?"

Uridge, G.A. Richards, and Jim Riddell met at the Detroit Athletic Club. Owen Uridge had a special interest in this meeting. From the lofty prosperity of his present situation at WJR, his memory of displacement at WXYZ by Allen Campbell had become slightly acidulous. He could gloat a little now about his move to a more powerful station.

"One thousand watts to 50,000 isn't a bad jump,." he said.

"Want to make it?" Richards asked Riddell.

"We're 5,000 now," Riddell reminded. Then, frankly, to Richards, "I don't know."

At that moment Trendle and Campbell walked into the grille. The five men exchanged nods.

"Three WXYZ sales managers in one room at one time," Uridge observed. "Trendle must feel like a mother hen."

"Jimmy, you've done one hell of a job at WXYZ," said Richards. "We know it; and we want you at WJR."

"Well, let's wait and see what happens," Riddell suggested.

"Do you belong to this club?" Richards demanded.

"No."

"If you work for me, you will. Do you own your own house?"

"No."

"You work for me; I'll see that you have a house."

"Dick, it sounds good," Riddell admitted. "But let's wait. Meanwhile, what's your proposition?"

Richards and Uridge went into some detail; the offer took definite shape.

Five minutes after Riddell returned to his office in the Stroh Building, he had a

James G. Riddell. Broadcast House was his idea.

buzz from Trendle.

"I want to see you."

Riddell signed a letter his secretary had had waiting for him, dropped it on her desk as he went by, and waved and smiled at Priaulx as he passed the little office next door.

"How's it going, Nick?"

Jimmy's face was a little less sunny than usual, but Riddell was not disturbed. His walk was relaxed, unrushed. He had the air of a man well on top of life.

"When I sold the station," said Trendle as soon as Riddell walked in, "I sold management as well. I sold a going concern. And you are the one who's been running the inside of the station for quite a while. What's going on with Richards?"

Jim's face broke into a sunny Scots smile.

"Just what you think," he admitted.

"What did you tell him?" Trendle demanded.

"I'm not going to stay on here and do all the work and have this Swanee Hickman come in and get all the credit," Riddell explained seriously. "We have this thing going pretty good and I'm just not going to do it."

"Well," said Trendle. "Don't make any move until I get back to New York."

Trendle, Campbell and Priaulx had a conference that afternoon. That evening, Campbell called Ed Noble in New York. That night, Trendle called Riddell at his home.

"It's all right," he said. "It's worked out. You're going to be the new general manager."

30
Year of the Giants

The sale of the station had no discernible effect on day-to-day activities at WXYZ. Charles Livingstone was obliged to abandon his roomy office and move into the writers' wing; and Trendle, instead of holding his daily meetings in what had been Mr. Mendelssohn's library, met with his more limited staff in the same wing.

Raymond Meurer was retained for some time as legal representative of the station. He did have contact with ABC in New York and persuaded the network not to dissolve the local identity of WXYZ lest union negotiations be raised to a national level; it would be cheaper to negotiate as a local station. Under his urging, he told us, WXYZ was incorporated as a Detroit company.

Meurer's tenure as legal counsel for this new company, however, was short-lived. Whatever friendship he and Riddell might have enjoyed had ended some time before the sale as the result of an incident which had occurred during one of Campbell's absences. It involved relations with the sponsor of Tiger baseball but none of the three could recall exactly what had happened and their memories of the hassle differed considerably. Whatever the details, Meurer and Riddell came out of the situation with bad feelings and soon after Riddell became manager, Meurer's relationship with WXYZ ended, and George Trendle, Junior, became the company lawyer. Meurer went with the new Trendle-Campbell Enterprises, Inc. It bothered him more than a little that he had not been included in the "Inc."

Dick Osgood had feared the move out of town into the mansion might curtail getting stars for his *Show World* interviews: but the celebrities still came. Once, however, El Brendel, a Philadelphian who affected a Swedish accent and became a top comic in the movies, arrived early for an interview and walked around Indian Village. He became so delighted with the neighborhood that he missed the interview.

During the Detroit engagement of a play in which Henry Fonda was starring, the actor stayed at the Whittier, a pleasant hotel on the Detroit River across Jefferson Avenue from WXYZ. One night, driving home, Osgood saw Fonda waiting for a street car, picked him up and drove him to the theater, marvelling that a major star would ride on a streetcar.

Paul Muni, after an interview, expressed a desire to meet Tonto and the Lone Ranger. Dan Beattie introduced him to bald John Todd.

"Mr. Muni, this is Tonto."

"Oh, my God!" Muni exclaimed. "A scalped Indian!"

But all was not laughs for Osgood. Behind the scenes, *Radio Schoolhouse* had gone sour. Osgood felt that Fred Rivard and the theater manager were trying to smudge the luster of his "stardom" and he was trying to muster the will to leave the show and give up the fee.

With the end of the war, WXYZ veterans began coming home.

Fred Flowerday: "George Trendle had the faculty of holding good qualified people. What the club was, I don't know. When I was in the army, I told myself, I will not go back to WXYZ. I was going someplace else, because...you know, when I wanted a $5 raise, I'd go down and meet Mr. Campbell, and Campbell would get the ledgers out and show me—I'm sure it was a second set of books—it was always in the red and I'd invariably sneak out feelin' sorry for the company because it couldn't give me a $5 raise.

"How they managed to keep—Chuck Livingstone, tremendous talent!—for so long—the actors. It wasn't money. Promises? Was a time when Chuck, Fran and myself were goin' to be given some stock—which was another little club—and so you stayed on and hoped—and pretty soon the whole damn thing was shot out from under you."

But Flowerday returned.

Ernie Winstanley: "I was captured in 1945 in Czechoslovakia by the Germans. I was with Patton's Third Army. I was a prisoner of war, close to the end. The Russians liberated us. I was sent back to a repatriation camp at Le Havre. I had shell shock and combat fatigue. I had to take, oh, almost nine months of recuperation. You had a choice; you could go wherever you wanted. So I went to Van Nuys, California, where my aunt and uncle lived.

"I was bound and determined that I was going to try to make Hollywood.

Line-up Sunday noon to watch Radio Schoolhouse broadcast

Behind the scenes of *Radio Schoolhouse*

FEDERAL
DEPT. STORES
PRESENT
WXYZs
RADIO
SCHOOLHOUSE

Johnny Hodiak had me all set up for a screen test. Two days before I was supposed to take the test, the entire motion picture industry went on strike. I had no job; I had no money; I had nothing. I worked in a gas station. I still wasn't in very good health. I stuck it out for six months and finally wrote a letter to Chuck and he said, come back anytime. And I packed up my bags and I came back."

Tom Dougall came out of the army a major. He stayed in New York for a while. His commanding officer in the army was now with Benton and Bowles, an advertising agency, and he got Tom a job writing *Buck Rogers* scripts and the *Prudential Family Hour*. Tom also wrote for a show called *Ladies, Be Seated*. But, he returned to Detroit for a vacation—and stayed. Again he wrote for the show he had helped to create, *Sergeant Preston of the Yukon*.

Trendle's nephew, Bill Hendricks, had come from the theaters and had worked in promotion and advertising as the assistant of Lambert Beeuwkes. At the end of the war, Beeuwkes left and Hendricks took his place.

Another to return to WXYZ after the war was the little Italian firebrand Jack Petruzzi. Soon after Pearl Harbor, Jack had followed Mike Wallace to Chicago and looked up Ted Robertson. As men went into service, it became increasingly difficult to find male voices and Ted used Petruzzi in some public service shows he was producing, but, as they faded out, so did Petruzzi's employment. He, too, returned to the security of the Trendle dramas.

Others had tried Chicago without success. Both Gille Shea, who played Axford in *The Green Hornet*, and Paul Hughes failed there.

Ted Robertson remembered that "They were totally non-competitive."

They both returned to Detroit. Robertson moved on, very successfully, to Hollywood.

Another soldier out of the army began acting at WXYZ: short, rotund, versatile Rube Weiss. He had studied at Wayne University and had acted on some of their radio dramas.

The sale of WXYZ to ABC was the beginning of the end of the Michigan Radio Network. Eventually, the stations which had been fed by and through WXYZ received their programs directly from the ABC network.

Elizabeth Ashton: "I couldn't tell the date but I know we kept that name for a long, long time. It was, of course, after ABC bought us. You see, when ABC... When ABC bought—of course—why, they dropped that and they put in the lines direct. Well, it was after some time and I don't recall the exact date."

The Pfeiffer Brewing Company sponsored Harry Heilman with *Tiger Training:* quarter hours transcribed in Florida and air-expressed to Detroit, reporting the progress of the team and its players. It had been the custom for free-lance producer Mort Neff to accompany Heilman to the training camp and produce the transcriptions. Neff was one of the first to switch from electrical transcriptions to the less cumbersome wire recording. It was an improvement. But in the Spring of 1946, Mort Neff was too busy producing shows around Detroit to go to Florida. This put Riddell and WXYZ in a bind. It was already March. Who could they find to send to Florida with Heilman?

Fred Foy

Mr. & Mrs. Phil Brestoff

Like a miracle, an answer materialized in the presence of Dick Osgood, who was still acting program director and occupied the office by the back door. One March morning, a powerful-looking man in Coast Guard uniform came through the back door lugging two heavy cases.

"Aren't you Dick Osgood?" he asked, peering around the office door.

"Yes. Come in."

"You interviewed Victor Mature when he was here with the Coast Guard show *Tars and Spars.*"

"Yeah."

"I arranged it. I handled publicity. John Pival."

"Oh." Osgood had forgotten. "Sit down."

Pival sat down, setting a case beside each leg. He kept on his heavy coat but removed his cap. He was bald, had a very short moustache, and a rather swarthy complexion.

"What are you selling?" Osgood smiled, looking at the cases.

"Those are wire recorders."

Osgood had never seen a wire recorder; WXYZ was still transcribing programs on disks.

"I'm getting out of service and I need a job," said Pival.

Osgood called Riddell.

"There's a man in my office with two wire recorders and he wants a job."

"Send him down."

Pival was hired, but there was a hitch: the wire recorders did not belong to him, and he had no money with which to buy one. He and his wife, Virginia, went to a loan company and borrowed $300. That would buy one recorder. The station expected him to have two.

Virginia Pival: "We hurried up and went to another loan company hoping we could get the other recorder without that first loan company knowing about it. By golly, by the time we got there, they knew it! So we borrowed it anyway, from somewhere, and he got the wire recorders."

John Pival inherited his appearance from his parents, who came from Yugoslavia. His nose was small; his face was broad; he was 33 and built like a weight lifter. Actually, he had been a track star, best at hurdles, at Detroit's Eastern High School. When he was 15, he was assistant manager of a theater; at 21, he was general manager of a chain of seven theaters. He emceed stage shows, amateur mostly, although he did occasionally introduce beginning professionals like Amos Jacobs and Joan Leslie.

Virginia Pival: "He'd have contests going on, like tryin' to catch a pig that was greased. Oh, he used to come up with some fantastic things for the audience to participate in."

Pival and Virginia met at the Senate Theater where she took a job as cashier. She attracted him. He dropped by "to see how she was doing." They discovered they lived only three blocks apart. He asked to walk her home. At first she turned him down because she was afraid of him. But he was "so nice, spoke so beautifully" that

John Pival

they dated. Before he went into the Coast Guard they were married and bought a little house in Dearborn.

Now it was 1946 and John Pival was beginning a career in broadcasting. He was to become one of the most controversial characters in the story of WXYZ.

Fred Wolf found radio a fantastic expansion of the sports world; to him, broadcasting was a wonderland. He remembers being introduced to a bald-headed guy in a Coast Guard uniform with two wire recorders who, on his very first day on the job, was going to Florida to cover Spring training with the Tigers. He also remembers with what awe he would watch Harry Wismer, the sportscaster whose network broadcast preceded his own by 15 minutes at 6:30 P.M.

Wismer was already sports director of the ABC network in New York. He was also an executive at WJR and would often be in Detroit in that connection at which times he would be delivered to WXYZ for his ABC broadcast.

Fred Wolf: "I used to sit there and see this guy—big, black limousine with a driver, and two, three guests that he hauled out o' the stands, and storm in there at 6:30 with A'Hearn standin' there in the door—or little Alexander—with all his scores sliced in little pieces. This guy would sit there and in 15 minutes he could take four pieces of goddam yellow stuff off the teletype and do a whole story! Fake it. I thought, why can't I do my own bowling stories that way? Why do I need all this copy—twelve pages! This guy amazed me."

In May, 1946, shortly after the station was sold, the bowling season ended. Bowling did not continue through the summers then. The *Ten Pins* series was a dead duck. And for *The Sandlotters* on Saturday nights, Judge Watts said, "Gee—what can we do instead of bowling?"

"I could do the baseball scores," Fred suggested. And that's what they did.

Fred was in the news room one Saturday night, tearing scores off the wires, when Tom A'Hearn, never one to enthuse, picked a piece of wire copy off his desk and held it out to Wolf. In a casual, flat tone of voice, he said, "Why don't you do this?"

Wolf read it. It was a 4-minute feature which came over the wires every Thursday night. A'Hearn saved it; and Fred made a trip to the station every Friday morning to pick it up. He re-wrote the stories after the style of Bill Stern.

Wolf: Bill Stern was my idol. I'd lay on the carpet on Friday nights when he did these—real ham—but to me—they made my goose bumps come up. I copied his style to where I made myself believe I was Bill Stern."

Larry McCann, his announcer, would find schmaltzy violin music and play the record under Fred as he read his story.

When Wolf did his first remote broadcast—a bowling match in Chicago—he met Guy LeBow. LeBow told him of a series of sports stories he could buy from Mac Davis in New York. Fred used to buy 26 at a time for $35.00

When the Spring training season was over and John Pival returned from Florida, his first studio assignment was to produce *The Sandlotters.*

Fred Wolf: "We hit it off just like THAT. (He snapped his fingers.) So now, he wants to make a little mark so he's goin' to take my stories and he's goin' to make a production of 'em. And boy, we did! Somebody drownin'—you'd hear the bubbles. We went the whole route. I mean, we'd do football stories and big band music and crowd cheers. He PRODUCED it. He'd take my stories that I bought from Mac Davis and he'd re-write 'em, and we were constantly getting more time on *The Sandlotters*. Now we're 7 minutes; now we're up to 8 minutes; and we've finally got half the show. That's when I started gettin' paid on *The Sandlotters*. Ten dollars."

One of the last to haul on the "civvies" was long-legged Jack McCarthy, looking handsomer, a little harder, and less boyish. He had been undecided whether to try Hollywood again—or New York. A conversation with the persuasive H. Allen Campbell had convinced him Detroit was his best bet. He returned to WXYZ as studio manager and Earl Moore who had held that position was moved out to the station in Flint.

"One of the first problems when you take over," he told McCarthy before leaving, "we can't seem to get announcers in here on time in the morning. Engineers are putting the station on—records without voice—you know."

So there was a little shake up. First to go was blonde Larry Kelley. McCarthy had moved him up into announcing before he joined the Marines, now he had to let him go for being late on the job.

McCarthy also took over the programming from Dick Osgood who moved back upstairs to the maid's quarters, between Lee Smits and the newsroom.

Hal Neal was not well when he came out of the Air Corps. In fact, he was a nervous wreck. He was announcing *The Lady of Charm* from her offices in the New Center Building. Edythe Fern Melrose found him one day lying on her couch with a cover pulled over his head.

"Hal!," she said. "You can't *do* this. My sponsors come in here. This is my office and you can't do this."

She was fond of him. Beyond his good looks and his zip as an announcer, she felt he had a potential for something more. He seemed restless, dissatisfied. For his part, he admired Edythe Fern Melrose. Her keen business ability set her apart from other women in radio. His best friend at the station, Bill Morgan, had announced her show occasionally and the two had talked about her.

Bill Morgan: "Edythe was a very positive-minded up-tempo individual—a great person, really. She liked Hal. Edythe liked anybody that gave her half a chance. Some of the guys were down on her and laughed at her and all that, but I never could buy that at all."

She was a hell of a business woman.

"Oh boy! She sure was! but she was more than that. I could tell you a lot about Edythe. She was a great person!"

Edythe noticed that Neal usually took a pill or two before the broadcast. One day, the thought of a talented young man in his late 20's leaning on pills haunted her throughout the broadcast. When it was over, and they relaxed in her studio-office over coffee, the ends of her mouth disappeared in her cheeks as she beamed at him, but there was an analytical concern in her eyes.

"That was a good one today," she commented. "Hal, you are a fine announcer.

Why do you take pills before the broadcast?"

"They're no big thing," he said. "They just calm me down a little, that's all. It's nothing important. I buy 'em in the drug store."

"You don't need pills. Hal, why don't you tell me what's bothering you?"

"Me, I guess."

"Of course. You. But what about you? You are talented, good-looking—you are—a very good announcer or I wouldn't have you on my show."

"I'd like to be more than just a good announcer, Edythe. I guess that's what's eating me. How did *you* get started on this track *you're on?"*

The pleasant chime of her laugh brightened the office. She opened a drawer of her desk, took out a book, and placed it in front of her. She still smiled, but she was serious.

"I read this book," she said. He couldn't see the title from where he sat. She studied him a moment. Then she said, "Hal, you've got to get into an executive capacity—and out of a salary—if you're going to think and grow rich."

"Think? And grow rich?"

She patted the book, *Think and Grow Rich* by Napoleon Hill. "I'm going to let you borrow it."

"I'll buy a copy."

And he did. No one knowing Hal Neal and Bill Morgan casually would have guessed the characters that existed deeper than appearance and speech. Morgan might be taken as small town with a gleeful humor, an infectious laugh, and beyond that perhaps even a little stodgy. Handsome Hal Neal was smooth, sober, and not overly likable—except to the opposite sex. The clue to the pressures deep in each was in their actions—Morgan's restless hopping from station to station in Texas and back; Neal a nervous wreck, popping drug store pills to help him hold together.

Think and Grow Rich by Napoleon Hill changed the direction of their lives as it had done with Edythe Fern Melrose. To understand it, one has to understand the monotony of being a booth announcer.

"If you go back to that kind of radio," Bill Morgan said, "a radio announcer worked an eight-hour shift and for one minute or 30 seconds out of every half hour, he did something. You'd do a station break and a spot—and go back to the network. And this went on ad infinitum. You could play poker. I mean, there were all kinds of devices for just killing time.

"Early in the game, when I was at WOEI, they had a nice grand piano and I taught myself to play piano in my spare half hours. Also, I was writing—and selling. I did some of my writing in Studio E, but it was hard. I could see that I was either going to have to quit everything and write, or else stop writing and get all my eggs in the basket, but I didn't want that one basket to be announcing. And neither did Hal. We were both the kind that we just didn't want to totally waste our time."

Reading Napoleon Hill's book released the steam in Hal. He said to Bill Morgan, "This is the greatest book I've ever seen and you've got to read it."

Morgan bought a copy and it produced the same result with him:

"It absolutely changed me all around. It gave me new ideas, bigger ideas, goals.

Lone Ranger rehearsal. Left to right: Harry Goldstein, Bill Sanders, John Todd, Bob Dale, Brace Beemer, Jay Michael, Ernie Winstanley, Jim Fletcher and announcer, Fred Foy. In the booth: Ray Skidmore (engineer), Fred Flowerday (producer) and Bill Hengstebeck (recorded musical interludes and recorded sound effects).

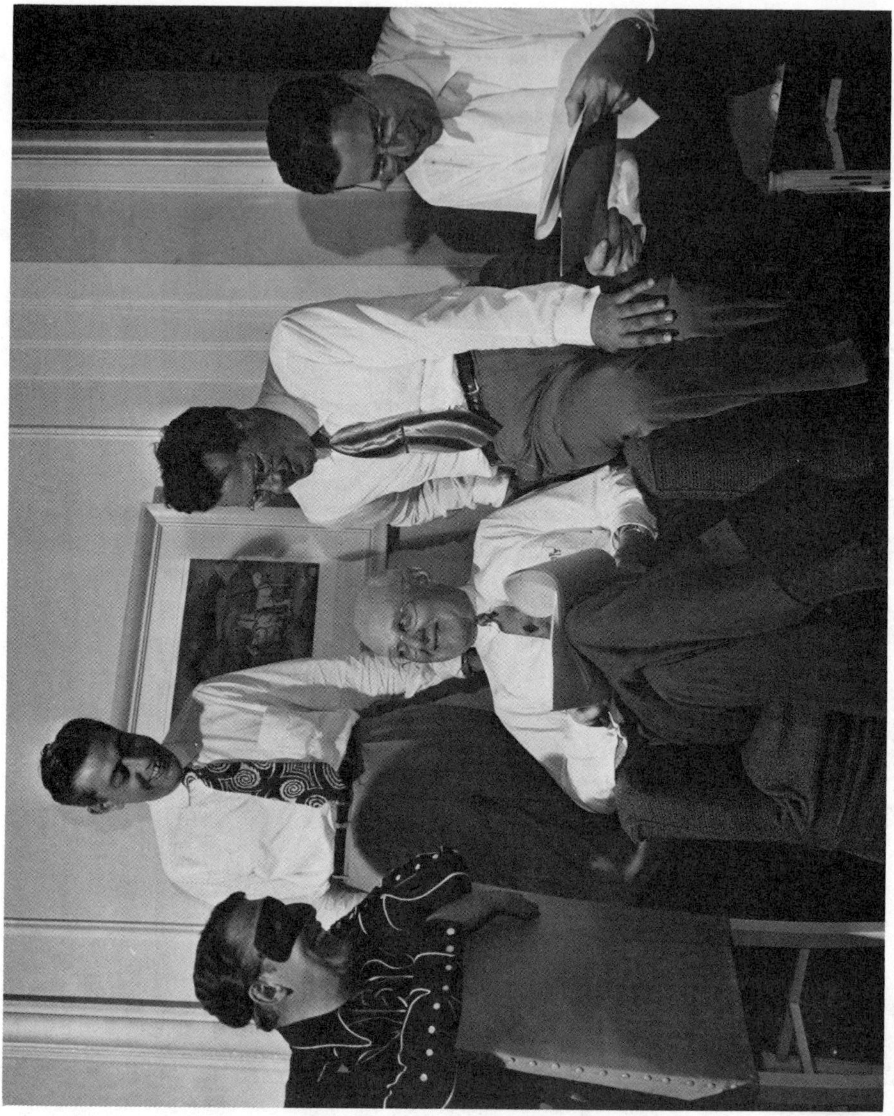

Story conference, left to right: Brace Beemar, Fran Striker, George W. Trendle, H. Allen Campbell (sales manager, later part owner of the Lone Ranger, Sergeant Preston, and his favorite, the Green Hornet) and Charles D. Livingstone who replaced James Jewell as director.

It straightened my thinking out. Okay. This really gave Hal and me burning ambition. I mean, we just wanted to get going on something."

Neal echoed these sentiments: "Bill was a pretty good writer and a fair announcer. I was a lousy writer and a better announcer. So anyway, we decided we were going to get in the business of making spots—recording spot announcements—and do 'em at a cut rate."

Morgan: "We continued on staff as announcers, and did this in our spare time—evenings—days off—and so on. We started in the basement of Hal Neal's parents in Redford." (Morgan broke out in his merry, infectious laugh as he recalled it.) "We started doin' air checks for all the kids on *Radio Schoolhouse* and for guys in the band that were going to do a solo on a show—and all this stuff."

His face squeezed up with glee as he laughed. Neal, on the other hand, was serious and not too sentimental as he remembered:

"I would announce Radio Schoolhouse and Bill Morgan would go out to Redford on Sunday mornings and cut air checks and we'd sell 'em to the kids and that's how we paid for the equipment."

Thus in 1946, James G. Riddell, one-time mail boy became President of WXYZ, Inc.: two wire recorders started a Coast Guard veteran on a 14-year push to top spot at WXYZ; and a book started two booth announcers on a long climb to heights not dreamed of before—one to be manager and part-owner of a prosperous, award-winning station; the other, president of a network.

31
A Rustle in the Leaves

Had Trendle talked with that tiny scarecrow of a fortune teller? Was Madame Weston still alive, tucked off in some tiny office where he could consult her? And had she told him of things to come? Was that why he sold WXYZ without hesitation? When Campbell commented, "you could see what the hell cooked," was he referring to bigger events than the sale of one radio station and a program or two? One thinks not.

But across the world of broadcasting a breeze was stirring. Those in the highest places with sensitive nostrils could smell great change in the air. But only in retrospect would the average veteran of the industry point to the late '40s as the time of the first positive stirring of the wind. At the time, the broadcaster was too involved with his own ego to note the stirring of the leaves. Signs of the tempest to come were unobserved in their total significance, even by the bright Riddell when they appeared in his very office.

Example: Himself! He was a symbol of the change. Trendle was a showman, and broadcasting had been a showman's business. Riddell was a salesman. He thought as a salesman. He "meddled" less in programming; he left that to the creative people who worked for him. Even later, when he would pioneer in a whole new concept in programming, the change he wrought would be inspired by salesmanship rather than showmanship.

Example: Ross Mulholland. The high strung announcer who loved music had deserted WXYZ to play records on WJR. His program was called *Music Hall* and his catch phrase was "How you gett'n' along with your music?" Consumed by ambition, he went to the independent producer-packager Mort Neff and said, "Gotta have more, Mort. More." Neff signed a contract to be his manager.

Example: Eddie Chase. His *Make Believe Ballroom*, a record show with spot announcements, was on WXYZ every afternoon for 45 minutes after Lady of Charm and Charm Kitchen. Chase was more salesman than performer.

Example: Ed McKenzie.

Ed McKenzie was one of the most fabulous characters in the story of Detroit radio. A blacksmith's son from Flushing, Michigan, he started in Detroit radio as an engineer on Station WJBK. He was assigned to remotes and "rode gain" on all the big bands of the later 30s.

Then came World War II and WJBK's chief engineer left to work on radar on ships; and McKenzie, who was exempt from service because of a double hernia, became chief engineer. After the war he started doing a little announcing. WJBK was then a small 250-watt station, had no union contract, and an engineer could go on the air from his position at the control panel. Ed introduced the *Polish Hour*, the *German Hour*, the *Italian Hour*, and when there was no foreign language program, the station would fill time with transcriptions, usually without announcements.

Ed's first contact with the outside world, after the big bands, was a station representative from New York named Joe Bloom. Bloom went around the country visiting advertising agencies and placing spot advertising on stations he represented, for which he received a commission. WJBK was one of his stations.

Through Bloom, McKenzie first learned of Martin Block in New York, and a record player down in Washington, D.C. named Arthur Godfrey. They were making a lot of money. McKenzie was making $65 a week.

"Gee," Ed said. "I'd sure like to hear those guys—what they sound like."

Bloom cut a 16-inch platter of the two and sent it to McKenzie. Ed listened.

"Look," he proposed to WJBK manager Fred Hopkins. "We got an hour after the Polish program in the afternoon from one to two. We've got another hour-and-a-half from 4: to 5:30. I'd like to play records in there and see what would happen."

"Give it a try," said Hopkins. "Pick some theme."

"Well—there's a fast little piano and drum thing that Lionel Hampton and Nat King Cole recorded. Sprightly, I guess you'd call it. I like it. It's called *'Jack, the Bellboy'*."

"Okay," said Hopkins. "We'll put your name on the schedules for the papers."

"Gee—wait," said McKenzie. "Don't use my name; I might be a big bomb. Why don't you just call *me* Jack, the Bellboy?"

McKenzie, a big young fellow with blonde hair, was naive. He didn't think anything of it when more and more commercial spots came along for him to read and record pluggers began calling on him.

One day a promoter brought a new platter to him—*Peg O' My Heart* played by

The Harmonicats. Ed loved it.

"Would you believe it, Ed?" said the plugger. "Ross Mulholland over at WJR won't play it. Do me a favor. See if you can get him to play this."

McKenzie called WJR.

"Dah," said Mulholland, a great egotist. "The other side is better."

They kidded around a little; then Ross said, "Hey, you know you're really getting the numbers."

"Numbers? What do you mean?"

"You're leading the whole parade in the afternoon in the Hoopers."

"Hoopers. What are they?"

"Are you kidding?"

"I never heard of 'em," McKenzie confessed. "I'm just a dumb engineer. Remember?"

"Ask 'em to show you the Hoopers. They're selling your time over there for triple the rate they get for anything else. I hope they're paying ya."

It was true; and they weren't. And with his new knowledge, McKenzie began to burn. He went to the Terrace Room at the Statler and cornered Phil Brestoff. Brestoff, besides being musical director of WXYZ, played luncheon music at the hotel. Brestoff knew who Jack, the Bellboy was; Ed was the talk of Detroit—an engineer on a 250-watt station capturing top rating in six months! McKenzie confided that he was just learning of his commercial pull and was not getting paid for it.

"Would you like to come to WXYZ?" Brestoff asked, his brown eyes calculating.

"Well gee. I'd sure like to talk about it."

When Brestoff was finished at the Statler, he saw Riddell.

"Do you listen to Hugh Holder on our station in the morning?" Brestoff asked.

"Sure. Why?"

"Is he commercial?"

"Nothing sensational."

"Any of these record fellers make any money?" Brestoff prodded.

"For themselves? Or for the stations?"

"The stations." Brestoff's watchful eyes waited for Riddell to ripen.

"Mulholland's doing all right for WJR," said Riddell. "But the hottest guy in town right now is Jack, the Bellboy, over at WJBK."

"I can get him for you. Want to talk to him?"

"Why? Is he unhappy?"

"They're paying him only $65 a week."

"Well—JBK is only 250 watts," Riddell pointed out. "But he's making a lot of money for the station. Sure, I'll talk to him."

Riddell made an offer. McKenzie went back to WJBK and told Fred Hopkins he was leaving.

"What's their offer?" Hopkins asked.

"$15,000 plus $7.50 for every spot they sell on my show."

"We'll meet it—or better it," Hopkins promised.

McKenzie stayed with WJBK. Within 12 months he had moved up from $65 a week to $65,000 a year! *The Saturday Evening Post* featured a story about him.

Ed McKenzie was incredulous: "I just couldn't believe what was happening. Some little voice inside said this—can't—last. I wasn't an entertainer; I knew nothing about show business; I really didn't know music. I wasn't interested; I really wasn't interested in music."

But didn't he choose his own records?

"Yeah, I picked my own music and I just seemed to have some kind of—I suppose—God-given faculty for pickin' the kind o' stuff that people wanted to hear. I think probably the thing that did more for me than anything else was—I was honest. If I thought something was lousy, I wouldn't say it was good. I used to break records— throw 'em in the waste basket. You know. And if I thought Guy Lombardo played out of tune, I'd say Guy Lombardo and his Tone-Deaf Fourteen, or whatever. Vaughn Monroe with muscles in his tonsils. Jan Garbage and his orchestra. I'd make really nasty remarks."

But people found them amusing; they didn't think they were nasty.

"Oh, yes they did! I beg your pardon. I used to get letters by the hundreds that called me a dirty, rotten S.O.B."

Meanwhile Ross Mulholland kept pushing his new manager, Mort Neff, for a better deal. Instead of promoting Ross where he was working, Neff went to Riddell at Wyxie. Riddell was interested; Neff and Riddell wrote up a contract; and Ross Mulholland moved to WXYZ to do a record show. No salary to speak of, but a very high commission deal. He received a good fee for every commercial sold on his program—a deal such as McKenzie might have had.

Yes, there was a rustle in the leaves.

But things seemed much the same to the unobserving at the Mendelssohn mansion, even though the station had new owners. Trendle continued to drop in every day. Actors showed up every afternoon and performed in the evening. *The Lone Ranger* was on Mondays, Wednesdays and Fridays; *The Green Hornet* played Tuesdays; *Sergeant Preston of the Yukon* had Thursdays.

There was another program for about a year called *Bob Barclay*. Jack McCarthy played the lead. The hero was a spy, posing as an international correspondent. This led to trouble.

"Over in Czechoslovakia," Trendle remembered, "they arrested a man for being a spy and he was an American newspaper correspondent. The government sent a man out to see me and said that foreign countries were claiming that all correspondents for American newspapers were undercover agents because of the program we had on the air. So they asked me if I would take the story off, and I said no. By the time we had had some 40 weeks of it, it was going very big—played opposite the swinging door show—*Inner Sanctum.*

"Well, about a month later a little fellow who was head of the foreign correspondents' association pleaded with me to take the show off. So at the end of 52 weeks we took the show off."

When *Bob Barclay* went off the air, Jack McCarthy took over *The Green Hornet* and curly-haired Bob Hall went to New York. He eventually did *Music 'Til Dawn* for American Airlines until he died, much too young.

After the war, Fran Striker bought a jeep and used it in business as a tax deduction. The income tax people could not believe his claim that he was supporting twelve people. They went to his house to investigate, found it was true, felt sympathetic, and wound up finding deductions he had not thought of.

Jan Striker still felt the pinch: "Even from his peak salary, he couldn't save very much. That house on University Place was a dog to heat. The utilities were awful in that house, but we had to have the room. And then—the farm."

One of Fran's writers, Bob Green, not only swiped a story for *The Lone Ranger* from a magazine but committed the blunder of matching his ending word for word with that of the original author. Trendle fired him. Rumor also reached the station that Trendle had fired Earl Moore because of "a fiasco in Flint." And Harry Golder lost his job because of his diet: whiskey and beer for breakfast, lunch and dinner. Golder followed his friend Malcolm McCoy to California and found a good job as newscaster on a station in the Los Angeles area. McCoy was now out of show business altogether; he was working in the office of a construction company.

When Golder left, an announcer was needed for *The Lone Ranger*. Chuck Livingstone offered the job to Bob Hite—for AFRA minimum pay. Hite was doing commercials in New York on Edward Murrow's broadcasts from London. He had bought a house and married again. He stayed in Manhattan.

But after James G. Riddell became manager, John Slagle, whose first marriage had disintegrated in New York, returned to WXYZ—and soon indulged in another unhappy marriage.

A remarkable loyalty and pride in one another's success existed through the years among people who worked together at WXYZ. Old friends were happy to learn through Tom Dougall, for instance, that Joan Vitez had helped raise $85,000 for the production of a musical called *Brigadoon*. It was a solid hit and Joan had a piece of the action, financially. Her old friends back in Detroit were proud of her prosperity.

It was also considered a marvelous coincidence that four radio shows were being produced simultaneously on four different networks by four ex-Detroiters: Fred Flowerday was producing *Sergeant Preston of the Yukon* in Detroit; James Jewell was producing *The Silver Eagle* from Chicago; Ted Robertson was producing *Straight Arrow* in Los Angeles; and Ernie Ricca was producing the fourth show in New York. All were on the air at the same hour!

In 1947, *The Lone Ranger* established a record for mail pull on a giveaway—more than 3 million letters and postcards.

There was another law suit that year. An Earl Currey claimed he was the original Lone Ranger, declared the radio series was patterned after an act he did in rodeo shows, and sued for $50,000. He charged he had been harassed and embarrassed while appearing in Muskegon, Michigan, when he was ordered to change his act or be "run out of town." At the end of the year, the *Detroit Free Press* reported that Currey had suffered a "temporary setback" in his suit.

Seven months later, Currey was in court again—in Scranton, Pennsylvania—but as a defendant. The judge observed that *The Lone Ranger* was now (1948) being carried over 246 radio stations; that the Ranger cartoon was now being published in 104 Sunday newspapers throughout the world, and 143 daily papers in the United States and foreign countries; that Lone Ranger books had been published regularly for ten years; and that various manufacturers, distributors and miscellaneous venders had been selling articles of merchandise identified as having some connection with *The Lone Ranger*; and that more than 4 million youngsters were enrolled in The Lone Ranger Safety Club for which six awards had been presented to the program. He noted that the trade mark was registered with the State of Michigan and that the trade mark and articles of incorporation were registered with the U.S. Bureau of Patents, etc., etc., etc. Muerer had done his work well. The court enjoined Earl Currey from further use of the name The Lone Ranger or use of the costume.

Currey did not give up. He brought a final suit in Detroit for $50,000. Judge Theodore Levin threw it out of court. That was the end of that.

As soon as the Tiger training season ended, besides producing *The Sandlotters,* John Pival went all over Detroit with Fred Flowerday recording sounds for the dramas. He also introduced what have come to be called "actualities"—inserts in newscasts of statements and comments by persons involved in news stories. It was a year of state elections, and Pival wire-recorded winning speeches by all important candidates *before* the votes were counted so that as soon as it became obvious who the winners would be, WXYZ had their acceptance speeches on the air.

When Fred Wolf's bowling show returned in the Fall of 1947, he and Pival drove around town picking up material on a wire recorder. They started a new show called *Tenth Frame* which was *all* recorded. Live broadcasting was becoming less important. (Another rustle in the leaves.)

Pival also created a show called *Down Our Street.* He picked up interviews and minor events of civic interest—an interview with a boat captain, a talk with a truck driver, a policeman, a visit to a fire station, etc. Dan Beattie wrote scripts around these things, devising a character called Old Timer to connect components of the program. Pival produced and directed the show.

Dick Osgood was connected: "I can't remember much about it except that I played the Old Timer and talked with a little girl. It was broadcast on Saturdays which meant I was on the air seven days a week; but I enjoyed performing the old man."

Hal Neal and Bill Morgan, still following the philosophy of Napoleon Hill, graduated from the Neal basement in Redford to a small recording studio on 2nd Avenue with the help of WXYZ engineer Ray Aiken. Business was picking up.

Charles Livingstone was teaching radio classes one day a week at Wayne University.

"I've found a perfect Mustang Mag," he announced to the writers one day.

Mustang Mag was a female hoyden type of character written into *The Lone Ranger* from time to time. Ruth Dean Rickaby had played her originally, followed by others through the years. But for a long time, she had not appeared in the series. Now, Livingstone suggested to Fran Striker that he revive the character.

"Wait'll you hear her, Dan," Livingstone enthused. "She's great!"

"Any experience?" asked Beattie.

"She'll be great," said Chuck.

She was a big woman—one of his students at Wayne. Mustang Mag all right, thought Beattie when he saw her. Chuck spent time with her, coaxing her into character; and she did sound good as they rehearsed. Livingstone beamed. He turned to Beattie.

"How's she doing?"

"She's doin' all right, Chuck."

"What do you mean—all right?" Chuck laughed, his mouth screwed up into a smug expression.

"She's doin' fine, Chuck! *You* know."

The broadcast, however, as Dan Beattie remembered, was a disaster: "Comes the show! Goes to pieces! Script's all mixed up. She hadn't put her pages back in order. Rollie Parker ran up to the mike with his script—and *he* played Mustang Mag!"

Yes, life went on much the same. The breeze seemed soft, the rustle in the leaves seemed gentle, and radio was still fun.

32
There Was a Monster in the Basement

In 1948, television was definitely coming alive. Pictures and sound were being transmitted from New York to Boston by microwave relays; a coaxial cable connected New York and Washington; another tentacle was reaching westward. The American Broadcasting Company managed to get Channel 7 in every city where it owned a station and Detroit's Channel 5 which it had purchased from Trendle was discarded. At WXYZ there was a monster in the basement.

Genial Jack Tierney had his office and repair shop under the master control room alongside the mailboy's headquarters. Scripts and transcriptions were stored in the basement; and somewhere Rollie Parker managed to keep a bottle hidden. Now, at the end of the underground hallway, there was something new.

It was a television camera. Whoever stood in front of it could be seen on a small screen in the back of it, and Osgood predicted that Larry McCann with his dark good looks would be a natural in the new medium. The camera was a depressing reminder that many were saying radio was passing its zenith.

Above the basement, however, there was no sign of decline. Live programming kept everyone as busy as ever, and no one was prospering more than Edythe Fern Melrose. She was investing $87,000 of her own money in the building of a six-level real-life House of Charm on land projecting into Lake St. Clair.

"Jimmy," she told Riddell. "I want to buy a thousand dollars worth of time."

"Why?"

"I want to mention the people who are participating and whose products I am using, because they are so good, and that's the only reason I'm using them."

Riddell wondered if "participation" meant special prices or maybe a donation here and there.

"I have combed all of the makers of brick," Edythe explained, her manner feminine yet forceful. "I am buying the finest brick. I'm putting in the electrical units that I think are the best, and I want to mention them."

"You don't have to buy any time," Riddell told her. "What you're giving us goes far beyond a thousand dollars because the way you refer to it, it sounds like the station is building the house."

Edythe moved from the New Center Building into her dream house with a staff of seven people, and did her broadcasting from there—live, until tape replaced wire for recording. Tape could be edited; wire could not.

Big, hearty Norman Lenhardt had to be removed from the Lady of Charm broadcasts. Letters addressed to the Lady which he was required to read on the air struck him as being hilariously silly, and he couldn't help breaking out in robust laughter. The Lady didn't find them that funny and asked for a new announcer.

Larry McCann inherited the post: "I was the last remaining Indian and I was assigned to the Lady of Charm. I was paid a dollar a week to drive back and forth from the studio to her house. That covered my gas. That was my talent fee for a one-hour radio program."

Show World celebrated its 10th anniversary in 1948. Osgood had kept a record of all the celebrity interviews and put two girls to work in their spare moments, counting the number. They got mixed up in their count after they had passed a thousand and asked if they should begin all over again.

"Forget it," said Osgood. "Nobody would believe it. Over a thousand celebrities? I wouldn't believe it myself."

The national trade weekly *Billboard* front-paged the anniversary because *Show World* was still unique. Detroit movie exhibitors were now quoting excerpts from Osgood's reviews in their advertising—years before broadcasters would be quoted in New York or Hollywood.

Announcer Ed Michael (Lessic) had switched to producing. He and John Pival took turns producing *Show World* and between them had arranged a surprise for the 10th anniversary broadcast. They recorded congratulatory messages by station executives, theater men, and all the touring celebrities who were in Detroit at the time. Early in the broadcast, announcer Hal Neal departed from Osgood's prepared script, interrupted Osgood's attempt to continue, and presented the surprise recording. Osgood, a sentimentalist, sat there with tears running down his cheeks—not only because of the anniversary.

Ed Michael was not there to hear the show he had helped prepare. That day, his wife, a beautiful English girl with a face like a cameo, had plugged in a sun lamp while wet from her shower, and was electrocuted. Ed left WXYZ, went to New York, and announced over ABC until he, too, died young.

In 1948, Hi-Speed's show, *Eyes on Tomorrow*, set an all-time record for the sale of Lee Tires over a 3-month period. Osgood was also broadcasting a Saturday night remote series of give-away shows called *Crack the Safe* in Kaiser-Frazer show rooms. The Sunday program *Radio Schoolhouse* had completely lost its *esprit de corps* and was steadily deteriorating. Osgood's ego was large as any, but it was non-competitive. He was convinced Rivard wanted his job and brooded over it, not himself for hours after a broadcast.

John Pival had moved into the office across from the Hammond Organ studio on the second floor. One evening as Osgood passed his door on his way up to the Hi-Speed rehearsal, Pival called him into his office, opened the deep drawer of his desk, pulled out a bottle, poured whiskey into a glass and held it out.

"You look like death warmed over."

"We still have another show to do."

"Drink it."

After some hesitation, Osgood obeyed. He had already been working twelve hours. Pival had a swig and smacked his lips.

"I came down the hall to see you this afternoon and your door was closed," he said.

"I was taking a nap on the top of my desk."

"You also had lunch on the top o' your desk— a can o' salmon and potato chips."

He poured another finger of whiskey into the glass and made Osgood drink it. Then they went upstairs and did the Hi-Speed show.

In January, 1948, for the first time, WXYZ-FM went on the air with separate programming. Jack McCarthy set up the operation in the old Mendelssohn library. He programmed classic music.

"Originally the intention was to replace AM with FM," Charles Kocher remembered. "Nobody intended an entirely different program service. During the war it could be seen that by the time the war was over, the first big development in communication would be FM transmitters to replace AM transmitters. That would simply take over as a superior means of transmitting. They would just take over the same radio programs; there was no thought of separate programming."

Ten WXYZ announcers, doubling between AM and FM, went to their union to fight for additional pay. The Detroit Local of AFRA had a new and forceful Executive Secretary, Boaz (Buzz) Siegal, a physically powerful, gentle-mannered, sharp-minded young labor lawyer who taught at Wayne University.

He met with James Riddell and Raymond Meurer. Riddell explained no money would be made with FM; it was not commercial; nothing would be sold on it.

"That's your problem, Jim—not ours." A friendly smile lifted Siegal's broad cheeks to narrow his laughing brown eyes. "The announcers work overtime and want to be paid."

"Well, of course FM is a separate operation and FM has no contract with

Radio Schoolhouse

AFRA," said Meurer, slick as oil. "We could probably solve this whole thing by hiring non-union announcers."

"No, Ray," said Siegal. "It's still WXYZ and we have a contract."

"How many hours are the announcers supposed to work in a week?" Riddell asked, restoring progress.

"Forty," said Siegal.

"That's easy," said Jimmy. "Jack McCarthy can work it out so that their schedules stay within forty hours. You don't want to dictate where we use 'em for the forty hours, do you?"

"Their complaint is they're working extra hours without pay."

"Well, Jack can work it out."

The announcers threatened to quit AFRA and join the International Brotherhod of Electrical Workers to which the engineers belonged. They said they might even join the CIO. But the threat faded away as McCarthy worked out scheduling that was intricate but satisfying. No new announcers were hired, but a new engineer was required. He was a little man named Paul Dudek who liked movies and wore French berets.

The 1948 FM experiment lasted less than a year, then FM went back to simulcasting, broadcasting the same programs that were on AM. Paul Dudek stayed on.

As preparations were made for television, Riddell or McCarthy talked with Charles Livingstone, Fred Flowerday and Dan Beattie about producing for WXYZ in the new medium. Both Livingstone and Flowerday felt they had too good a thing going with the Trendle dramas to leave them for $75 a week they were offered. Beattie would have switched, but when he tried to make a deal with Trendle to continue writing one drama a week (instead of two) to add to what he could make on TV, Trendle turned a cold, deaf ear. Beattie had just signed a new contract with him for two scripts a week and he would not have his people working for anybody but himself; so Beattie remained with Trendle.

Riddell also gave some thought to Fred Rivard who produced the kid show, but his mind finally zero'd in on John Pival. Both Beattie and Livingstone had recommended him, and it was true, he was working out extremely well in radio. Pival was ready, eager and willing to learn. So it was announced: John Pival would be production head of Channel 7. Jack McCarthy now had to hire two new radio producers.

When radio moved to the Mendelssohn mansion, the Alcove was replaced by the Van Dyke Club as the hang-out. The Van Dyke had been a plush and exclusive gambling club, but by the time WXYZ invaded it, the "club" had become a public restaurant with a bar, owned by John Phillips. It had more class than the Alcove—carpeting, dim amber lighting, table cloths, and dark wood paneling. The bar was long and beatifully carved, and an impressive wooden chandelier hung over it.

One night, after the last Lone Ranger broadcast, several Wyxy-ites were at a table in the bar. Announcer Fred Foy was sitting beside Brace Beemer who had been chasing martinis with beer. Alcohol was now chasing blood through his cheeks. He turned to Foy.

"You think you're going to get my job?"

"What?"

"I heard about your talk with Trendle."

"Brace! Let me put you straight on that. Number One: he called me down there. I didn't know what the hell it was about and he said, 'You ride?'

"And I said, 'No.' Well, he said he'd been having some kind of problem with you and he told me to go out and take some riding lessons. So I did. Hell, I just work here."

"You couldn't ride a pony cart on Belle Isle," said Beemer, laughing derisively, grabbing at his next martini.

"Right!" Foy laughed. "I hate horses. But I went out. I think I rode two days and I said oh forget it! This is ridiculous! And that was it. So then, nothing more was said about it. Apparently you and he patched it up."

"Well I heard—"

"Forget it!" said Foy. "I'm not a rider. Forget it! I wouldn't want that job, believe me. Forget it! Look!"

Beemer and Foy were big men. They sat looking each other in the eye, blue to blue. One side of Beemer's forehead was wrinkled and his smile was a little one-sided. Foy's eyes were clear, his face unflushed, and his open smile was honest.

"Let me buy you a drink," said Beemer.

"Right," said Foy. They became good friends.

Beemer was now making many personal appearances and sometimes he had to pack his right arm in ice to recover from hand-shaking. When he went out on a p.a. assignment, his scenes would be recorded in advance and integrated into the broadcast. Foy was signed as understudy and would read his part in rehearsals.

Foy remembered the working arrangement: "I did the show—*once*. I played the Lone Ranger. Brace developed laryngitis and at the last minute he couldn't talk. So they had Jay Michael do the announcing and I played the Ranger. One time! God! Why didn't I get a copy of that one show for my own!"

In 1948, the Stroh Building changed its name to the Mutual Building. In it were the offices of the United-Detroit Theaters, Trendle-Campbell Enterprises, and the executive offices of WXYZ.

Campbell and Ray Muerer had luncheon together. It was obvious Meurer wanted to be part of the company.

"When you talk about interest what are you talkin'?" Campbell wanted to know.

"O-oh," said Meurer. "I'm not—choosy about what the interest moight be but I think I should be part of the company."

"I got Trendle a pretty good deal out of King in getting the stock back, so I don't believe that he would be too opposed to workin' out some kind of a partial arrangement with you. Supposin' I could work out a deal for you on a 5% basis with $300 a week salary. Would you be happy?"

"I think so, Allen."

Campbell got Meurer 10% instead of 5.

Campbell's success was a mixed blessing: "Trendle objected to that to beat hell. And later I was sorry about it, too, because Meurer was tough to work with—real tough. He fought us on a lot o' things."

Furthermore, he was still not satisfied. He was in on Trendle-Campbell-Meurer Enterprises (the dramas) for 10%, but he still had no share in the Trendle-Campbell Broadcasting Corporation which owned WTAC in Flint. Now he received an offer from Jules Stein, then President of the Music Corporation of America. Using this for a wedge, he went to Trendle while Campbell was vacationing in Florida. Trendle relayed Meurer's new suggestion to Campbell in a letter. Campbell blew up. He phoned Trendle.

"I want you to know that I am terrifically miffed at Meurer talkin' to you about this while I'm in Florida," he said. "I am not goin' to write you a lettuh, and I'm not goin' to talk about it—until you get down here."

Their vacations overlapped. Campbell greeted Trendle in Miami with his usual golf-course tan and strong white smile. They talked of other things while he waited for Trendle to bring up the subject of Meurer's latest suggestion. Trendle enjoyed playing one person against another, knew well what was really on Campbell's mind, but waited for him to open up the subject. Campbell gave in at last.

"All right. What is this thing with Ray Meurer?"

"Oh," Trendle said, casually, "he wants to get into the Trendle-Campbell Broadcasting Corporation."

"Your lettuh said 10% of *my* stock. WTAC is not makin' a dime. I'm not makin' any out of it; *you're* not makin' any out of it."

"I know."

"He's askin' for 10% of nothin'! And you tell him that if he gets 10% I oughta give him that 10% outa *my* stock."

"Well, I thought as long as I gave him 10% of *my* stock in Enterprises—"

Campbell leaned over the table. His voice was even and relaxed, but one finger tapped on the table with certain words.

"Here's a bird that I interceded for with you," he said, "and I've made him a hell of a lot o' dough. He's going to make a hell of a lot o' dough with us. And you're goin' to be the guy that contributes to it. But now (tap) you turn around and tell me that (tap) I oughta give him 10% of (tap) mine that I've worked for all these years. He can go screw himself. I'll (tap) never give it to him."

Campbell was still angry when he got back to Detroit and called Meurer to his office. "I can't i*mag*ine your havin' the guts—!"

Raymond Meurer declared nothing of the sort ever happened.

Radio was going to broadcast the Gold Cup boat races in the Detroit River. John Pival would produce the broadcast.

"You can use Johnny Slagle," Riddell insisted. "He knows boats."

"Don Wattrick," said Pival.

"Oh hell yes—you have to use Wattrick; he's Sports Director. How many men do you need?"

"We ought to have two out on the Belle Isle side," said Pival. "I figure Wattrick for the Yacht Club and Slagle at the Boat Club."

"You ought to have somebody on the Belle Isle Bridge."

"El Prough is sports-minded."

Riddell nodded approval.

"And I want a man on the Detroit Towers," said Pival. This was an 18-story apartment house on the Detroit side of the river.

"Who've you got in mind for that?"

"Fred Wolf," said Pival.

"Wolf! He's a bowling announcer. What's he know about boats?"

"He'll be all right."

Wolf was nervous.

"Don't worry, Fred," said jolly John Slagle. "If you get stuck you can throw it back to me."

On the day of the race, Fred got to the roof of the Detroit Towers.

But he was not without problems: "Man, I am ready! I am studying and studying. I got stacks o' stuff. I got it on cardboards 'cause I know it's gotta be windy up there. I'm 18 stories up. All the positions are identified. You know—'Here's Fred Wolf. Let's go to Fred Wolf.' And I've got the opening copy all typed out a la Osgood, and I'm readin' that thing right on cue. 'Back to Wattrick.'

"*Bang*! The race starts and now I'm—you know, it's runnin' down my pants legs. When is he gonna give it to me and what in hell am I gonna say? I got a pair of binoculars an' you sure need 'em. I'm so high I'm gettin' air sick."

The boat everyone was interested in that year was Miss Golden Gate because its two young drivers had caught the fancy of all race enthusiasts. Directly below Wolf's position, the Golden Gate broke down. In his ear phones, Wolf heard Wattrick broadcast, "Okay, Fred Wolf. What's going on with the Golden Gate?"

Wolf couldn't think of anything more dramatic to say than, "Well—the boat's down there—not goin'. Uh—take it away John Slagle."

But Fred redeemed himself later that year. The 1948 national annual competition of the American Bowling Congress was held in the Coliseum on the State Fair Grounds in Detroit. The E. & B. Brewing Company, the *Detroit Free Press* and Station WXYZ picked up the tab for the most complete radio coverage ever given such a tournament. Wolf succeeded in getting the Congress to build a radio booth in the bowling area. This had never been done before. In the afternoon, Wolf, and again Pival with his wire recorder, went to the Coliseum and recorded the minor events—doubles and singles. The team events at 7:00 and 9:30 P.M. were covered live. What was wire-recorded in the afternoon was intergrated in the night shows. This was the only ABC tournament that was broadcast for 80 consecutive nights with ball-by-ball coverage.

A couple of times the wire on the recorder broke, and Wolf had to sit ad libbing until it could be repaired.

Emily Wolf: "It can't be as much fun now."

Tall Jack McCarthy appeared at the door of Osgood's office with a short man with black hair, black eyes and a screwed up face.

"Dick, meet Boyd Lawlor—one of our new producers."

Lawlor had been fired from the east. He had worked at WCAU in Philadelphia.

The same day, McCarthy introduced the other new producer—short, sturdy, curly-haired and pleasant. He had grown up in Sault Saint Marie, Michigan; had played football; had just returned from working with a former Detroiter, Axel Gruenberg, in west coast radio. His name was John F. Rellis.

Later, McCarthy returned to Osgood's office alone. He had not assigned the new men to any programs.

"What do you think?" he asked.

"I'll take the second one," said Osgood, meaning Rellis. "The first guy seems to be talking down his nose at us."

McCarthy laughed. "Lawlor *does* carry a kind of truculent atmosphere."

"He's too important for me. Give me Rellis."

So Rellis was assigned to *Show World* and the Hi-Speed show. Osgood and Rellis went to the Van Dyke Club for lunch to get acquainted. Rellis was so impressed he let Osgood sign the check, thinking the "club" was private. Rellis immediately became "one of the gang."

Rellis was impressed: "I was thrilled. That mansion was such a beautiful place. There was never such an atmosphere to work in—that grand staircase, the elegance, the privacy. And we had such beautiful people! And of course—the theater carpet in the master bedroom."

"New York calling."

It was Ed Noble, President of ABC.

"Jim, I had a call from Bing Crosby. Hollywood. You got an announcer named Callum?"

"Yes. Don Callum," said Riddell.

"Well, it's the damnedest thing," said Noble. "Bing said, 'Mr. Noble, I have your telegram here before me recommending, I believe it's a Detroit announcer, Don Callum of WXYZ, as the best man to be my announcer on my new ABC show.' And he says, 'Ken Carpenter has always been my announcer and I don't understand.' And I said, 'I don't understand, either, Bing.' Can you figure it out, Jimmy?"

"No."

"I didn't send Bing any telegram."

"I'll look into it," Riddell promised.

Don Callum had a large family and wanted to make more money. He had gone through the Detroit telephone book and had found the name Edward J. Noble. The Detroit Noble turned out to be a used car salesman. Callum took him to lunch and explained that he wanted to play a joke, and asked permission to use his name. The salesman agreed. Callum had sent the telegram of recommendation to Bing Crosby from Detroit using the Detroit Noble's name. He was summoned to Riddell's office. Riddell got the President of ABC on the phone.

"I've got Don Callum here," Riddell explained. "He admits he sent the wire to

Bing Crosby."

"He forged my name," said Noble.

"He says he didn't," said Riddell. "He says he has a friend in Detroit named Ed Noble."

There was a long silence at the New York end. Riddell, minus his usual sunny grin, sat looking at Callum, drumming on his desk with the eraser end of his pencil.

"Ask him if he was drunk," Noble demanded, finally.

"Were you drunk?"

"No. I wasn't drunk."

"He says, no, he wasn't drunk."

"Well, then, there's only one thing you can do," said Noble.

"Right," Riddell agreed, and hung up. "Okay, Don, I don't know what your reasons are but—I'll give you a couple o' weeks and that's it."

According to gossip, the last anybody heard, Don was driving a delivery truck.

33
New Times Like Old Times

The monster moved out of the Mendelssohn mansion, but the threat remained. The placid waters of radio were roughed up. The wind came from the direction of Woodward and Putnam. Something new moved into the Maccabees Building. It was Channel 7. Lynn Smeby's transmitter had served AM and FM; now it would send out pictures. Into what had been the old radio studios in the penthouse, now, besides microphones, came cameras, cables, booms, dollies, scenery, props, lights and—confusion. It was like the early days of old radio, only more so. Everything was rule of thumb. Stage hands, electricians and one cameraman came from theaters. Charles Kocher also hired students. Curiously, Wes Gover, the man who had had television experience found himself stuck in radio because a shortage developed there of engineers who could handle a control board.

Radio sound engineers glued inexperienced eyes to the rear ends of cameras and practiced photography. Some men stayed with audio but in the new medium hung their microphones on long metal arms ("booms") and learned how to extend the arms and twist the mikes and roll ("dolly") the whole units along the floors on carriages. Everything was new to everybody. And in the middle of it all was John Pival.

WXYZ was ABC's third television station. The network owned one in New York and operated one in Chicago. Pival had gone to New York to learn how things were done and now he was back charging into the new project like a bull. "Knock that stage down. Take this out. Put this in here. Put that there. Get that camera unpacked. Find a settee. Paint a flat. Get the dust off that table. Don't dolly the camera over the cable, you jiggle the picture; move the cable."

Times had changed. Eighteen years before, George Trendle had had to buy ads in the newspapers to let the city know of his new radio station. Now, the *Detroit Free Press* published a special section of 30 pages hailing Detroit's second video station.

Television was referred to then as "video."

The first TV station in Detroit had been WWJ, owned by the *Detroit News*. As it had ignored the debut of WXYZ-Radio 18 years before, the newspaper ignored Channel 7's opening as completely as possible. Only WWJ programs were listed, and radio-TV columnist Hershel Hart made no mention of the event.

The main attraction of the opening day was a variety program produced locally for which Paul Whiteman, Frances Langford and the dancing Hartmans were imported; and there were speeches by Michigan Governor Kim Sigler, Detroit Mayor Van Antwerp, ABC President Mark Woods, Edward J. Noble, Chairman of the Board, and General Manager of WXYZ, James G. Riddell.

What Detroit first saw officially from Channel 7, at 12:15 P.M., Saturday, October 9, 1948, was a view of downtown Detroit from the roof of the Maccabees Building. This cross-faded into a close-up of the round and dimpled 47-year old face of Dick Osgood. The camera dollied back to show Frances Langford on his right and Paul Whiteman on his left—all three uncomfortably crammed between the arms of a very small settee. Osgood welcomed the city to the new channel, interviewed the celebrities, and gave a build-up to that night's variety show.

This quarter hour was followed by the Chevrolet Glee Club, a World Series baseball game between Cleveland and Boston, and the second half of a football game between Notre Dame and the University of Michigan. The big variety show with its speeches began at 6:15 P.M. and lasted 45 minutes. At 7:30, a sportscast by Don Wattrick.

A photograph accompanying an article by J. Dorsey Callaghan on the first page of the special section in the *Free Press* showed how much more manpower it took to put TV on the air than was required in radio. The picture showed Don Wattrick at the microphone surrounded by announcer, producer, director, program assistant, production manager, floor manager, two stage hands, one screening engineer, audio engineer, technical director, boom mike operator, three cameramen, film screening engineer, film projection engineer and transmitter engineer—eighteen men behind one on the air.

When Detroit's 3rd video station, WJBK-TV, was launched fifteen days later, there were just over 9,000 TV sets in Detroit living rooms. As of November 1, 1948, there were 42 television stations in 22 of the larger cities; 87 others were under construction; and 197 different models of TV receiving sets were on the market costing anywhere from $99 to $5,000—average price, $375. A woman who was divorcing her husband in Chicago asked for and won custody—of their TV set. Walter Abel, Chairman of the TV Committee of the Screen Actors Guild, was complaining that actors wanted to be paid if movies they appeared in were shown on television more than once.

Programming was far behind technical progress. It had become obvious to those who had invested $1 million to get a TV station equipped, manned and on the air that the prime requirement now was to sell goods via TV in order to raise money for better programming. TV would have to pay its own way.

A coast-to-coast television network did not exist. What came to Detroit from

New York had to be sent through space from Toledo by Microwave. Not until the next year would a co-axial cable extend as far west as St. Louis. Television-wise, the west coast was in limbo. No coast-to-coast connection would be made, even by microwave relay, until 3 years later in 1951. Clearly, John Pival was going to have to dream up some programs. He loved it; within a month he had Riddell's okay for regular afternoon programming. WXYZ was one of the first stations in the country to televise during afternoon hours on a daily basis.

Pival's first move was to lure WXYZ's leading established radio personalities before the cameras. One of the first to make it was Osgood. Two days after his appearance with Whiteman and Langford, he received a letter from Riddell:

> Dear Dick:
> I thought you should know that everyone thought you did an excellent job on the opening television show Saturday. I watched it with Messrs. Noble, Kintner and Woods, and they were very much impressed with your work. I think you should pretty definitely consider television work to supplement your present AM, because television is finally and definitely here. It is just going to be a matter of time until TV will be more important than radio. In addition to all this it is a way of augmenting your income now. Kindest regards.
>
> Sincerely,
> Jimmy

Soon, Pival and his executive producer, Pete Strand, had Osgood on TV with a new amateur show once a week. The program was called *Starlit Stairway*. In quality, it could not compare with *Radio Schoolhouse,* but that classic was rapidly falling apart.

That wheeler-dealer Ross Mulholland had seen what was coming and by opening day had an iron already hot and ready. A popular and well-established lecture series in Detroit was *World Adventure*. Travel films with their makers appeared Sunday afternoons at the Detroit Institute of Arts. Films and speakers were booked and introduced by a stocky gourmet-traveller, George Pierrot, who also managed all the business of the series.

Mulholland had gone to Pierrot as soon as the monster appeared in the basement of the Mendelssohn mansion and had suggested that the appearance on television of parts of the travel films with the speakers and George himself would be splendid promotion for the lecture series. Pierrot thought Mulholland represented the station and Ross did not disillusion him. It was arranged. On Sunday, October 10, 1948, the day after Channel 7 went on the air, Pierrot started a regular series.

Pierrot did not get rich: "The celebrity was paid $50, and I was given $25 for expenses. Not for a year-and-a-half did I learn that Mulholland was pocketing $125 a week as my 'agent'."

John Slagle, with his fresh complexion, sparkling deep-set eyes and jolly

Graduation day at Radio Schoolhouse

Graduation day at Radio Schoolhouse

personality, was a TV natural. By February of '49 he had a regular program called *Wax Wackies* on which he played records. With him appeared three pantomimists, *The Gay Deceivers*, who burlesqued what was heard on the records. In December, he and a tall, attractive, swingy blonde girl, Pat Tobin, inaugurated what became one of the first real hit shows on Detroit television, *Pat 'n' Johnny*. There were pets in the studio, well-known visitors, and it was an engaging combination of chatter and personality.

Phil Brestoff was not going to be benched by television, especially with a pretty and talented wife whose ambition he served with devotion. In March, 1949, for the first time, Detroit became a network point of origination in television for a regular series. It was a show called *Rehearsal Call*, and it went over ABC's eastern network every Sunday night at 9:15. John Pival produced it. A critic in *Variety* found Dee Parker pleasing and the camera work on Leonard Stanley's piano solos "outstanding."

In May, the Motor City Open Golf Tournament was telecast over WXYZ from the Meadowbrook Country Club. Showman Pival had a 70-foot tower built to give the viewers a birds'-eye view of the 1st and 10th tees and the 9th and 18th greens. He added 4 walkie-talkie units to the cameras for complete coverage. Chris Brinke, a golf announcer, did play-by-play. Don Wattrick, of course, was on hand. And so was Fred Wolf.

"Fred Wolf!" exclaimed Riddell."What does he know about golf?"

"What did he know about boats?" said Pival.

"Not much." Riddell laughed and turned to Wolf. "You play golf?"

"I play once in a while," Wolf admitted, modestly.

"What's a dormie?" Riddell demanded.

"Well—you're talkin' about matched play," said Wolf. "Dormie is when you've got me down with two to play."

"John, I told you he doesn't know anything about golf. Dormie is when *I* have *you* down with two to play."

"Well—what the hell did I say? Anyhow, what's the difference? Somebody's down and all they do is tie."

Everybody laughed.

"Okay," said Riddell.

So Wolf did the Motor City Open with Wattrick and Brinke.

In the summer of 1949, an auto dealer, Floyd Rice, bought Fred Wolf on a hot rods show.

Fred Wolf: "One night we had the midgets and boy! These guys were really goin'. And they had an accident. Three of 'em. They didn't make the turn. They skidded right at Bruno's camera."

Bruno was a big red-head.

Wolf: "One went up on its end and the fire started immediately and it fell right into the platform an' Bruno wouldn't leave his camera. I remember the picture. The picture just went all over hell. I guess the camera went up in the air but he got it back down. In the meantime, he's shoot'n' down an' one guy landin' out o' the thing an'

Covering the 1959 Motor City Open at Meadowbrook Country Club in Detroit, Wolf interviews, from left, fellow sportscaster Don Watrick and noted golfers Marty Furgol and Cary "Doc" Middlecoff

you know—seriously hurt, two of 'em. And the ambulance. An' they finally put the fire out. I wanna tell ya, that was one of the greatest things that ever happened in television."

Emily: "Fantastic."

Two months before WXYZ-TV went on the air, the national trade magazine *Advertising Age* published an article about Edythe Fern Melrose's House of Charm. Pictures were shown of the house, including the kitchen and its equipment. Melrose had chosen Gibson appliances because the company "was the first to come out with a refrigerator with a top freezer." General Electric saw the article and called Riddell.

"We want into that house. How can we buy her?"

"She's sold out," said Riddell. "Besides, Edyth won't sell rival products on one show and she already has Gibson on contract."

"That's radio. How about television?"

Melrose refused to go on television. Some famous ladies of radio had ventured into the new medium with disastrous results. Edythe held out for eight weeks. Then Riddell called her down to the Mutual Building. He led her into the Board Room. The table was surrounded by men.

"Edythe, we've signed a contract with General Electric."

The argument went on for two hours. Finally, with a deep sigh and a tremulous voice, Edythe Fern Melrose said, "I'll try."

Unlike Harriet Livingstone years before, the Lady of Charm knew her way around a kitchen. One was built for her in a studio in the Maccabees Building. Furthermore, Edythe had trained for the stage. This carried her through her TV debut. Contributing to her ease was the floor man, Jack Flechsig, who crouched under the cameras and repeatedly made a gesture holding thumb and first finger together, indicating perfection, and pointed at which of the two cameras was the one to work to.

Christian, Riddell and the salesmen gathered around a receiving set in the Mutual Building. When the program was over they looked around at each other happily.

"She made it!" one exclaimed. "She *made* it!"

The Lady of Charm sold $10,000 worth of General Electric kitchens in one show. She designed and built three more houses.

Larry McCann went from radio into television with her. In fact, Edythe became his mainstay at the station for a time. He was paid $200 a week to do her program and it was understood that anything he did over that, he would be paid for.

McCann retained his sense of humor: "So we went on the air with the television program. We were also doing an hour a day on radio. That was two hours a day of The Lady of Charm. I was so charming when I got home I kicked the dog."

To radio people who doubled in video, radio continued to be more important. One month after Channel 7 made its debut, Hi-Speed's *Eyes on Tomorrow* series celebrated its 1000th broadcast. Floyd Tracy was still in the advertising saddle at Hickok Oil, avoided TV and still did all his advertising in radio from WXYZ and the regional network.

Tracy was bullish: "We had the biggest volume of business at grand openings. We would open a station and in three days sell better than 35,000 gallons of gasoline starting from scratch. The Pure Oil Company heard about it, and they sent a feller up from Atlanta, Georgia, and one from Memphis, Tennessee, to see how we conducted our grand openings."

Jack McCarthy was doing a noontime show with either Jack Franz or Marguerite Werner at the Hammond organ, in which he would make telephone calls with a give-away format. He climbed into first place in the ratings. Famous Furniture sponsored the program which was called *I've Got Your Number.*

There were several such shows during which people were called and rewarded in some way if they were listening and/or knew the answer to a simple question. These were the idea of Lambert Beeuwkes, and had been quite successful in promoting listenership. John Slagle had one. Hal Neal had one. They lasted several years after Beeuwkes left WXYZ and returned to the east in 1945. They were copied all over the country.

Bill Morgan and Hal Neal were still announcing, but their recording business had grown to the point where they had to move to larger quarters in the Michigan Theater building.

Radio Schoolhouse succumbed to a series of blows. First, Osgood left it to escape its antagonisms. Besides, he had married again and wanted his weekends. Then the show suffered flack from the Detroit Federation of Musicians on some contractual technicality. Next, the sponsor cancelled. Then it lost the theater and moved into the television studios, although it was not televised; and Fred Rivard got his heart's desire: he became master of ceremonies. Two or three months later the show went off the air. In one form or another, it had been on the air for more than 20 years.

In January, 1949, the Hi-Speed broadcasts underwent a radical change of format. The accent was shifted from stories by Osgood to songs by Dee Parker. Osgood was kept on to do the commercials and to read dialogue with the new star. One night a week, the show moved to the Maccabees Building and was broadcast from a TV studio because time between it and the rehearsal for *Starlit Stairway* did not allow for commutation. Osgood, Brestoff and the orchestra were all on the television show. Brestoff, ordered his balder musicians to get hair pieces for the new medium, and there was some talk with Osgood along these lines. His hair was thin and TV lighting was aimed at the top of the head. But he balked.

"It's no secret that my hair is thin," he argued. "People have seem me on the stage of the Broadway Capitol for years. I've been on the stage of almost every theater in town to introduce celebrities. Who would I be kidding with a hair piece?"

The subject was dropped.

An important new personality was added to the radio roster in June, 1949. Dick Femmel was a Chicagoan. He had won an Evans scholarship to Northwestern University and had graduated from the Medill School of Journalism. James G. Riddell was a member of the Western Golf Association which sponsored the Evans

Scholarship and was active in the scholarship program in Michigan. Femmel learned of this and wrote to him for a job. Riddell invited him to call on him in Detroit, and he would see what he could do.

Femmel, shoes shined, suit pressed, shirt clean, tie carefully tied, and hair as neatly combed as a resistant cowlick would permit, sat waiting in the reception room when Riddell walked in from lunch carrying his wife's stole over one arm. He was taking Fadellis to the race track after keeping his date with Femmel.

Dick Femmel: "I thought he was very stylishly dressed. He had a sun-tan and seemed to be man-about-town-sportsman. He was a charming individual and handled himself very well—bright—sparkling. He was great with people."

Riddell's first question after they shook hands and sat down in his office was, "How's the golf game?"

Femmel laughed. His teeth were small and poorly placed. Otherwise, he was a handsome young man, tall, with a strong nose, strong chin, and large, observing blue-green eyes.

"I'm shooting—well—around 88 to 95."

Riddell's game was 15 strokes better than that. They got down to business.

"Have you had any actual experience?"

"Yes. I have worked in journalism in Chicago. I worked for a community newspaper—editing and reporting—while I was in school at Northwestern. I also worked at Station WEAW, an FM station in Evanston, doing newscasts, and I worked briefly at WGN-TV while I was still in school."

"Television, eh?"

"Spencer Allen was doing the news and Clifton Utley was another one."

"Did you know Paul Harvey?"

"We were down the hall from Paul Harvey." Femmel was amused. "He used to wander around the newsroom, mulling things in his mind. This was probably around the 24th story and there was a roof out there and he would open the window and he would walk out on the roof and look around the Chicago skyline, muttering to himself, and would come back in and go to his typewriter and write some great prose."

"Such as—'PAGE TWO!!' " Riddell commented, giving a good imitation of Harvey's staccato style. They laughed.

"Harvey is an entertainer," said Femmel. "You don't come away with a comprehension of the news after you've heard Paul Harvey but you come away with a chuckle and a feeling that you've heard something."

"He can sell." They dismissed Paul Harvey and turned to Femmel's reason for visiting Detroit, as Riddell said, "Well, you're in luck. Austin Grant—he does news in the morning—he's leaving, going over to CKLW in Windsor. Now, I've got a man, Dick Osgood. He's never done news but he's done everything else. I'm going to put him on the news in the morning and he'll need writers. Henry Alexander is head of our news department. You can work it out with him."

So Femmel was hired. On July 4, 1949, Osgood started doing four newscasts every morning, three 15-minutes, 1 five-minutes. They were prepared by Alexander,

Marguerite Werner and Johnny Slagle. Telephone Show. Photo courtesy of Al Schaub.

bushy-browed Tom A'Hearn and the new Dick Femmel. During their vacations their places were filled by Harvey Taylor and Jerry Chiapetta, then of United Press, later a popular telecaster of outdoors hunting and fishing shows.

Meanwhile, the Trendle dramas continued and *The Lone Ranger* moved into television besides, in September, 1949. Trendle took Fran Striker with him to Hollywood to help find the right actor. They chose a man who had been a trapeze artist, had played in movie serials, and appeared willing to perform under the anonymity of hiding half his face behind the mask. His name was Clayton Moore. *The Lone Ranger* was the first western series on television.

In October, 1949, the movie company, 20th Century Fox, offered $20,000,000 for the American Broadcasting Company. Edward J. Noble would not take less than $25,000,000. So the deal fell through. At the end of the year, television in general had gone into the red to the tune of $25,300,000. Radio was still hugely profitable. But, as Allen Campbell would say, "you could see what the hell cooked."

What cooked was a cyclonic upheaval.

34
New Directions

The gathering storm broke loose and the winds blew in all directions. Talent and crews were buffeted between Mendelssohn and Maccabees, both ways. A man would work the controls in a radio studio one day, and a camera in television the next. Genial Jack Tierney scheduled engineers for both radio and television for a year or two, until the storm blew over. Engineers were temperamental. Some wanted to stay in TV; it was new; a challenge; it was where the action was. Others, usually the older men, preferred the safe shelter and security of radio.

Charles Kocher: "It levelled off after a while so that we had about 50% senior men and 50% new men in radio and, of course, a great deal more new men in television."

Television revenues, nationally, tripled in 1950 from $34,300,000 to $105,800,000; but costs were high and TV, generally, was still losing money. ABC was losing more than $21,500 a *day*. In 1950, there were 107 television stations in the United States and only 54 of them showed a profit. There was still no coast-coast network.

All was confusion in the Maccabees Building. Everybody was learning, doing his best, but results were sloppy. Once a camera stayed too long on an announcer and caught him spitting out with a grimace the beer he had just praised and tasted.

Nothing ever seemed ready; everything was uncomfortable. Osgood occasionally hosted a TV round-table of four people to promote a new theater movie. He and his guests would gather outside the studio door and wait for the program ahead of theirs to end. Then there would be a 30-second scramble to get in place between programs—people coming out, people going in, crews wondering, "What now?"

There were claps of temperamental thunder under stress. Even the usually sanguine Edythe Fern Melrose lost her charm on occasion. Once a prop man was slow preparing a set for her. She needed a table for a display of cosmetics. The table was not in sight and the seconds were ticking away. Her secretary, Dorothy, went to the prop man, clasped her hands, and begged: "Please! Please! Won't you bring in that table so we can set up our cosmetics?"

The prop man had had a bad morning. He got the table, but instead of placing it, he slid it across the floor with a show of contempt. The Lady of Charm strode across the polished floor shaking her finger.

"If you ever dare do that again, I'm going to slap you right in the face, because I am helping you earn *your* salary. You're working for *me* and not just WXYZ."

Over all was bald and bullish John Pival. He lived and breathed television, awake and asleep. His enthusiasm was unflagging.

Virginia Pival: "No matter where we'd go or what we'd be doing, he'd talk about some idea—and some of them seemed pretty far-fetched to me because they seemed like they were almost impossible, but by golly, we'd come back, and he'd try it, and nine times out of ten it would work."

Directly under Pival at the studios was John Lee. Once, in a conference with Riddell, there was discussion of cutting down expense by letting a man go.

"We *can't* let him go," Lee declared. "We can't get along without him. We *have* to have him."

Riddell grinned. "Pretend he died."

That ended the discussion, and the man's job.

Charles Kocher: "I learned from Jimmy how to handle people. He used to cut right to the heart of the matter."

While television was trying to land on its feet in the storm, radio seemed to be in a kind of vacuum. It was deceiving. Actually, tremendous change was in the wind, nationally and locally, but on the talent level things seemed the same. Osgood was riding high. In radio he was doing *Show World, Around the Town, The All-American Review,* the Hi-Speed show and four morning newscasts—36 broadcasts a week, plus *Starlit Stairway* and the occasional "round-tables" on television.

Dick Femmel had an idea for one of the morning newscasts—a series of in-depth reports on local situations. Osgood welcomed it because it added stature to one of the newscasts and he paid Femmel for the extra work out of his own pocket. Femmel investigated a lamentable housing situation in Detroit and wrote a sharp, observing piece about it. Detroit's Mayor Albert Cobo heard it and was not happy. He called Riddell and Riddell ordered Femmel and Osgood to cease and desist. So they went back to straight reporting.

Shortly after Femmel's arrival, WXYZ came up with a new gimmick. The Trans-Lux Sign Corporation erected its up-to-then largest fixture on top of the Maccabees Building—two signs, each 7 feet high and 90 feet long, one facing south, the other facing north. For 14 hours a day, news, advertising, station promotion, time and weather reports rolled across the surfaces of the two signs in lights. Messages were fed to them from the radio newsroom in the Mendelssohn mansion.

This newsroom, once largest of the maids' rooms, was crowded. Four flat-top desks were jammed against one another; the wire services were lined up against one wall tapping out the news 24 hours a day; a large air conditioner occupied space against another wall; and there was the mechanism into which tape was fed for the Trans-Lux sign.

It was a busy little room.

One day Hal Neal sat down in Osgood's office. Neal had the same clean, tailored look characteristic of Bob Hite. Both had an indescribable electric quality, physically. The difference was that where success came to Hite with no apparent effort on his part, Neal would have to dedicate himself and fight for it.

"I'm fed up with Studio E," he told Osgood. "All those guys do is gripe."

He was talking about the announcers.

"I've got to get away from it," said Neal. "Riddell has offered me a job downtown for about a third of what I'm getting and I'm trying to make up my mind."

Osgood advised him to take it. "At your age you can afford to gamble. You're a helluva good announcer and if you don't like it downtown, you can always go back to announcing."

Neal took the job downtown.

"You're going to make a lot o' mistakes," Riddell told him. "Just listen to people and I'll try to help you. The only thing I ask is don't make the same mistake twice."

Hal Neal broke into the business end of broadcasting as Advertising and Sales Promotion Manager of WXYZ Radio *and* TV. He and Bill Morgan continued their recording business. They recorded the Hudson Motor Car Company's 40th anniversary program which was fed to the Mutual network. They had lucrative assignments from the Ford Motor Company.

Morgan felt confident: "We were on our way. We could have gone ahead with this and had a real business."

Hal's concentration on his new job was a handicap, however; and Morgan continued at WXYZ as a staff announcer.

No one ever came to the station under better auspices or was more naive about it than big Bob Baker. He was a Canadian from Fort Williams on Thunder Bay in Ontario. He breezed through school like a genius and began his radio career as program director of a very small station. By World War II he was in Toronto where he occasionally wrote newscasts for Canada's then most listened to newscaster, Lorne Greene.

In 1950 Bob's mother was visiting his uncle in Detroit. She didn't like to travel alone; so they called Bob to pick her up. The uncle was Clarence Hackwell, once one of the biggest Buick dealers in the United States.

"So you're in radio, Bob," he said. "I've got a friend in radio—Jimmy Riddell. He runs WXYZ here. Would you like to meet him?" Bob would.

Baker: "Next morning we go down to the Mutual Building, Clarence breezes past the receptionist and breezes past Margaret Roach (Riddell's secretary), and we go in and there's this guy sittin' behind this big desk and he's on the telephone and I won't tell you the language he was using."

Riddell's fresh complexion was flushed. Seeing Hackwell and Baker, he stood up and waved them to enter, continuing on the phone.

"I don't give a goddam *what* that sonofabitch does. He's not telling *me* how to run a radio station. Tell him to take his fifty thousand dollars and—"

He hung up the phone and, unruffled, broke into his toothy white grin as he shook hands with Hackwell and Baker and invited them to sit down.

Riddell had been talking with a salesman about Harry Wismer. Wismer had offered an account, but with strings attached. Riddell had long been unfond of Wismer. Wismer had been hired by Campbell and when crossed by Riddell had developed a habit of going to Campbell over his head. Now, Riddell was the head and he was taking no strings attached from a sportscaster.

Baker regarded Riddell with awe. Riddell was mildly impressed with Baker.

As Riddell remembered him: "Bob Baker. I had to help him get in from Canada. His uncle, Clarence Hackwell, was my best friend, and he asked me, he said, 'This kid wants to get into radio badly, and he's got to have a job to get into the country.' So we worked it out. And of course, I thought Bob became a gem; I thought he was a *very* good man.

Baker started at WXYZ as a record assistant. Riddell apologized for offering him only $65 a week. Baker didn't tell him he'd been working in Toronto for $45.

When he moved to Detroit, he lived at the downtown YMCA, and as payday approached he usually found himself with the choice of riding out Jefferson on a street car without breakfast or having toast and coffee and walking. It was a four-mile walk. Sometimes he would walk both ways.

Once a *Radio Schoolhouse* graduate now a sound man, Bill Hengstebeck was driving down Jefferson Avenue when he saw Baker, picked him up, and took him home for dinner.

Baker remembered this poignant occasion: "I remember to this day—I'm positive Nikki (Bill's wife) had two hamburgers—one for her and one for—and somewhere along the—some bread crumbs—she broke 'em up and ended up with three patties and I remember to this day I wouldn't have had any dinner because I remember I had ten cents in my pocket. That's how Bill and I became friends."

Despite his experience in Toronto, Baker remained naive and small-townish. He was impressed with Detroit and excited to know the Trendle actors. It thrilled him to be with the station where TV's Captain Video, Al Hodge, had been the original Green Hornet, and whose Michael Tolan of Broadway had once played Kato. WXYZ, he thought, was a prestigious and glamorous station and he found being part of it a joyful experience.

Harold Christian's unique merchandising service continued to thrive.

Christian: "Each year we would survey four different products that we were interested in putting on the station. We would interview over 1,000 women in each survey, using our food demonstrators to get answers to many of the questions. We interviewed wholesalers to find out the sales volume of comparative products, and such things as price structure and any pertinent information that gave a manufacturer a true picture of his standing in this market."

Impressed with the work that had been done on his behalf and with the facts presented, the manufacturer would often, of course, be easily persuaded to improve his position by advertising on WXYZ.

Small indications of major change to come passed unobserved, except as they applied to the moment. *I've Got Your Number* slipped a little in the ratings and lost Famous Furniture. Without a thought of how it would affect the program, Riddell allowed it to be cut in half and sold for seven-and-a-half minutes. Jack McCarthy was upset; it ruined the show. He called Riddell. Riddell referred him to the sales staff. McCarthy made his plea to the salesmen. His argument fell upon deaf ears and cold eyes. After a brief contract at half length, the program blew away in the storm.

Changing winds blew the career of Fred Wolf into the unrecognized trend. In the summer of 1949, the ever ambitious Ross Mulholland operated a drive-in movie theater in the northern vacation land of Michigan. Wolf substituted for him on the morning record show when he was away on Friday or Monday. The theater adventure turned into a minor fiasco, in addition to which Mort Neff concluded a final session of complaint by extracting from a desk drawer his contract as Mulholland's manager, and tearing it up.

In June, 1950, the restless Mulholland went to California for three weeks to court Dick Richards, owner of WJR, and owner of the Los Angeles "station of the stars," KMPC. Mark Haas, another old friend from WJR, was KMPC Vice President in charge of programs. Pacific prospects looked bright when Mulholland returned to Detroit, but so did the money he was making at WXYZ. Characteristically, he didn't want to give up one for the other; he wanted both.

"How you gett'n' along with your music?" he grinned when he saw Fred Wolf who had been substituting for him.

"Purty good," Fred told him.

"How do you like it?"

"Purty good."

"Okay, old buddy." said Mulholland. "We'll make a deal. I'll hire you to stay on the show. I've got something going in Los Angeles."

His proposal to Riddell was that he would go to the west coast for six months, but hire Wolf to substitute for him on WXYZ.

"You either go on the air—or forget it," said Riddell. He called Wolf on a Saturday.

"Fred, you do the Mulholland show Monday morning."

"What's goin' on?" Wolf asked. "I talked to Ross yesterday; he's back."

"Never mind," said Riddell. "Be there Monday morning."

Wolf continued as a substitute for two or three weeks; then Riddell called him.

He had a contract on his desk, waiting.

"If you don't sign this, you've got to have a hole in your head."

"What happened?"

Mulholland was out. He had been making close to $100,000 a year with WXYZ. But he wanted more. The salesmen were down on him; they couldn't talk to him.

"They're pitching for you," said Riddell to Wolf. "Now I'm not going to pay you what we paid Mulholland—at least, to begin with. But this is a good deal; you can make a lot o' money. Read it over. Go take it to your attorney. Check with your wife."

"I don't want a record program," said Wolf. "I want to do sports."

"If you turn down this kind o' money, you're not as smart as I think you are. Talk it over with Emily."

Emily Wolf was a small blonde woman with a thin little-girl voice, but her mind was firm and her devotion to her husband without limit. Fred was still working for Chrysler for something like $550 a month. The contract Riddell had drawn up offered more money than the Wolfs had ever dreamed of. They had to accept it.

Wolf was given an office in the Mutual Building—and a secretary. Every day, Emily Wolf listened to him and made notes. When he got home, she would have pages of shorthand pointing to improvement.

What were his main faults?

Emily: " 'I seen.' His pronunciation. And like most sports-minded men, he didn't read well. I had always been very interested in my English classes and I think I helped him a little to develop a flow—and emphasis—not just a monotonous—"

She also went over his commercial logs every week and billed the station.

Emily: "I wasn't a little housewife any more. I was a big-time secretary."

At the end of six months, Fred was "doin' purty good." So was the station. Wolf with the recordings and Osgood with the news turned out to be good chemistry. The two men had little in common other than mutual respect, integrity and the morning hours. Behind both was the newsroom. News writer Tom A'Hearn who had prepared humorous material for Mulholland, now wrote gags for Wolf.

Wolf, Neal and Osgood had been pushed into unthought-of directions by James G. Riddell, and his decisions were bearing fruit. It became increasingly difficult to buy a spot with Fred Wolf; all of Osgood's newscasts were sold; and Neal's application and grasp of promotion and public relations were impressive.

Riddell now found his own mind riding with the wind in a new direction.

35

Turbulent Times.

Word was around that Edward J. Noble would like to dump his controlling interest in the network. In April, 1951, *Business Week* reported that the International Telephone and Telegraph Corporation was dickering for ABC. In May, Detroit heard that CBS was interested. If CBS took over ABC, four stations would be affected in the motor city: WJR-Radio (CBS), WJBK-TV (CBS), WXYZ-Radio (ABC) and WXYZ-TV (ABC). In June, James G. Riddell went to New York to be briefed on still a third possible merger, this one with the United Paramount Theaters. The last would be no small wedding. 294 stations were now affiliated with the American Broadcasting Company; and there were 950 Paramount Theaters, 600 of them wholly owned.

The man who would come out on top in the third merger would be a Harvard Law School graduate, now 45, and a theater man from way back, Leonard Goldenson. He had been lawyer for the theaters which had been forced into a divorce from Paramount Pictures by the United States Supreme Court; and he had been made President of the new United Paramount Theaters Corporation in 1940. If the ABC-Paramount Theater merger became effective he would be head of the combined companies, Noble would become head of the "Finance Committee" with 9% of the stock, and an ex newspaper columnist, Robert Kintner, would be President of ABC.

Such a merger would create an interesting situation in Chicago. Paramount Theaters owned Station WBKB. ABC owned WENR. One would have to be dropped in compliance with the FCC ruling against single ownership of two stations in one city. It was believed that CBS would be the tommy tucker who would stick its thumb into the conglomerate pudding and pull out the plum WBKB.

In June, Goldenson, Kintner and the Secretary-Treasurer of Paramount Theaters, Robert H. O'Brien, all came to Detroit and inspected WXYZ facilities in the Maccabees Building, the Mendelssohn mansion and the Mutual Building. During the visit in the Mutual Building, Goldenson shook hands with George Trendle and with another old friend, Earl Hudson. Hudson was now head of United-Detroit Theaters which were part of the Paramount chain.

WXYZ would be no small acquisition in the proposed merger, not only because Detroit was a major market and home of the automobile industry but because Pival programming and Riddell salesmanship had already steered the ABC-owned Detroit television outlet into the black. WXYZ-TV was the only ABC-owned station and one of the very few television operations in the country to be showing a profit at that time. And of course, WXYZ-Radio was a well-established gold mine besides having international prestige because of *The Lone Ranger* which was now being broadcast all over the world.

In September, 1951, the proposed marriage had the blessings of the stockholders of both companies, but faced some opposition from the Federal Communications Commission. The opposition faded somewhat as it became obvious that advertisers looked upon the wedding with drooling anticipation. However, the FCC announced there would be no formal consideration of the proposal until January.

A problem had developed with Boyd Lawlor. Early one afternoon, he burst into McCarthy's office, his face red with rage.

"McCarthy," he cried, "I just bumped into Riddell at the Statler."

"Oh?"

"And he says, 'McCarthy tells me he's having a problem with you because of your drinking.' What the hell's the idea of going to Riddell?"

"I go to Riddell on a lot of things," said McCarthy coolly. "There *have* been a couple of rather serious episodes involving you and your drinking."

It was true. The climax came the morning after a card party at Larry McCann's house. A dozen men had been invited, including Lawlor and McCarthy. As Boyd progressed toward intoxication, McCarthy began to worry.

"He's got a rehearsal for *The Sandlotters* at 9 o'clock tomorrow morning. Judge Watts will have guests. This guy's got to be there. Somebody better get him home and sober him up. He's had enough."

When McCarthy arrived at the studio Saturday morning, Lawlor had not shown up. An announcer was conducting the Watts rehearsal. That did it; Boyd Lawlor was fired. His producing job eventually went to Bob Baker who became assistant to Jack Rellis.

Rellis was working out very well. His writing was not distinguished but he was imaginative, creative and enthusiastic. He was also frustrated. He needed praise and praise was rare at Wyxie.

Rellis remembered getting none: "I used to have an hour show on the network, radio, *Speaking of Songs*, where we dramatized the stories behind the songs. Dee Parker, Phil Brestoff, the twin pianos—George Primo and Leonard Stanley—and the full orchestra. And I could write the whole show. I could never get anyone to read the scripts; nobody paid any attention. The only credit I got was the line at the end of the show: 'written and produced by Jack Rellis'."

Jack was something of a show-off. One time he had Bob Baker watch him do the Hi-Speed show.

Baker remembered what happened: "Jack was a very dramatic producer which is great because that was old-time radio, but—but—Jack was throwing a cue, and because I was there, he dra-*mat*-ically wound up, misjudged, and nearly broke his finger against the window. He backed up a little and got a little embarrassed."

The engineer, Ray Skidmore, gave Baker some advice.

"When you produce this show, just tell me when Brestoff wants the music louder and I'll give it to his earphones."

Skidmore would raise the level of the music in Brestoff's earphones only, enough to burst his eardrums. It didn't get any louder anywhere else but Brestoff was happy.

Big, burly Bob Baker was still surprisingly naive. What first impressed him when he got to the Van Dyke Club was the sight of the folk hero, the Lone Ranger, chasing martinis with beer. And there was Jay Michael, once Sergeant Preston on *Challenge of the Yukon* and villain extraordinaire on the Hornet and Ranger shows. He had had some toes removed because of some kind of foot trouble and walked even more heavily than his 200-pound-plus.

Baker: "Jay Michael would arrive at the Van Dyke Club and they'd have about three or four shot glasses of booze sitting there, and he'd walk in—shlurp, shlurp, shlurp. I remember Jay walking in and—he said to Phillips, 'Give me my check. I might as well give ya,' he says, 'I just got my—.' And they'd give him a bill like—for one month—of seven hundred and some dollars. It could have been $600 or $800.

And he gave them a check for *you* know, the Lone Ranger thing, like a thousand dollars. Jay was doing nine and ten shows a week then. And I can remember sittin' there and I'm makin' $65, $75 a week. And the only reason I'm in the Van Dyke is because Jack Rellis or Jay Michael said come in and have a drink, 'cause I sure as hell—I just wasn't makin' that kind o' money."

Jolly, easy-going John Slagle had a record show at 11 P.M.

Baker: "Nick, the bartender, would say, 'It's gettin' close, John.' And they'd turn up the radio behind the bar. The 10:59 show would sign off and back in the studio they'd put on Johnny's theme record, *Cherokee*, and Johnny'd yell, 'Gotta GO—O' And he'd get up from the bar stool and whip down the two blocks to the studio in three minutes and go on the air."

Johnny also had a program on Sundays. His theme served one Sunday as music for a wedding. Jack and Theda Rellis were married in B Studio, once Mr. Mendelssohn's library. Slagle's *Cherokee* filtered through the studio B control panel as Judge Watts conducted the ceremony.

Trendle's writers still occupied offices over the sound studio, and there was a fabulous new character among them—Steve McCarthy—no relation to Jack but Irish and part leprechaun. Once he was sitting beside Tom Dougall at the bar in the Van Dyke. Tom was well on his way with the drinking and had become, as usual, shrill.

"Steve," he cried. "You don't look well. You look awful. You look *sick*."

"Well, Tom," Steve, who never felt better, replied, "I don't feel well, either."

Steve removed himself from his stool and sank slowly to the floor. Freddy Tojer, a waiter in tuxedo, with a napkin over one arm, came from the dining room to pick up some drinks, saw Steve on the floor, and said, "Is there anything I can do, Mr. McCarthy?"

"Yes," said McCarthy, gasping for breath. "Pull up your pant leg."

Trojer obliged. Steve bit him on the ankle, got up, and returned to his drink. Trojer went about his business with a smile; the Wyxie gang amused him as it did everyone else.

Jack Rellis: "There'd be about 20 people on the Ranger or the Hornet or Preston, and about 20 more on the third floor on a musical show. Hi-Speed could be going to one network and the drama to another, and both would get off the air at 7:29:30 P.M. At 7:29:45, about thirty cars would go careening out of the parking lot, race two blocks, slam on their brakes—and suddenly the long bar in the Van Dyke that had been empty would be filled with radio people. The public got to know this after a while and they used to come in just to watch the frolic.

Ernie Winstanley: "There'd be a couple of very strange young ladies sitting at the bar and Steve McCarthy would walk in, see them, and get down on his knees and take the hand of one, or kiss the hem of the skirt, and look up into her eyes and say, 'Oh my God! I have never in my life seen anything so beautiful as you. Oh, what a fragile beauty. *So* beautiful. *So*—may I buy you a drink?' "

Steve McCarthy lived in an old downtown hotel called the Barlum, and he never forgot to carry bread from the Van Dyke to the pigeons which came to his window sill.

One night he had had more than usual to drink, but he remembered to take a loaf of bread to his room. It was summer and he opened his window and started back to his dresser to break up some crumbs for the sill. But he didn't make it; he dropped onto his bed and went to sleep. When he woke up, there were fifty pigeons in his room, and no way to get them out. The more he flailed his arms, the more panicky the birds became. He finally opened his door into the hotel corridor and the stupid pigeons, instead of flying out the open window, flew into the hotel hallway.

A less engaging character among the writers was an oldish, fat, sloppy gent named Galt. He worked mostly at night, and he smelled. Trendle found him asleep at his desk one morning in his shirt sleeves, and told Fran Striker to fire him.

Jan Striker: "Fran tried hard to save him and save his soul. He was alcoholic. He'd bring him out to the house for meals and I didn't particularly appreciate that. Fran liked characters. I mean—I used to wonder about him a little bit."

Her laugh is gentle, loving.

Tom Dougall had no office at the station now, wrote free lance, and acted in the dramas. Betty Joyce wrote less and worked more as Striker's secretary. Another woman, Mickey Merrill, wrote Yukons only. Beatty was writing two scripts a week. Steve McCarthy wrote mostly Hornets; he was fond of the Axford character and fed him Irish jokes as far as Trendle would allow. Fran Striker was still writing, but the actors noticed that his scripts were less and less original; they seemed to be re-writes of stories from years before. He still worked nights and worked hard but he was getting tired and a little deaf. His hair was greying, and although he still walked with a strong display of German energy, his shoulders sagged a little.

Fran's devotion to his family was impressive, never dimmed by fame.

Jan Striker: "I never really thought of him as being famous. He was—Dad."

Bob Striker: "I don't think any of us were conscious of his fame. He certainly was not an egotist."

Jan: "I think in later years he *under*estimated himself."

Janet: "He needed a lot of building up."

Bob: "I think a little more conceit when he got into middle age would have been extremely helpful to him, especially if it'd given him enough confidence in himself to seek something beyond working for Trendle."

Janet: "Trendle never did anything to boost his morale."

Bob: "There was a time when I thought I'd like to follow in his footsteps and I think Dad realized long before I did that I wasn't really cut out for that business. He encouraged me to study engineering, and it's kind of ironic, because I wound up writing in the engineering field."

Would Fran Striker have been happier to sit down without the pressure of daily deadlines and write deeply, rather than hacking out radio scripts day after day?

Bob: "If he had not had to worry about the economic aspects, I think the thing he would have liked most would have been to do something like the book he worked on five or six years."

Jan: "I think so." (The book has never been published.)

Bob: "At the same time, I don't really believe that dad could have supported the family unless he had somebody to crack the whip."

Jan: "Or unless he had a deadline."

Janet: "Remember when he wrote a couple of cartoon strips in the back of the car while you (Jan) drove us back to Detroit from the farm?"

Jan (laughing): "To get them in on time."

Janet: "So you'd have a free evening."

Bob: "I think he knew that he needed the pressure to get something done."

Jan: "He was a procrastinator: He'd get the rubber bands out and all the pencils and sharpen them, and paper clips—."

Janet: "Clean the typewriter."

Jan: "—Typical of many writers. But he knew eventually he had to stop this and get to work."

Bob: "He definitely did have a lack of confidence, though."

Janet remembers when she was six and Fran cut down a tree and decorated it for her very own on Christmas. Once he and her brothers built her a beautiful doll house—"big as a television set."

A brook ran through their land in Arcade. Fran and the boys dammed it up and had a swimming pool. A huge hen house had come with the farm. This, Fran and his three sons—Bob, Don and young Fran—turned into a comfortable bunk house. They also made their own fireworks for the 4th of July.

Bob: "Dad was always a great reader and one of the books he read was on the history of fireworks; so he went right back to the Chinese and started doing it the way they did."

Fran and the boys spent months preparing elaborate set pieces. They took them to the farm which was on the top of a high hill; and people would come from miles around to see the display. If the weather turned bad, Fran would fill the house with strangers who had gathered for the fireworks, and Mommy would be kept bustling with the coffee pot and the house would be "eaten out of pastry." Mommy ruled the roost; she was jealous of Jan; and Striker's gentle wife preferred peace for her family to rising against her dreadnought of a mother-in-law.

Mommy Striker's meek little husband was the first to die; then Uncle Gus, leaving Auntie Bess alone with Mommy on the farm. Auntie Bess was something of a secret tippler, but it was only her way of putting up with Mommy. Bess was always gentle, and sweet, and a lady.

The Ranger broadcasts were now carried on 221 stations with an estimated audience of 12,000,000 listeners a week. The television audience at this time was estimated at 5,000,000. Seven million copies of the Lone Ranger comic books, all written by Striker, were being sold every month.

In 1951, *The Lone Ranger* was voted the "best network program for children" by the General Federation of Women's Clubs; it won the Academy of Arts and Sciences Award, taking first place in children's shows on television; the TV Forecast Magazine presented it with the statuette for being the most popular network TV program in western New York: and it won the TV-Radio Mirror Award as the "best western."

Striker kept a list of the awards; Trendle kept the awards.

Although Clayton Moore wore the mask as the Lone Ranger on television, radio's Brace Beemer was still making personal appearances. He left no doubt whatever that he was the star.

James Jewell remembered one appearance: "One time he came over to Chicago dressed in the full regalia. He shook hands with me, sat me down at the same banquet table at which he was being honored, and completely disregarded any connection I might have had with *The Lone Ranger. He* was Mr. Lone Ranger. He was the entire thing."

Leta Beemer remembered another occasion: "Brace, at one time, was making a personal appearance in Chicago with a circus and—you know—there's all these wires and guide lines and things, and somebody opened a newspaper right in front of the horse as he was going around a corner, and of course the horse went sideways and jumped up in a ring.

"Instead of trying to save himself, Brace tried to save the horse when he went down; so he stuck his leg out and pulled all the ligaments loose. He lived in a hospital that week. They'd shoot him full of some kind of Novacain or something to dull the pain, and he'd go back and ride the horse and then go back to the hospital and go back to bed."

In October, 1951, Beemer made personal appearances at the Madison Square Garden in New York. Bob Hite took one of his daughters to see it.

Hite: "He was a great Lone Ranger because he had the physical set-up for it, and the voice, and the ability to ride the horse. It was wonderful, his appearance. It was most impressive. He rode around the whole runway, and then he dismounted and walked to the microphone, and he was all alone in this huge arena, with a white light down on him, with his white hat on, the white horse, the silver guns—very impressive. The kids just went wild."

Hite took his daughter, Bobbie, to meet Brace Beemer backstage and in contrast with Jewell's experience in Chicago, Brace was cordial to the former Lone Ranger announcer and invited him to visit in his hotel suite the next day.

Trendle denied that Beemer's horse, Silver Pride, ever made any personal appearances. Leta Beemer told us more convincingly that he did. There is no question that Beemer's horse was a magnificent stallion. And beautifully trained. Brace delighted in showing him off.

Leta Beemer: "We used to bring him in the house. People would want to see him, and Brace would go get the horse and bring him into the house and he'd do his tricks right there in the living room. He'd look down that long patrician nose of his, and these people were climbing the walls and stairs and everything else. The horse thought they were crazy."

That occupational disease was making grabs at Beemer.

Jack Rellis: "I never saw Brace drink during rehearsals or before a show but I certainly remember the Van Dyke afterward."

Bob Baker: "I was just aghast. Brace would sit at the far end of the bar and they knew what he wanted—one martini and a bottle of beer as a chaser and they'd put another martini in front of him and he'd finish that bottle. One bottle of beer and two martinis. And I'd sit there and I thought, I don't believe this! And then, sure enough, another bottle of beer and another martini and—shlurp. He'd do this over and over and he'd get up and walk out of the Van Dyke Club. And he wasn't sober—I'm not saying he was sober—but he damn well could navigate. And he'd get in his car and drive out to Oxford."

Charles D. Livingstone admitted: "Brace and I had our times, believe me. He wanted things his way and there were times when I wouldn't let him have things his way and it caused a lot of tension. You didn't know when there was going to be a blow-up. I finally changed over to a certain thing that he wanted and everything was fine from then on."

What certain thing?

"Oh—about who would play certain parts or things like that."

Women's parts, maybe?

"Maybe. It's possible."

He and Harriet laughed together.

Fred Flowerday: "I always worked easier with Brace than I did with anyone else. I think Brace wanted me to do well and he did his damnedest because we personally liked each other. Brace knew I was kind of on the spot. It was damn hard for me."

James Jewell: "He did a lot of work with children in hospitals. He'd go in and talk with youngsters."

Leta Beemer: "Those are things that Brace never mentioned. Lots o' times he would go and visit children who were very, very sick—only on the promise that there would be no cameras or anything there for publicity. He used to go to old people's homes—and nobody knew this, nobody but me. I'd find somebody that I thought would appreciate a visit from him and he'd go and see them. He was real wonderful about doing things that he didn't want publicized."

Was Brace an easy man to live with?

"Oh, heavens yes! Oh yes! Brace was the head man. There was no doubt about that, ever. He wore the pants always and that was the way I wanted it. By the same token, he was the most considerate man. If I reprimanded the kids, if he thought I was over zealous, he would never say anything in front of them. He'd wait until later, and he'd say, 'Don't you think you were a little—uh—.' Oh, he was a doll. He was always doing something nice."

Once the Osgoods played bridge with the Beemers. The time came to pick up two of the sons who were attending a Boy Scout meeting. The Beemers lived in the country.

"Take the gun, Leta," Brace ordered.

"Oh—Brace!" Leta objected.

"Take the gun."

Leta carried the pistol without further question.

John Todd, the original Tonto and still playing him after 18 years, was aging.

As Herschell Mayall had done before him, John would now occasionally fall asleep between his scenes and one of the sound men or actors would keep an eye on him.

One night, the script called for the Ranger and Tonto to search the crook's room on the second floor of a hotel. Fred Foy, the narrator, described their creeping up the creaking stairs, forcing the door open, and rummaging through dresser drawers—all accompanied by appropriate sounds.

In the script as written, the Ranger finally says, "Well, Tonto, there is no evidence here. Let us get our horses and go back to camp."

Tonto grunts and says, "Mm. We do. Kee-Moh-Sah-Bee."

Music comes up briefly to indicate their return to the ground floor and the hitching post, then fades as the Ranger says, "Come on, Silver. Come on, big fellow," and Tonto says, "Get 'em up, Scout"—and there follows the sound of galloping.

During the broadcast, while Foy described the long climb up the stairs, and the search of the room, John Todd fell asleep. Brace was reading the line, "Well, Tonto, there is no evidence here," etc. when Ernie Winstanley roused poor John. As he jumped up to the mike, his script fell from his lap to the floor. Ernie held his script up, pointing to the line, "Mm. We do. Kee-Moh-Sah-Bee." But John was rattled, didn't see the line, and jumped ahead to "Git 'em up, Scout"—still on the second floor of the hotel. Brace was very hard to break up, but he laughed all through his "Come on, Silver. Come on, big fellow." And the galloping was slightly ragged and delayed.

Another night, one of the characters was a tinware peddler. The sound men hung a number of pots and pans from a rod in the sound room, and gently jiggled the rig to get the effect of the metal as the peddler's wagon jounced slowly over an old dirt road. At one end of this paraphernalia was a box of gravel in which a sound man marked time for footsteps.

The story has the old peddler attacked by outlaws and left to die. He is found by the Lone Ranger and there is a very quiet scene between the two.

During the performance, John was standing in the sound room. He started to tip-toe back to the actor's studio, stumbled over the box of gravel, and knocked over the hanging array of frying pans and stew pots, creating a sudden clatter and crash that drowned out the actors. Quickly, the director brought up some music. It was half a minute before that show settled back into the rut.

Trendle fired John Todd once, because "he was too old." They found a real Indian to play Tonto, but he was college-bred and refused to say "ugh" and "git 'em up, Scout." Besides, there was an immediate flood of mail complaining, "Where is Tonto? Who is this new Indian?" Trendle had to admit defeat; John Todd returned to his loyal fans, his pals, his little cat-naps, and the role he had created.

By September, 1951, Detroit, with WWJ-TV, WJBK-TV and WXYZ-TV, ranked 6th among the "Ten Top Cities" in television: New York: Los Angeles, Chicago, Philadelphia, Boston, Detroit, Cleveland, Pittsburgh, Baltimore and St. Louis.

The tentacles of co-axial cable had reached across the nation. The first transcontinental telecast occurred on Tuesday, September 4, 1951. It was President Truman's opening address at the Japanese Peace Treaty meeting in San Francisco.

It reached Chicago by radio relay and was carried east from there by cable. All three Detroit stations carried it.

Regular coast-to-coast television got the gun 25 days later. On Saturday, September 29, Paul Whiteman's ABC show was the first to go from east to west; and on Sunday, September 30, Eddie Cantor was the first regularly scheduled TV show to originate on the west coast and arrive in the east via NBC.

In October, 1951, the first *international* coast-to-coast telecast was fed to the nation from WXYZ. The station picked up the royal visit of Queen Elizabeth II and Prince Philip to Windsor, Ontario. Dick Osgood was assigned to this pick-up but had no time to prepare and in a fit of exhaustion and temperament begged off at the last moment. The honor went to John Slagle.

In the turbulence of the times, personalities came and went on Channel 7. Bill Hendricks who had switched from promotion to TV sales, sold *Starlit Stairway* to an automobile dealer and the dealer did not want Osgood as emcee. Tom Waber took over briefly, then the assignment went to a new man named Ted Johnson, not to be confused with the actor Ted Johnstone. By July of 1951, it had run through five masters of ceremonies and Johnny "Scat" Davis was the sixth. Davis had appeared in movies during the '30s, was a "scat" singer and orchestra leader, and was imported to Detroit by John Pival to do a morning TV show called *Coffee and Cakes.*

John Pival had introduced a number of freakish characters to preside over a showing of movies. There were the White Camellia who played it sexy, Jean Loach of *Heart Throb Theater,* and Tom Dougall whose thin face materialized by trick photography from a skull to be Mr. X of horror movies. Probably the most popular movie host of the time was Bob Dale who played himself on a series ambitiously called *Academy Theater.* Dale made the news when a girl admirer climbed through a window of his apartment and killed herself.

1951 was a big year for Fred Wolf. He was a complete success as morning disc jockey on radio. In September, he did TV wrestling matches sponsored by an auto dealer; and in November, he started a TV boxing show for the Pfeifer Brewing Company. The last made history. Again, Bill Hendricks was the salesman. The bouts, for television only, were arranged by the Motor City Gymnasium Company and approved by WXYZ. They were submitted to the state boxing commissioner for his okay. There were two 6-round bouts every Saturday night and there were standbys in case of a knock-out. The show was televised from the Madison Ballroom at Woodward and Forest.

Meanwhile, John Pival, Pat Tobin and Johnny Slagle were learning, as was everyone else, one great difference between TV and radio: in radio people lasted; in television, they did not. By July, 1951 the *Pat 'n' Johnny Show,* which had been so popular, wore out its welcome and was discontinued. The next month, Johnny started *River Boat Jazz* on WXYZ-Radio, Monday, Wednesday and Friday at 6:30 P.M. He was a real jazz buff and used music from his own library of 1500 jazz records. Pat Tobin moved away.

Pival was promoted from Manager of TV Programming to Manager of WXYZ-TV, and to an office in the Mutual Building. Under him, Pete Strand and John Lee took over the programming.

Riddell, Pival and the salesmen in these early days of TV have been described as "the gas house gang." There was a lot of drinking and the gang was rough. The motto was: *Sell! Make Money!* Sponsors were not checked; they were accepted blind. Their copy might not be in good taste and it might be very, very long. There were give-aways and gimmicks. Anything went as long as the money came it. Larry McCann, besides accompanying Edythe in her Charm Kitchen, moved out onto the used car lots and ad libbed commercials that sold old automobiles like cotton candy at a carnival.

Into this high pressure turbulence moved a sober young idealist. The gas house gang decided it was time for Channel 7 to have a news department. Golfer Riddell thought of golfer Dick Femmel and called Hank Alexander. Hank passed along the information that TV was in the market for someone to set up a news department. Was Dick interested? He was.

Although they had both worked in the Mendelssohn mansion at the same time, Femmel and Pival had not met. They were now introduced to each other in the Mutual Building; and they crossed swords from the beginning. Pival had no interest in news; he was all show biz. Femmel had studied journalism; he had worked at television news in Chicago; he had definite ideas as to how news should be handled.

"We'll need camera crews to go out and cover stories. We'll need a set. We'll want visual materials. And we'll want to hire a bona fide news man for our air personality."

"Hold everything," said Pival. "I don't have money for all that. Besides, we have some good announcers out there—George Hunter, Bob Mabry, Mike Master, Tom Waber—. You just prepare the news. They can read it."

"What about camera crews—sound men—a reporter to go out on stories?" Femmel asked, very much let down.

"Nope. I'm not going to spend all that on news. We've got Mike Halley on the staff out there. He's got a 16 millimeter movie camera. I'll give you a little money and you can slip him something extra to go out and film a story, if it's important enough."

"What about a reporter to go out with him?"

"Naw. You can do a voice-over. And don't send Mike out unless the story is really big."

So, Channel 7 news got off to a very poor start. Dick Femmel wrote from wire copy and telephone conversations, staff announcers read what he wrote, and occasionally there would be an "actuality" filmed by a one-time commercial photographer with a 16 mm camera. Twenty years later, Channel 7 news was still fighting for an audience it did not have.

Meanwhile, the winds were blowing from another direction.

36

Wolf on the Porch

The most successful network shows and the top national stars began to move away from radio into television. When Eddie Cantor ventured into the new medium, NBC had no problem finding a sponsor for the series. But Cantor wanted to continue in radio, too, and for the first time, the network had to scrounge around for an advertiser. Ted Mack was on both radio and TV in 1951 with *The Original Amateur Hour*. Asked if there was any difference in response between the two, Mack replied, "Oh brother! Don't ask me that. The answer would break radio's heart."

But radio was fighting. Somebody at CKLW in Windsor (Art Gloster who was handling publicity for the Canadian station is given the credit) thought up the slogan, "Wherever You Go, There's Radio." It was seized by the entire Detroit area radio industry and promoted in a noteworthy campaign by Hal Neal who was now Assistant Sales Manager of WXYZ. In late September, a song with that title and theme was written by Claude Moye and Joe Troyan who appeared on WJR as Pie Plant Pete and Bashful Joe. WJR's Don Large made a glossy arrangement of the song and introduced it to the nation over CBS from WJR on his *Make Way for Youth* program. The campaign was picked up all over the country. By Christmas, "Wherever You Go, There's Radio" bumper stickers were ready for automobiles. The car radio, ignored up to then, became one of the most treasured assets of the industry.

Some talent was still prospering in old-fashioned radio and clung to it. Some sought jobs in it as a step toward television. Among these was Fred Weiss—short, solid, round-faced, light blue-eyed and very light blond. Fred was eager, painfully honest and even more naive than the Bobs Hite and Baker. He attended the University of Wisconsin and was working on an efficient little NBC station at Madison.

He read all the trade papers and came to the conclusion that the only future in the broadcasting industry would be in TV. There was no television yet in Madison, so during his summer vacation in July, 1951, he auditioned in Dayton, Columbus, Toledo, Cincinnati and Detroit where there was TV as well as radio.

At WXYZ he found not just one but *two* receptionists inside the grand entrance hall of a beautiful mansion. He was led up an impressive stairway and into the sumptuous office of a young man as short and solid as he who smoked his cigarette in a holder and had a rather leisurely air of elegance and importance. It was Jack Rellis. He was extracting some audition copy from a desk drawer when another short man, trim, with short, curly well-greased hair came into the office. Rellis introduced him.

"Fred Weiss. This is Fred Wolf."

"I want a great big banner with the letters WXYZ on it," said Wolf.

"What do you want that for?" Rellis asked.

"I want to hang it on the porch so everybody can see it when I do my show out there."

"On the porch?" Rellis gave him a wide gray-green stare.

"Where've you bin?" Wolf muttered. And walked out.

Rellis was friendly and warm as he led the announcer back down the stairway and into a studio.

"I want you to know," Rellis announced grandly, handing Weiss the audition copy, "that you are now standing in the studio where *The Lone Ranger* originates."

It was enough to scare the auditionee to death.

"A-And *The Green Hornet*," Rellis added with an upward inflection.

Fred Weiss could only do his best. His audition was transcribed. (The tape recorder had not come into its use.) At the conclusion of the audition, his face flushed, his hands trembling, Weiss returned the copy he had read.

"You're pretty good," Rellis drawled as though a little surprised, but smiling in a friendly way.

He led Weiss back up the grand stairway and along the carpeted corridor past his door and into another big square office. A tall, good-looking man in his late thirties sat behind a big desk, looking grim. He was introduced as Jack McCarthy. He used a cigarette holder even longer than Rellis'. Rellis put the transcription of Weiss' audition on McCarthy's desk.

"What station did you say you were with?" McCarthy asked.

"The NBC station in Madison, Wisconsin—WIBA. Of course, its nothing like *this*. It's a nice station but it's *small*. It isn't show biz like this. Gee!"

Weiss chuckled modestly.

"They have *Morning Melodies*, and then they have *Noon Melodies*, and then they have *Evening Melodies* and they have *Serenade for Monday* and *Serenade for Tuesday—you* know. And they're on 1310 on the dial, so they have a *Thirteen-Ten Club*—that's a *Bi-ig* record show at night. But—gee."

Weiss looked around admiring the office.

"Well, we'll let you know," said McCarthy.

What Rellis didn't know about Wolf and the cause of the grim expression worn by McCarthy was that Fred Wolf had done his record show that morning from the brick-and-granite porch of the Mendelssohn mansion!

It was the first step toward an innovation that would make history. Wolf was convinced that one way for radio to fight television was to get out and meet the people. So that morning in July, 1951, he had gone to his engineer and said, "I can't work in the studio. It's too hot. I'm goin to work on the porch."

"You can't do that," said Harold Winter.

"The hell I can't," said Wolf. "Get a mike out there and a card table and chair. That's all I need. We'll move out there while Osgood's doin' the 8 o'clock news."

Winter gave in reluctantly. The change was not as simple as it sounds. Fred's records were run by Bob Baker in the director's booth of the dramatic studio and the turntables could not be moved. As soon as a record started playing, Wolf would remove his headset, race into the studio, and find out from Bob what the next record would be. The two also hollered back and forth through the open windows. Meanwhile, the porch microphone picked up all the sounds of the morning traffic. Electric street cars still ran on Jefferson Avenue and there was a safety zone and stop in front of the Mendelssohn mansion.

At 9 o'clock when he was off the air, Fred lit a cigar, collected his copy and headset and was stepping inside as Winter came out to restore the card table to a lounge in the basement and retrieve microphone and cable.

"You're wanted upstairs," said Winter.

McCarthy was livid; his Irish was up.

"Where in the hell do you get the idea that you can go outside and do your broadcast?"

"I thought it was a good idea," said Wolf.

"Well, the problems you caused with our engineering—of course, you know what's going to happen downtown."

A meeting was set up for 2 P.M. in Riddell's office.

"Oh Christ," thought Wolf. "Here goes. I've blown the whole thing."

He had lunch with some of the salesmen. They had heard the outdoor broadcast and liked it.

"I think I'm in trouble," Wolf confided.

Riddell listened to McCarthy first. Jack's argument was not only that it created problems in the engineering department and separated Wolf from his producer and engineer back in the studio, but that it looked bad and sounded bad.

"It sounds like 50 watts," he concluded.

Riddell turned to Wolf.

"I just thought it was a good idea," Wolf shrugged. "Poor man's television. *You* know—guy can't afford a television set—hell!—listen to WXYZ. Come on down and watch it. What da hell? Let's try it."

Riddell was becoming a little disenchanted with Jack McCarthy. He felt that Jack was an excellent detail man but that he lacked imagination. Besides, Wolf was where the money was.

"I'm sure the engineers can rig up something so Fred can talk with the guys inside without having to yell through the windows," said Riddell. "Can't Jack Tierney work out some two-way thing?"

"I suppose he could," McCarthy admitted.

"Okay," said Riddell. "Let's try it."

Norman Lenhardt of the rich and oily laugh ran afoul of his alarm clock that summer. It worked, but he didn't. The tall announcer whose thinking, Tom Dougall used to say, got lost in his vocal cords, turned off the alarm, rolled over and went back to sleep. This time he was fired. On September 1, 1951 Fred Weiss replaced him as a staff announcer.

Fred Weiss:"I suddenly fell into show biz. Oh! I talked to these guys—*you* know—the big tall guy in *Challenge of the Yukon* who played Sergeant Preston—Paul Sutton. I can hear his voice. And I thought, gee, I got in the wrong place, somebody made a mistake or sumpn. I figured I wouldn't last around that crowd. They were all—it was very show bizzy there. *You* know."

But he was good. He went on the night shift immediately and was assigned to the important 11 P.M. news.

Weiss: "Here I am on a network-owned station, *you* know, *Big City*—scared to death and I had glasses on and I wanted to make sure I could read my copy so I shined up my gl—and broke 'em in half—and—the very first night! And I had to go on and read this news without my glasses—and I found out I didn't need glasses, and I have never worn glasses since!"

One of Jack Rellis' important shows was broadcast Saturday mornings from the Detroit Institute of Arts. It was called *Quizdown*. Tom Waber was the quizmaster, and the new blond from Milwaukee got the announcer's job on it. After a year, *Quizdown* was simulcast over radio and television. But Weiss was considered primarily radio and although he wanted TV assignments, John Lee thought his hair was too light.

He was doing well enough in radio, however.

Weiss: "There's a show called *Quittin' Time*. Johnny Slagle was the personality and I was the announcer. Jack Rellis was the producer. Bi-i-ig—*you* know— this is radio with things goin'. We had Phil Brestoff and a *band*. Phil played his violin and Jack Franz played the organ and it was done in the ballroom up on the top floor. And we read corny poetry and—*you* know—it was *drive* time and we're up doin' this terrible stuff! Oh, it was awful! But it was fun and it was live and there was a kind of rapport among all these people that was really kind o' nice, you know? There's nothin' live like that any more."

Fred even got to do the Cheerios commercials on *The Lone Ranger*.

The second step toward the innovation which would bring station representatives from all over the country to observe was taken in October when Fred Wolf discovered that it was cold on the stone porch at six o'clock in the morning. He went to Bill Hendricks who was sales-minded, close to Riddell, and a neighbor in Grosse Pointe.

"Bill," Wolf told him, "I want to build a glass house—some little eight-by-ten thing, three sides o' glass, and I want a lot o' lights inside, in fact, lights *out*side—at night. Helluva thing! *You* know what that mansion is. Drive by it—Christ! It's a mausoleum!"

"I've got a friend," said Hendricks. "Joe Decker. Joe's one of the top men with the Peterson Glass Company.

They took the idea to Riddell.

"All we want," Wolf explained, "is a little—shingle roof—three walls o'glass, three feet off the ground so you can see it above the hedge when you come around the bend on Jefferson. And we gotta have a heater in there."

"Find out how much it would cost," Riddell told them.

Fred suggested the phrase: "If you want class with your glass get Peterson glass." If Fred would just say that phrase on the air three times a morning, Joe Decker said he'd give them the glass. Another friend, Chaplow of Chaplow Lumber Company, agreed to build the house for about $1200, but it could be paid for in plugs. So the little studio was built on the front lawn. Fred called it "Wolf's Wacky Wigloo."

It was an insult to the classic architecture of the Mendelssohn mansion. Jack McCarthy asked that it at least be built to one side of the mansion, but it was erected smack-dab in front of the entrance. An advertising agency, B.B.D.& O., which

Mendelssohn Mansion. Wolf's Wackie Wigloo in front. Fran Striker's light on as he writes. Photo courtesy of Fred Wolf.

occupied another old house on the same block, threatened to sue on the grounds that the "eye-sore" projected beyond what was allowed in the building code, but Jimmy Riddell got with the boys at B.B.D.& O. and turned on the power of his own personal sun over a few drinks, and the agency men gradually came around to agreement that Wolf's Wacky Wigloo was "commercial as hell."

Jack Tierney tied the little shack into master control and the turntables of the drama studio so that Wolf and Baker could communicate through mouthpieces and ear phones; and an extra microphone which Fred could control was placed outside the Wigloo to pick up street sounds, including the electric street cars.

Occasionally, Wolf would say, "Look at all the cars goin' by. Wonder if anybody's listenin'!"

He would turn on the outside microphone and listeners all over Detroit would hear the horns blowing. Kids stopped by on their way to school.

Wolf: "I had ten or twelve, but two of 'em got me—Joan and Davey. Joan had two braids and she was loaded with freckles and was sharp as a tack. And Davey lisped and there's nothin' better in radio than to have a voice that is different, and this kid—people still ask me whatever happened to little Davey. Well, he was in the service, and he's married and he's six-foot-three."

Eye-sore or not, the Wigloo was a success, and led to something even better. Wolf's ratings began a slow but steady climb.

The management of the *Detroit Free Press* listened to *Show World* and wanted a column in the same style but about radio and television. One of its top editors, Frank Angelo, drove out to the mansion and proposed to Dick Osgood that he write a column about the business he was in and the people in it. Riddell gave his permission and the first column appeared in April 1951.

Later that year, in November, Osgood—not to mention stations in Indiana, Ohio and Michigan, including WXYZ—lost Hi-Speed. The Hickok Oil Corporation was purchased by Pure Oil. Pure Oil sponsored H.V. Kaltenborn nationally and cancelled out WXYZ and the regional network ending an association that had lasted fifteen years and five months.

Radio programming—local and network—had been resting on its laurels, and its laurels were withering and blowing away. Something new was in the wind.

37
All Hail! The Deejay!

Twelve years after Bill Morgan looked up Woodward Avenue at the WXYZ sign on top of the Maccabees Building thrilled to be in the big city of Detroit, he became one of the "gas house gang" in the Mutual Building. When Hal Neal was promoted to Assistant Sales Manager, Bill had gone after the job he vacated, Sales and Promotion Manager—and won it. It ended his career as an announcer and the existence of the recording company he and Neal had started. They were both too busy now as station executives.

Many nights there were meetings. Nobody called them; they were impromptu. After the day's work, Pival, Morgan, Neal, Christian, Riddell and the salesmen would just sit around and talk about the station. There was gossip. There were ideas. There was liquor.

"The midnight never came when John Pival wasn't dead drunk," one recalls. "He consumed a minimum of one fifth of Cutty a night. And many a night he was into his second fifth."

But another remembers, "After the roughest night in the world where John would be stoned after 5 o'clock in the morning, at 9:00 A.M. he would be at his desk clean shaven, neatly dressed; and his eyes would look like two cherries. He had unbelievable strength."

Head man, of course, was James Riddell. He had a built-in enthusiasm. His judgments were usually quick, his action immediate and open. In one sentence he could clear an issue of all confusion. He was blunt, yet kindly. He had a knack for exposing humorously what was phony. Resistance melted before him. Trendle had been austere and Teutonic. Campbell had been vigorous, quiet and shrewd. Riddell was gregarious and fun-loving. But he was no less creative than the others in his own time and in his own way. In fact, he started something in Detroit that spread across the country.

Riddell: "Everybody does it now; it became the format of radio. As far as I know, WXYZ was the first one in the country to do it. And it was something that I sort of backed into."

What he "backed into" was a new format—the *Disc Jockey.*

Insecurity still gnawed at WXYZ because of uncertainty as to whether the long-delayed merger with Paramount Theaters would ever be consummated. The FCC didn't get around to lending its ear to the proposal until April 1952; and the hearings dragged on.

Meanwhile, the winds blew and radio's old house was being demolished. Still, CKLW of Windsor told the world it had had its best year, WJR declared a nice dividend, and WWJ was building new studios. Radio networks suffered the most damage. Many favorite programs blew away in the wind. Some that sought the refuge of television suffered for it. TV's Amos and Andy were not the Amos 'n' Andy we had lived with and loved so many years. When they saw it on TV, many viewers felt that *One Man's Family* had turned into *Another Man's Family.*

After establishing Fred Wolf with records in the mornings, Riddell "backed into" Jack, the Bellboy. This came about through a one-time publicity man whom Riddell had helped into the industry.

"Dick Jones just wanted to get in so badly," Riddell remembered. "Ted Campeau ran CKLW and I called Ted, and Dick got a job there."

Jones made a strong showing at CKLW as Sales Manager and when George B. Storer purchased WJBK, he hired Dick to manage his station. Early in his tenure there Jones called Ed McKenzie into his office and, tapping papers spread before him on the desk top, said, "I've been going over these figures. You have been cheated. I want you to know that's going to stop. As long as I'm here, you'll get what's coming to you."

The two became buddies.

McKenzie: "Dick could drink and cut it off any time he wanted to. I'd get to drinking with him and I couldn't stop, so eventually I had to cut it off completely. Dick and I hit if off real well, but he got to raising hell and wouldn't do what Storer wanted him to, so eventually he quit. So—when he quit, I quit right along with him."

Jones went to New York to manage the Du Mont station. McKenzie went again to James G. Riddell. On February 15, 1952, Ed McKenzie moved over to WXYZ, and a $100,000-worth of business moved with him.

Riddell gave McKenzie a real star set-up at Wyxie. Instead of broadcasting from the Mendelssohn mansion, he had Jack Tierney build him a studio in the Mutual Building. A pretty brunette with sharp features, Joyce Chapman, had been doing publicity for WJBK, had released the story that Jack, the Bellboy was leaving that station to go to WXYZ, and had been promptly fired for it. She now followed McKenzie to WXYZ and became his Girl Friday. McKenzie was given the same kind of deal Mulholland had had and which Fred Wolf had now: he was paid a guarantee plus so much per spot. It was not long before he was again in the big money.

He and Riddell made one mistake, however: they used the name Jack, the Bellboy. George Storer sued and the court ruled that McKenzie could not use the name; it belonged to WJBK.

Riddell: "George Storer was my friend and we appeared in court against each other and that was one of my regrets. I was quoted in the paper as saying, after we lost the decision, that 'as long as there are people like Storer around, these kinds of things will happen.' It was a loser's statement. We have become good friends since. Strangely enough, I almost went to work for Storer after that; he wanted me to move over with him and run JBK and WJBK—TV." (Twenty-odd years later, Riddell became a consultant for Storer.)

Even without the name he was known by, Ed McKenzie was a good buy. The station made a lot of money with him, and he with the station.

Riddell now had two men running record programs, and although another had not yet reached his maximum potential, Jim sensed what was happening. You could pay these men a lot of money and still come out ahead because you didn't have to maintain an orchestra of twelve to fourteen men, and you could eliminate all the overhead of actors and writers and directors and sound men. Furthermore, personalities like Wolf and McKenzie were easier to sell than big program packages. Moreover, the advertiser was not confined to one program; he could spread his commercials throughout a 24-hour period.

In March, 1952, two months after McKenzie went on the air at Wyxie, Bob Baker had a phone call from Riddell.

"Bob, I've got an audition record here. I've been told it's good. Doug Campbell has a client—aluminum windows. They want a one-shot half-hour show and are talking about having the kid do it. I'm sending the transcription out to you. Tell me

Dick Osgood and Ed McKenzie, original "Jack, the Bellboy"

what you think."

The audition disc was labelled *Winter & Company.* On it, a teacher of philosophy at the University of Michigan performed not only as Winter but as the "company." He did a number of characterizations in skits and sketches which he had written himself. It was a remarkable display of talent.

Winter's real name was Saul Wineman. The "Winter" was inspired by a brother who enjoyed knocking about town but didn't want their very orthodox father to know it, so he got himself on various mailing lists as David Winter. The "Paul" evolved from Saul.

Paul Winter had majored in history and German at the University of Michigan and had played classical records over the U. of M. radio station. He had done a couple of years of social work, then returned to the university for graduate work in philosophy. He had taught physics as well as philosophy; he had auditioned for Chuck Livingstone who had never hired him; and he had done some acting on WWJ and WJR. He was now 29, and married.

A university acquaintance and neighbor, Adele Hager, had gone into advertising in Detroit and Paul had told her that he enjoyed performing on the university radio station so much that he was thinking seriously of giving up his academic career to become a professional broadcaster.

"Why don't you put together an audition disc and give it to me?" Adele suggested. "I'll give it to a friend of mine who works at WXYZ in Detroit. He's a salesman there, Doug Campbell. Let's see what happens."

Among her clients was the man of the aluminum windows. She took him and the *Winter & Company* audition record to Doug Campbell. The result was the profesional debut of Paul Winter. After the broadcast, he received a call from Hal Neal.

"I liked that show," said the new Sales Manager. "I'm going to try to sell Jim Riddell on putting you on staff."

But nothing happened for several months.

Baker received another phone call from Riddell.

"Bob, I want you to audition a man named Jack Surrell. Make a tape for us to listen to. The Stroh Brewing Company will buy six nights a week, three spots a night, if we come up with a good black disc jockey. Surrell is black. Ed McKenzie found him in a night club playing piano. Maybe we could use piano with records. Work something out with him and let us hear him."

Jack Surrell was thin haired, thick lipped, quiet, well-spoken and unusually talented. He did his late night show for Stroh's from the Lone Ranger studio, and Bob Baker produced it. Jack used *Penthouse Serenade* for his opening theme and *Night Train* to sign off with. He had a very pleasant air personality; and his special gimmick was segwaying on the piano from the key one record ended in to the key of the opening of the next record while he spoke easily and quietly.

Hal Neal: "That was long, long before the days of Civil Rights pushes and things as we know them today. Doug Campbell had the Stroh Brewery account and they were interested in developing their sales in the black community. It was a big 52-week contract to give them exclusivity in the program for beer. They did many things with it—merchandised him and did counter card signs that were put up in the bars and so forth of the black community. Harold Christian and his merchandising department were deeply involved.

"We didn't hire Jack *because* of the Stroh thing; it developed out of it. They were the first to jump on the bandwagon. Went on for a number of years."

The important thing was, programs were being replaced by record shows and the men who ran them were not staff announcers but "personalities" called disc jockeys. Wolf, McKenzie, Surrell, Winter—there would be more.

The year Jack Surrell came in, two other Jacks went out.

Jack McCarthy was not happy. His feelings about broadcasting from the porch and building a hut on the lawn had been ignored. Winter and Surrell had been sent to Baker over his head. His highly rated telephone program had been first cut, then discontinued. Pival was not giving him a crack at television. He unburdened himself in a letter to his one-time predecessor, now vice president and TV director of Knox Reeves Advertising Company in Mineapolis, Russell E. Neff. Somehow, the letter was misdirected and returned to WXYZ. It landed in the hands of Riddell.

Jack McCarthy resigned the job of studio manager but continued playing the Green Hornet until it ceased being broadcast live, in December, 1952.

One would have expected the job of studio manager to go to Jack Rellis, or perhaps even to his assistant, Bob Baker. Neither one got it.

When Hi-Speed went off the air, WXYZ radio had no commercial need for the orchestra. It had moved to television in the Maccabees Building. Phil Brestoff and his wife, Dee Parker, now spent all their time in television. "Auntie" Dee did a show with children which by year's end was the only locally produced television program to land in the Top Ten rating by Pulse: *Howdy Doody,* Kate Smith, Perry Como, Auntie Dee, *Strike it Rich,* CBS News (Doug Edwards), Camel News, *The Big Payoff* and *Captain Video* (Al Hodge).

Brestoff haunted the executive offices, observed what was going on and seized opportunities. It was he who took over the job of studio manager in radio. Bernardine Ryan quit and there was a new secretary. One of Brestoff's first actions was to send for Paul Winter. Paul went on staff in September, 1952. His first regular program went from 6:30 to 7 in the evening and was called *Winter & Company.* It consisted of recorded music and skits written around topical matters. Paul played many different characters. But not without some trepidations: "I was scared stiff," he confessed. "I want you to know. Absolutely and positively terrified."

Three or four months after he had started, Muntz TV purchased his time—but with Ed McKenzie. This squeezed Winter off the air, for there was no other available time. Gloria Swanson, Tennessee Ernie Ford, the "soaps" and other remnants of network programming consumed all the time not used by already established local shows. However, Riddell at the insistence of Phil Brestoff, kept Winter on staff, but pressed the studio to find something for him to do. One of Paul's intriguing characters was "Lonesome Luke Borgia." A record program created around that character was put on from 5:30 to 6:15 AM, ahead of Fred Wolf. Winter did that program for years.

The second Jack to go was Rellis. By the fall of 1952, his assignments had petered out to one little live hill-billy program at noon. It was about all that was left outside the Trendle dramas, news, *Show World, The Lady of Charm* and those rising stars of radio, the disc jockeys.

Jack Rellis was the son of a practicing Christian Scientist, and a born optimist. He bought Theda a mink stole for Christmas and hid it in one of the walk-in vaults in the Mendelssohn basement. He went to see the head of public relations of the Hudson Motor Car Company about being their PR man on the west coast. The job would have consisted mostly of putting celebrities in Hudson automobiles and getting them into parades.

Rellis: "Fortunately, they didn't hire me because you know what happened to Hudson. The same thing that happened to radio."

When he returned to his office, Rellis found that the door connecting his office with the one now occupied by Phil Brestoff had been locked. He found out why when he sat down at his desk and discovered the reason tucked into the pocket at one end of his desk pad. It was a note of few words explaining that since the station was producing no more programs, it had no further need of a producer.

Two of Rellis' best friends, Dick Osgood and Mark Beltaire who wrote the *Town Crier* column in the *Free Press*, advised him to get a refund on the mink stole, but such was the cheery confidence of Jack Rellis that he gave his wife the stole anyway. Within three days, he went to a new job with the Ross Roy advertising agency.

Stray items from *Observations by Osgood* in the *Detroit Free Press*, 1952:

Harry Wismer voted Sportscaster of the Year by radio and TV editors polled by Radio Daily.

Seven hundred American Women in Radio convened in Detroit in April and Edythe Fern Melrose was elected chairman. In September, Edythe was named Radio and Television's Outstanding Woman of the Year by the Cleveland Women's Advertising Club.

Norman Lenhardt goes to WJBK to announce Ralph Bingé and Joe Gentile who followed Dick Jones over from CKLW.

Don Karle, who joined WXYZ radio as an announcer from a smaller station, switches to the announcers' booth on Channel 7.

Don Wattrick buys a drive-in restaurant.

Ross Mulholland came a cropper on the west coast, now back in Detroit on WWJ.

Rosemary Clooney's sister, Betty, a regular on Channel 7.

Johnny Desmond, once of Wyxie, in his 4th year on Don McNeil's *Breakfast Club*.

Irv Immerman appearing as The Black Spider on Sunday afternoon movies, Channel 7; and a girl named Loree Marks, asked how she makes herself so sexy as the White Camellia hostess, says, "I lower my voice and raise my eyebrows."

Announcer Bob Longwell, on a television commercial, was supposed to prove the unbreakability of Permaglass by whamming it with a hammer. The head of the hammer flew off. Permaglass was delighted but there was a complaint from the company that made the hammer.

Brace Beemer asked John Pival to visit the Beemer estate in Oxford and meet his son, Bob. Bob, Beemer thought, was a good bet for TV.

"Go saddle up Bob's horse," the father ordered, looking at his younger son, J.D. This was unusual; each boy ordinarily saddled his own horse; but today Beemer was giving Bob the star treatment for his "audition." J.D. obeyed; but he also harnessed his own horse. Beemer took Pival to the edge of a field to watch Bob display his riding ability. While Bob "auditioned," J.D. at the far end of the field, put on his own display.

Pival decided against Bob in favor of J.D. They gave J.D. the name of Justice Colt, and by March, 1952, he was so popular that he was mobbed by his fans when he awarded puppies to youngsters who had excelled in Safety Patrol work. Brace and Leta Beemer watched from a box.

"Brace, J.D.'s wearing out," said Leta. "Get him out o' there."

"Let him sweat," said Beemer, beaming with pride. "Do him good to know what I went through for years."

But Brace took pity eventually and yelled out: "Justice Colt, your car is waiting. You have a show to do. Come on! Hurry!"

The son looked up from the autographs he was signing and yelled back, "What keptcha?"

Jack McCarthy took a trip to New York between Green Hornet shows. There he found the original Hornet, Al Hodge, doing *Captain Video* and another Hornet, Bob Hall, announcing at CBS. Former WXYZ announcer Hugh Holder was announcing Edward R. Murrow's *See It Now*, and another, Gunnar Back, was moderator on *America's Town Meeting of the Air*. Bob Hite was announcing Waring's Pennsylvanians. Mike Wallace had moved to New York from Chicago and was doing a series with Buff Cobb. Douglas Edwards had become one of the most important news men at CBS. And McCarthy found three ex-Detroiters rehearsing *The Romance of Helen Trent*: Ernie Ricca from WWJ, directing; Fielden Farrington, announcing; and a former Trendle actor and taxi driver, Chuck Webster, acting a part in it.

The next month, McCarthy visited Hollywood and reported that more Detroiters failed than succeeded there but he did find Amos Jacobs, George Seaton, Martha Scott, John Hodiak and Tom Harmon all eating very well. WXYZ's former sound man, Ted Robertson, was directing *Dr. Christian*; Forest Owen, once a Wyxie announcer, was head of television activities of the Wade Advertising Agency, producing the Curt Massey-Martha Tilton series; and another ex-Detroiter, Earl Ebi (WWJ) was connected with the *Lux Radio Theater*.

By the end of 1952, when *The Green Hornet* went off the air, Jack McCarthy had no prospects on either coast, but he had an idea for a television show.

As Advertising and Sales Promotion Manager of WXYZ, keen-witted Bill Morgan was discovering the difference between talent and executives.

Morgan: "Once I got used to the business operation and saw how things worked, I felt that talent was very unrealistic—not necessarily trivial—but kind o' living in a dream world.

"When I was an announcer, the announcers would gripe like hell: 'We're doin' this good show, why the hell can't the salesmen get out and sell it?' Well—then you get downtown and you hear the salesmen saying, 'Those goddam announcers! Why can't they do a better job so we can sell!' "

Morgan broke out in a merry, infectious laugh as he recalled the vagaries of human nature, his nose wrinkling, his eyes crinkled.

No longer station breaks and idle time in between for Bill Morgan. There was a great deal of promotion for the new television operation, and material to prepare for sales presentations and trade publications. He had several girls working for him. Added to everything else at the end of 1951, he was put under pressure by the network to enter a promotional contest conducted by Gruen Watch, sponsor of a *Theater Guild* series which was being carried by some hundred stations. The promotion manager who did the best job, documented, for the Gruen program, would receive $150 in cash and a beautiful Gruen Autowind. There were other prizes. WXYZ being a key station, Bill was under special pressure from ABC. But he was so busy with the everyday burden that he didn't have a chance to assemble his material for the contest until very late. He finally threw up his hands and went to Riddell.

"Listen, Jim, I don't think we're going to make a real good showing in this thing. I—I don't think we got enough stuff. Should we—shall we go ahead and submit it anyway—or—?"

"Oh yeah," Riddell ruled. "We don't want the cheese; we just want out of the trap. We've got to submit it."

Bill won the contest! He split the $150 among the people in his department but kept the Gruen Autowind.

Morgan: "Riddell was the complete leader. I mean, we loved him. He was fun to work for. I think probably, next to H. Allen Campbell, this is the best salesman I have ever seen."

Unlike Campbell, however, Riddell did not do any direct selling. Neither did Hal Neal.

Riddell: "Sometimes you can be a good sales manager without ever being a salesman, and sometimes you can be a very bad salesmanager after having been a good salesman. The reason, I think, that we had a successful sales department was that we left the salesmen alone. I believed in letting a man earn his money, so I never personally sold anything, but I was always very close to a lot of the accounts, and of course, I knew most of the agencies. But I would never go out and see an account without the salesman being with me because it wasn't fair. He's entitled to not just money, but self-respect."

Riddell loved people; he loved golf; he loved fun. But in 1952, he was well aware of the storm that was raging and of the winds that were blowing, and that the old house was being blown down. Radio would never be the same. He had to think of some way to create a new WXYZ. He had help but, as with George Trendle before him, the decisions and much of the hard thinking were his own. He himself was stimulated by the enthusiasm he inspired in others. Hal Neal was "hot as a pistol." Hal, in turn, stimulated, and found stimulating, two young radio salesmen, John Lyon, short, florid with piercing blue eyes, and Doug Campbell, short, happy and smartly progressive.

Neal: "The three of us were all young and full of get-up-and-go, and lots of ideas. And—uh—hungry. The three of us really did a lot of thinking and talking and working together."

One thing that was surviving the storm was a stubborn faith in various kinds of audience measurement—the ratings.

Bessie Ashton: "You have to have the figures—this business—sets in use—or listeners—or something else. They're all figures—strictly figures. Well, my dear, I started—I started putting ratings on spots—spot offerings—in radio. I started listing the ratings on spots. They had never *bean* listed before."

Bess pronounced "been" "bean." What she was saying was that if a salesman had a client who would buy spot announcements, he learned from her what times were available and what the audience measurement was said to be at those times.

Ashton: "I would list them. I mean, when I sent out, sent out—weekly availabilities—weekly availability lists—why—list the spot and the rating."

James G. Riddell took radio beyond the disc jockey. His new idea came to light during a meeting of the gas house gang. He stood in the window behind his desk looking down at the changing lights of the fountain 17 floors below in Grand Circus Park.

"Doug brought in Sinclair Oil today," Hal Neal announced behind him.

Riddell turned into the room and looked at Doug Campbell who was comfortable on the couch cushions.

"Don Wattrick?" Riddell asked pleasantly.

"Three 5-minute sportscasts a day, 4, 5, and 6 P.M." said Campbell, grinning. "I've also got First Federal."

"Savings & Loan? What did they buy?"

"They want a newscast at 11 o'clock."

Riddell looked at Neal, a little surprised. "Did Cunningham's drop it?"

"Not at night. In the morning," said Neal.

"11 *A.* M.," Campbell confirmed, waiting for praise.

"Doug," Riddell said after a moment. "I'm going to turn it down."

"First Federal Savings and Loan!" cried Campbell as though Riddell had just cancelled the Second Coming.

"A newscast at 11 o'clock in the morning would just clutter up the—"

"I agree," said Neal. "Look. We've got Fred Wolf on—a nice island of music and talk. Then you've got Don McNeil and the Breakfast Club—a whole hodge-podge—True Story Hour, another give-away telephone show—"

"I'm going to try something," Riddell announced. "It used to be that you'd produce a program in radio. My idea is to produce a DAY. Where you used to produce an hour show, now you produce a whole schedule. You program the station."

"The afternoon is a hodge-podge, too," said Neal.

John Lyon, his chair tilted back against the wall, looked from one to the other, blue eyes bright but confused.

"You could keep Wolf on all day, or McKenzie, and make money," he said, trying to keep up.

"What we have to do," Riddell went on, "is figure out how the audience flows from one show to the next show to the next show. We know that the only thing that you can sell is whatever has a rating, whatever is being listened to. What we have to find out is where the people *want* news—not where the advertisers want it, but where the *people* want it. That's what I mean by programming the station."

"Network shows are on the way out," Neal observed.

"You mean you would go to an advertiser and say, 'Here is our schedule. What do you want to buy on that schedule?' " Campbell asked.

"What happens if somebody doesn't want to buy into your format?" Lyon asked dubiously.

"No matter how much money somebody has, if they spoil your flow of audience, it's going to hurt the station," Riddell reasoned. "Anyone that—came in with a show—. For example, Chevrolet's *Musical Moments*—quarter hour—beautiful things—sponsored all over the country. Well, under this plan, we wouldn't accept that show because it wouldn't fit our format."

"Foreign to your audience," Lyon agreed, catching on.

"Right. It would interrupt the flow."

This was a switch for Riddell—turning down money.

"How do you know it'll work, Jimmy?" Campbell demanded. "Has anybody tried it?"

"Not to my knowledge. No." Riddell grinned. "What do you want to do? Run for cover?"

"I like it," Neal said flatly.

"I'm just wondering what to do with First Federal," Campbell admitted.

"Is Osgood sold out with the newscasts in the morning?" Riddell suggested.

"No sign of anything cancelling," said Neal. "And that's a clutter, too. Too many of 'em, and they're too long. Fifteen minutes is too long."

"Well, we can't touch those yet," Riddell warned. "Sell Federal on the format idea; sell 'em spots." Riddell's face broke into a wide grin as he gave the advertiser a nickname: "First Format Federal."

The tension was broken.

"When are you going to start this?" Lyon asked.

Riddell chuckled.

"We've already backed into it," he said. "Wolf is picking up in the morning; we're going to get a lot o' listeners and make a lot o' money with Ed McKenzie; and we have Paul Winter and Jack Surrell. Your 5-minute sportscasts with Wattrick fit right into the flow."

Thus, at the end of 1952, James G. Riddell and WXYZ were planning a new edifice. As the storm died down, much silt and clutter remained to be cleared away. And caught in the ruins of the old house of radio were Trendle's dramatic treasures. As the sun rose on a new day, they slowly came apart and were cast aside.

38
A Merger

The wedding of Paramount Theaters and the American Broadcasting Company, first proposed in 1951, finally took place two years later. The merger was completed in February, 1953. It was no small affair. The participants represented combined assets of $140 million. There were now reported to be 651 theaters in the Paramount chain. In 10 years, from 1942 through 1952, ABC had increased its billings more than $36 million—from $15.7 million to $52 million.

For ABC it was an especially good marriage. CBS, aggressive and profitable, and immensely wealthy NBC had always been out in front because ABC had lacked money to develop competitive programs. Edward J. Noble, now Chairman of the Finance Committee, revealed that Paramount Theaters brought a dowry with them of up to $40 million. Certainly part of this would be available for improved programming. Indeed, only eighteen days after the wedding, Robert Kintner, President of the American Broadcasting Company, announced that $100,000 would be spent immediately on developing "family serials" for TV and $2 million on getting increased power for its five owned-and-operated stations. In New York, the key station call letters were changed from WJZ to WABC, and across the nation 14 new affiliates expanded the ABC television network to 95 stations; and by the end of March there were 362 outlets for ABC-Radio.

The optimistic Mr. Kintner seemed to have high hopes for the survival of network radio. In February, he announced a new concept in radio for weekends made up of music and news—Margaret Whiting as hostess Saturday nights. Burgess Meredith hosting on Sundays, and on Mondays, Jan Peerce. In September Kintner bravely launched a big hour-long live radio program with George Jessel as emcee, celebrity guests and music by Paul Whiteman; also *Horatio Hornblower,* a dramatic series, starring Michael Redgrave.

Early in March, top AB-PT brass descended upon Detroit. Among the giants, Robert Weitman, vice president in charge of programming and talent in the TV division. In the studios, he spotted Johnny "Scat" Davis who had tooted a horn in his brother's band when Weitman was manager of the Paramount Theater in New York, and Dee Parker who had sung at the Paramount with Jimmy Dorsey's band. One he did not recognize was actor-announcer-disc jockey Stan Dale who proudly revealed that he, too, had worked for Weitman at the Paramount—as an usher.

As the mighty giants of AB-PT whooshed out of Detroit to return to Manhattan, an invitation from Miami to Jim Riddell was sucked into the vacuum they left behind. George B. Storer was in the market for a manager for his Detroit station, WJBK. Riddell flew to Florida and the two talked aboard Storer's boat. Storer was

Edythe Fern Melrose and staff preparing for a Christmas broadcast from the House of Charm

obviously fed up with the transitory nature of his top personnel in Detroit and wanted a man who would satisfy him and stay to build up his station. Even though the two had fought in court, Storer liked and admired Riddell and his pitch was strong. He pointed out that WXYZ was no longer a Detroit-anchored enterprise. There was no George Trendle upstairs, nor an H. Allen Campbell between him and the remote corporate ownership in New York.

Riddell remembered the readiness of New York to install its own man in Detroit when Noble first bought the station. New York could do it again.

Storer also suggested that AB-PT might want to move him to another city. How would he like that?

He wouldn't. Jimmy Riddell loved Detroit; it was home. He and his family were happy; he had many friends of whom he was fond; and he had prestige.

Storer's final clincher was an offer that was not cheap; and it was solid.

"All right, George," Riddell said at last. "We've got a deal."

Duplicity was not part of his nature and upon his return to Detroit, Jim called Robert Kintner in New York and told him of his decision to leave WXYZ. Kintner persuaded him to go to New York to talk it over. He went. He spent time with Kintner; with Leonard Goldenson; with Ed Noble. They would not let him go; they gave him a better deal; they prevailed upon him to stay with ABC.

Riddell: "And I want to tell ya, the next few days—it was just hell. Ducking. I finally had to call him (Storer) and say, 'George, I can't do it.' I said, 'I know I gave you my word, but I can't do it.' And he was understanding enough."

What Riddell did not want to have happen to him happened almost immediately to Earl Hudson. United-Detroit Theaters was part of the Paramount chain. Leonard Goldenson persuaded Earl to leave Detroit and become vice president in charge of the western division of AB-PT with headquarters in Los Angeles.

The corporate shuffle had begun.

Between radio and television, the Lady of Charm made $100,000 at WXYZ in 1952. Many of the advertisers who bought her were national. They placed business with her through national agencies. Sponsors would say "this is what we want to sell; you sell it in your own way"; and she did. At the end of '52, Riddell called her down to his office.

"Edythe," he said, "this year we have paid you more than ABC paid their president."

"I didn't tell it," said Edythe, the ends of her mouth disappearing into her round cheeks in a happy smile. "You're the one who released that."

"I know it. I'm proud of it."

"So am I," said Edythe. "Young and Rubicam wants me to move to New York."

"That so?" Riddell showed no apprehension; he was genuinely pleased and interested.

"The manager of General Foods was there. Young and Rubicam buys the time on my television show for General Foods, and he said, 'We should be very happy if you would come to New York'."

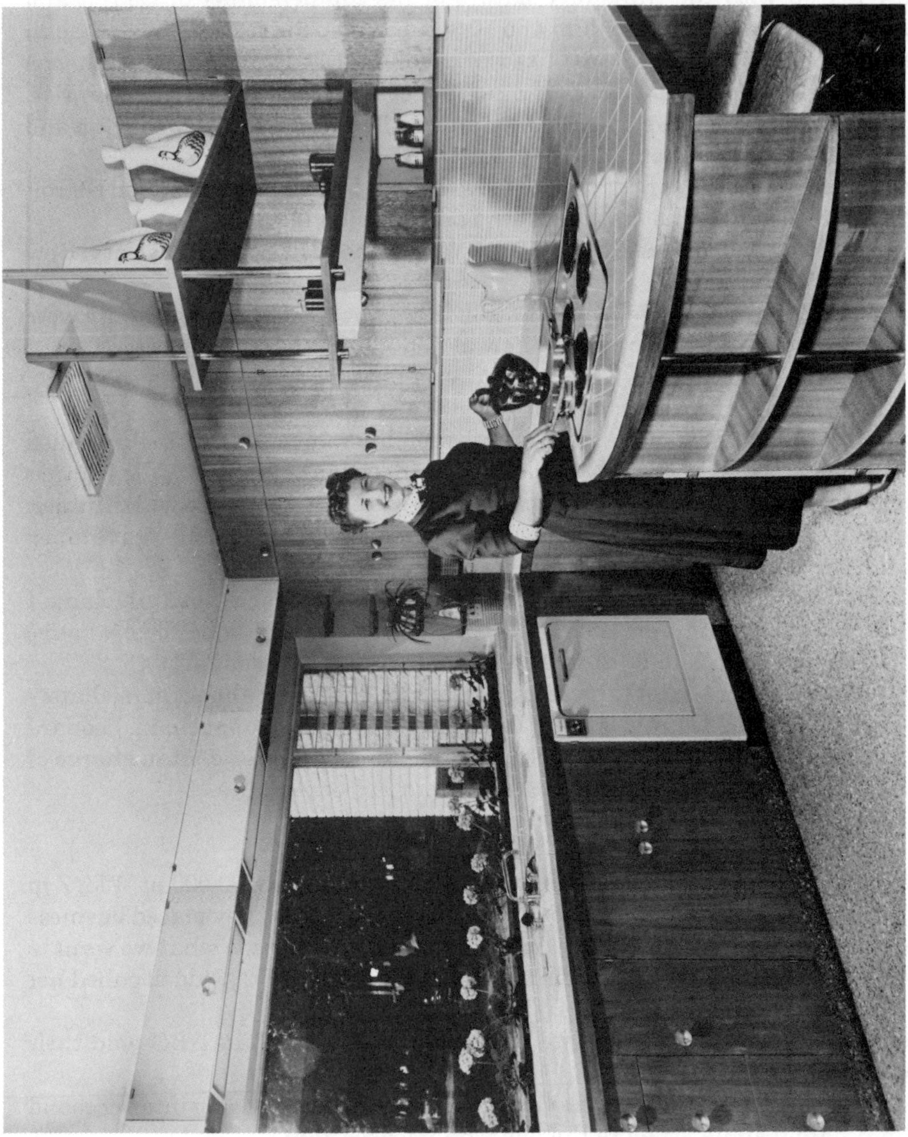

The kitchen. . . House O'Charm. . . every inch especially designed for ease in food preparation with each area away from any traffic lane.

Edythe Fern Melrose, in center, came to WXYZ in 1941 with an account she had sold—the Chrysler Dealers. As The Lady of Charm, she was popular and very "commercial." Building an actual House of Charm was her idea. She built three. This one, the first, was on Lake St. Clair just north of Grosse Pointe. The other girls were on her staff. The man on the end was probably the builder. The younger man behind Edythe is Larry McCann, her announcer after Hal Neal. Both moved into TV; both were with WXYZ many years. Edythe's death notice should appear in a later book. McCann became associated with the Mutual Network, first in Florida, then in Washington, D.C. (House built in 1948).

The House O'Charm...first home ever built for testing of all products before recommendation on WXYZ's Charm Kitchen...mikes were in each room for direct broadcasts where home furnishings, decoration, cooking and entertaining were broadcast 5 days per week.

"What did you tell him?"

"I said, 'I would be very *un*happy.' New York would not let me be autonomous as I am here; it would no longer be my show."

"What can we do to put on a big promotion for you?" Riddell offered.

"Well, Jimmy, do you want to make anything of the fact that I came to this station 10 years ago?"

"That's it! Our 10th anniversary."

WXYZ splurged in New York with a big party. Twice as many agency people showed up as had been invited, and Harold Christian had to rush around Manhattan in a cab to pick up additional liquor.

Again in 1953, Edythe Fern Melrose was chosen Detroit's Advertising Woman of the Year.

One day early in 1953, Phil Brestoff and Dick Osgood encountered each other outside Phil's office.

"Sometime, if ever one of my shows goes off," said Osgood, "I have an idea for a record show I'd like to do."

"Come into my office," said Brestoff.

"Phil, I can't. I'm on my way downtown to a screening. Besides, I couldn't handle anything more right now anyway."

It was true. Besides all the movie screenings, the morning newscasts, *Hollywood Headlines* and *Show World*, Osgood was broadcasting and telecasting from theater lobbies. Also, he was touring Kiwanis and Rotary clubs in the Detroit area, doing speeches on the motion picture industry—a promotion of Alice Gorham. It kept him out several nights a week for many months.

"Well, what's the idea?" Brestoff asked.

"Well—some day I'd like to do a show of theater and movie music. I'd call it *Reprise.*

"Reprise. Nobody'd know what it meant." But Brestoff's little black eyes were absorbing.

"Well—*Reprise, Encore*, Something," said Osgood. "It's just an idea for the future, like money in the bank."

"Tape an audition."

"Phil, I just do not have the time right now."

Not long after this encounter, Hal Neal was talking with Paul Winter. A half hour had opened up in the late morning.

"Paul, there isn't any show music program in the city and I think it might go with women in the morning. Why don't you and Bob Baker work something out?"

The show was called *Curtain Calls* and was extremely successful.

According to the ratings in May, 1953, the top five television shows being produced in Detroit were: *Auntie Dee* (WXYZ); 11 P.M. News (WJBK), *Motor City Fights* (Fred Wolf—WXYZ), *World Adventure* (George Pierrot—WXYZ) and *Playschool* (WWJ). *Auntie Dee's* ratings even topped Kate Smith's! Wolf's *Motor City Fights* were on the ABC network.

Betty Clooney, Rosemary's sister, was on Channel 7, and so was an ambitious young comic from Cleveland. He had been brought to John Pival's attention by the aggressive young announcer Stan Dale.

"I knew him in Cleveland," said Dale. "His real name is Milton Supman. They call him Soupy."

"Soupy Hines. I've seen him," said Pival. "My wife and I were visiting my folks in Akron and we made a side trip to Cincinnati and saw him in a club there. Yeah. Get him up here. I'll talk to him."

Soupy Hines had auditioned for radio in Detroit three months before with no luck. He was now appearing in a night club in Cleveland. He returned to Detroit and met Pival. He was 27, married, and had a 17-months old son. He was a veteran of the Navy, had taken part in the invasion of Okinawa and was on the first U.S. ship in Korea. More recently his luck had run out. He was hard up. Someone pointed out to Pival a hole in the comic's shoe.

"So what!" said Pival. "That's show biz. It's got nothing to do with the guy's talents."

Pival, John Lee, Pete Strand and Soupy created a noon-time show for kids. There would be movie comedies and Soupy would make up a luncheon menu each day with producers Jack Schwartz and George Manno.

"I'll wear a funny silk hat," said Soupy.

"And a great big bow tie," Pival suggested. "A yard long."

"Yeah," said Soupy.

"But drop the Hines," Pival ordered. "C.F. Smith Foods will buy this show, but they won't buy a guy whose name sound like Heinz pickles."

"What'll we call him?"

"Just call him Soupy."

The show was an immediate hit. The audience wanted to know more about Soupy. Pival called another conference.

"We gotta get a last name."

Somebody read from a phone book. "Sachs—Saddler—Sage—Soupy Sage—no—Saint—there are a million Saints in here—how's Soupy Saint?"

"Hell—no!" said Pival.

"Salansky—Sale—"

Pival pounced. "Sale! Sales! Chic Sales. Remember Chic Sales? He was great. That's it—Soupy Sales."

Soupy Sales worked happily for eleven months for AFTRA minimum without a contract. Then Pival gave him a contract for a year—$11,000. The second year, 1954, he signed for $28,000; $59,000, the third; $84,000, the fourth; and on his fifth contract at WXYZ, he topped $100,000.

He also legalized the name John Pival had given him.

In September there was a shake-up. New York had big plans for the new season, and Riddell announced local changes in line with the national push for bigger audiences in television. Johnny "Scat" Davis was let go; and Betty Clooney was

Soupy Sales

through in Detroit. There would be a new morning show with Warren Michael Kelly; and such was the success of Soupy Sales that another night club comic named Marv Welch was hired for a morning show for children.

Marv Welch, 29, was an energetic and fortissimo extrovert. He had grown up on smoke and the smell of beer. His parents had been too poor to hire a sitter, and he had gone with them to the night clubs where his father sang and his mother played the piano. He had served three years with Patton's army in World War II, and had worked his way through Wayne (State) University as a stand-up comic in the clubs. When John Lee sent for him and told him they wanted him for a show for children in the morning, he couldn't believe it.

"What if I forget and tell a dirty night club joke?"

He, Lee, Pete Strand and John Pival worked out a show and Welch named it *Wyxie Wonderland*. He played Wyxie.

(WXYZ was first called Wyxie on *Show World*. Dick Osgood asked George Trendle if he objected and Trendle made a little grimace but said, "Sounds all right when *you* say it.")

By the time WXYZ-TV celebrated its 5th anniversary in October, 1957, Soupy Sales had added a show at 11 P.M. for adults; and a cute pianist from cocktail lounges, Diane Dale, was doing a show in the morning. Already television was outgrowing its space in the Maccabees Building. What had served as a men's make-up room was now so cluttered as a storage room for props that there was no room in it for a person. In November, Riddell began looking around for a place where he could build facilities for both radio and television.

Incredibly, *The Lone Ranger* had weathered the storm. In January, 1953, with Script Number 3,128, this hardy program celebrated its 20th birthday. It was now being broadcast three times a week over 249 stations. Michigan Senator Homer Ferguson pointed out, in paying it honor on the floor of the U.S. Senate, that it was now the oldest continuous half-hour program in radio history. It had collected 19 radio and TV awards for "distinguished service."

During the Year, George Trendle took Chuck Livingstone to the west coast with him in connection with the television of the show which was now being seen on 90 stations. It was sponsored in 40 states by General Mills, and by American Bakeries in 8 states in the southwest. Trendle-Campbell-Muerer had 60 license agreements for manufacture, distribution and sale of toys, novelties, books, records, musical instruments and other items tying in with their programs.

The Lone Ranger comic strips were appearing in 117 daily newspapers, 28 of which were published in foreign countries. They were also being printed in 119 Sunday papers, including 33 in Norway, Argentina, Hong Kong, South Africa, Cuba and Canada. Lone Ranger comic books had a circulation of some 2 million a month. Fifteen Lone Ranger novels were in distribution. The comic books, novels and copy for the comic strips were all written by Fran Striker.

Observations by Osgood in the *Free Press* was keeping up with people who had worked at WXYZ.

Jack McCarthy's program, *Hit a Homer*, was produced on WWJ.

Johnny Desmond came home to Detroit with Don McNeil and the *Breakfast Club* to do a week of broadcasts from Masonic Auditorium.

Mary Lou Harrington, another graduate of *Radio Schoolhouse*, was a regular on *One Man's Family.*

And Jack Rellis, producing commercials for *Wide, Wide World*, introduced Fibber McGee and Molly to television on a commercial, adapting their famous cluttered closet gag to the cameras. When Fibber opened the closet, everything fell out, including, last of all, a set of NBC chimes. Half the stations on the network thought it was a cue and cut out of the program.

Osgood went on his first junket to the west coast in 1953—a one-man jaunt. He spent a half day each for the first five days on different movie sets. On one set, the director said, "Watch that kid who's boxing the star in the ring. He's a natural. You'll be hearing about him. He's another John Garfield."

It was Charles Bronson. The "star" is forgotten.

He visited Leslie Caron in her modest home in Laurel Canyon, and found her painfully shy and uncommunicative. There was a half day on the set of Elizabeth Taylor's *Rhapsody*; a half day with Richard Widmark, and a luncheon with Cyd Charisse.

He spent the weekend as a Detroit delegate to the national convention of AFTRA.

The second week he visited TV sets of Ray Milland, Joan Caulfield, Dick Powell and others, and had luncheon with the author of *One Man's Family*, Carleton E. Morse.

On his last night in Hollywood, too late for dinner and too early for supper, he was sitting at the bar of what had been the beach home of Marion Davies with an old friend from New York theater days. Except for the man behind the bar they were the only two there. The friend (unemployed) talked so *big time* that Osgood's ego was dragging. The bartender who had been polishing glasses asked him if he happened to come from the middle west.

"Yuh."

"Detroit?"

"Yeah!"

"Aren't you in radio?"

"Yes."

"You're Dick Osgood."

Osgood's ego picked up.

"My wife and I used to listen to you all the time back there," said the bartender. "She wouldn't go to a movie unless it was listed on your Five Best Bets. I got tired of that. One night, I picked a movie I wanted to see and I told her, this is what we'll see.

"She said, 'Dick Osgood doesn't recommend it.'

"I said, 'To hell with Dick Osgood.'

"So we went to see the movie I picked. It was lousy."

" 'Serves us right,' my wife said. 'It wasn't on Dick Osgood's Five Best Bets'."

The bartender laughed.

Don Wattrick had praise from out-of-town in 1953, too. Sherley Uhl who wrote a column in the *Louisville Times*, after finding fault with sportscasters in general, wrote: "Among those gridcasters heard and seen here this Fall, my top rating goes to Don Wattrick who calls the Cleveland Browns games each Sunday over WAVE—TV.

"Don's commentary is just humorous enough to keep the proceedings in proper perspective. He never foams at the mouth like Harry Wismer and while it's his job to talk up pro-football, he doesn't forget that it's sport and not a diversionary skirmish in the Korean War."

By the end of 1953, sales of radio and television sets (not counting car radios) became equal for the first time; Eastman reported that 68% of its raw film sold in Hollywood was now used for television shows; WXYZ-TV was in the black; the disc jockey had become the most important factor in radio; and Jim Riddell's idea of audience flow and formatting a station was a success.

Harold Christian's comment: "We had the best year we ever had in '53. And I mean *Radio!*"

39

A Sad Year for Actors

Dick Osgood looked up from his desk and saw Fran Striker grinning in the doorway. His teeth were large, separated and a little yellow, but the grin was warm and friendly. Osgood returned the smile and for a moment nothing was said. Twenty years of friendship filled the air; both were now in their fifties. The flesh below Fran's eyes was a little puffed and the whites of his eyes were yellow.

"You look tired." Osgood remarked.

"Wha'?" Fran turned his head a little to one side.

"Work all night again?" Osgood asked a little louder. It was 9:30 in the morning.

"Yes," Fran sighed in a resigned kind of way, his grin remaining. "A television script. They sent it back for some changes."

"You're writing the TV things, too?"

"Yes." He chuckled uncomplainingly. "People I never heard of—telling me how to write *The Lone Ranger*."

"Get tired of the grind?" Osgood asked, raising his voice.

Fran shrugged and nodded a little. His oldest son, Bob, was already home from Korea and since hostilities had ended, Don would soon be back, too. That worry was over. Daughter Janet and youngest boy Fran were still in school. The family would be together again, close and happy.

But Fran was tired. It was becoming increasingly difficult for him to knock out scripts; it was labor. But after twenty years of prolific output and supporting a large family, he still needed money; he still had to work.

"All I ask," he said, grinning, "is for one more year—maybe two."

A number of executives in the corporate structure of the American Broadcasting-Paramount Theaters disliked ABC President, Robert Kintner, but he and James G. Riddell were congenial.

"Which would you rather be," Kintner asked, "President of WXYZ or Vice President of ABC?"

"I never counted the number of Vice Presidents in ABC," said Riddell. "But you can't turn around without bumping into one."

Kintner's round face wrinkled when he smiled. "There are eighteen."

"That many."

"And there is only one President of WXYZ."

"Well, I think you're right."

So, in 1954, Jim Riddell, one-time mail clerk, became President of WXYZ, Inc.

Bill Morgan: "We got totally sold out on Channel 7. I mean we had no time left for sale; every hour of the evening was sold out. And we had all the beers going and it was a problem keeping them apart where we'd get a half hour intervening between them. We had a boxing match sponsored by Pfeiffer's Brewing every Saturday night. Okay. Now bear in mind we have no more time available.

"Now! The network pre-empts Pfeiffer's fights and puts in *Budweiser* fights. We have no other place to put Pfeiffer's. We're losing the hour and a major advertiser. *and* a contract with the promoter of those fights—I mean, a long contract. Everything is just a mess. The only way we could resolve this would be by pre-empting the E. & B. Brewing Company's program to accommodate Pfeifer. Pfeiffer is up in arms; E. & B. is up in arms; they are both furious. This all ends up in Jimmy's lap."

Riddell, with his positive optimism and sunny nature, accompanied the salesmen involved into a series of meetings—first with the Pfeiffer people, then with E. & B. Finally, in his office, there was a joint conference with WXYZ salesmen, Pfeiffer men, E. & B. men, and representatives of the advertising agencies of each. Everybody was ruffled except Riddell. He was not altogether relaxed; he had two fat accounts to save; but his balance and humor were not lost in the flush of argument.

At one point an E. & B. executive made a disparaging aside about the Stroh Brewing Company which had nothing to do with the problem at hand.

"Are you kidding?" laughed Riddell. "Stroh spills more on the floor than you bottle."

The Pfeiffer people laughed and relaxed a little.

But everybody tensed up again, frozen in their varying points of view. Nobody would give an inch. One of the E. & B. men was old and deaf and he had to go to the john. As he left the room, Jimmy jumped up and did an imitation of him. It was funny and everybody laughed and the laughter broke the mental log jam; everyone realized at last that the end of the world was not at hand, and began to reason.

The final solution was that one show was shortened and Pfeiffer and E. & B. were scheduled back to back. One agreed not to run a commercial at the end of his program and the other promised not to open his show with advertising.

Shortly after this, Bill Morgan switched from advertising and sales promotion to selling television. It was a mistake. He had absolutely no empathy for John Pival.

Neither did Bess Ashton who was handling traffic for both radio and television. When Riddell made John Pival Vice President, she expressed herself.

Ashton: "I told Riddell—at the time, I said, that's one man I wouldn't—I wouldn't trust as far as I could spit. Yeah. Jimmy had a very strange look on his face. I said I certainly wouldn't trust him. The man was driven by something or other. I mean, Pival—he was just driven by something. He was a driven man. 'Course alchohol probably had something to do with it. But he had delusions.

"Ja-Jo-John was always trying to pit one person against another. Trendle did it but in a different type of way; you knew it. But John—uh—you wouldn't know it because he was double-faced. Sure. Sure. Yah. Yah."

Late in the winter of 1954, George Trendle sent for Charles Livingstone. Trendle wore his years remarkably well. There was hardly an exterior sign yet of his aging—only the beginning of a sort of puckered look around his mouth and chin. His eyes were clear and he maintained that fresh just-showered appearance. There was no perceptible sag, physically, and his mind was as keen as ever.

"I don't like the way things are going on the west coast," he said with a little grimace. "They don't know what they're doing out there. Do you think you could direct films for television?"

"Why, I don't know, Mr. Trendle. I—"

"Radio is changing. It would be good for you to be out there where they're making stuff for television."

"I don't understand, Mr. Trendle. You're talking about *The Lone Ranger?"*

"They keep making little changes that I don't like. I just need somebody who knows the Ranger out there to supervise it."

"Me?"

"Radio is not going to last. You could give it a try if you want to."

Livingstone thought it over. He discussed it with his wife. They decided to move to California. On the last Saturday in March, 1954, there was a farewell party for the Livingstones. Brace and Leta Beemer were there, the Paul Suttons, the Gillie Sheas, the Fred Flowerdays, the Fred Foys, Tom Dougall, the Dan Beatties, Lee Allman and her husband, the son of the late Dewey Cole who had followed his father into sound effects, and Ted Johnstone who played Sergeant Preston's dog, King.

The sentimental party reached a lump-in-throat stage when Jay Michael read some doggerel he had written during Chuck's last rehearsal of the radio Ranger the night before:

Some 20 years of throwing cues,
Of ranting, raving and blowing a fuse
One score plus two of Malemute manges
Chlorophyl hornets and riding the ranges,

A lot of years of riding herd
On actors with the IQ of Mortimer Snerd,
A lot of time and deep frustration
Over radio operas that cover the nation.

A few gray hairs and an ulcer or two,
Acute indigestion and air that is blue.
But when it's all over and the show's put to bed,
And everyone leaves, heading home to be fed,

A warm sort of feeling settles over the frame
And we start to forget who's right, who's to blame.
It's then that we know a director's life
Is pretty much rugged. Just ask his wife.

He had his problems and a big job to do,
And doing it right means a job for you.
So tonight, Chuck, our hair's really down:
We're sorry to see you leave this town.

We've gone through a lot in the past several years,
We've argued, we've laughed, and we've shed a few tears.
But right now, buddy, the time's drawing near
When you and your family will be leaving here.
So from all of us—God bless you, Chuck,
May you leave in good health and with the best of luck.

Fred Flowerday became dramatic director of WXYZ-Radio; Tom Dougall was his assistant.

Harold Christian was now Commercial Manager of WXYZ-Radio. Early in February, he announced "the biggest hunk o' programming we've accomplished in three years." WXYZ-Radio had purchased 260 transcribed half-hour shows of Red Skelton through Frederic W. Ziv, a transcription and recording company with offices in New York and Cincinnati. There would be a 30-minute Skelton show, Monday through Friday at 7 P.M. Bill Stern's network sportscast would be transcribed off the air and moved to another time; *Show World* sank into the 8:15 P.M. slot. Then there would be an additional one-hour Saturday morning show, *The Best of Red Skelton.*

The complicated job of putting the show together went to big, burly Bob Baker. Ziv supplied three special transcriptions for each show. One bore the opening theme with Skelton's announcer, Rod O'Connor. Another was the standard closing with Skelton saying good-bye, the theme music, and O'Connor's signoff; the third transcription was Skelton's part of the broadcast with four different cuts. The first was usually straight monologue and the other three were scenes involving Red's characterizations—Clem Kadiddlehopper, Cauliflower McPug, the Mean Widdle

Kid, etc. Altogether, these four Skelton cuts ran approximately fifteen minutes.

Baker chose three musical selections from the record library for each program. One would be instrumental music by David Rose who played the opening and closing themes; the other two would be a popular vocalist. Three commercials for each program were written by a Detroit agency and recorded by Bob Maxwell, an announcer from WWJ.

Now, with his opening and closing transcriptions, his four cuts of Skelton material with cue sheet from Ziv, his commercials transcribed separately and locally, Baker wrote a three-page script tying everything together. WXYZ announcer read the Baker copy—Don Karle the first three days of the week, Chuck Dougherty the last two. There was also an applause record and a musical fill.

After a rehearsal with the engineer and the announcer, Baker would give the signal and they would plunge into the half-hour taping—two tapes, one for stand-by in case the air tape failed. For thirty minutes, Baker changed records seventeen times and operated two switches and two faders on the turntables. It was characteristic of him to do everything himself.

This Skelton mish-mash was virtually the last effort to produce an individual program on WXYZ-Radio. It lasted one year, then the time went to disc jockeys—a show called *Stars at Seven* with Paul Winter on Mondays, Jack Surrell on Tuesday, Ed McKenzie on Wednesdays, John Slagle on Thursdays and Fred Wolf on Fridays.

John Pival sent for Ed McKenzie.

"You've got to go on television," he said.

"You've gotta be kiddin'!"

"I want a two-hour show on Saturdays," said Pival. "A variety show—for teenagers—dancing to recording outfits. I think you're the guy to do it."

"Jesus Christ, John! *Two hours*? Up there before a camera?"

But he did it, *The Ed McKenzie Saturday Party*.

McKenzie: "I was scared to death. Just scared—out of my wits. Terrible. But I did it—uh—caught on. It got reasonably decent ratings."

The high spot of the annual Spring Carnival at the University of Detroit in 1954 was presentation of the American-of-the-Year Award to Danny Thomas. This star would be appearing at the Chez Paris in Chicago, and Marte Mogge, a pleasantly aggressive young sound man who was handling publicity for the Carnival prevailed upon Thomas to come to Detroit for the presentation.

Danny Thomas had signed a long-term contract with ABC and filmed his first television show in Hollywood. John Pival made plans for a big variety program on Channel 7 honoring Thomas—and, not so incidentally, cashing in on his talent and prestige.

Thomas would have none of it. He agreed to appear on TV with a U. of D. program, but he insisted that it must be an all-student variety show, presided over by the King and Queen of the Carnival. Mike Masters, WXYZ-TV announcer, who was a graduate of the university could serve as emcee, but Thomas stipulated that the program must originate on the campus and was to be worked out by the U. of D. television workshop with WXYZ's Pete Strand.

AFTRA was planning its annual ball. Dick Osgood, President of the Local, was credited with snagging Thomas for the ball, but most responsible was Marte Mogge working with Thomas through the Grant Advertising Agency and the Dodge Division of Chrysler which sponsored Thomas on TV. At the ball, Danny Thomas quieted the noisy, happily inebriated mob and maintained complete control. He reminisced about his early days in Detroit—hilariously, and sentimentally, too. He recalled old friends by name and remembered appearing in theaters for a dollar a show. He switched from electric moments of sincerity to shocks of wit.

"Actors," he said, referring to the purpose of the ball which was to raise money for the Detroit Local's welfare fund, "are the only people in the world who are always ready to *give* away the only thing they have to sell."

For his closer, he stood in the spotlight at the front of the projecting stage in the huge Latin Quarter and had his audience sing his last song with him, a sentimental old-time kind of thing. All the lights were extinguished except the spot he stood in. At a given cue, even that was killed and everyone in the huge, crowded room held up a burning match for the song's last phrase. It was a moving moment, for the star and his non-star contemporaries were all conscious of sharing the ups and downs of a harzardous profession.

Most of the crowd in the Latin Quarter that memorable night did not know that this generous, powerful entertainer was the same man who had saved the very first AFRA party over protest fifteen years before when he was Amos Jacobs. For those who remembered, it was doubly moving.

In California, the Livingstones met a new George W. Trendle. He went to the coast on television business and visited them.

Livingstone: "I got to know George Trendle in a way that I'd never gotten to know him before. As an individual. As a person. He was quite different, because all I knew about him before was—he was the boss, and when the boss came around everybody immediately stopped smoking, and did things—

"He was a very, very sensitive—terrifically sensitive man."

Harriet: "Very charming to me."

Livingstone: "Very charming to be with."

Harriet: "He was out to the house. He liked the real plain, simple—I don't know—maybe I was sort of small-town Ohio—oh—funny to him. Maybe I reminded him of an aunt or sister or something. Daughter was entranced with him."

Livingstone: "He didn't drink. Rhine wine—that's all."

Harriet: "A lonesome kind of man."

Livingstone: "This is something that most of us who worked for him never knew. Away from the office, from the studio, he was entirely different."

But even in those relaxed, friendly moments, George Trendle did not confide that he was selling *The Lone Ranger*. When it was sold, Chuck learned of it from

Signing papers on sale of Lone Ranger. Left to right: H. Allen Campbell, George W. Trendle, Jack Wrather, Jr., and Raymond Meurer.

Jack Chertok who was producing the Ranger films for television.

The news broke in Detroit in July, 1954, the month of Trendle's 70th birthday. Negotiations had been going on for some time with Jack Wrather, president of Wrather Television Productions, Inc. Wrather, a Texan, came from an oil family and the University of Texas. He had owned stations in Tulsa and San Diego, married the movie actresss Bonita Granville and was a true Lone Ranger fan.

Campbell handled all the negotiations.

Trendle: "I sold the shows because I thought I might die anytime and these damn things—they stay in estates for years and you can't fix any price on them. So I made up my mind I'd fix my estate up and here were two on which I could get a flat price and I sold 'em. The Hornet was questionable and I'd just hang onto it and put no value on it and if I die, why people can fiddle around with it if they want it.

"They wanted *The Lone Ranger* and uh—*Sergeant Preston* was sold on the air and had a wonderful background to it. And *The Green Hornet*—we neglected the damn thing. It was on and off and on and off. And uh—Campbell interested them in *Sergeant Preston* and they didn't want—they picked *Sergeant Preston*—if my memory is right."

When he said this seventeen years later, his memory was not clear. Campbell's however, was crisp and "right on."

Campbell: "We had the Hornet all sold to Wrather at the same time, and somehow it fell through. I actually offered Wrather in 1954 when he bought *the Lone Ranger*—I offered him *Green Hornet* and *Sergeant Preston* for the same amount of money that he paid for *Sergeant Preston* three years later. So he wouldn't pay another half million for *The Green Hornet.* So we wouldn't give it to him."

Trendle-Campbell-Meurer sold only *The Lone Ranger* in 1954 for $3,000,000—said to be the highest price paid for a single radio property up to that time. Fran and Jan Striker each received "bonus" checks for $4,000. In California, the Livingstones also received checks.

It was a sad blow for the actors. Beginning in September the radio Lone Ranger would no longer be broadcast live. There would be 52 weeks of transcribed repeats. The actors would be paid residuals, but their income would be cut in half.

The last live broadcast of *The Lone Ranger* occurred on the 3rd of September, 1954, one month after Charles Livingstone directed his first TV film. Fred Flowerday directed the final live broadcast, and in his cast were Beemer, John Todd, Rollon Parker, Harry Goldstein, Bill Saunders, Ernie Winstanley, Paul Sutton, Jim Fletcher and Jack Petruzzi.

There was no let-down in the last performance. The voices were dramatic; there were action and spirit. But as the actors sat on a bench between their scenes, or stood in a corner, script in hand, they had a feeling of unreality and apprehension. None thought back to the first performance of this classic series 21 1/2 years before. Only one of them had been there and that was John Todd.

He was the most philosophic of all. "It's just as well," he said. "We were getting into a rut."

When the show was over, there was a sudden emptiness. Some went to the Van

Dyke Club and got drunk. Others went to their homes, still feeling vaguely that something impossible had happened. All concealed their foreboding. One, as he snuggled into bed that night thought, "Oh God, let me sleep tonight and not worry. Something will turn up. Sleep, Lone Ranger, sleep."

It was not really the end of *The Lone Ranger,* but it was the end of the radio *live* Lone Ranger, and it marked the end of an era in radio history—and the beginning of a new one.

40.
Radio In Exile.

Driving home from golf one afternoon in 1954, James G. Riddell's mind skipped around the two-headed station he was operating.

Harold Christian had ended two years as commercial manager with a breakdown. But he was over it, back in merchandising and functioning beautifully, helping John Blair & Co. capture accounts for the station in New York and through Blair branches in Chicago, San Francisco, Los Angeles, Dallas, St. Louis and Boston. Christian and WXYZ introduced Lark and Kent cigarettes to the Detroit market.

Driving along, Riddell whistled softly to himself thinking of his salesmen. They were younger now. Tom O'Leary and Joe Molin were dead; Chuck Gunn was with CKLW; Garde Wallace had moved to Toledo; Irv Stoll had disappeared. Today, in TV, Bill Hendricks was gutsy, excellent, down to business; Ben Hoberman was top man in number of sales. In radio, Doug Campbell and John Lyon, young and aggressive, complained that there was no room for local clients on the Wolf morning show, but they had Ed McKenzie in the afternoon; Surrell was commercial; and Paul Winter was coming along with a new record show called *Winter Wonderland.* The Pulse ratings for July and August put Wolf with the records and Osgood with the news ahead of all opposition in every time period between 6 and 9 in the morning.

TV was in good shape, too. He had arranged for the first regularly scheduled commercial network TV show to originate live in Detroit. Soupy Sales would be summer replacement for Kukla, Fran & Ollie on 39 stations, Pete Strand directing.

And WXYZ was getting a new 50-kilowatt transmitter designed to send out programs in *color.*

As he waited for a traffic light to turn green, drumming on the steering wheel with his fingers, John Pival came to mind. He had got very drunk at the last Christmas party, nearly choked Jim Washburn, a producer, and had gone into a characteristic alcoholic rage, ripped off his shirt, flexed his muscles, challenged anyone to come near him. Riddell had sat watching the melee thinking how bad it was for an executive to make such a show of himself in front of the men who worked for him. The next day he had sent out a memo: no more Christmas parties.

Riddell was turning into his street when a thought came back that he had had before: *Why not dump the Mendelssohn mansion*? Soon, he had heard, *Sergeant Preston* would be transcribed repeats like the Ranger. Radio didn't need all that room any more. Why not sell the mansion and move radio into smaller quarters?

In the back of his mind, a dream was taking shape: all divisions—radio, television, traffic, sales, executives—all under a single roof. That sure would give Bob Baldrica something to crow about. Baldrica had replaced Bill Morgan in Promotion and Public Relations. He was a young man with a happy face, Italian brown eyes, straight brown hair and a constant smile.

Yes, Riddell thought, as he turned into his driveway, the station was nearly up to his golf game—and that was good.

His hole-in-one—was Hal Neal.

Napoleon Hill's "think rich" disciple with the fiery blue eyes had no negative thoughts about radio. Handsome Hal had replaced Christian as commercial manager in 1955. He was determined to pull radio out of the wreckage left by the storm and nurse it back to life. He developed a flair for improving the ideas of others and making them work for Wyxie.

In the beginning his pipeline to other parts of the nation was the manager of the Detroit office of John Blair & Co., Charles D. Fritz. Chuck Fritz was tall, had a patrician profile, thin hair, steel-cold blue eyes and a rather hesitant way of talking. He was a Detroiter, devoted to his family, and much admired by those who survived the initial impact of his icy facade.

Through Chuck Fritz, Hal Neal learned of the jingle. Singing jingles were common tie-ins with commercial products now; but to do a station break in music was new. The first station to use a singing jingle for station identification was WFBR in Baltimore.

WFBR was represented nationally, like WXYZ, by John Blair & Co. On the trail of a Detroit automotive account, Robert B. Jones, Vice President and General Manager of WFBR visited Fritz in the Book Building of Detroit, and together they went to an important advertising agency—Brooke, Smith, French and Dorrance—to make a station presentation to the Agency's media director. The media director decides how much of an advertiser's money will be spent in print, how much on billboards, and what percentage in radio and television. Part of WFBR's presentation was its singing station break. When he heard it, the agency man's eyes popped.

"Get out!" he cried. "I wouldn't buy any time on that station."

The man was angry.

"This is incredible," said Chuck Fritz, "to have this reaction. That's good sounding music. It's beautiful."

"Forget it!" The media director waved them away.

"Why?" Jones asked.

"Because you're competing with my commercials. I think that's terrible that the station is putting musical spots on. You listen to those and you forget our commercials."

He practically rushed them out of the office. Fritz told Hal Neal about it afterwards when the two had lunch together at Berman's Chop House. The jingle station break idea was trapped in Neal's mind. Because of him, WXYZ was the first

Harold J. Neal, Jr.

Detroit station to sing its way through a station break.

As the station developed, so did Neal, and New York took note. When, in 1955, Neal became the youngest member of the Board of directors of the Michigan Association of Radio and TV Broadcasters, ABC offered him a job as manager of WABC-Radio in New York. But he fought it off; he was not yet ready.

By November, 1954, it was known that WXYZ-Radio would move away from the Mendelssohn mansion. Riddell decided to move radio several miles from the opposite side of the city to the caretaker's cottage at the base of the transmitter on Joy Road near Greenfield. Bob Baker and Jack Tierney worked on the remodelling of the interior.

Fred Wolf: "I am sick. I don't want to drive from Grosse Pointe way out to Joy road. And I'm goin' to lose my—my image. I mean, my poor man's television's goin' to hell. They're goin' to stick me in some room an'—we're fight'n' television in those days. 'Course now, they've more or less given up, but radio is still radio. But in those days there was a fear that radio was goin' to die."

One of Wolf's accounts was McDonald Trailer Sales, Inc., maker of mobile homes. Wolf had met Frank McDonald through the salesman on the account and they had become friends.

"Mac," Fred said one day, "you know we're leavin' the Taj Mahal an' I'm gonna lose my little glass shack. I don' wanna."

"I don't blame you."

"Well look. I got an idea. Why couldn't we take a normal little 30-foot trailer and knock out the back end and put in three sides o' glass?"

"Easy," said McDonald. "But give me something to go on."

Wolf talked with Riddell and he talked with Hal Neal.

Wolf: "Hal grabbed it quick. He could visualize. Here we got X local sponsors. We put the thing in front of their business—car dealers—for a week. If they'll give us 26 weeks and give us 5 spots a day, we'll give you Fred Wolf and the Wigloo. This was the way it was sold."

McDonald built what came to be known as Wolf's Wandering Wigloo. It was no simple job wiring and equipping the trailer as Fred wanted it. He and an engineer would be ten to fiften miles away from the station. His producer—at that time, Bob Baker—would be back at the home studio with records, turn-tables and recorded commercials. Fred wanted to be able to talk with his producer while a record was playing so they could keep ahead of their show and know definitely what was going to happen next. Fred believed in keeping his show organized and moving. So he had Jack Tierney install a second off-the-air microphone through which he could talk with his producer. Both wore head sets. In one ear, Fred heard his producer; in the other, he heard what was on the air.

For his first remote in the Wandering Wigloo, contact was made with the University of Detroit. Bill Rabe who handled public relations for the university was an imaginative, aggressive promoter.The university even removed some hedges to make room for the mobile studio on a busy corner.

Wolf persuaded another sponsor, DeSoto dealer Glenn Walker, to furnish him with a white convertible with Fred's name all over it. This was used to haul the Wigloo; also to serve as pace setter at the Hot Rod races.

Wolf: "So now they take this mobile unit, an' I wanna tell ya, they created more good will—! Each salesman that had a local contact—we had a schedule every day. We had a waiter and we had this thing loaded with booze, and we had a girl as hostess, and we drove this thing from one client to another—parked it right in front of his joint. The dealer and his salesman came in, shook hands with everybody. We did this after I got off the air.

"This was the start. Now! We got a mobile studio *available*. But there are a few little handles. You can't just get it for fun. Ya gotta buy X. *You* want it, *you* gotta buy X. All these car dealers: 'You stand in front of my place three hours in the morning.' This can't hurt. Gas station openings—I don't know how many things we got involved in."

Here was the Lone Ranger on Belle Isle all over again; this was new radio.

When WXYZ moved its radio operation to the transmitter cottage, Fred Wolf was prepared not to go into exile with it.

Fred Weiss: "In the beginning, it just started out weekends and we worked in the *hallway* of this guy's home. the caretaker lived in that building; and he lived there when we began working out there only on weekends. During the week, we worked at the estate on Jefferson Avenue. On the weekends, in order to cut down on heat and overhead—radio fallin' apart—the announcers went out to the transmitter site and we had one table and a set o' earphones and we sat in the hallway and did the broadcasting. And then the caretaker would get mad because the cigarette butts would wind up on the floor. And it was really kind o' tragic because this guy was havin' these people invade his house to do their broadcasting."

Melvin Morehouse, a short, pudgy little engineer with popped eyes, who had transferred to television, was an organ buff. He asked Riddell if he could have the pipe organ when radio moved out of the mansion; he wanted it for his church. Riddell gave it to him.

Osgood: "It was a sad day for me when I saw the old pipe organ being hauled out of the carriage house and loaded onto a truck, piece by piece. As far as I was concerned, that old organ had been as important a part of WXYZ as *The Lone Ranger.*"

Morehouse rebuilt the organ for the church. Soon after he had finished it, the church burned down and the organ was completely destroyed.

The network made another effort, investing a huge sum of money in night programming with Herb Oscar Anderson, Merv Griffin and one or two others. It had been an expensive flop. The year before, ABC radio had produced only 3 shows to ABC-TV's 33. The three were something called *Harvest Years,* TV's *That's My Boy* without the picture, and a show with Harry James and Betty Grable. They all failed.

Kintner had said in June, 1954: "Morning radio is thriving; afternoon radio is not as healthy; with nighttime radio we're still experimenting."

He was speaking of network radio, of course, and the network was healthy in the morning because of Don McNeil's *Breakfast Club*. Individual stations wished it dead, because they could make more money selling the time locally, but it was always commercially loaded and the network insisted it be carried.

In October, Osgood observed in the *Free Press* that network radio was "very, very sick, maybe on its deathbed."

Figures showed network billings down by millions of dollars. Radio time sales, nationally, in 1954, had amounted to $453 million; but only $76 million of that was network; the rest was local. As Riddell had proved with Wolf and the others, there was still business to be had on the local level.

In September, 1955, Kintner made one more bold attempt. He tried another night-time network format made up of five 30-minute segments, each cut up into 5- and 10-minute sections. The first half hour was called *Events of the Day*; the second, *The World Around Us;* the third, *A Better Tomorrow;* the fourth, *Sound Mirror*; the last, *Off Beat.*

"We are programming with the individual listener in mind," he said, "rather than the mass audience."

Neither mass nor individual wanted it. It folded.

The last live *Sergeant Preston* broadcast took place on June 9, 1955. In the cast: Paul Sutton, Harry Goldstein, a fine old ham named Warren Colston, the ex-policeman Bill Saunders, Jack Petruzzi, TV's popular Bob Dale and Ted Johnstone. Fred Foy announced it.

The Trendle actors were now without work in dramas but they could collect residuals for a year because *The Lone Ranger* would continue five nights a week on NBC by transcription, and transcribed *Sergeant Preston*s were sold five times a week on the Mutual network.

Fred Weiss: "All you ever heard was that radio was *dead —dyin' —goin'*. Whshoo! *The Lone Ranger* and everything quit and all that stopped. And that beautiful panelled studio was left empty with the dead mike hanging off the ceiling and you could walk around there in the sound effects room and there was *no*body in there and the stuff just sat—and dirty—and it was kind o' sad."

Fred Flowerday: "The shows went off but I continued for the next two years via tape. I would tailor these shows—take out the old commercials, put in the new ones, and send five tapes to NBC in New York, or Mutual, for network broadcasts. But they were repeats. Jay Michael would come in and do new narration to cover a deletion or a new commercial. And then that finally phased out."

Overweight and alcoholism had all but downed Jay Michael. He was living in a clinic.

In March, 1955, WXYZ-Radio moved to the remodelled caretaker's cottage at the transmitter. It might as well have moved to Siberia, but it was cozy and intimate once it was reached. Brestoff's office was quite large and pleasant. His secretary, Marian Vedder, sat outside in a hallway. Osgood's office was next to Brestoff's, faced south, and he was happy in it. Next to him was Ed McKenzie who shared his office with his Girl Friday, Joyce Chapman. Tall, aging Lee Smits, who celebrated his 15th year of broadcasting a month after the move, had a small office at the back of the building and trained his dog to fetch copy from the newsroom next door. (Tom A'Hearn was now the only man in the news room. Seymour Kapetansky had been lured to California; Hank Alexander accepted an offer to handle public relations for Michigan Blue Cross-Blue Shield; and Femmel was in television.)

One small office was shared by all the record personalities—not yet called disc jockeys; and there was a kitchen which was a pleasant place where the small staff could enjoy coke, coffee and lunches together.

There were only two studios, both quite small. One accommodated Jack Surrell's piano and was also used by Smits. The other, equipped with turntables for records, was used by Osgood and the record players.

The control room was cluttered and noisy. Tapes were edited in it, and it was a confusing tower of Babel with tape sounds, engineers' conversation and radio programs all mixed up together.

Reduced quarters opened with a feeble little party, attended by Riddell and Hal Neal. Jack Surrell's small upright was hauled into the hallway. He played and sang and the liquor flowed.

When Riddell walked in, he beamed at Osgood.

"How do you like it?"

"Only one thing wrong," said Osgood. "My name isn't on my door."

"Can't take a chance," Riddell declared, grinning. "You may not be here much longer. How many years is it now?"

Osgood laughed, hiding his apprehension. He and Edythe Fern Melrose were the lone survivors of special programs. All that remained were record men, announcers and the dying network.

Jimmy Riddell sold the magnificent Mendelssohn mansion for $100,000. Some of its interior beauty was sold at auction. Then it was torn down to make way for a supermarket.

41

The Luckiest People.

In 1955, the AFTRA Ball was held on the last night of April in the huge Latin Quarter. Detroit now had the 5th largest Local with 300 members. Ernie Winstanley was President. Vice President, Paul Hughes, was out west somewhere prospecting for "radio-active rock." The Recording Secretary was a graduate from Radio Schoolhouse, Maureen Bailey, who would eventually go to Broadway and play Wendy to Marty Martin's *Peter Pan.*

Everybody was dressed in his best, lapping up the liquor, table-hopping, greeting old friends, rubbing egos. When AFRA became AFTRA, the addition of the "T" for television introduced a subtle new rivalry into the ranks of the union. The TV performers' sense of superiority was only beginning to be felt in 1955 because many had come from radio and were still working in both media, and the record player—now called a disc jockey—was not sensitive to the possibility that anyone existed who might be more important than he. But the veterans of radio had become a little defensive.

Tom Waber smiled up at Dick Osgood as they shook hands. *Cunningham's News Ace,* created 14 years before by H. Allen Campbell, was now being broadcast on radio by Osgood and on television by Waber.

"We're still zooming into their homes by propeller plane," said Osgood. "Isn't it about time we arrived by jet?"

"Where've *you* been?" Waber exclaimed. "We've been flying an F-80 jet for a *long* time in *television*."

Osgood walked away laughing, but feeling a little sour.

The slick program of the AFTRA Ball was filled with ads bought by agencies, sponsors, stations, and the "artists" themselves. The money—with all net proceeds of the annual Ball—went into the Welfare Fund of the Local. Changing times were symbolized in 1955 by the ads made up by the Beemers. Brace Beemer's ad consisted of two drawings, side by side. One showed a rather grim-faced Sergeant Preston standing idle, holding his horse, with King, the husky, behind him. In the other, the Lone Ranger's hat, spurs, rope, holsters and boots were hanging unused on the wall of a tack house. This was a half-page ad. In contrast, a full-page ad inserted by Brace's son, J.D., showed J.D. as TV's Justice Colt, leaning forward, guns drawn and full of action.

At one table sat Bertha Forman, a plain little character actress who spoke with an English accent when she wasn't acting. She was studying a page in the program. It was solid black except for a small white circle in the middle of the page. In the circle, names of members who had died, among them Earle Graser, Herschel Mayall, Fred Fry, Fred Reto, Dewey Cole—and Harry Heilman.

Bertha and John Todd were now appearing quite regularly on Soupy Sales' 11 P.M. show—a loosely prepared slapstick program which John Pival figured would siphon off thousands who preferred entertainment to news.

Soupy was master of ceremonies at the '55 Ball, and the entertainers were Sam Levenson, the Fontaine Sisters, and Four Stepbrothers, and that long-ago WXYZ kid, Johnny Desmond, now in his thirties and singing with surprising vigor.

But most Aftrans went to the Ball to drink, talk and visit.

"Where's Stan Dale now?"

"He's at WAIT."

"Where's that?"

"Chicago."

"There's Fred Foy."

"Handsome bastard. He's doing all right."

"Still got the Rambler dealers?"

"Yeah. Channel 4. And he's the voice of Goebel Beer on baseball and football."

"Who needs the Trendle dramas, eh?"

"He doesn't."

"Hi, Bob Dale. Congratulations, professor."

"Thanks."

"Why—'professor'?"

"He got his M.A. degree from Wayne last June."

"Let's see. How many Green Hornets were there?"

"Well—Al Hodge was the first."

"He's doing television in New York."

"Then there was Don Faust."

"He's managing WJR-TV up in Flint—if it ever opens."

"After Don, it was Bob Hall."

"Bob's doing *Music 'Til Dawn* for American Airlines on a New York station."

"And Jack McCarthy."

"That makes four."

"Jack's doing news on WPIX-TV in New York. He's calling himself John E. McCarthy. There was already another Jack McCarthy in New York."

"Where isn't there a Jack McCarthy?"

Among the new faces at the Ball that year was Leon McNew. He was young but his hair was white. He was now hosting old western movies as Captain Flint, for which he wore a mask and boots. He had been on the Trendle-Campbell station in Flint.

"I replaced Bob Reynolds up there when Bob came to Detroit. I did sports up there. John Lee came up to Flint one time to see Rube Weiss and Jim Washburn and he saw me on television doing a football game and he invited me to come down and talk and I did and so John Lee said, 'I want you down here on Channel 7.' So I came down last October and my first job was with Fred Wolf out at Motor City Speedway. I did the pit interviews and Fred did the races."

Rube Weiss (no relation of Fred Weiss) had managed the Flint station. He was a multi-faceted, many-talented man with kind jewish eyes, thinning hair and a distinctive voice. He had acted in the Trendle dramas and was married to Liz Weiss.

They had met in the east. Liz had auditioned for Livingstone without results; so she had gone back into the theater when Livingstone decided to use her. She lost no time returning to Detroit, because she was in love with Rube.

"What do you want for your birthday?" Rube asked her one year.

"An engagement ring," Liz replied without hesitation.

She got it, and they were married, and had several children. Their talent and versatility were in great demand for commercial movies, radio and TV commercials and slide film.

A small group of men had gathered around Budd Lynch near the foyer. Budd broadcast the Red Wings hockey games. He was a Canadian and had lost one arm fighting with the Black Watch contingent in World War II. The men shook their heads in amazement as he explained how picture and sound travelled separate routes from Detroit to Toronto and Montreal in the telecast of a recent game between the Red Wings and the Canadiens.

"The sound went by London, Ontario," he said. "But the picture had to go by cable to Toledo, then to Cleveland and Buffalo and up through Hamilton, Ontario. By some miracle both picture and sound arrived on receiving sets in Toronto and Montreal at the same time."

Disk jockeys had a lot to talk about at the Ball of '55. Sixteen had become charter members in a new organization called The Detroit Association of Disk Jockeys. Their purpose was to do some housecleaning. Young men leaned forward on the tables and emphasized their words with fists and fingers dropping into the white light on table cloths from under pink shades.

"I don't like the stuff we have to play any more than the public does. We're out to reform the methods by which these records are picked."

"If a jock thinks owning stock in a record company is a good investment, and he has the money to invest, what's wrong with that? It's his own business and he's smart to plug his records."

"All I know is, the record companies are sending some of their top men into Detroit to find out what we're up to. We are not after any individual. We just want to get rid of payola."

Each faction of AFTRA had its own line of conversation. The actors, now reduced to residuals, talked of the past, boosted their own morale.

"Isn't it great about Johnny Hodiak? He's on Broadway in 'Caine Mutiny Court Martial'."

"That makes two from Wyxie. Mike Wallace is in a play, too. *The Reclining Figure.*"

"What do you hear from Chuck Livingstone?"

"I hear he's not happy in California."

"I was just talking with Jack Petruzzi. He's going out west, too."

"Might as well. There's nothing here."

(Petruzzi did go west and wound up delivering mail.)

On the television side of the Latin Quarter:

"Hay, Axford! Gillie. Gille Shea! Saw you on *Night Court*—Great!"

"Tom Dougall, you're drunk!"

"I'm drunk and you're frustrated. You live a very frustrated life."

"It took two policemen to hold back the crowd in the lobby of the Maccabees Building last Saturday after *Ed McKenzie's Saturday Party.*"

"Ed's building a new home."

"Marv Welch has bought his own night club."

"*Wyxie Wonderland* is in the Top Ten in the new ratings."

"So's Auntie Dee."

"I can't figure it."

"We have a better show and we don't get the ratings."

"Oh well—the ratings don't mean anything."

"Sure."

Jack Petruzzi

Johnny Desmond

After midnight. Brace Beemer really rolling, bore down on Dorothy Davenport. Bob Baker intervened. Beemer insulted him good-naturedly. Baker outstayed him. Beemer finally returned to his table.

"Thanks," said Dorothy Davenport. Husband Don was off somewhere, leaning forward, brows together, listening earnestly, glass in hand, to a smiling friend. Don was now Fred Wolf's producer, and Dorothy had given up her secretarial job to raise children. After four kids, the marriage collapsed.

One thirty in the morning. The band packed up to leave. Several hundred individuals and parties in the broadcasting industry separated into the night—people who needed people—shoulders to cry on, ears to brag into, backs to slap, arms to touch, cheeks to kiss—some flushed with success, some fighting for recognition, some scared. They all met once a year at the AFTRA Ball.

42
Old Faces; New Faces.

Despite the sense of having been benched to which radio succumbed, there was reason for hope. The A.C. Nielsen company, which measured audiences by attaching gadgets to receiving sets, submitted a report in mid-1955 in which it declared, "Radio '55 is of such scope and dimensions as have never before been fully realized." Nielsen backed up its statement with interesting figures. In the northeast, 19% of all homes could now be reached by television, 96% by radio. In the east-central section of the country, 76% of the homes had TV, but 96% were receiving radio. Radio was listened to by 96% of the homes in the west-central part of the United States, 63% watched TV. In the south, only 49% of the homes had television while 95% of the people still listened to radio.

In Detroit Fred Wolf's Wandering Wigloo had already proved such a success that a second trailer studio had been built to be used by Paul Winter. A permanent broadcast booth had been installed at Sears in suburban Lincoln Park. And a third trailer was rented for use at a shopping center in Dearborn. WXYZ engineers were "on remote" from 6 A.M. until 10 o'clock at night. Station executives from many other cities visited Detroit to observe WXYZ's "mobile studio operation."

Riddell and Hal Neal decided to give FM another try. To head the effort, they made a deal with an optimist named Larry Gentile. Larry had had an all-night record show on CKLW in Windsor where his brother Joe had enjoyed great popularity teamed with Ralph Bingé. Larry was as commercial in his way as Edythe Fern Melrose was in hers. He sold and serviced his own accounts and had been successful for a long time.

The second effort of WXYZ to establish a separate broadcasting entity of its FM station opened in June, 1955. There were no extra studio facilities, and the FM programs had to be taped at odd times. It was an uphill struggle for Gentile but he never lost his optimism. However, this second attempt was only slightly more successful than the first and the experiment ended in October, 1956, after a year and four months.

On October 19, 1955, John Hodiak died of a heart attack. He was only 41. His marriage to Anne Baxter had fallen apart. His contract with Metro-Goldwyn-Mayer had expired. He was bitter about the fickle friendships of Hollywood. He knew that in Detroit his friends still loved him, but he was too famous to turn back. Besides, what was left for him in Detroit? He had recovered his prestige with a strong performance on the New York stage in *Caine Mutiny Court Martial* and he was to have repeated his performance on television.

Detroit friends recalled his clowning during rehearsals of the Trendle dramas—always spontaneous, never malicious, rarely rough, and welling up from within to turn an essentially somber personality into something radiantly funny. It was a priceless facet of the Hodiak personality never used in Hollywood.

Trendle-Campbell-Meurer, Inc. still owned *The Green Hornet* and *Sergeant Preston of the Yukon.* There was no action on the Hornet, but Sergeant Preston was filmed for television and Fran Striker went to Hollywood to indoctrinate the screen writers, edit the scripts and help in the adaptation. It was a temporary absence.

In Detroit, Fred Flowerday continued editing and tailoring the Ranger and Preston radio tapes for Wrather and the networks. Ray Meurer, representing Trendle, kept in close contact with Flowerday. Both saw the potential in recordings and Meurer turned his town house near downtown Detroit into a posh recording studio. He became a partner. They named the company Special Recordings after the statement made at the end of each show: "This program is brought to you by special recording."

Two ex-Trendle sound men joined the company—Tony Caminita (who eventually became manager) and Ernie Winstanley as Sales Manager.

(Seventeen years after both *The Lone Ranger* and *The Green Hornet* went off the air as live programs, in 1973, Special Recordings was cutting Dodge commercials using the Lone Ranger character: and American Motors was producing a car called the Hornet and using *Flight of the Bumble Bee* in its commercials. Both characters had originated nearly 40 years before.)

During the theater season of 1955-56, Dick Osgood pulled a stunt which caught national attention in show business. The touring company of *The King and I* with Patricia Morrison and Leonard Graves played Detroit at the Shubert-Lafayette Theater. By arrangements with David Nederlander who operated the theater with his sons Jim and Joe, Osgood invited high school students who had never seen a stage play to be his guests in the second balcony for a matinee.

The balcony was jammed with young people whose enthusiasm was caught and reflected by the actors in a performance even better than the one they had given on opening night. During intermission, Osgood taped questions and answers with his young guests, and that night used parts of the tape on Show World. The questions asked by the teen-agers were valid and intelligent, and the experience corrected Osgood's thinking as to what people wanted to hear on the radio about show business.

A permanent result of the experiment was special prices for student admission to the theater.

ABC-TV network sales crept up like a tide, covering more and more local sound. In the Fall of 1955, the Auntie Dee program was lost in the flood. Dee Parker and Phil Brestoff did some "tall thinking."

Phil was still Musical Director of WXYZ, but there was less and less live music on television locally, and virtually none on radio. Phil was not happy as Studio Manager of radio; important decisions were made downtown and he felt ostracized in the caretaker's cottage 6.8 miles away.

In September, 1955, the Brestoffs spent their vacation "casing" Los Angeles; in February, 1956, they bought a home in Beverly Hills; in May, they left WXYZ for the big gamble on the west coast.

Within days of their departure, James G. Riddell announced some promotions. Hal Gordon who played the piano for Soupy Sales was named Musical Director. Bob Baker moved into Brestoff's old office as radio's Studio Manager. Bob Baldrica of the constant smile would be Assistant Sales Manager in radio; Harold Christian now became Vice President in charge of merchandising for both radio and television; and Harold L. Neal, Jr., his blue eyes ablaze with ambition, was raised to Vice President in charge of radio. John Pival was Vice President in charge of television.

"Fred Wolf? My name is Sidney Goltz. I'm from Chicago. I just watched your show and I'd like to talk to you."

Wolf had just finished a telecast of *Bowling Champions* which he owned and produced over Channel 7 for National Food Stores. Olympic Recreation where the bowling took place was on the second floor of a building in downtown Detroit and, as usual after the telecast, he, his producer, and others who had worked on or watched the show went downstairs to a bar.

"I'm sure you have seen *Championship Bowling*," said the man from Chicago.

"Well I sure have!"

"I produce it. Can we talk?"

The two men lifted their beer from the bar and moved away from the others to a small table. *Championship Bowling* was a syndicated show produced in Chicago and sold to 148 stations.

"Your show is *Bowling Champions,* ours is *Championship Bowling.* People in Detroit must get confused."

"No," Fred assured him. "Not the bowlers anyway. My show was on—hell I've been doin' bowling shows around here since 1949. You've only been on two years."

"Right. No problem. The bowlers probably watch both shows." Goltz swallowed some beer. "You say you've watched *Championship Bowling*?"

"I've watched every one," said Wolf. "Your first year I did all the commercials. We'd cut out of your film at the end of each game and I'd do some scores and stuff and a couple commercials; then they'd roll the film for game number two."

Goltz tipped his glass back and forth watching the beer hold its level. Then he looked up at Wolf and said, "Chris Schenkel has been doing our show."

"Yeah."

"We want to make a change next season."

"Oh?"

"Would you be interested?"

Fred looked at the Chicagoan and took a drag of his beer.

"We tape the whole season in two or three weeks in the summer," said Goltz.

"This presents a problem," Wolf told him. "I'd have to be away from my morning radio show."

"I'm sure something could be worked out."

"I'll have to see Jim Riddell."

Riddell was impressed. He was all for it. It was prestige for WXYZ. He and John Pival negotiated a contract for Wolf.

Fred Wolf: "It was a pretty good piece o' money. However, I received no residual thing; I just got a flat amount of money which was a big mistake as the years run on. So okay! So we're all set to do this thing in Chicago. So Jimmy said, 'No,' he said, 'there's no way. You'll have to take your morning radio show with you.' He don't want me away from here for two weeks 'cause we're flying' pretty good in this stage o' the game with the morning show."

Emily Wolf: "Can you believe this? How could a man do four hours of radio and then go into three television shows? How could he do it?"

Wolf: "Three shows a day we did—three one-hour shows. So. I said, 'Okay—what the hell'."

On the 11th of June, 1956, Fred went to Chicago and stayed at a motel on Lake Shore Drive. He did his morning radio program under a canopy on the grounds of the motel. After four hours of that, he'd take a shower and race to Chicago Heights where *Championship Bowling* was filmed.

Wolf: "Mr. John Pival was the first station manager to put *Championship Bowling* on the air at 12 noon. They told him, 'You gotta be out of your mind! Who's goin' to watch a sports show at 12 o'clock on Sunday; everybody's in church.'

"John said, 'No. Everybody's *not* in church, an' this is a good time on my station.'

"An' he started it. And over the years there were 150 stations that put it on at 12 noon Sunday."

Emily: "Amazing."

A novice announcer at this time discovered almost immediately that Fred Wolf was important. Hugh Copland was given an early morning trick preceding Fred, and was instructed to keep reminding impatient listeners that the god would soon be heard by his waiting worshippers. Hugh was so busy plugging Wolf on his first day that he forgot to report the weather.

Fred phoned the station.

"Tell that new kid to tell the world it's foggy."

This made the new announcer even more nervous. Copland had observed that Fred Weiss who had played records before Wolf's arrival invariably introduced him with some kind of humorous remark. So when he turned the air over to Wolf, the nervous man said, "And here is the only man in Detroit who didn't have trouble getting to work in the fog. He followed one of Detroit's biggest landmarks—his nose."

Hugh Copland: "The minute I said it, I knew I was in trouble. Wolf was outraged, didn't get over it for quite a spell."

Hugh moved into television the next year. In 1958 he joined Don Faust at WJRT in Flint. Eventually he went to Channel 4 in Detroit.

Another new face that would be around longer belonged to Gene Avram. He had sung with Wayne King and on Channel 7. He had been a member of a male chorus which had supported Dee Parker in an act written for her by Doris Fisher and performed at the Statler Terrace Room. Even after seeing him, it was impossible to imagine Gene Avram tripping a heavy fantastic toe in a chorus. Built like a pick-up truck, he was the sort of man who would build his own fence or finish his own basement. As a radio announcer, he did not possess the old Trendle timber but he was thoroughly efficient and a man of good humor with a broad face and a kind smile, and strong integrity.

1956 saw the passing of an old face and familiar voice from radio. Lee Smits had been doing the evening news for the Michigan Consolidated Gas Company for 15 years. He wrote every word of his newscasts and spent all day doing it. It showed. He had a remarkable ability to remove hysteria from the headlines and see everything that happened as an expression of human nature.

Dick Osgood: "Lee was a great character. We were office neighbors for years. Once he appeared at my door with an exaggerated display of self-control and the expression of a man who was in the middle of a frightening experience. With an elaborate show of preventing himself from exploding, he asked if I believed in the supernatural. It seems his scissors had disappeared.

"At length he explained how they had been in his hand—then were not in his hand, nor on the desk, nor even in his office. Yes, he had looked everywhere—and he said so in the manner of a martyr in the presence of a skeptic. He expounded on the marvels of invisible gremlins capable of making a hard, cold, steely pair of shears evaporate.

"I suggested that perhaps he had dropped them into his deep trash basket filled with cuttings from the newswires. With the despairing air of a man who describes a miracle to an infidel, he declared that he had searched through the basket three times.

"He borrowed my scissors and returned to his office, resigned vociferously to the

Lee Smits, woman unknown, Earl Hudson and Dick Osgood on TV

amazing supernatural power of inanimate objects to make themselves simply non-existent.

"Some time later, I needed my scissors and found Lee quietly snipping the news with his own, mine lying idle on his desk. I asked him where he had found his. He pointed his recovered shears toward the trash basket with a smile of wonder, as though his faith in the supernatural had been confirmed. And with a happy sigh, he said, 'Isn't it marvelous?' "

Lee Smits could roar like a lion or bestow a cracker upon a child like a doting grandfather. He knew all about the prohibition days and the bawdy houses of old Detroit. He may easily have been the best hunter living in Detroit. He raised a wolf in his office.

When he left WXYZ in 1956, a bit of badly needed personality faded from broadcasting. However, the same year radio lost Lee Smits, it won Dick Kernen.

Dick Kernen was waxing his car which his father was paying for when his mother came running out the back door with a newspaper in her hand.

"Richard, there's a classified ad in the Dearborn paper here that I want you to answer."

Sloppily dressed, unshaven and uncombed, her 18-year-old son looked up at her with absolutely no interest.

"They want a young man with an automobile that has insurance on it," she told him.

"Who does?"

"WXYZ. The radio station. Richard, I want you to go after this job."

"You're laying an unhappy word on me."

"Before your father throws you out," she added.

Dick followed his mother into the house, showered, shaved and put on a short-sleeve shirt and a snug pair of flannel slacks. He was 5 feet 8 inches tall, weighed 145 pounds and had the same clean, firm look that distinguished Bob Hite and Hal Neal.

Beatrice Kernen watched her son drive off, and worried. Of all four sons, Dick was the strange one. She had to face it, he was what people called a "juvenile delinquent." Robert, her husband, had had to get him out of jail a couple of times when Richard and some other boys had been caught breaking down market doors to steal liquor. It was just a prank. Richard was really not much worse than his brothers; it was just that they were smarter somehow and didn't get caught. No, that wasn't right. They were not any smarter; Dick had the best mind of all if he would only apply it, give it some direction. It wasn't that he didn't want a job; he had gone after it willingly enough. But he didn't have one ounce of ambition.

Young Kernen told the girl at the desk just inside the back door that he had come in answer to the ad. Marian Vedder ushered him into Bob Baker's office. Baker liked his looks immediately. One brown eye turned in just a little and he wore glasses, but he was a trim looking kid, obviously came from a good family and had a brain.

"How come you're not in college?" Baker asked.

"I have a kind of negative attitude toward college," Kernen admitted frankly.

"Well, I'll tell you what the job is," said Baker. "Our executive offices are downtown in the Mutual Building. What we need is a guy with a car to run between here and downtown mostly—once in a while maybe to television in the Maccabees Building—but mostly between here and downtown—mail, scripts, logs, pay checks, newspapers, inter-office stuff."

Dick Kernen went to work for WXYZ in July, 1956. On one of his first deliveries, he lost the seat of his pants to Lee Smits' dog. Around November, he had "the unhappy word" laid on him that the U.S. Navy desired the advantage of his services for two years. He put up a little fight but there was no contest. He went into the Navy. His impression of WXYZ-Radio as it was in 1956 remains vivid:

"It was like a corner store. Everyone had his own little thing going. Some people played a certain kind of music and some people played another kind of music and it was all very weird. Saul (Paul Winter) played show music and McKenzie did a kind of pop show in the afternoon; and everyone was into whatever they were into.

"Jack Surrell, as I recall, was into a little more sophisticated music than the rest of the station. I don't think it was necessarily black music 'cause at the time Jack was on the air there was no black consciousness. Most people thought Surrell was white; most people were astounded to find he was black.

"The rest of the station was into a pop—uh—you know—Doris Day and that whole kind of thing. Jack tended to use a little better class of music, I always thought, than the other people on the air, which I think is probably significant."

Kernen was right. It was significant. Surrell's music was ahead of its time. But time was catching up with it.

Television had added studios on the ground floor of the Maccabees Building. When Leon McNew walked in one afternoon he was grabbed by Tommy Davis, a stage hand.

"Lee, Larry McCann isn't here. Will you stand in for him with 'The Lady of Charm'? You've seen the show."

McNew was "doing booth" then—station breaks and commercials between shows from the announcers' booth.

"Well, yes," he said. "I've seen it. Where's McCann?"

"Well, it seems Larry took off with Justice Colt, Eddie Chase and some of the boys for a week of hunting up in Canada without notifying Edythe that he was going."

Toward the end of the week, Edythe Fern Melrose went to John Pival with a demand: "I want Larry McCann off my show and I want Lee McNew on it."

She had her way. Edythe Fern Melrose had now built her second "house of charm," tropical style, near the Grosse Pointe Yacht Club. It had cost her $120,000. She was planning a third. In 1956, for the 4th time, she was chosen by the Women's Advertising Club of Detroit as "Detroit's Advertising Woman of the Year," and she was elected national president of American Women in Radio and Television, Inc.

But she was being phased out of radio.

James G. Riddell: "You just couldn't leave a disc jockey and go into an Edythe Fern Melrose with her loyal but limited audiene, and then pick the big audience back up again."

Trendle-Campbell-Meurer sold *Sergeant Preston* to Jack Wrather in June, 1957. The 17th anniversary of its first broadcast was celebrated, not in Detroit where the program was created but in Wrather's publicity from Hollywood. The month it was sold, Frank Russell who had first played the French Canadian, died. In August, John Todd, the original Tonto, died. One year later, in August, *The Lone Ranger* faded from television.

Brace Beemer who played the role longer than anybody else was living quietly and not altogether happily on his 300 acres in Oxford, Michigan, wearing twill trousers, flannel shirts and western riding boots. He was now 56 years old. He and Leta were developing a section of their property as Lake Shore Estates.

Fred Flowerday called Beemer from Special Recordings occasionally, and he did a few commercial announcements for a dairy; but actually, he was through with broadcasting.

And he drank. Heavily. His son Bob tells of his being hospitalized and suffering from delusion that "people were after him." His sons extracted him from Jennings Hospital and took him to a "dry-out" clinic in Brighton. He sent for Ed McKenzie who belonged to Alcoholics Anonymous, and the two had a serious talk. McKenzie "laid it on the line." Beemer did not join Alcoholics Anonymous, but he stopped drinking completely. Both McKenzie and Brace's sons confirm it: Brace Beemer became a teetotaler.

Meanwhile Tom Dougall, the first writer of *Sergeant Preston* when it was *Challenge of the Yukon*, was researching and co-ordinating a new program in television. *Night Court* began on Channel 7 in October, 1957. On it, actual cases were reenacted by actors who ad libbed after they were briefed on what actually happened at court. Often, the original policemen played their original roles. Working steadily on this show as the court bailiff was the one-time Axford of *The Green Hornet*, Gillie Shea. The program was on the air five nights a week.

George Pierrot's *World Adventure* series was such a success that Pival created a second series called *The Explorers*. It was sold to Michigan Bell Telephone which bought a special network to carry the program throughout the state. So many were auditioned for the announcer's job on this program that even Fred Weiss was summoned. Much as he had wanted to do television, Pival and John Lee had never called him because they thought his hair was too blond, his face too cherubic.

He was caught on a morning when he was doing the morning trick with Wolf and Osgood, shabby and unshaven. But as soon as he was free, he obeyed orders, and raced to the audition.

Fred Weiss: "Everybody was there—Rube Weiss and Fred Foy—you name it—Rollon Parker—all the actors from the Ranger—Bob Leslie and all those people were there. And I didn't have time to memorize the commercial. We sat on a sort of straw box or something. And I did this commercial. I just ad libbed it as best I could.

"And then they had about four or five of us stand up and give our backgrounds—where you went to school and all this stuff. I was scared to death and here I am in this dirty suit and I stood up and I said to 'em. 'I went to the University of Wisconsin but I can't see what in the world *any* of that has to do with any of this stuff. All that I can think that should be of interest to you is the fact that here's the job the guy does, and this is the way he looks in public, and that's *it*. What else could interest you'?"

The sponsor's number one choice was white-haired Leon McNew, but Leon had been associated with beer commmercials and that ruined his image for the telephone company. Weiss was new, fresh. He got the job. Things began to roll for the modest blond from Wisconsin.

Tom Waber replaced Lee Smits on the 6:15 news. He also broadcast news on radio at 3:30, 4:30 and 5:30 P.M. He referred to himself as the "featured newscaster," but station publicity still featured Dick Osgood.

It was a peak era for Osgood, even without Hi-Speed and *Radio Schoolhouse*. He returned from a movie junket to find a note from Bob Baldrica which read: "I thought it might be a welcome back: the Pulse ratings for May and June show *Show World* with a 3.9 rating while Thomas has a 3.8—putting *Show World* into the first spot."

Lowell Thomas was on WJR from New York.

Observations by Osgood was discontinued in the *Detroit Free Press* in December, 1958. The paper had asked for more gossip and Osgood felt that in that regard he lived in a glass house. Besides he was doing extra newscasts now at 10:30 and 11:30 A.M. and 12:30 P.M. So the columns were discontinued by mutual desire and Bette Lou Peterson became radio and TV writer for the *Free Press*.

Bessie Ashton had moved her traffic department from the Mutual Building to the 14th floor of the Maccabees. She had made up the logs for both radio and TV until 1956.

Bessie: "It reached the point where it was—ooh—getting a little hairy—doing this sort of thing—and so I—we went out there—14th floor—yeah. Yeah. M-hm."

Harold Christian was given an honorary Life Membership in the Detroit Association of Grocery Manufacturers "in recognition of loyal service to the Association for over 20 years." He was the only man in broadcasting to be so honored.

Christian: "For two years we (WXYZ) selected a panel of six supermarket buyers and asked them to tell us what they thought was the best product to come on the market during that particular month, the best new package, and the best promotion for a new or old product. We published the results in a trade paper with the photos of these products. This brought us many inquiries from manufacturers who wanted to enter this promotion. It was not unusual for us to receive 40 or more entries. And here again, it produced new business for WXYZ."

In addition to radio shows, Paul Winter was doing a teen-age dance show on

WXYZ
5000 WATTS • 1270 KILOCYCLES
DETROIT
1700 MUTUAL BLDG. WO 3-8321
James G. Riddell, President
RADIO
Program schedule for
NOVEMBER, 1958
Owned by
American Broadcasting Company
A Division of
American Broadcasting-Paramount Theatres, Incorporated
AMERICAN BROADCASTING NETWORK
Now! on the Station of the Stars...
Survey-proved
DETROIT'S TOP NEWSCASTS
as done by veteran reporters
DICK OSGOOD & TOM WABER
during peak listening hours
Monday thru Friday
DICK OSGOOD MORNING
6:45
7:00
7:30
8:00
8:55
TOM WABER AFTERNOON
3:30
4:30
5:30
6:15
DICK OSGOOD
TOM WABER
NEW YORK
CHICAGO
520 N. Michigan Ave.
SUperior 7-8659
ST. LOUIS
1037 Paul Brown Bldg.
Chestnut 5687
DETROIT
617 Book Bldg.
WOodward 1-6030
DALLAS
Rio Grande Nat. Bldg.
Riverside 4228
LOS ANGELES
3460 Wilshire Blvd.
DUnkirk 7-1333
SAN FRANCISCO
2502 Russ Bldg.
DOuglas 2-3188
BOSTON
6 St. James Ave.
HUbbard 2-3163
ATLANTA
101 Marietta Street Bldg.
Alpine 6482
JOHN BLAIR & Company
Representing leading radio stations

television; and in December, 1957, an album he recorded of satirical songs and sketches, called *A Winter's Tale*, sold surprisingly well around the eastern campuses.

In April, 1958, *Life Magazine* did an article on Soupy Sales. This year, he also had his own network show, *Lunch with Soupy*, on Saturdays.

In 1958, with partners, Fred Wolf opened his own bowling establishment, which would eventually become his main interest, Fred Wolf's Eastland Bowling Center.

The Strikers sold their home in Grosse Pointe and bought an old house in Arcade, the nearest town to their farm in New York state. It was a sad time for Fran Striker. He did a little teaching, a bit of public speaking, wrote a few TV plays, and spent time working on a long, historical novel.

But when he left Detroit, he suffered a double blow. He lost his job; and he was losing a son. Don was only 24, and in love.

Jan Striker: "Don had leukemia. That was a very bad year-and-a-half—a terrible year-and-a-half. We were told to let him do anything he wanted to do, go any place he wanted to go, try to make each day a good one, a happy one, not to let him know we were worried. And yet—we were living with this awful thing; so—there was a terrible strain.

"And we took him to Arcade. He was married. In June."

The girl Don married knew he was dying when she married him.

Jan: "He lived seven weeks after that. Prior to that, we took him to Arcade in an ambulance because that's where he wanted to be. And we decided we'd never be able to take him back; so we bought a house in Arcade. Then he did die—he was in the hospital at that time. He died—he never got into the house.

"Well—then Fran—after all this—was very depressed. It was very hard for him to work."

Janet (the daughter): "Took four-and-a-half years for him to—"

Jan: "We two just literally watched this man go to pieces—almost. He would just sit—sit and stare. He'd try to work and he couldn't work."

But he still had to keep ahead of *The Lone Ranger* comic strip which was appearing in 286 newspapers.

James G. Riddell was working on his dream. Several sites for new studios were considered, including an old church in Highland Park, but Riddell saw future in land northwest of the city. The largest shopping center in the world up to that time had been built in Southfield just beyond the northwestern edge of Detroit. Riddell went scouting.

One day he gathered his favorites around him—Ben Hoberman, Bill Hendricks, John Pival, Hal Neal, Charlie Kocher, the chief engineer, and Bud Reynolds who had been comptroller since ABC's purchase of the station. Riddell's face was pink and tan; he was neat, plump and enthusiastic.

"I found it," he announced. "I found the place. It's a big farm—a big farm at Ten Mile and Northwestern Highway."

"Jesus," someone breathed. "That far out?"

"That's going to be the hub of things," Riddell insisted. "They're already extending Lodge Expressway out to the Northland Shopping Center, and they'll go out beyond that. There'll be theaters out there, businesses out there. Oh—this is where we have to be."

Residents, however, were violently opposed to building a broadcasting station in the neighborhood. Riddell drafted everybody—his salesmen, Neal, Pival, Bob Baker,Pete Strand, John Lee, Kocher—and they carried on a door-to-door campaign to win support for re-zoning the farm so they could build a station on it.

Charles Kocher: "We had to convince them that the way things were growing, they could choose between an airport and transmitter towers. They were mutually exclusive; if you had one, you didn't have the other. We went after the people who didn't care and we got an amazing number of signatures. I think it was a Friday and we met after canvassing at nine—ten o'clock and had dinner some place."

That was the beginning. There were several hearings. But there was less objection from the Southfield Council than expected. Also, the FCC approved the move. So the plans were made.

The Austin Company of architects, engineers and contractors worked out the plans with Riddell. He was particularly impressed with the design of Station WSB in Atlanta—a Georgian mansion complete with a front lobby two stories high with a white stairway. ABC people came in from New York to confer with Riddell, but it is generally agreed that the project of building Broadcast House was his all the way.

Riddell: "We tried to make it a model broadcasting station. We tried to make the inside of that building *something*. And John (Pival) had a lot to do with that, too. We tried to work it so that everything would flow right from where the shipping came in to where it went on stage. And it was done well."

Virginia Pival: "John and I used to go over there and walk along the foundations and he'd say,'Now this is going here and this is going there but I know we don't have enough storage room right now before the thing is even built.' And he tried to explain this and wanted the company to do something about it, but apparently, they didn't."

Construction began in July, 1958.

Then what George Storer warned Riddell would happen happened: New York wanted Riddell—in New York.

43
Birth of the Top 40

Harold L. Neal, Jr. sat alone on the deck. All the others were below in the cabin enjoying the cocktail hour. The boat belonged to the head of the Bratton Tool Company and Hal and Shirley Neal and Chuck and Barbara Fritz were among the guests. The sun was losing its heat and sinking over Michigan and a warm, pleasant breeze was stirring. Hal Neal tilted back in his chair, his feet on the rail, feeling no breeze and not seeing the reddening sun nor the sparkling water of Lake St. Clair.

Hal Neal was now 34 years old and obsessed by his determination to revive radio. He was not thinking now of his triumph of last December, but the satisfaction from it did color his confidence. He had had Bob Baker create a tape with Wolf, Waber, Surrell, Winter, McKenzie, Shorr and Osgood, show-casing the WXYZ radio personalities and displaying how the Detroiters did what they did. He had musical station breaks and different program identifications worked cleverly into the presentation. No asset of WXYZ was omitted. It paid off. WXYZ had stolen the meeting of ABC station managers in New York. Everybody, right up to Leonard Goldenson, had been impressed. It was a decided victory for Hal Neal.

But when the quick and hearty laugh of Shirley Neal rose from the party below deck and interrupted the churn of his brain, Neal was not thinking back to December, 1957. He was thinking ahead, trying to determine how he might bring to Detroit a radio renaissance such as had been effected in Texas, Nebraska and Louisiana by three men: Gordon McLendon, "the wild man of Dallas"; incisive Todd Storz of Omaha; and Bill Stewart from Boston. What these three had done was simply fantastic.

Hal Neal knew of it through Wells Barnett, an extremely nervous little man who perspired a lot, whose hands trembled and who was based with the John Blair Company in New York. "Scoop" Barnett travelled all over the country for Blair, picking the brains of the best-managed stations and analyzing the reasons for their success. Hal had become close to Barnett and Chuck Fritz when he was sales manager because the Blair company represented WXYZ nationally. But he was now becoming more and more interested in the programming end of things, and he was fascinated by what Scoop Barnett had told him about those three radio rebels in the south and midwest.

Gordon McLendon was a born showman; his father, B.R. McLendon, owned a string of theaters in Texas. Gordon had published a newspaper before he was 20, graduated from Yale, spoke 5 languages including Chinese and Japanese, and served in Naval Intelligence during World War II. He was now in his late 30s, had a stocky build and a round face with deceptively sleepy eyes, deceptive because behind them were a jumping mind, an active imagination and an ever ready sense of humor.

In 1947, McLendon had taken over Station KLIF in Dallas with his father's money. He had won listeners with the wildest stunts imaginable. He promoted a treasure hunt with a hidden check, good for $50,000. He paid Lloyds of London to insure himself in case anybody did find the check. Then he had clues presented over the air as to where the check might be found. A listener did find it, Lloyds had paid the $50,000 and the million-dollar promotion had cost McLendon the mere $5,000 for insurance.

He maintained, however, that ideas to catch the fancy of listeners were of greater value than money. Once he offered to give 25 words or less to any listener who would send in $125,000. Promotion brings revenue, he said, and he had proved it.

The second man, Bill Stewart, went to work for McLendon in 1954. He saw KLIF zoom from 10th place in Dallas to 1st—in just 60 days! Gimmicks, promotion and "the McLendon sound" had done it. The McLendon sound was created chiefly by a tight, fast pace. Fred Wolf's "moving along" was a crawl by comparison.

Bill Stewart was a Bostonian. He had studied and taught classical music. He went around the country studying success in radio. McLendon's father-in-law owned a station in New Orleans. Stewart went there from KLIF and pulled off the stunt that brought him to the attention of Todd Storz.

Stewart had Al Jarvis write a song. It was an awful thing called *Shtiggy Boom*. Stewart had this song recorded by three janitors, and he had that record and *only* that one record played on the station for 4 straight days. Announcers would announce something else, but the same awful record went on—and on—and on. Newspapers made news of it. Petitions were circulated to stop it. Finally, Al Jarvis himself was imported to beg the station to stop playing his number. The stunt had worked. The station zoomed from 20th place in New Orleans to 2nd place—in 2 weeks.

Todd Storz, whose Station WTIZ held first place despite the stunt, hired Stewart away from the McLendons and took him to his home office in Omaha where he operated Station KOWH. Eighteen years later, in 1973, Stewart told Claude Hall of *Billboard*, how he and Todd Storz discovered, or invented, what came to be known as "the Top 40."

One night they went to a bar in Omaha and sat there 4 or 5 hours, discussing business. It gradually dawned on them that the same record was being played over and over again on the juke box. When the bar was closing, even the waitress put money in the machine to hear the same record 2 or 3 more times.

Shtiggy Boom had driven everybody up the wall because nobody liked it. But here, a record was being played repeatedly by *choice*. What had happened on that juke box in the Omaha bar was more important than anything the two men had discussed and they were quick-witted enough to realize it. People would rather hear again and again a record they liked than others picked by disc jockeys. Todd Storz and Bill Stewart applied their discovery to the programming of Station KOWH in Omaha, Nebraska, and the Top 40 format was born.

"Hal. The station will survive without you for a few hours. Come on down and join the party."

Chuck Fritz had climbed the companionway with a drink in his hand and stood smiling down at Hal Neal.

"Chuck, I was talking with Scoop Barnett yesterday. These guys down in Texas and out in Nebraska are really something."

"Gordon McLendon? He has the guts of a cast-iron monkey."

"All three of 'em have," said Neal. "But it pays off."

"I guess." Fritz watched the sun sink over the horizon. "We're in an era of 30- and 60-day wonders. In 60 days a radio station can jump into first place."

"McLendon and Todd Storz are sure proving it in Dallas and Omaha," said Neal. "And you know what? I'm going to prove it in Detroit."

"You won."

"I lost!" cried Fred Weiss. "I don't want to be a disc jockey."

It was 10 o'clock in the morning and they were in Bob Baker's office. Hal Neal was sitting at Baker's desk. He was as steady and earnest as Baker was ebullient. Neal wore a dark suit; his tie was in harmony; his curly, close-cropped hair was neatly brushed. Baker wore a silky gray suit; his tie looked as though he had chosen it before he was awake; his light brown hair seemed to have been forced straight back with a pocket comb. Fred Weiss looked flushed and a little wilted.

"Listen," said Neal. "You're making 140 a week as an announcer. I'll give you a 35-dollar-a-week raise. $175 a week. How's that?"

"You'll be a star," Baker tossed in, his big face bright and happy.

"I don't want to be a star," Weiss moaned without animosity. "I don't care about the money. I like what I'm doing and I don't know anything about this music. I couldn't ad lib my way out of a paper bag and I think you've got the wrong guy."

"I think we got the right guy," Hal Neal said firmly.

"You'll only have to work 2 hours a day instead of 8," Baker pointed out.

"You'll be on the air from 7:15 to 10 P.M.," Neal corrected. "I'm moving Mickey Shorr up to noon-to-3, ahead of McKenzie. You'll follow the hour of news, sports, Osgood and Edward P. Morgan."

"7:15 to 10! That's 2 hours and 45 minutes! What'll I do?"

"What you did on your audition record—more or less."

"I didn't want to make that record."

"I know," said Hal Neal.

"He stalled for 2 weeks," Baker yelled, laughing exuberantly. His own little conflict with Weiss was over now and he could make a joke of it, his language as usual trailing behind the rapid pace of his brain. "Everybody else—and he—couldn't cut it."

Fred Weiss picked up the joke as though by telling it he might ease away from a monstrous assignment. "Every morning I'd see this note sayin', 'Cut that.' And I never did."

Baker looked at Neal, laughing. "I told him, I said, 'You of all people better cut a good one'."

He looked back at Weiss, laughing.

"After 2 weeks, I saw it there, that note, and I thought, gee, I better do this or they'll crawl down my neck," Weiss chuckled. "So I went in and I recorded 10 minutes of—terrible—you know."

He and Baker laughed again. Baker's laugh was surprisingly small and high-pitched for so large a man, reminding one somehow of the yap of a Pekinese pup. Weiss' chuckle was gutteral. Hal Neal had sat there without laughing.

"I heard it. It isn't great, but it isn't lousy, either," Neal said. Then he added firmly, "I know you can do it, Fred, and you're it. Starting Monday."

This sobered everybody.

"I'll tell you what," Weiss said at last. "If you guarantee me 6 months, Hal—you know, if it's terrible and you leave it on, I'll guarantee that at the end of 6 months, you can live with it. But don't stick me on for 2 weeks and say, oh my God, I guess I made a mistake. You know?"

Hal Neal's fiery blue eyes were steady. "Okay. I'll guarantee you 6 months."

"Then I'll try it," said Weiss.

Fred Weiss: "Oh, it was terrible. The very first day I was on, I played a thing by the Eberly Brothers. Their first big hit. And I was the guy who sat at home and listened to the symphonies and turned on the opera. And *this* stuff! I knew *nothing* about it. And that night I got on and played the Eberly Brothers thing and I said, 'Boy, if this is music, you can have it!'

"I got threatening letters in the mail from the Eberly Brothers Fan Club—real threats! They were goin' to come and kill me! So one thing I learned from the very first night was that, gee, there are people who listen to this and really believe somethin' about this. So I didn't play *that* funny game any more."

Fred and Carrell Weiss were soon in a position to invite friends to a house warming in Grosse Pointe.

The farewell bash held for James G. Riddell at the Lochmoor Golf Club in Grosse Pointe began with cocktails and luncheon and continued through golf, highballs, swimming, cocktails, a big dinner, cards, dice and liquor until early morning. Fifty friends came and went. When Riddell was presented with oil paintings of his family, he was so genuinely moved he couldn't speak.

His embarrassment was relieved by "newsboys" racing through the room hawking an extra edition of the *Redford Observer*, printed for the occasion. A headline screamed across the front page: "TV EXEC HIT WITH CHEATING CHARGE." The executive was accused in bold print below of cheating at golf. The executive, of course, was Jimmie Riddell and among the accusers were John Pival, Don Wattrick and H. Allen Campbell.

Riddell's equilibrium was restored and he gave a warm and witty speech, needling some of his friends around the room and passing along some amusing memories of his days as an office boy under George Trendle.

Riddell had once bawled out a salesman because he told of running short of cash while entertaining a client at lunch. Riddell had ordered every salesman to carry at least $100 cash at all times. John Pival went beyond this. He was not a salesman but he was a gambler, and he always had $1,000 cash on him. This was bad news to crapshooters because he could outlast any streak of bad luck he might have. Cards and dice appeared at Riddell's farewell party.

Working at WXYZ-TV now, and present at the party, was a flamboyant news commentator named Lou Gordon, short, dapper, keen-eyed and well-heeled. He had inherited the White Stag sportswear franchise in Michigan from his father and had returned from Washington to take it over. He had done news work in the nation's capitol, and was a rather sensational addition to the staff at WXYZ.

Lou Gordon walked over to the crap game, bet $100 a throw, made 6 passes, coolly reached down, picked up $600, and said, "Thank you, John. Good night."

He turned around and walked away. John Pival was drunk and he was dumbfounded.

Lou Gordon

"Hey, wait a minute," he called out finally. "Come back here."

Lou Gordon turned around, waved, smiling, and continued to walk away. He was through. Pival stood there a moment swaying a little. He was angry. Suddenly, the successor of George Harrison Phelps, George B. Storer, George W. Trendle and James G. Riddell, wheeled around, picked up the table on which the dice had been rolling, and smashed it against the wall.

James G. Riddell: "I never wanted to go to New York. But you see, I organized the owned-station group, and any time we had meetings, I made all the arrangements and I conducted the meetings. And I guess we did the best job in Detroit; we made more money than any other station. So they prevailed upon me to come to New York as Executive Vice President of the company. I went with great trepidation."

It was a sad day for WXYZ and everybody felt it even as they whooped it up at an historical farewell party.

On a bleak morning in 1958, 6 disc jockeys sat or stood in Bob Baker's office facing Hal Neal. Sleek and trim as usual, Neal had tried to engage in friendly small talk with "the boys" as they arrived, but friendly small talk did not come easily to Hal Neal at any time and even less easily this morning. He had made a decision, knew it would not be popular and was on the verge of a showdown.

"We're going to start a new thing here," he announced as soon as they were all together. "We're going to go with the top hundred."

Ed McKenzie's moan, "Oh Jesus," was barely audible.

"Wha—what is—what does that mean?" Fred Weiss asked, his chuckle rattling in the back of his throat.

"We're going to play the same hundred records all week on every show," said Neal.

"Not—*only* that!" cried the unbelieving Weiss.

"That's all," said Neal.

The air was thick with a sense of shock. An uncertain white smile gleamed in the black face of Jack Surrell and was aimed vaguely in the direction of Paul Winter. Winter's reaction was a giggle, falsetto but unquestionably masculine. Mickey Shorr, like McKenzie, had already heard or read about this type of format and he was less opposed. Fred Wolf slouched in a chair—quiet, removed, unapproving.

Baker's voice cut through the silence in shrill defense of Hal Neal, his words tumbling over one another in their effort to keep up with his mind. "It's *very*—in Dallas and Omaha—fantastic! Very successful. Top 40—you—wouldn't believe!"

"I've heard about it," McKenzie scoffed.

"Well—I don't believe it'll work in Detroit," Fred Wolf commented quietly.

"I realize it's a big change, Fred, but it'll work," Neal declared. "*And*—another thing. Everybody's going to cut down on the talk. What people want to hear is the music. They tune in for their favorite records."

"Fiddlesticks!" McKenzie exclaimed. "They—music's part of it—but they tune in to a deejay because they like him."

"Personality is important," Neal agreed. "It's not to put personality down at all. But they don't tune in to listen to a disc jockey talk about inane, ridiculous things that *he* and only a minor segment of the audience might be interested in. The object here is to get a mass appeal and a consistence of sound going, so that the listener—no matter what time of day he tunes in—he can depend upon *that* station for *that* type of sound. And WXYZ is going to be *that* station with *that* sound."

Neal was not weakening, not arguing; he was firm; he was telling. Mickey Shorr, who was believed to have an interest in a record company, broke the short silence that followed.

"Who's going to pick the hundred?"

"Baker and I," said Neal. "We'll make up a list and those will be the records you play."

"You mean—the same records over and over—*all day*?" Fred Weiss still couldn't believe it.

"All day and all night," said Neal. "And the most popular ones will go on the air once every hour. We'll mark those on the list."

"We'll mark those on the list," Baker echoed. "Each of you will get a copy of the list and there'll be a new list every week."

"When does this start?" Wolf asked, quietly.

"Monday," said Neal. "Less talk between records, more music. We're going to tighten things up."

As the meeting broke up, Neal watched his 6 men leave the office. One or two seemed embarrassed. Two—McKenzie and Wolf—showed signs of instant antagonism.

"It won't be easy," Neal declared when he and Baker were alone. "But this is the way it's going to be."

"Well!" Baker shrugged, his little blue eyes wide and serious. "You slapped their little egos."

"They have a right to them," Neal observed. "Fred Wolf will go down as one of the most important and prominent personalities in the radio scene. And Ed McKenzie has been a very important personality in this market. Very successful. Very big salary and at one time in ratings, in popularity. I can understand what they're thinking; here's this young upstart—me—coming in with controls and saying this is the way it's going to be. They resent me and they're going to resent controls."

"Well!" Baker exclaimed, defending his disc jockeys. "They'll come around. They'll compromise."

"No compromise, Bob," Neal declared flatly. "It's got to be my way—all the way."

Thus in 1954, Hal Neal was the first to introduce the tight, fast "McLendon sound" to Detroit. And he transferred choice of records from disc jockey to management—as it turned out, not a month too soon.

44
All Together Now

Dick Kernen, now 20, was released by the Navy late in 1958. He hung around the family for a couple of months, trying to decide what to do, then finally took the easiest route.

Kernen: "It's always seemed to me that people who left the station always kind of regretted leaving. There is just something about WXYZ that attracts people."

Bob Baker rehired Kernen with the promise that if Dick would be patient for a month or two, he'd try to move him up into the "production" department—which now meant running records for disc jockeys. In the Spring of 1959, Kernen was promoted from mail runner to record runner. He also enrolled at Wayne State University and Baker arranged his hours so he could go to school full time and still work an 8-hour day. But the important thing to Kernen was the station.

Kernen: "When I went into the Navy, everyone had their own little thing going. When I got back, there was a tremendous difference. The whole station tended to sound more cohesive. Everyone was basically doing the same thing with the exception of Fred Wolf. It was a whole new thing and I loved it."

The disc jockeys had all been fitted out with red jackets and told to wear them whenever they appeared before the public. The record lists had been shortened from 100 to the Top 40. Neal was now allowing the disc jockeys no more than 7 seconds of talk between records. They had slogans to repeat at set intervals. The current one was "WXYZ Sounds Good Like a Radio Station Should."

Some things had not changed. The same engineers were around—Al Schaub, Ray Spejna, Ray Skidmore, Elmer Immel and Wes Govier. Jack Tierney was installing new studios in the new building.

Porter Bourne did less announcing now because his job was to schedule announcers on the Log. Gene Avram was announcing, and big Chuck Daugherty, thin George Hoffman, and Don Potter who later became a Michigan State Senator.

Dick Osgood was still doing the morning newscasts; all were commercial although they had been reduced to 5 minutes each. He was still doing Hollywood Headlines for the United-Detroit Theaters, and Show World in the evening. He was now 58, told everybody he was 90, and had a boring way of talking about former Wyxyites nobody ever heard of.

Music, however, was "where it was at" today. Fred Wolf was given a little more leeway than the others because he was sold out, his ratings were high, and Neal conceded that the "morning drive" audience was a little different. But Wolf did feel restricted and he was fighting "the new sound." It was a losing battle; forces other than Hal Neal were lined up against him.

Fred Wolf: "All the New York brass came in for a big meeting at the Statler. We were all told to be down there in our red jackets. This is when the big thing started about 'Young America.' Nobody's over 35 years of age at this time, an' you just gotta play music for Young America. An' I can remember everybody' yessirin' and yessirin'—but me. An' I know I embarrassed whoever the hell my boss was—I guess it was still Neal. But—I couldn't go with it. I fought it. I stood in front of this guy from

New York and I said, 'I think we're doin' very well. Our ratin's are just—'

"An' this guy sits there an' said, 'Fred, how do you know how much *higher* the ratin's could be if you didn't play your kind o' music?' "

Don Davenport continued to be Wolf's "producer." Dick Kernen worked for all the others. The first he remembers was Fred Weiss.

Weiss was adapting very well to the role of the disc jockey but it was not easy.

Weiss: "I used to come home absolutely sick. My stomach was all churned up. And partly it was nerves because I didn't know what I was doing; and secondly it—it disturbed me to think I was playing this *corrupting* kind of music. It really disturbed me for a long, long time—until I finally began to realize that most of the kids that listened to this were very young and didn't really know what the lyrics were all about. There were a couple o' words that would stand out and fit the rhythms but this was all. And if you asked them what it was all about, they really didn't know."

There were those who, if they didn't like the music they had to play, still admired the man who made them play it.

Paul Winter: "Hal Neal had a remarkable intuition about mass audiences. Hal was not always an easy guy to work with. But at least where I was concerned, he was always a marvelously fair guy. I never fought him as much as some people did. On the other hand, I don't think he gave me as hard a time as he might have other people. Whatever he did, I think he was a splendid program guy. He knew what he was doing. He knew how to do it. And he knew—this is most important—how to tell people what it was he wanted."

In the Maccabees days, the hang-out had been The Alcove. From the Mendelssohn mansion, everyone went to the Van Dyke Club or Dad Buslip's. Now, the special place was The Ranch House. Like many restaurants built in the late '50s, it had the outside look of an oversize septic tank, but inside, the bar was friendly, the food was good, and the Wyxie gang appropriated a big round table—especially on Fridays.

One Friday noon in February, 1959, gathered around the table for lunch were Baker, Ed McKenzie, Osgood, Weiss, Winter, Kernen, Don Davenport and Sonny Pryor who was McKenzie's Girl Friday when Joyce Chapman left to become a police woman. Sonny seemed to have separated from her husband, spoke often of her father who was an artist, was herself interested in art and choral music, was Nordic, attractive, and a solid personality.

There was not a great deal of drinking. McKenzie, of course, did not touch liquor. One or two beers would be enough for Don Davenport who was bald, thin, and military. Also for Dick Kernen. Sonny ordered something fancy. Baker slurped up a number of martinis and Osgood kept him company. Winter was a very amateurish drinker and ordered Cutty Sark because he'd heard it was good. The group had gossiped and drunk briefly and was now settling down to food.

"Michael Tolan opened at the Ethel Barrymore Theater in New York last night with Gertrude Berg in *A Majority of One*," Osgood was saying. "He used to play Kato in *The Green Hornet.*"

"Real name's Seymour Tuchow," said Baker.

"I'm hoping *Maggie* will come to the Northland next summer," Osgood went on. In 1959, the Northland Theater was a tent which booked summer attractions.

"Who's Maggie?" Kernen asked.

"*Maggie*—it's a musical," said Osgood. "Billy Roy wrote the music. He used to be on the kid show we had. *Maggie* flopped in New York but Ella Logan took it into the strawhats last summer."

"Fred tells me the new ratings are in," said Davenport. Baker had a way of pouncing into a conversation.

"Yes!" he cried. "Hal and I—I was looking at 'em with Hal—and would you believe it? Fred Wolf tops J.P. McCarthy."

J.P. McCarthy was Wolf's popular and formidable opposition on WJR.

"Fred's up a little over J.P.," Baker repeated. "Well, it's Hal Neal—his—he's—it's Neal."

"Oh come on, Bob," McKenzie protested. "It's Wolf. The guy has built an audience."

"He still won't play Elvis Presley," Davenport said.

"Neither does J.P. McCarthy," McKenzie pointed out.

"WJR is like crackerjack," Paul Winter observed. "It never changes."

"Why should it?" Osgood demanded. "Radio is habit and JR's got its audience addicted."

"Yeah—well—." Baker's tone was a put-down of the opinion, colored by devotion to Hal Neal.

"I hear JR won't let Don Faust into the radio station," Osgood went on. "Don used to announce with us but now he's manager of JR's TV station in Flint, and he wanted to get Bud Guest before the cameras, but Worth Kramer told him to keep his cotton-pickin' fingers off the radio personnel, won't even let Don into the studios."

Nobody seemed interested.

"Don Faust used to play the Green Hornet," Osgood finished, lamely.

"TV. Who needs it?" said Ed McKenzie. Mickey Shorr had taken over Ed's *Saturday Party* on Channel 7 while Ed had surgery. McKenzie continued, "When I got out of the hospital, I went to Pival and I said, 'John, I don't want to do that show any more.' 'Cause I could see, you know, it's going' down hill. So, he kept Mickey on."

"Didn't last long," said Weiss.

"The show just folded. You know. Nothin'."

"I was watching Perry Como last night," said Osgood. "And I heard the announcer. I knew his voice, and then I realized it was John Cannon. He used to announce here. They tell me he's just going great in New York. And Elaine Alpert. She's part owner of *Crossroads*."

Crossroads was a network program. Nobody was interested.

"Elaine used to act in *The Lone Ranger,*" Osgood murmured.

"Jack Surrell's wife on his trail again?" McKenzie asked, looking at Weiss who was on the air just ahead of Surrell every night.

"Oh God *yes*!" Baker cried.

"What's the trouble?" Davenport asked.

"Who knows?" Baker shrilled. "I guess Jack's giving her a bad time."

Baker's face flushed as he broke out in his little Pekinese laugh. "She wrote to Jack Rellis once. She said, 'Everybody in the world should be like Bob Baker. He doesn't like girls, doesn't like boys, doesn't like *anybody*—just keeps his nose clean.' "

Winter broke out with his masculine falsetto giggle which was so infectious everybody laughed. There was a happy congeniality in the Ranch House—a momentary revival of the spirit of the Alcove and the Van Dyke—and one which was soon to pass.

"This tight format is tough on Surrell with his piano segways," McKenzie said seriously. "They don't go with the new music. It's murder."

"Well," Baker pounced. "He won't last! You know that. He—he— this isn't going to last."

"I'd rather listen to his segways than the records," said Fred Weiss, chuckling.

"You have to be out of your mind," declared Dick Kernen. "Do you mean you still want the June-moon stuff?"

"No," Weiss replied. "I admit that the lyrics might be better if you could hear 'em. I took a 45 this morning and slowed it down to 33 tryin' to figure out what the hell they were singin'. I'll never know."

"Oh happy day," McKenzie intoned sadly.

Sonny laughed. "Oh, Ed—tell them about that."

"You know the one where the Branches sing 'oh happy day' like it was a funeral dirge," said McKenzie. "I played it for a psychiatrist friend of mine an' I asked him what makes 'em sing a bright idea like that oh happy day so mournfully, so sad?"

"What did he say?" Winter asked.

"I forget. What *he* said didn't make any sense either."

The gutteral chuckle of Fred Weiss sparked a little laugh around the table. McKenzie went on, seriously, "I wouldn't mind workin' in some of this new stuff if they'd let me pick what I played on the air. But this top 40 crap—"

"You better get used to it," Baker warned. "Hal's going to narrow down the list even more. I predict within a month it'll be down to the Top 20."

"Jesus," said McKenzie. "All the fun is gone. The job is no longer fun. I don't like somebody who never made it as a disc jockey tellin' me how the hell to run a show."

"Hal never *tried* to make it as a disc jockey," Baker reminded them. "But you better go along with him. We *all* better go along with him because what he wants he's going to get. Besides, the ratin's are on the way up, and you can't argue—you know—you can't—with that you can't argue."

"I don't see how you can argue with the new music, either," said young Dick Kernen. "Regardless of whether you like it or not, this is what is happening in music."

"I can stand the music," Winter agreed. "First, as a performer, it gives me great satisfaction to change my style to meet the challenge. I enjoy it. It's like acting a part. Secondly, it's my job to do it. If I didn't do it, I could do something else."

"Teach," Kernen suggested. "I think Paul would make a hell of a teacher."

"Paul's a performer—a pro," Osgood observed.

"How do you define a 'pro'?" Sonny asked.

"My definition of a pro," said Fred Weiss, "is to get from one end of the show to the other without quittin'. That's all I try to do. I just try to make it from one end to the other. I have no idea what I'm doin'."

For no reason, Osgood said, "I see Stan Dale made the Top Ten Disc Jockey list. WAIT, Chicago. He used to be an annuncer here at—"

"Well!" Baker cut in."I got to get back."

The plates were empty, the coffee had been drunk, and chairs were pushed back. Another Friday lunch at the Ranch House broke up. It was one of the last.

Dick Kernen rode back to the station with Baker.

"So you're dropping school?"

Kernen admitted he had not signed up for the new semester.

"I have a whole thing about education and what it does to you," he told Baker. "The education, for instance, in this country, taught two of my brothers beyond any point of refutation—it's impossible to convince them otherwise—that they're stupid. That's the only thing they took away from high school—that they were stupid, and they have never forgotten it. It's the whole system. It's pretty weird."

Clearly Hal Neal was running into trouble with his disc jockeys. Only two were openly antagonistic, but not one of them was really in key with what was happening in radio. And as the music got more and more into rock, they seemed to be less and less compatible.

Neal was absolutely determined that the station was going to sound as he wanted it to sound. He was as finicky as George W. Trendle had been about every last detail. He listened to the station constantly in his office and in his home. He carried a tape recorder with him in his car and as he heard something he didn't like on the car radio, he made a note of it on his tape recorder.

The disc jockeys received a flurry of inter-office memos. "You talked 13 seconds between 2 records yesterday afternoon. Keep it down to 7 seconds *maximum*" "You gave the slogan only once in the quarter hour; it must go on the air twice very 15 minutes." "I want station identification after *every* record." "Your voice pitch is too high. Learn to talk fast without letting your voice go up. Keep it down."

The conflict between Hal Neal and Ed McKenzie finally arrived at a confrontation.

McKenzie: "Hal Neal was on me all the time. 'You didn't say WXYZ 2 times with every record.' 'You didn't stick to the format'."

Neal: "Ed just became *so* undisciplined that —there had to be something done."

McKenzie: "I just couldn't take it any more. Neal drove me crazy. He got to be a real tyrant."

Neal: "It was one of my first functions as a manager after Jimmy Riddell had left town and I was on my own and I had to make a decision."

McKenzie: "We kept gettin' those memos, you know, Li-ife is too short you know."

Neal: "I was ill one Friday afternoon and was listening to the station at home, and Ed—he sounded like he was really ready to flip. So I called up. He had recorded a show for Saturday morning and I had that pulled and asked him to be in my office Monday morning. We sat down and had a talk and—we agreed to disagree."

McKenzie: "Finally, I said *shove* it."

Neal: "I had to ask McKenzie to resign. That was the first decision of the kind I had to make and I never will forget it."

It was a lonely fight for Neal. He and John Pival were not especially congenial, and he and Bill Morgan had drifted apart. He did go to lunch with Charles D. Fritz of Blair, and he could talk to Baker. Actually, he was winning the battle; the ratings were up; and the station was doing a lot of business. But he was discovering the isolation of top position. Most of all, he missed the sunny optimism and support of James G. Riddell.

As for the Riddells, they hated New York. They stood it for about 8 months; then Jim went to Leonard Goldenson and suggested that Earl Hudson be retired and he be sent west to replace him. This was done.

Riddell was picked up at the airport near Los Angeles by Phil Brestoff. Things had not gone as the Brestoffs had hoped. Dee Parker's one or two shows had not lasted and Phil had spent most of his savings and inheritance from his father.

Riddell gave him a tip he had been given by a friend on the stock market. Brestoff put every dollar he had into the stock, and recouped all his losses. He was so happy he sent Riddell some cashmere sweaters. But he couldn't get a job.

The Riddells were happy almost immediately in California. Jim was now top man with KABC radio and television in Los Angeles and KGO radio and TV in San Francisco. He imported the one-time WXYZ salesman Ben Hoberman to manage KABC for him, and persuaded him to hire Phil Brestoff as sales manager at $35,000 a year.

It didn't work out. Brestoff was unable to get along with the salesmen.

"I think he's sick," Hoberman told Riddell.

Riddell went to Brestoff's doctor.

"Frankly, I think his job is too much for him," Riddell explained to the doctor. "He's a pretty good salesman, but he is not a good executive. Now, I'd appreciate it if you would tell Phil that he's got to give up this job for his health. I'll give him a job making the same kind of money as a salesman."

The doctor collaborated. Brestoff, a very sensitive man, was hurt by the demotion. In a few months he quit. He ran a little station in Burbank for a while; then he suffered heart trouble and died. Dee eventually married again, moved to San Francisco. Bill Bixby who played the father in *Courtship of Eddie's Father* on television, moved into the old Brestoff home in Beverly Hills.

Meanwhile, back in Detroit, WXYZ moved into Jimmy Riddell's dream house.

"American's newest and most modern broadcasting facility" opened officially with a simulcast on WXYZ radio and television, 6:30-7 P.M., Thursday, June 4, 1959. As with most milestones of WXYZ history, newspapers made no report of the

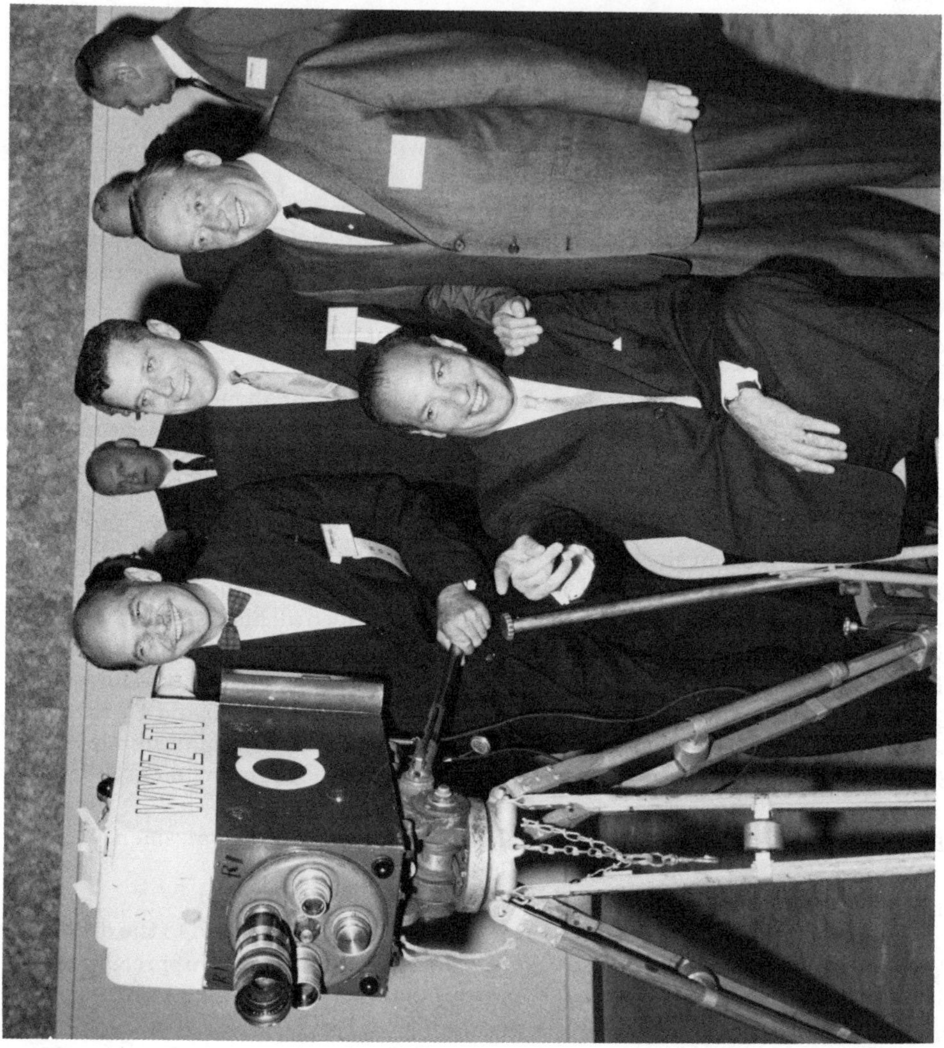

At opening of Broadcast House. Foreground, left to right, John Pival, Harold L. Neal, Jr. and James G. Riddell. Sitting, Leonard Goldenson

Radio stuck in a corner at Broadcast House. Osgood at console, Al Schaub, engineer at left. Bruce Miller at right

opening. *The News* did list the "Broadcast House Dedication" among the night's Best Bets and promised appearances by Governor G. Mennen Williams, Mayor Louis Miriani of Detroit, Mayor Swanson of Southfield, and the Very Reverend Celestin J. Steiner, S.J., President of the University of Detroit.

The Free Press front-paged a Big Four parley in which the United States, Britain and France proposed a Berlin truce plan whereby Russia would guarantee freedom to isolated West Berlin in exchange for a freeze on western troops there. The front-page banner in *The News* announced that a 17-year-old high school outfielder from Youngstown, Ohio, Andy Kosco, had been purchased by the Detroit Tigers as a bonus player for $60,000. Both papers reported that Governor Williams seemed to be losing his grip on the State Legislature; and four mice had perished in an abortive attempt to put Discoverer III into orbit.

There had been rioting in Ecuador; Fidel Castro agreed to give American sugar interests a chance to harvest the 1960 crop before taking over the industry; Ingrid Bergman and Roberto Rosselini were fighting over possession of their 3 children; and Mickey Rooney was married again, this time to Barbara Thomason. Banks were paying 3% on savings; Harry Suffrin was selling men's Palm Beach suits for $39.95 and Demery's advertised a 3-piece "Hop-Silk" ensemble for ladies at $17.98. Eighteen inches of hail had fallen in Kansas and a sure-fire cure for sunburn was announced on the front page of the *News*.

WXYZ's new TV transmitter, tallest structure in the Detroit area—1,073 feet, was equipped with an elevator to carry engineers to the top. It had gone into operation in January. Broadcast House was completed in May. In June, Jimmy Riddell flew in from California and Leonard Goldenson with others including Oliver Treyz, President of ABC-TV, flew in from New York to participate in the grand opening of the 4-million-dollar facility. Advertisers and agency people were invited to the opening festivities which were arranged by Bill Hendricks and the former manager of the Fox Theater, Jack Hurford, who was now producing in television. Dick Osgood was chosen to be master of ceremonies. Lou Gordon and Leon McNew served as hosts.

Early in the evening, McNew was walking along a back hallway, handsome in his tuxedo, nervous, trying to get things organized in his mind. He stepped into one of the offices and found a little man standing there, as nervous as he.

"Hi."

"Hi."

"I have to make a speech," the little stranger said, "and I'm as nervous as I can be."

"Oh, there's nothing to it," said McNew to comfort him.

"Maybe you're used to making speeches."

"Well—no," said McNew, "But—just have one or two things in mind, and make it brief, because all the bigwigs will be making speeches—the Governor and the brass from New York—even Leonard Goldenson."

The nervous little stranger was Leonard Goldenson.

Broadcast House was not only the most modern broadcasting facility in the country, it was also the most glamorous. Set several hundred feet back from Ten Mile Road, the brick-and-white Georgian front was backed by a two-story building which looked much smaller on the outside than it turned out to be on the inside. Apple trees remained on the landscaped lawn to blossom in the Spring. A sweeping hard-top driveway led around the building to a parking lot with space for 175 cars. Top executives could park under a roof built onto the main structure.

The original farm house, connected with the main building by a covered walk, now housed a cafeteria and, on the second floor, small lounges and dining rooms. In each lounge was a bar where a salesman could entertain clients until lunch was served in the adjoining dining room. On the first floor, in two carpeted and wall-papered rooms, 80 employees could enjoy the food at one time.

Beyond the cafeteria building was a cottage for the caretaker and at the far end of the paved parking lot was a 10-car garage for company autos and other motorized equipment. The TV transmitter stood behind the garage, and almost lost at the far end of the 110 acres were 2 radio transmitters.

Behind the square-paned windows along the front were offices for the top executives. The most luxurious, originally intended for Riddell, now went to John Pival. On the opposite end was space for the comptroller and his staff. Bud Reynolds moved into the choice corner office in this financial suite. Occupying front offices between Pival and Reynolds were Harold L. Neal, Jr., Earle Moore, Harold Christian, Bob Baldrica and Bill Hendricks—and their secretaries.

Behind this elite row was a catacomb of offices and cubicles for radio and TV traffic, promotion and advertising and salesmen. On the other side of this maze, running from a reception desk at the back door to the financial suite was a hallway which also had doorways into the three TV studios. Each of these studios measured 40 by 60 feet, and each was equipped with 90 overhead lights. In one, a complete permanent and practical kitchen had been installed for Edythe Fern Melrose.

On the far side of the studios was a large areaway plus 77,000 cubic feet of space for storage, receiving clerk's offices, the heating and air conditioning plant and the transmitter room. There was a side corridor for offices which looked out on radiators of the executive cars under the shed. These were assigned to John Lee, Pete Strand and the art department of former floor man, Jack Flechsig. The mail and supply department was on the opposite side of this corridor, and dressing rooms for men and women.

On the second floor was the film department with Kenny Sim's office, storage room for 2,000,000 feet of film, and three screening rooms. Telecine—TV equivalent of a movie theater projection booth—was conveniently situated in front of the film section; and there was a spacious repair room in this area.

The real show place of the building was TV's control room, 120 feet long and about 40 feet wide—the largest, most modern control center in the world. The first impression was one of great space. Then one observed three large directors' consoles, one for each of the studios below, and behind them, a big master control area which relayed chosen pictures and sound to the transmitter room.

"This is really something," marvelled Leonard Goldenson, standing behind the master control console and looking around.

"The reason this is so large," Riddell explained, "is your chief engineer. Frank Marx had a lot to do with the designing of the building. I think it's too large really."

"Don't let any of the men from our other stations see this," Goldenson warned. "They'll want something just like it."

Aside from its sales and traffic departments which were lost in the inner catacomb of the first floor, radio had been crammed into a total floor space equalling about two-thirds of TV's spectacular control room. It was hard to find. In fact, the installation of television had been well underway before anyone thought of radio.

Jack Tierney: "All of a sudden they say, it's your turn. You know? I came out here and Harvey Mellion from ABC was here and TV was being put in. I was introduced to him. I said, well, how do you do? and here I am—uh—what would you like me to do?

"He says, 'Put it in'."

"I says—*you* know—friendly—it's only a room here now—a big, bare room'."

"He says, 'Yeah. Go ahead. What you got over *there* (meaning Joy and Greenfield), bring over *here*'."

"So that I proceeded to do. It was decided that they needed a chief engineer for radio as well as a chief engineer for television; so Paul Dudek, being Charley's assistant, he came over and was chief engineer, and I continued to do what I'd been doing. We had crews of half a dozen guys all workin' on it and I was supplying the information for them to put it together."

Television got all new equipment; radio used the equipment it already had.

Charles Kocher: "It had been new three years earlier when they built the plant at Joy and Greenfield; so they didn't feel they needed new equipment."

Fred Weiss: "They did everything for television and stuck us in a couple of cement rooms on the second floor."

Even so, radio managed to install a record library, an office for Bob Baker, a small secretary's office between that and the hallway, one office for production men to share with disc jockeys, a combination control and taping room, and 3 very small studios into the largest of which Jack Surrell managed to cram a small upright piano.

A large newsroom shared by both radio and television was fringed with offices for Dick Femmel, Tom A'Hearn, Dick Osgood, Don Wattrick, Tom Waber and Porter Bourne.

Tours conducted through Broadcast House concentrated almost entirely on television. Following the breath-taking sight of the mammoth control room, visitors were herded rapidly along the corridor past the little radio section and, hardly slowing down, the guide would apologetically wave one hand in passing and say, "And this is radio." nobody ever stopped to look or seemed for that matter to be interested."

The important thing was that now, for the first time in its 34 years, all divisions of WXYZ were under one roof. It is generally agreed that James G. Riddell's greatest contribution to the station was Broadcast House. He thought of it in the beginning, persuaded the New York powers to build it, chose the site for it, planned it and started construction.

Radio's old site at Joy and Greenfield was sold. On it now is a shopping center.

John Pival was a very happy man. He had a golf green installed on the back lawn, had his office decorated to his taste and chose all the furniture. Television was making a lot of money and he was king of the palace—a mighty Potemkin.

Hal Neal, on the other hand, was disappointed. Radio was making a lot of money, too, and the ratings were up. But radio had obviously received the short end of the deal in the building of Broadcast House, and almost immediately he demanded changes. The first thing he ordered was a separate room for taping and editing.

The sound of the station did not suit him. What was happening in music was not happening to his disc jockeys; the jocks and the music simply did not belong together.

Then, like a clap of thunder, in November, 1959, 5 months after the move into Broadcast House, *Life Magazine* printed an article, ghost-written for Ed McKenzie, which fanned into the nostrils of the public an odor that had long filled the nose of the industry. He told the world he had quit radio (meaning WXYZ) to escape the stink of—PAYOLA.

45
Men of Low Estate

The payola scandal was only one plant in a patch of skunk cabbage. The first smell-releasing kick had come from a stand-by contestant on a TV quiz program called *Dotto*. *Dotto* was sponsored by Colgate-Palmolive on CBS in the daytime and on NBC at night.

The stand-by contestant was a part-time butler and night club comic named Edward Hilgemeier, Jr. Before one telecast, Mr. Hilgemeier observed a contestant studying some notes. She had survived the questioning of several programs in the series, and when she was called to the cameras this time she made the mistake of leaving her notes behind her. Mr. Hilgemeier was curious and examined them. They consisted of questions the lady "might" be asked—and the answers.

Mr. Hilgemeier reportedly accepted $1500 to keep his mouth shut, but he brooded, and his conscience won. He finally took his information to the Federal Communications Commission. The FCC promptly got in touch with those concerned, and within hours, *Dotto* was dropped by NBC, CBS and Colgate-Palmolive. This happened in August, 1958.

New York District Attorney Frank S. Hogan caught the smell and began investigating. He learned that two years before, in 1956, a New York college student, Herbert Stempel, had been invited to appear on another TV quiz program called *Twenty-One*. He had been told how to dress, coached on how to perform in the "isolation booth," and had been forced to lose to Charles Van Doren, 33-year-old doctor of philosophy and a teacher at Columbia University who later admitted that he had won a $129,000 by seeing a script in advance. Loser Herbert Stempel had offered his story to two New York newspapers in 1957. Both had turned it down.

Attorney Hogan's curiosity kicked up a lot of cabbage. He, and the world, learned that Xavier Cugat had picked up an easy $16,000 on another show called *$64,000 Challenge,* a companion show to *$64,000 Question,* which had a smell all its own. Patty Duke, a little 12-year-old actress appearing in a stage play called *The Miracle Worker* had been maneuvered onto *Challenge* by her manager and had won a tidy $32,000 in a fixed match.

The stench stirred up in New York reached a special House Subcommittee on Legislative Oversight in Washington. This committee, chaired by Representative Oren Harris of Arkansas, heard producers testify how quiz programs were controlled.

A producer might "calm the nerves" of a contestant before going on the air, and discover in the "relaxing" conversation what special interests the contestant might have in which he might be well versed. Or the producer might say, "Well—and what shall we talk about tonight? You're interested in history. What if we talked about World War One? I'm sure you would know that Kaiser Wilhelm of Germany led the forces against the Allies." And sure enough, a question would be, Who led the forces against the Allies in World War One? And the contestant would answer Kaiser Wilhelm—after acting out the agony of trying to think.

A producer could control the length of time a contestant would be on. A contestant with a dud personality could be asked an extremely difficult question and bumped off. If audience reaction to a contestant was good, questions could be made easy. Often, sponsors would tell the producer after a show which contestants to keep nursing along, and which to get rid of.

There was also what the producers called "plotting." They could plan the order in which contestants would appear for cumulative or dramatic effect, or for contrast, or according to how much time was left for the telecast. Plotting established an overall program tempo. Contestants were told to miss an easy question occasionally because this was good audience psychology: the viewer would say, oh, I knew that, and feel smart. It was also customary not to let anybody fail on the first question. For some reason, this was believed to be bad audience psychology.

"There were the lambs, patsies and champions," Herbert Stempel observed. James Snodgrass, who had won $4,000 on *Twenty-One,* was told to have his teeth polished by a dentist before appearing before the cameras. He admitted he was given his answers in advance, and told when to lose to Hank Bloomgarden.

The winner of "a glorious vacation in Paris" testified that the glorious vacation had consisted of a plane ride to France one day and back home the next, with no hotel accommodations provided for the one-night stop-over.

Televiewers questioned on street broadcasts across the land seemed not to mind very much that the programs were rigged; they enjoyed watching them anyway. President Eisenhower, however, expressed dismay and hoped the air would be cleared; the Justice Department examined the matter to see if any law had been broken; the Securities and Exchange Commission investigated a block of stock given as a prize; the Internal Revenue Service began to review certain tax returns; the Federal Trade Commission cracked down on deceptive commercials; and the television networks got busy preparing a flood of situation comedies for the next season.

The kick that stirred up the stinkweed called payola was delivered under the nose of Robert Lishman, counsel for the House Committee. It was a letter from the president of the American Guild of Authors and Composers, Burton Lane, who wrote the score of *Finian's Rainbow*:

"The practice of audience deception in broadcasting which has been revealed in the testimony adduced by your committee," Lane wrote to counsel Lishman, "is by no means limited to quiz programs. It has a counterpart in the promotion of music and musical products. There is no doubt that commercial bribery has become a prime factor in determining what music is played on many broadcast programs and what musical records the public is surreptitiously induced to buy."

Lane declared that the Federal Trade Commission and the Federal Communications Commission had both known of this as far back as 1956, and had taken no action. Also, in September, 1956, the Antitrust Subcommittee of the House Judiciary Committee had held a hearing in New York. At that time, Representative Emanuel Celler had commented on a number of articles which had been printed in various trade papers about a practice in which men who were in charge of selecting singers, bands and other artists to make records, and what songs to record, would surreptitiously accept cash or gifts to promote songs or give preference to tunes that emanated from Broadcast Music, Inc. This caused a great deal of irritation among songwriters not affiliated with BMI, namely the members of Mr. Lane's American Guild of Authors and Composers and of BMI's old enemy, ASCAP.

The American Society of Composers, Authors and Publishers maintained that rock-and-roll music, originating largely through BMI, would never have been played on the air, hence would never have become popular, without payola. ASCAP claimed 75% of the Top 50 list got there by payola. Some disc jockeys were inclined to agree.

Now, in 1959, the stench of the rigged television quiz was clearing and the radio industry had to hold its nose over payola. It could not have been kicked up at a worse time for radio. Individual stations had survived the shock of television and were still fighting. Recorded music was now the main substance of their existence, and the men who played the records on the air were caught in their own whirlwind of discontent.

At the first disc jockey convention, held in Kansas City in the summer of 1958 under the auspices of Todd Storz, the emotional state of the disc jockey was very low. Deejay morale had been shot down by the Top 40 format. Top men were quitting all over the country. Six men had left one station alone—KLAC in Hollywood: Peter Potter, Dick Haynes, Gene Norman, Earl McDaniel, Duke Norton and Jack Smith. In New York, Alan Freed left WINS, Art Ford left WNEW, and Ted Steele left WOR. In Detroit, soon after Ed McKenzie was through at WXYZ, Eddie Chase ended a short marriage to CKLW.

At the 1958 convention, Todd Storz called the disc jockey "radio's most celebrated star." Robin Seymour of Detroit's WKMH, Reed Farrell of KWK, St. Louis, and Bob E. Lloyd of WHEC, Rochester, took part in a panel discussion of the topic: "Is Rock and Roll a Bad Influence on Today's Teen Agers?" They all agreed that it was not harmful.

But many jocks were rebelling against both rock music and the Top 40 format, and when Mitch Miller attacked both in a speech they stood up and cheered. Miller accused the disc jockeys of abdicating to the "pre-shave crowd" which, he pointed out, represented only 12% of the population and "zero in buying power." Station managers resented his speech but the disc jockeys loved it.

Another disc jockey convention the next year in Miami was a disaster. Those who went spent more time on the sand in the sun than listening to speeches. Again the Storz stations hosted, with a lot of help from record companies. A newspaper that covered the activities headlined its piece, "Booze, Broads and Bribes."

So Ed McKenzie's attack in *Life* Magazine in November, 1959, came a little late; but it did serve to fan the smell into the nostrils of the public. Broadcasters finally went into action.

NBC sent a questionaire to its personnel probing into the matter of gifts and bribes, and employees had to take an oath that their answers were honest. Frank Stanton, CBS president, declared the industry was in the worst mess he had known in 30 years of broadcasting, and admitted, "We've been lax in our responsibility." ABC gave its most popular record man, Dick Clark, the choice of remaining on the air, or devoting full time to his record-pressing plant, his recording company and his music publishing firm. He decided to abandon his outside interests and stay on the air.

This roused Burton Lane to comment that ABC should obey its own dictum and divest itself of its own recording company. Lane pointed out there were subtle forms of payola beyond a plugger paying a disc jockey to push a record. He decried the network practice of promoting their own records and their own recording stars. He declared even the Top 40 charts could be rigged: a girl who made up the charts could be wined, dined and who-knows-what. Many nefarious variations on the payola theme came to light.

The stations themselves were not entirely innocent. One Los Angeles station had charged $225 to play one record 8 times a day, but limited such deals to 3 records a week. This was not called payola; it was called the "test record plan." A station in New York was reported to be offering 6 "plays" a day for $600. One music publisher was paying a weekly fee to have his music played on a TV show. When the cabbage was kicked, the agreement was quickly and quietly cancelled.

There was also the business of "slicing up an artist." In Detroit, where he first became popular, Johnnie Ray was said to have been sliced 3 ways. He eventually had to purchase his freedom from a record player, a night club owner and a local disc jockey.

Because of Ed McKenzie's article in *Life* Magazine, the Detroit newspapers aired the smell on their front pages. There were some firings and resignations at

WJBK, and one of Detroit's most popular disc jockeys, Tom Clay, admitted taking a few thousand dollars in a year-and-a-half. No one at WXYZ suffered, although Mickey Shorr's aura was clouded.

He became involved in a financial squabble between a quartet known as *The Royal Tones* and their former manager, Stuart L. Gorelick. The singers wanted an accounting on a record they had cut called *Poor Boy*. Gorelick admitted that 10% of the money the quartet wanted had gone to Mickey Shorr to pay off a "loan."

Harold L. Neal, Jr., issued a statement: "Information available to WXYZ affirms that no disc jockey now performing on the station, no person who participates in the selection of music for broadcast over WXYZ, nor I, this WXYZ station manager, has either solicited nor accepted any personal consideration, money or otherwise, to have any record played on WXYZ."

Hal Neal: "We at WXYZ were in great shape because I had put in a music control system which put the selection of records in the hands of management and did not allow disc jockeys to have any great amount of influence. We already by that time had in effect our committee for record selection."

How did this work? How many people were on the committee? Who were they?

"It varied from time to time. For a while, it was basically Bob Baker and myself who were doing it. Eventually, as I recall, we ended up with a committee consisting of myself, Baker, and a couple of record spinners, production people like Dick Kernen.

"The talent would recommend records. Fred (Wolf), I remember, would always find a record or two and had a submission to the committee. We'd take an hour-and-a-half or 2 hours on a Tuesday or a Wednesday and listen to new releases and go over the releases we were currently playing and come up with a list approved for airplay.

"There were some things going on in some stations, but WXYZ was scrupulously clean."

Fred Wolf agreed with ASCAP that rock and roll would never have got started without some form of payola, but he believed payola started way above the disc jockey level.

Wolf: "I believe many people in the business of making records were paid off, all the way along the line from the man who made it first to the man who took a carload of 'em and went to record distributors where that label was supposed to be distributed and said, 'Here are 400 of this recording that is going well in Bixwell, New Jersey. Now hustle it!'

"Now the record distributor would find a buck in his pocket and take the record to the music stores and radio stations in his district. When he went to the music store, the trend was, 'Here's a hot record. You pay'—whatever the price is—if the records are selling at 80¢ or 70¢, prob'ly the dealer's price was 40¢—he'd say, 'Here. You take 200 copies instead of 50 and I'll give 'em to you for 20¢.'

"So the guy buys 'em for 20¢, and he immediately has a 'big hit' posted, and the kids come in and they all buy the 'big hit'."

Fred Weiss admits he came close to payola once in an enthusiastically naive kind of way characteristic of him. He encountered a record pusher who was plugging a record with a foreign title.

"What's it mean?" Weiss asked.

"Blue on Blue," the pusher explained.

"Hay, if you want the thing played," Weiss chuckled, kidding, "why don't you give the guys a couple of blue-on-blue shirts?"

To his surprise, the plugger took him seriously.

"I can't do this for everybody because they're pretty expensive," he said, "but why don't you go down to Capper & Capper and get three of 'em made for yourself."

Weiss almost fainted. Of course he didn't go, but his attitude toward payola was not uncommon. He felt that all the music was awful and it didn't make much difference what was played; nobody seemed to know the difference.

Weiss: "So he gets paid for playing a lousy one 10 times. So they got *that* lousy one going instead of a different lousy one going."

Weiss was not alone in thinking the smell was that of a normal backyard rather than a patch of skunk cabbage. Jack LeGoff, popular newscaster, commented that payola was just a part of American business in general. He got fired for saying so.

Abel Green, in the same edition of *Variety* that announced the firing of LeGoff, wrote of payola: "It's almost as part and parcel of the (record) business as the very vinylite is to the platter."

Green did not get fired. As a matter of fact, Jack LeGoff was not long idle. CKLW grabbed him. Eventually, he returned to the station that had let him go and was Detroit's top-rated newsman.

The breezes blew, the odors passed, and WXYZ was clean. But Hal Neal still had his problems.

46
The Gemini Twins

Greek astrologers sometimes referred to the twins of the third sign of the zodiac, the Gemini twins, as Apollo and Hercules. In Broadcast House, the Gemini twins were Harold L. Neal, Jr. and John Pival, and they were as unlike as the gods they resembled.

Apollo was an averter of evil; Hal Neal had saved WXYZ from the stigma of payola. Apollo was the god of music; Neal chose the records his disc jockeys could play. Apollo restored health; Hal Neal restored profit and high ratings to WXYZ radio. Like Apollo, Neal was dedicated to victory, and remained essentially remote from man.

John Pival, like Hercules, was basically human. Hercules wrestled with a giant; Pival's giant was television. Both Hercules and John Pival were enormously strong, consumers of much liquor, amorous, and generally kindly but given to occasional outbursts of brutal rage.

Apollo and Hercules fought each other for the sacred tripod at Delphi; Neal and Pival sought favor with the upper echelons of ABC; and the struggle was put to an end by Zeus (New York) throwing a thunderbolt between them.

James G. Riddell: "They were different types. Totally different. Completely. I wish that Hal Neal had more of the confidence that John had. I tried to get him to stand up more. He was a shrinking violet. Sometimes John was a bull in a china shop."

Charles Kocher: "John, underneath it all, was a good human being. His own rules were that you must be ruthless in business; but when you got right down to it, he was the softest of the lot. Sometimes I would recommend that a guy ought to be fired and he'd say, 'We-ell, just tell him to take it easy'."

Pival enjoyed prosperity. He had not known it when he was a boy. Now, although he was not a good golfer, he bought clubs enough to fill three golf bags. He installed a workshop in his basement. When he was a child, his mother could keep him occupied and out of her way by giving him a hammer, a block of wood and some nails. Now he bought every kind of tool there was, and was good at carpentry around the house. Photography was a hobby.

He did not neglect Channel 7 for hobbies, however. ABC ranked 3rd among the networks in TV, but Pival kept WXYZ the leader in local programming. He thought his own programs were better than what came over the networks, and often he was right—especially in the area of sports. He televised bowling, boxing, wrestling and hot rods. When the novelty of the Skull, the White Camellia and other local movie hosts wore thin, he imported Herbert Marshall and Don Ameche to introduce the movies. He purchased a batch of comedies starring Edgar Kennedy. Warren Michael Kelly had a musical show. There were Soupy Sales and Marv Welch and Johnny Ginger; there were Edythe Fern Melrose and Rita Bell. There were *Starlit Stairway,* George Pierrot with *World Adventure* and *Adventure, Ho!* and Jerry Chiapetta with a hunting and fishing show. Least of all—was news.

News did not interest John Pival. This frustrated Dick Femmel. In the beginning, he simply wrote newscasts from wire services, and they were read on the air by announcers George Hunter, Bob Mayberry, Tom Waber and Mike Masters.

When radio and television moved into Broadcast House and shared the same news room, Tom A'Hearn was ostensibly head of radio and Femmel presided over TV. But shortly after the move-in, A'Hearn suffered an emergency attack of emphysema. It was then decided Dick Femmel should head both radio and TV news. Tom continued to write news for Osgood and gags for Fred Wolf.

It was some time before Pival would let Femmel hire a full-time photographer. WXYZ's first full-time movie camera man for news was a square-dance enthusiast, with white hair and spectacles, Art Mazur. If a story was important enough, Femmel might go out with him, but usually Dick told him where to go and what to capture in his camera. Mazur would stop on his way back to have the film developed by a man who had a lab in his basement and could develop in a hurry. His chemicals were not always fresh which created problems. It was a crude operation.

Femmel: "We never had a budget to do anything. It was always clinging to the news by your fingernails."

Pival's first step into news was the hiring of Lou Gordon. Gordon dug into his stories with outside help. He had no office at the station but came in with his material which was usually original. He rooted for the sensational and his style was flamboyant. Pival liked it because it had show biz pizzaz; Femmel was not as pleased because to his mind it was not journalistic.

Less grounded in news were three people chosen by Pival, not Femmel, for a program called *Detroit Deadline*. One of the three was an attractive blonde, a diminutive Joan Blondell-type, whose best claim to journalism was her husband, Marc Beltaire. Marc had fallen heir to *The Town Crier* column in *The Detroit Free Press*. Beverly Beltaire later developed into a first class publicity woman.

Femmel: "Pival was a showman and he always went out and got show people but with no news background and I was the guy who had to make these people look and sound like reporters."

Pival was under some pressure from New York to program news at 11 P.M. So he decided to give it a try.

"Hey, Lee, they're auditioning men for news. Why don't you go in there?"

Leon McNew happened to be walking by the studio. He was no longer hosting movies as *Captain Flint*. Soupy's night show, on which he had occasionally played bits, was going off and Leon's main job outside the announcers' booth was announcing *The Lady of Charm*.

"What have I got to lose?" he grinned.

Somebody called Pival who was watching auditions in his office with Dick Femmel.

"Lee McNew just walked in. As long as we're auditioning people, he'd like a crack at it."

"Okay," said Pival. He hung up and looked at Femmel. "McNew."

Femmel didn't argue, but not only did McNew have no background in news, he had even played a comic reporter with Soupy Sales called Mike Walters (for Mike Wallace).

McNew was handed some copy and read it. His face wore a natural expression of relaxed vigor, his white hair gave an impression of authority, and he was not pompous.

"Well—he looks good," Femmel conceded.

Pival picked up the phone and dialed control.

"Tell Lee he's our news man."

Edythe Fern Melrose tried to give him a gracious send-off. Leon McNew had been her announcer 2 1/2 years. Her cast gathered around a dining table, as usual, for the sign-off.

"Ladies," she said sweetly to the camera. "It's my great displeasure to have to tell you that my right arm is not going to be with me any more. The announcer we all loved so long is leaving us to become a newscaster right here on Channel 7. I'm sure you all join me in wishing him the best of success. So let's all say good-bye to my dear friend—Lee—uh—Lee—Lee New."

Pival wanted no straight newscast. He wanted it jazzed up. A show.

"We'll have a man down in police headquarters and if a story breaks while we're on the air he'll phone it in as it happens."

"John, you stage a lot of other things," Femmel declared, "but you can't stage news. You can't say, 'All right, we're on the air, stage a crime'."

"It'll work out," said Pival.

It didn't. News did not break to suit Pival. The man from the police department was obliged to review past happenings—minor muggings, a drunk or two, a juvenile delinquent, a domestic squabble—hardly news of vital interest to the viewer. Femmel laughingly called it *Pival's Telephone Hour*. It delivered no great jolt to the competition.

Gathered in the board room were Warren Michael Kelly, Johnny Ginger, Larry McCann, Leon McNew, Soupy Sales—all the air personalities of WXYZ television. Pat Morris, the weather girl, was smoking in the lounge. Edythe Fern Melrose arrived from her show with Harold Christian. Rita Bell came in through the kitchenette from the back reception hall and greeted everyone with sparkling brown eyes and a wide, open smile. She had sung on the kid show in the days of James Jewell, and had now taken over the Auntie Dee program and was doing *Starlit Stairway*.

Floor-to-ceiling drapes covered one side of the board room, blocking out daylight at the windows. A couch, chairs and smoke stands were set along the opposite wood-panelled wall, and at one end of the room were screen and speakers for auditions. In one corner, a door was open into a bar-kitchenette where hot coffee was being prepared. At the opposite corner another door led into an entrance from the elite corridor and commuted with Pival's office.

John Pival came in wearing tailored dark blue and looking fresh as he usually did in the morning. He motioned everyone to gather around the huge table that dominated the room. Ten people could sit uncrowded along each side. He sat at one end. Earl Moore served the coffee.

"I've called you in here," said Pival, "because things are happening that I want you to know about. ABC is grabbing more and more of our time. The faces around this table may not be here another year, because there won't be any time for local programming. I wanted to tip you off so you can be looking to your future."

Sentences flew in all directions for half an hour. As the meeting broke up, Pival said, "Edythe. Come into my office. Harold. You, too."

They followed him. He gestured to couch and chairs as he sat down behind his desk.

"Another big week for you, hey Harold?" Pival said. He looked glum.

"Oh, yes," said Christian brightly in his scratchy little voice. "Our annual WXYZ-advertised Brand Week. Next month, we've got Wrigley Supermarkets, Kroger's, First National—several chains are in on it, and we're getting radios into some of the stores with an agreement to turn them on when our commercials come on."

"That's radio," Pival commented without enthusiasm.

"It all helps," Christian maintained. "These promotions produce exceptional volume for the products we feature."

Edythe Fern Melrose sat watching John Pival and thinking how much she missed the smile of Jimmy Riddell. The sun was shining on the trees and lawn outside Pival's windows, but inside there was no brightness. The fun of early days had given way to competitive tension. She was thinking, however, that it was friendly of Pival to give those poor people warning of impending change.

"Edythe, I have bad news for you," said Pival.

"There is no such thing as bad news, John," she said happily. "Bad news is nothing but a new challenge."

"We've got to cut your show—either altogether, or down to so little it won't be worth your time."

"John, if you cut me off," Edythe pointed out, smiling confidently, "you're cutting off your own record, financially, with this station."

"I can't help it, Edythe."

Christian sat as a silent witness. He had worked closely with Edythe Fern Melrose for 19 years and had tremendous admiration for her as a business woman, a broadcaster and a woman. She still brought a lot of business to the station.

"My sponsors will go with me to another station," she was saying. "Can you afford to do that for yourself?" Pival was not saying much; the woman was making all the points. She turned to Christian for confirmation of one statement and as he nodded his head in agreement, it seemed to her his eyes were wet. Edythe Fern Melrose proved her persuasiveness at this meeting. The Lady of Charm emerged from the conference with 38 minutes a day, 5 days a week, all locked up on Channel 7.

As she and Christian left his office, Pival saw Larry McCann waiting outside his door to talk to him, and motioned him in. McCann had developed into one of the smoothest ad lib pitchmen in the business. Without a script, he could make second-hard cars seem irresistable as they rolled into view on a car lot. He wore good looking suits and was paying for a very expensive home in an exclusive suburb. On a signal from Pival, he closed the door.

"I have to give you credit," John observed. "You're the first one that's come to me."

"It just so happens my vacation starts today," McCann said, smiling a little wryly at disaster in the air. "I'm going up into Canada with Eddie Chase, J.D. Beemer and a couple o' the boys for some fishing. Maybe I'll hook a job."

Pival leaned back

"Larry, how'd you like to go into sales?"

"Sales!"

"Bill Morgan is leaving and we'll need somebody to take his place."

"Where's Bill going?"

"Adam Young, Incorporated—the agency."

Larry McCann thought of his mortgage, and went into sales.

Bill Morgan quit the station primarily because he did not like John Pival. Soupy Sales left WXYZ and went to Hollywood. There, Soupy repeated his Detroit success on a larger scale. The custard pie gimmick he had revived successfully in Detroit after watching some old movie comedies in a screening room became his trade-mark nationally. In only a year, he became so popular in Hollywood that police had to hold back the crowds when he made a personal appearance; and at Christmas time, 1961, an album he recorded was top seller in the Los Angeles area.

Both Pival and Neal went around the country committing what Neal called "creative thievery." They would lock themselves in hotel rooms in other cities and spy on the hottest radio stations or TV channels. Neal taped hours on end, returned to Detroit, analyzed what he heard, appropriated what he wanted, improved on it, and put it on the air.

"What was fantastic," Fred Weiss recalls, "was that *all* of it worked."

More and more rock and roll was moving onto the popularity charts. More and more hard rock was working its way onto the lists of records to be played by WXYZ disc jockeys. On other stations in Detroit and throughout the country, "star personalities" were being replaced by "format disc jockeys." There were opposite reactions to this development. One view was that the day of the old-time disc jockey with his "innocuous fore-and-aft banter" was just about over. Those opposed put it that free wheeling individualistic deejays were giving way to "faceless facsimiles."

Hal Neal: "The trick was to find a guy who could have a personality, or project a personality, in a few words."

Dick Kernen: "The music outgrew the people we had on the air with it."

As pop music became more and more raucous, the soft approach of Jack Surrell became inconsistent with the sound Neal was after, and Jack was "phased out." Mickey Shorr was the next to go. His departure was less because of his irrelevancy than because he had moved in a gray area from the moment payola had broken into the open. He persisted in violating the new ABC policy against outside involvement.

Hal Neal: "Mickey was a bit of an entrepreneur and he had outside interests, and he just couldn't live with our type policy. So we had to part company."

Mickey was replaced by a musical announcer, Chuck Daugherty. Another new announcer was added to the staff in February, 1960—a young Pole named Donald Zamenski. He had thinning blond hair and was undersize, but he possessed vitality and a very sly humor. His name was changed to Don Zee, and 6 months after he joined the station he went into the late night spot as a disc jockey.

Fred Wolf was having more and more of his suggestions turned down, and although he continued in television which paid him less, he was increasingly unhappy in morning radio. He and Bob Baker began to yell at each other.

Paul Winter continued to "perform." He was a miscast actor playing his role as well as he could. He didn't relish his work beyond the accomplishment or the performance, but he had no quarrel with Hal Neal.

Winter: "It's no overstatement to say that in numbers terms (ratings), we ran the city of Detroit for the 4 or 5 years Hal instituted and managed the formula that we were doing, and I don't care how anybody feels about it; it happens to be the case. Whether we liked it or not is a different question. But I notice that we all took it, and we all did it, and we were all rewarded for it, and I don't have any complaints about it myself."

The most extraneous person on the staff seemed to Dick Osgood. Neither an announcer nor a disc jockey, he somehow maintained his identity on the morning news and with *Hollywood Headlines* and *Show World.* Dick Jones who had offered him a job at CKLW long ago offered him one now at WQTE, but Osgood was approaching 60 and thought it wiser to stay where he was. Larry Gentile had told him during FM days at Joy and Greenfield that he was so firmly established he could coast along for years, and he was proving it. But he worked hard while coasting.

In 1959, Hal Neal had an idea. Public service announcements were a necessary drag. Stations were obliged to show a certain percentage of public service programming at license renewal time. Hal Neal decided that instead of dragging anchor with it, he'd give it an outboard style of presentation. He would combine events into fast, frequent, brief reports around the theme, "it's fun to live in Detroit"; and he decided the best man to handle it would be the one already identified with entertainment; so Osgood gathered the material, wrote and recorded the feature, and it went on the air 13 times a day.

These *Fun Alerts* not only created good will; they even went commercial! Bob Koch sold them to a company that operated filling stations throughout the area, the Marathon Oil Company.

Zeus threw the thunderbolt in the Spring of 1960 and a shaken Apollo sat in Berman's Chop House with his friend from Blair, Charles D. Fritz. The noontime buzz of business talk at Berman's was punctuated by the sound of a knife on a wooden block as potatoes were rapidly and miraculously shaved into thin slices by the white-capped vegetable cook who was famous for it. The crisp, fried potato shavings were a speciality at Berman's. And the steaks. Hal Neal and Chuck Fritz were on their 3rd martini and in no hurry to eat. Neal had just returned from New York with the news.

"Hal," Fritz said, "you've just got to live with it. What else can you expect? You have made WXYZ far and away the most successful station in the ABC-owned company."

"I can't make up my mind."

"Who talked to you?"

"Goldenson, Si Siegel and Steve Riddleberger."

"Riddleberger. He's in charge of both the radio and television owned stations."

"Yes. They are taking Ralph Beaudin out of Pittsburgh and putting him in Chicago to run WLS."

"Beaudin has run a very successful operation in Pittsburgh."

Neal drew on his martini. When he looked up at Fritz his eyes were a little red from the weight of the decision he had to make and the pupils were tense.

"I admit I have the hots for New York," he said. "I think it's the most exciting place I've ever been."

"So what's the problem?" Chuck's lips curled into a friendly smile.

"WABC," said Neal. "WABC is the problem."

"What's so different about New York?"

"It's the Number One market in the country. And—ABC doesn't understand radio. They want to build it up but they don't understand what you have to do."

"*You* do," Fritz insisted. "You've proved it here."

"But Chuck, nobody has been able to turn WABC around. About 6 managers—a string of 'em. Nobody's been able to turn it around."

"Hal, I'm tellin' ya, you can rock that market like it's never been rocked before."

"I'd be right under the nose of management there."

"They won't bother you. They want you to do there what you've done here. They'll leave you alone."

"If I go, will you take over WXYZ?"

Both finished their drinks as the steaks were put before them with the crisp paper-thin slices of potato.

"How about it?" Hal asked.

"No," said Fritz. "It's still pretty exciting to be out on the street."

They ate in silence. Hal Neal was thinking hard.

"I'd have to go through the same thing with Martin Block in New York," he said at last, "that I did with Ed McKenzie here."

"Look at it this way, Hal. You've got your practice. Now's the time to either get into the game or get out."

Hal Neal decided to go to New York. In May, 1960, there was a farewell party for him and he was given some expensive luggage. He wrote notes to everyone, thanking them for their cooperation and friendliness.

Apollo was off to meet a new challenge in "the Number One market."

47
One Hundred and Seventeen Enemies

The center of power was no longer in downtown Detroit; it was in New York; and already New York had grabbed two local powers, Riddell and Neal. Now came a new experience. For the first time a top executive was hired in New York and sent out to manage WXYZ Radio.

John Gilbert was a native Michigander; his family had been respected throughout the state as makers of Gilbert's Chocolates; but when New York hired him to manage WXYZ Radio, he had been working in Boston at WBZ. As far as Wyxyites were concerned he was as foreign as a trapeze performer from Denmark. Certainly he was handsome—young, dapper, blonde, hale and hail-fellow. He came on like May lightning and his enthusiasm was reflected in the overall chemistry of the operation.

"Listen, Harold," he said to Christian one day. "This merchandising of yours can't be beat! It's the greatest! Why don't we get up a new presentation for food brokers—with slides?"

From their very first presentation they landed a 39-week contract with Knorr Soups—a superior European-type soup that was not canned. Paul Winter did a series on the "uncanny" soups.

John O. Gilbert ran into a couple of road blocks. One of them was John Pival. When Neal left for New York, Pival was named President of WXYZ, Inc., over both TV and radio; and in selling television, he was using radio as a fringe benefit.

"The son-of-a-bitch has done it again," Gilbert wailed to Chuck Fritz one noon when they lunched together. "He's made a deal with Elias Brothers and he's thrown in $45,000-worth of radio, and radio won't get a dime of it!"

Fritz was extremely prosperous as the result of a stock bonus deal with the John Blair company, and he was relaxed and affable. He grinned sympathetically.

"What am I gonna do, Chuck?"

"John is runnin' WXYZ like stations were run years ago," Fritz commented. "Like he owned it himself."

"I bet we have $100,000-worth of business on radio right now, and not a dime of it on our books."

"Is John drinkin' a lot?"

"Oh sure. He sits in his office drinking at night and his wife sits at home drinking. It's a sad house. But hell! That's not my problem."

"He does too much drinkin' out at Oakland Hills."

Fritz and Pival were members of the Oakland Hills Country Club in Bloomfield Township.

"He gets drunk to the point he falls off the stool and gets abusive," Fritz continued. "He fights and throws things. I'm as embarrassed as hell."

"A great image for somebody in the broadcasting business," Gilbert moaned, shaking his head.

The other cloud that hung over the head of John Gilbert called itself The Listeners Lobby. This threat was made up of 117 Detroiters, 1 New Yorker, and members of the Tau Nu Epsilon Sorority. It claimed to have 172 members, mostly professional. In December, 1959, under the direction of its executive secretary, Mrs. Dvora Marcia, and with the school authorities in Detroit, Berkley, Dearborn, Ferndale, Highland Park and Royal Oak, questionaires had been mailed to 66,837 students in grades 9 through 12. Replies had been received from 7,658, and a 30-page report of the survey was made public in February, 1960.

High schoolers, it seemed to indicate, preferred pop and rock-and-roll music, disliked commercials, and half of those questioned condoned the practice of payola. Rock and roll, the report stated, was generally found to be more popular with "less proficient students and the poorer class." *The New York Times* quoted the group as saying, "the disc jockey tends to relegate the whole teen-age group to an inferior position in the social class structure—then keep it there." The Lobby protested that disc jockey programs were not designed to "cultivate musical tastes, to set examples of vocabulary or diction, or in any other way to educate their audience."

The Lobby claimed to have operated with a light touch heretofore, working more through praise for things well done than otherwise. John Gilbert felt the light touch being discarded as the Listeners Lobby zero'd in on WXYZ.

"Hey. I want to talk to you."

First ladies morning matinee (at Michigan Theater). Line-up to get in.

A brown-eyed, black-haired, good-looking young man took Fred Wolf's arm and led him aside at the weekly Sports Broadcasters luncheon in the Press Club. Dave Diles was Michigan sports editor of Associated Press. He had known Wolf for his bowling as far back as 1948.

"Who is this guy Pival?" he demanded.

"Who *is* he? I don't know what his title is, but he's Number One on the television side."

"Well, I was toastmaster at a banquet," said Diles, "and Pival was there and he came up to me afterwards and said, 'How would you like to do sports on television?' He offered me a lot of money; I can't make that kind of money with AP for years. Is this a man of his word?"

"You better believe it," said Wolf. "If Pival says he's gonna do somethin', he'll do it."

"I'm supposed to call him."

"Call him," said Wolf.

The next week, Diles approached Wolf again.

"Hey, Fred. I talked with Pival and he's offered me a bundle. But I don't know. Television."

"What the hell are you afraid of? You know sports."

"But—television!"

"You can talk your ears off anywhere you go," said Wolf. "You're a big emcee. You're everything else—toastmaster—television ain't any diff'rent. Just sit down and talk. Ain' nothing worry about."

Dave Diles had a sharp mind, a lot of courage, and a vicious wit. He joined WXYZ-TV in the early summer of 1960. Don Wattrick continued on radio, but he had less to do, and eventually he left WXYZ, first became the voice of the Detroit Pistons, then manager of the team, until he died of a heart attack. When he left, Dave Diles expanded into radio.

The Channel 7 news team now consisted of Leon McNew on news, Diles on sports, and Pat Morris on the weather. They just couldn't make a dent in the competition. Channel 4 benefited from ownership by *The Detroit News* and a feisty little comic named Sonny Eliot who brought a fun personality to the weather reports. Jack LeGoff, however, on WJBK—TV, Channel 2, was the most popular news man in Detroit.

On the radio side news had gone into a frantic era.

Dick Femmel: "Our news was a combination of clangs, whistles, buzzers, and echo chambers—call letters every other word. Date lines, machine gun clip. It was ridiculous."

Tom Waber's style was considered too pedantic for the new style, his voice too heavy. He left radio and became a Lutheran minister. He was replaced in the WXYZ news room by a wild man from Rhode Island who used the name Jack Powers. If news was not there, he made it. It was said that he once urged police to shoot at an escaped convict so the shots could pep up his newscast. He could make Red Cross help in a flood sound like a Communist invasion.

Femmel: "He had no real news judgment and he was dramatic as hell. And unstable. He was always late for his deadlines. He would look up and see that he had like 30 seconds to get into the studio. He hadn't prepared a darn thing for his newscast. He'd run over to a wire, tear off 25 feet of copy, race into the studio, sit down, and read it cold but with these dramatic gestures. Funniest thing you ever saw. But disastrous as far as any kind of newscasting was concerned."

Osgood was doing his best to adapt to the times. He rewrote every newscast, shortening Tom A'Hearn's sentences, eliminating details, and adjusting his style to the staccato effect of the frantic era. His 5-minute newscasts were still all sold, but he felt that his prestige was slipping away; his morale was low; and his lips turned down at the corners.

"What are you sore about?" Baker would say.

"Nothing. I'm getting old and my face sags when I'm tired," he would answer.

United-Detroit Theaters still sponsored *Hollywood Headlines* in the morning and wanted to promote a new movie in some new kind of way. Marie Meyer came up with the idea of inviting Osgood's listeners to a morning preview at the mighty Michigan Theater. Only listeners writing in for invitations would receive a card admitting 2 people. The card was printed to read:

DICK "WXYZ" OSGOOD'S
EXCLUSIVE INVITATIONAL PREVIEW
OF Warner Bros. Wonderful New Picture
"THE DARK AT THE TOP OF THE STAIRS"
in Technicolor—starring
ROBERT PRESTON—DOROTHY MCGUIRE—EVE ARDEN
at the MICHIGAN THEATER
TUESDAY, OCTOBER 4, 1960—10 A.M. SHARP
Admit Two

Woodrow Praught, ex-Navy officer who now headed the United-Detroit Theaters, doubted that the invitational preview would be a success. First, WXYZ was obviously playing to a teenage audience, and the teenage audience should be in school at 10 o'clock on an October weekday morning. Further, Praught doubted that Osgood had much of an audience. So did everybody else, including Osgood.

But Marie Meyer was confident. And the station could not deny the pervasive influence of an account now bringing in $80,000 a year. The disc jockeys were therefore instructed to bait listeners into listening to *Show World* for the invitation and the theaters put ads in the newspapers teasing people into hearing the invitation which would be extended exclusively on *Show World*.

The turnout surprised even Marie Meyer. No provision had been made for ushers and Osgood found himself finding seats for his guests before he went on stage to bid them welcome. Even the balcony had to be opened up. Nobody from the station saw the miracle except Osgood; but the theaters had hired a photographer

Dick Osgood hosts first movie invitational preview for *Show World* listeners who took pictures of the crowd lined up around the block waiting to get in, and of the mob in the lobby and the filled auditorium. What surprised theater and station most was the character of the audience—mostly women, of course, but well-dressed and middle-aged.

"Fantastic!" John Gilbert exploded when he saw the photographs. A surprise luncheon was arranged to celebrate Osgood's 25th anniversary with the station. It was attended by all the important theater men of Detroit. The salesmen found the results of the preview—which was the first of its kind and led eventually to a city-wide Ladies Day in theaters one morning every week—so useful that they tossed Osgood a second anniversary luncheon of their own, and presented him with a loving cup—which leaked.

On the night of Thursday, September 8, 1960, 3 men stepped forward into the lights to face the camera and a panel of 4. The master of ceremonies read the following:

"I, Fran Striker, am a writer for radio and television. I started my writing career in radio in 1930 and among my early assignments were *Covered Wagon Days* and *Robin Hood*. I helped in the creation of both *The Green Hornet* and *Sergeant Preston of the Yukon*. My major effort, however, was the creation of a character who has been a leading figure in network radio and in television and in comic strips for more than 25 years. I am the creator of *The Lone Ranger*. Signed, Fran Striker."

The 3 men turned, walked to chairs behind a prop desk and sat down. It was now up to the panel to guess which of the 3 was "the *real* Fran Striker." This was the network show *To Tell the Truth*.

The sentence "I am the creator of The Lone Ranger" brought a telephone call to Fran Striker the next day. It was George W. Trendle. The two must have a talk immediately. Trendle would fly to Buffalo and Fran must meet him at the airport.

Jan Striker: "I was a little bit—you know—I was cooking dinner for the *Great George W.*"

The pretty, quiet blonde grandmother laughs gently as she recalls the night. Was she nervous?

"Well, a little bit. I don't know what he came to talk about, exactly; but it was about the retraction, I guess, or the threat of a law suit because of *To Tell the Truth*. And this is where Bob thought that Trendle knew exactly how to handle dad."

Bob Striker: "Trendle was a lawyer; he was a sharp business man. Dad was not. Dad was a dreamer—a creator."

Jan: "Trendle told dad *The Green Hornet* was probably going to go on again and there might be a job for him."

Bob: "Trendle had a habit of dropping little bits of bait to take advantage of a situation. He knew dad was trying to get back on his feet."

What kind of a guest did Mr. Trendle make?

Jan: "He was very nice, very congenial, very courteous. I met him only once in all the 25 years we were in Detroit; and so—having heard about him, I was really not too happy that he was coming to dinner."

She laughs again, gently, almost apologetically.

"But he was a nice guest. He really was."

Whatever conversation took place when Trendle was picked up at the airport, and driven back again that night, Fran Striker wrote the following letter to Gil Fatos, then executive producer of *To Tell the Truth*:

Dear Mr. Fatos:

With embarrassment, I find it necessary to clarify for you my connection with *The Lone Ranger*.

During my 25 years of employment by George Trendle and his radio station, WXYZ in Detroit—to whom I assigned my interests in *The Lone Ranger*—I was under the impression that this dramatic show was my own creation. This impression was founded on correspondence between myself and Mr. James Jewell, WXYZ's dramatic director who subscribed to many of the drama series that I was syndicating from my office in Buffalo.

Mr. Jewell asked me to 'write up three or four wild west thrillers.' (Letter of Dec. 28, 1932.) I wrote them, sent them to Detroit, and charged the same price that I was charging for the syndicated shows that were being used by that station—$4. per half-hour script for use in the Detroit area.

After sending Mr. Jewell a few of the 'western thrillers,' I received a letter (January 21, 1935) suggesting some changes. Mr. Jewell wrote: 'I hope the above suggestions won't cramp your style. I realize that they have changed the character you created, but only in a minor way.'

This early correspondence, copies of which I am enclosing, was the basis for my belief that I created *The Lone Ranger*.

Mr. Trendle feels that this claim is not justified. Apparently, and I have no reason to doubt this, Mr. Trendle had the character and the format of the program well established before he asked Mr. Jewell to find the scripts.

If you will read Mr. Trendle's letters to me (enclosed) you may feel as I do, that his vexation is justified. Also, if you read the photo copies of the correspondence with Mr. Jewell, you will probably see a basis for the claim I made.

I'm sorry that we've run into this situation. I do hope you can understand, as Mr. Trendle does, how it could have evolved.

Mr. Trendle is still one of my best friends. He has no malice, bears no ill will, has no intention of making an issue. He simply wants me to clarify the situation as far as your office is concerned.

If you will acknowledge this letter, as suggested by Mr. Trendle in his letter of October 12, the matter will, I'm sure, be closed.

Yours sincerely,
Fran Striker.

Parts of this letter read as though they had been written by Raymond Meurer. At any rate, the matter was closed as far as *To Tell the Truth* was concerned. But the old controversy would flare up again.

In 1961, John Pival sent Leon McNew to Africa to do a documentary. The people of the Congo had declared their independence. The country was fragmented with interior political and military strife, and the United Nations sent in troops to prevent war.

McNew stopped off in New York where ABC provided him with visas, arranged for him to use an ABC camera man from Europe, and set up an exchange of material between himself and an ABC correspondent who was already in Africa.

He found conflict too hot for comfort in Leopoldville.

McNew: "The U.N. troops were on one side and General Mobutu's troops from the Congo were on the other and every night it was bang, bang, bang."

ABC's camera man from France refused to go into the native section of the city. McNew hired a Russian who luckily spoke Swahili and they spent 9 days shooting 1700 feet of film.

Bruce MacDonald, who wrote for television at WXYZ, was already preparing a script for the documentary when McNew was met at the plane in Detroit. Lee was exhausted and had dysentery besides, but he was given no time to rest. As soon as he was unpacked, he went to Broadcast House and stayed up all night with Bruce, Dean Erskine who edited the film, and Pete Strand who put the whole thing together. The program went on the air in Detroit that day. The ABC network ran it later on a series, and it won an award for WXYZ.

On September 15, 1961, WXYZ-TV put into operation another "Instant News" bulletin sign, erected on the building of The Detroit Bank and Trust Company at State and Griswold. It was 80 feet long, contained 90 miles of wire and 1120 light bulbs. It flashed the news from 11 A.M. to midnight.

Neither documentary nor sign lifted the ratings of Channel 7 news.

Work "dried off" for Fred Foy in Detroit. His TV shows were discontinued. When he tried for news he was not considered because he lacked a "journalistic" background. He was not a disc jockey. And nobody was hiring announcers. He didn't want to try Hollywood again; Hollywood was an actor's town and actors were having troubles. The only thing left was New York. He went to the big city and began knocking on doors.

Fred Foy: "It was almost a year before something happened and I thought, oh God, what have I done! Everything down the drain!"

He thought of calling Hal Neal but other ex-Detroiters warned him that Hal was not helping old friends and he thought why call him up just to get thrown out the door?

He saw others from Detroit prospering. Fielden Farrington was prospering as a novelist. Mike Wallace and Douglas Edwards were fixtures at CBS. Bob Hite was prospering mightily and was married again. Al Hodge seemed a little over his crest but was still managing. And Jack McCarthy quit broadcasting to go into an interesting and unusual business of analyzing executives who were't quite "cutting it," improving their traits and images, and preparing them for better breaks. Bob Maxwell from WWJ was in demand and John Cannon from WXYZ was one of the busiest commercial announcers in New York. But all doors seemed closed to Fred Foy.

At last, in desperation, figuring he had absolutely nothing to lose, he called Hal Neal's office. Hal was on the phone almost immediately and was surprisingly cordial.

"Fred! How are ya?"

"Fine."

"Somebody told me you were in town. Let's have lunch."

They met the next day. Foy told Neal he had been pounding the pavements for a year.

"Want a job at ABC?" Neal asked immediately.

"Hell! You're damn right I would."

"We've got our summer relief stuff coming up. Let me call."

He did. Foy started as summer relief announcer. Then came a newspaper strike and ABC inaugurated FM news around the clock and kept AM relief men on, and then came summer again and continued need for relief announcers; and a year later Fred Foy was hired permanently. Hal Neal had opened the door.

Dick Femmel and Lou Gordon sat facing each other. Both men were smiling but their faces were flushed. Louis Miriani, the Mayor of Detroit, was running for re-election and Lou Gordon was convinced the mayor was dishonest.

"It may be true, but you can't broadcast it," Femmel maintained.

"Dick, I have the information. I *know*."

"But, Lou, you can't prove it in court."

"I don't have to prove it."

"You would have to produce proof if we landed in court."

"He wouldn't dare haul me in court. He knows that I know."

But WXYZ would not allow direct charges to be made in his broadcasts. Pival had discontinued Gordon on television, but John Gilbert hired him for radio and placed him in an evening news block called *Assignment the World*. Jack Powers led off for 10 minutes at 6:15, followed by a 5-minute commentary by Lou Gordon. Alex Dreier broadcast national news from 6:30 to 6:40, John Cameron Swayze did 5 minutes of weather, Howard Cosell followed with 5 minutes of sports, Osgood did a 5-minute *Show World* report, Arthur Van Horn was on for 5, then Edward P. Morgan climaxed the hour with 15 minutes of real journalistic class at 7 o'clock.

Lou Gordon, although barred from direct statement, managed by innuendo to discredit Miriani. He threw his influence, which proved to be considerable, into the support of an engaging young Irish lawyer, Jerome P. Cavanagh. Cavanagh won the election and credit for his victory was shared to a very substantial degree by WXYZ's fiery and controversial little commentator.

Even so, Femmel held the reins and continued to check over every Gordon script to make certain nobody was slandered and that Lou was able to justify what he put on the air.

Fred Wolf was still fighting rock music.

Wolf: "I didn't believe it should be bombed at adults getting up in the morning. You have a radio clock and there's some guy—he wakes up and he don't know whether somebody's getting murdered in his room or not. Whether it was my fighting that did it, I don't know. I think it was half-and-half."

At any rate, John Gilbert, and Hal Neal who continued to visit Detroit often, agreed to "pull" the 10 most raucous records on the list and let Fred substitute selections of his own

"Which was an indication to me," says Wolf, "they didn't believe in it in the first place."

One of the other disc jockeys, tall, thin, blond Joel Sebastian who had left WXYZ for better things in Chicago, found them no better, and returned with great fanfare. That squeezed Fred Weiss right off the air. Weiss' TV assignments had come to an end also. The management kept him on because it liked him, he was honest, made no trouble, and really tried to adapt to any assignment. He was offered a chance to be a newscaster, but felt it would be "a come-down" and refused.

"How would you like to be in public relations?"

"I just don't know any of the things that would be required."

That idea was dropped. He walked around another month collecting his pay check, doing nothing. Then John Gilbert called him in again.

"How'd you like to do the weekend? Saturday morning, Saturday night, and noon to 6 on Sunday."

"This would cost you more money even than I'm getting now."

"How come?"

"Well, if I work Saturday morning and Saturday night, that would be a 12 or 14-hour day. That would amount to a lot of overtime. Then on Sunday, I wouldn't have any lunch hour, see? And you'd have to pay me extra for that."

"Okay!" exclaimed the affable Gilbert, spreading his hands a little and beaming. "No problem. Terrific!"

Weiss won high ratings with his weekend shows, but he still was not happy.

Weiss: "In the 5 years that I was a disc jockey I don't think I ever really enjoyed it at all."

He began looking around. He had no confidence that he could make it in New York, and didn't like the big city anyway. But he had a real desire to go to Washington, D.C. He wrote to station WMAL there, and was turned down.

In December, 1961, Bette Lou Peterson reported in her TV column in the *Detroit Free Press* that all Detroit stations had quietly had their licenses renewed by the Federal Communications Commission, except WXYZ. The Listeners Lobby had filed a petition with the FCC against renewing WXYZ's license because the station's programming, it declared, was "without merit or taste," and WXYZ had refused in spite of promises to "recognize its civic and public responsibilty in the community."

This petition put to a test a 1960 law which gave "parties of interest" a right to protest licensing of a station. Whether the "parties of interest" might include the listening audience was now tested for the first time.

The Lobby asked the FCC to conduct hearings in Detroit. WXYZ plunged into a gung-ho search for proof of public service. All personnel were commanded to dredge every possible good deed from memory; and all the daily logs for a year back were put through a sieve for every possible shred of helpful evidence.

Despite its cultural lapse, the station was maintaining its reputation as the most commercially successful of all ABC-owned stations. John Gilbert followed along in the Neal pattern with promotional gimmicks. There was the Why-I-Like-My-Mother-in-Law contest which netted 1269 essays from listeners. And there was the campaign to have listeners mail in jokes. This ended with a nicely prepared 10-page book of corn containing *The Worst Jokes of 1960* which was mailed to anyone who asked for it, and to advertising agencies, clients and station reps who didn't. Lively salesmen—Bob Koch, Jim Christie, John Lyon and Doug Campbell—were keeping radio loaded with commercials. Spot salesmen from all over the country visited the station and nodded their heads unbelievingly over the continued triumphs of Harold Christian's unique merchandising department which was introducing new products to Detroit as it had for years. There was still nothing like it anywhere else in the country.

In one disc jockey's case, ironically, the cultural activity of WXYZ had been turned, as it were, outside in. Even as Paul Winter sent forth music "without merit or taste" he sharpened on the whetstone of his vigorous intellect the thinking of young production men around him.

Dick Kernen: "It became interesting to me how much he knew. Probably Paul (Winter) had the most influence on me of anybody except my father in my entire life."

Paul Winter: "All of them were marvelous to work with—Don Davenport and Billy Hengstebeck, Richard Kernen, John Dew, Jim Neil, and then Jim Hampton and Larry O'Brien. I would throw something—dialect, style, or some kind of kookie idea—at them, and they would respond immediately and in an improvisatory way with some recorded gimmick or idea, and it just worked superbly. It never bothered me to admit that I was dependent upon them in a very special way."

Winter, who loved to talk, was unhappy about being cut down to a few seconds between records, and even then told what to say; but while the records played, he stimulated the minds of the boys who ran them.

Kernen: "We'd be in that studio 3 or 4 hours a day. We would talk about things and he would give me—like—opinions, slants that came out of a whole different thing than I'd ever realized before. It got to the point where I was just reading books and reading books and reading books."

Winter himself is modest about his influence, but "his boys" all did well. Jim

Neil became sales manager of a major TV station; Kernen and John Dew became program directors; Hengstebeck is part owner of a company producing commmercial slide film—and so on.

In 1961 and '62 WXYZ Radio rocked hard. John Gilbert rose to its defense before the Listeners Lobby and the FCC. He maintained that—survey or no survey—the Lobby's stand represented primarily nothing more than personal taste and opinion. In the showdown, his dynamic personality made his arguments sound better than they really were. The license was renewed.

Gilbert crowned the station's victory with a full-page ad made up by the new public relations man, Sam Semora, stating that WXYZ was still Detroit's top-rated radio station! The jewels in the crown were the ratings that proved it.

48
Man Alive! Man Dead

A lonely young man sat bored in a motel room in Cincinnati. He slouched in a chair and glanced without interest through the motel literature. The phone rang.

"Lee Alan?"

"Yes."

"Dick. Dot Records."

Lee Alan sat up.

"Dick. I thought you were in Buffalo."

"I get around. Are you feeling as badly as I am?"

"Yes I am!" Lee Alan exclaimed. This was his 2nd day as disc jockey on WCPO. "Doggone it, I don't know a soul in Cincinnati, except a guy or two at the station."

"I hear you're going back to Detroit," said the record pusher. "You're going to work for ABC?"

"WXYZ? Don't I wish! That would be super, but there's no way."

"Listen. I hear you're going to work for ABC," the other insisted.

Lee Alan laughed. "You guys are better than Western Union."

One hour later, Alan received a call from John Gilbert.

"Will you be in Detroit this weekend?"

"I hadn't planned to." Alan listened hard to convince himself this *was* John Gilbert. He had heard his voice only once before.

"Well, if you were in town, I'd like to talk with ya."

"I've changed my plans," said Alan. "I'm going back to Detroit this weekend."

Even as he drove back, he couldn't believe what was happening. He had listened to *The Lone Ranger* as a kid, and had since read everything he could find about the program. Today his idol was Fred Wolf; Wolf was a god. And here he was—on his way to talk about a job on that very same station.

John Gilbert gripped his hand and pulled him into the office. It was a cold, sparkling morning in January. Gilbert poured coffee in a closet at the end of his cheery office. Alan chose a chair near the desk and sat down.

"How long have you been broadcasting?" Gilbert asked.

"My first job was with WCAR when they were in Pontiac. I went to Michigan State and then Wayne. I just worked Sundays. Then Mr. Levinson called me and said he didn't need me any more because I sounded like a kid. I was about 18. Then I went to WJLB and ran records.

Gilbert liked Alan's looks. His clear brown eyes were alert and aggressive. There was confidence behind his eagerness.

"How old are you?"

"Twenty-nine," said Alan.

"Service?"

"I was drafted in 1957. Came out of the service in '59 and went to work for JLB. I was only there a coupla months. I was hired by WJBK when Don McLeod was there. Clark Reid was there and Tom George was there and Tom Clay. I was there during the incident when Tom Clay left. Then—I left WJBK and went to WKMH."

John Gilbert was not really interested in Lee Alan's past. His personality was too dominating to become a passive audience for long. What the hell! Baker was enthusiastic. Baker told how this guy got the numbers on a little station you couldn't even hear on the far side of town; he had to be good.

"How would you like to come to WXYZ and do 7 to midnight?"

"How much do you want me to pay you to do it?"

Gilbert laughed.

"There's no sense in me playing hard to get," Alan went on. "This is something I've wanted all my life. This station is king. I've had the opportunity to watch a lot of people. I watched and listened to Tom Clay. I've had 4 or 5 people influence me when I was on the air. Paul Winter—I'm amazed about the way the man speaks. But probably the biggest influence in my life has been Fred Wolf. Fred has something that very few people ever have: he has the ability to go on the radio and speak as if he's talkin' to one or two people."

"Well?" Gilbert prodded. "How much do you want?"

"You just pay me scale and see if I work, and if I work then I'll come in and hitcha for a raise."

"There's only one thing," said Gilbert. "Joel Sebastian is doing these Club 182 record hops Friday and Saturday nights. They are sponsored by 3 men and they buy a lot o' time on the 7 to midnight period, and—they don't like—you. Do you have any idea why?"

Lee Alan's spirit had groaned within him at mention of these 3 men, but he would not let them defeat him this time.

"We have to go back 2 years when I was with WKMH," he explained. "I was lucky enough to have some ratings over there."

"I've heard."

"Three Redford Township policemen came to me. They had acquired the use of a roller skating arena in Livonia and they said, 'Look. We're going to have a dance at this arena and we're going to have some record artists and so forth out there. Will you promote it and—uh—in return, we'll pay you some money.' Well—I said, 'No—I don't think so.' Well, they said, 'We'd like you to do it.' So I thought, I'll give 'em a figure that they won't be able to accept and then they'll thank me and go away. I told them I would have to have $600 to do that dance."

"Terrific!" Gilbert laughed and lit a cigarette, offering one to Alan who accepted.

"So what happened."

"So—they bought some time on KMH and I promoted the dance, and I went to the dance, and there were about 2,000 people there."

"Fantastic!"

"I was surprised. I didn't know that many people listened."

"Too much!"

"Well—when the dance was over, I went to a little room in the back to collect by $600. And there sat one of the policemen at a desk, and the other two were sitting *on* the desk. And he had a gun on the desk."

"These are the same guys who are running Sebastian's hops at UAW Hall 182—that's why we call it Club 182," said Gilbert. "I still don't see why they don't like you."

"Well—that gun," Lee resumed. "I supposed it was there because they had a lot o' money in the place. They had charged a dollar to get in—and the concessions—and so they had a lot o' money there, and I said, 'Okay. Everything went great and thank you very much.' And he gave me $100."

"A *hundred*!"

"One hundred. And I said, 'This isn't what we agreed upon.' Well—he put his hand on his gun. He says, 'You better take that and these fellers will help you to your car so nobody'll take it away from you.' An' he—and he turned the gun over in the palm of his hand, and I decided I better just go ahead and leave. So I did."

"Jesus!" Gilbert laughed. "Maybe you won't want this job, because—you'd have to go out to Club 182 Friday and Saturday nights and do your broadcasts from the lobby while Joel does his thing inside. You'd have to promote his hops."

There was silence. Finally, Gilbert leaned forward and pressed out his cigarette.

"I'll do it," said Lee Alan.

He joined WXYZ in February, 1962. Lee Alan had a lot of savvy. One of his gimmicks was a horn. It had started in college. He was a member of the Phi Delta Theta fraternity. As a pledge, he had to go through what was called "Hell Week." One of the things he had to do was read a long sheet every time the telephone rang. It took him 3 minutes to read the copy into the phone, ending with the words, "This is Lee Alan so forlorn at your service on the horn."

In signing off one night at WKMH, he had said, "This is Lee Alan on the horn." It had become a sort of by-word: Lee Alan on the horn. A record distributor called on him at WXYZ and presented him with a genuine camel horn from Pakistan. When he blew it, it sounded like a moose. People began to react. He used the horn more and more, making up stories about it—how it was insured by Lloyds of London for over $1 million, how it was guarded 24 hours a day by 17 Pinkerton guards, etc.

Sixty days after he joined WXYZ, he was called down to John Gilbert's office. When he arrived, Gilbert was showing Charles Kocher out the door, his arm around the chief engineer's shoulders.

"Best goddamn engineer in America!" he exclaimed as he shoved Kocher on his way and pulled Alan into the office and to his desk. "You sonofabitch, you've *done* it!"

A book was open on the desk. It contained the new ratings. Gilbert pointed to some figures on one page with a strong, well-manicured finger, Alan looking over his shoulder.

"You are Number One!" cried Gilbert. "In spite of that goddam horn of yours."

The two men studied the figures intently. Then Gilbert slapped the book shut. "Lee Alan you have to be the greatest!"

Alan straightened up, laughing, walked to the couch and sat down. Gilbert sat straight in his big chair.

"You know something?" he said. "I worked as a salesman on a little station, and then I went to Boston, and then you know what? They called me to come here and run this station. *This* station! The *Lone Ranger* station! And I never walk into this office without wondering what I'm doing here. And being amazed! And I hope that I'm going to be successful. Now the thing that I can't understand is that this rating book has come out and you've been here only 60 days. Why is it you're carrying a 26 and the next station is an 18? I don't understand that. Can you explain this?"

"No. I don't know," Alan admitted.

"You and your goddam horn!"

Record hops were now a big thing. Don Zee was doing them. Joel Sebastian was doing more than anyone else. And Fred Weiss did hops until he left WXYZ. Weiss went to his favorite city—Washington, D.C. He eventually landed on the station that had once turned him down—WMAL—TV. He finally made it to television.

Fred Weiss was replaced at WXYZ by a friend of Lee Alan from WKMH—dapper, 31, thin face, strong nose, very black eyes—mike name Dave Prince; real name, Pringle. Dave also did record hops.

Except for Club 182 most of the hops were put on by schools and churches Friday and Saturday nights.

Dick Kernen: "That's how I bought a house. I think I was carting home the grand sum of, like, $70 a week from the station, and I was making probably that much again in doing the record hops, booking them, getting bands and all that.

"The big schtick you had at a hop was you had a decent sound system and you had *all* the records. *Nobody* could afford to buy *all* Top 40. *You* had 'em all, plus you had all the show stuff and all the old stuff. It got very sophisticated.

"You would book in one, two, three bands—mostly local—depending on where the hop was and what the particulars were. If you went to, say, a Notre Dame High School which always had just *killer* record hops, you'd book not only maybe 2 or 3 bands but you would also try and get an artist that was in town, and they'd come out and lip-sinc their records and sign autographs."

In the Spring, Lee Alan had a call.

"I'm a partner of the Kramers," the voice said. "Walled Lake Casino. We've had everybody out here at Walled Lake and we don't get a lot o' people. Could you come

Lee Alan

Lee Alan

Lee Alan (on right) introducing guest star group to crowd at Walled Lake Casino "record hop"

out on a Friday or Saturday night?"

"No," said Lee Alan. "I can't do hops. I'm on the air at night."

But Lee Alan was interested. He saw what was going on at Club 182 with Sebastian, and he felt the whole thing could be handled better. Besides, he needed money. He had luncheon with Dick Kernen.

"What do you know about Walled Lake Casino?"

"You wouldn't believe it!" said Kernen. "All the big bands used to play out there, they say. The first time I walked in, I said, my God! What an incredible place!"

Walled Lake is northwest of Detroit beyond Broadcast House. The northwest suburbs were growing rapidly, as Jim Riddell had predicted—Southfield, Livonia, Farmington, Bloomfield, Pontiac, Brighton, Howell. High schools were crowded, more were being built. Walled Lake was something of a resort; there was an amusement park across from the Casino. It was also a year-round community.

Lee Alan phoned the minority partner who had first contacted him.

"What are you doing Sunday afternoon at Walled Lake Casino," he asked.

"Nothing."

"Well I have an idea."

"Talk to Mr. Kramer. He's right here."

Red and Cleo Kramer were the new owners of the Casino. Lee could hear the minor partner say, "It's Lee Alan, WXYZ," as the phone was handed from one man to the other. Another voice came on.

"Hello, Lee. Red Kramer here."

"I have an idea. We're getting into summer; schools are letting out. You're out there on a lake. What will it cost you to open up on Sunday afternoon?"

"O-oh—maybe—hundred dollars."

"Well I have a deal for you. You spend the hundred dollars to open up and another hundred on the radio and I'll come out."

"What's *your* take?"

"I'll gamble with you."

Announcements were purchased on Alan's shows Thursday and Friday nights that week. That Sunday there were no more than 300 people at the first Lee Alan record hop.

"That's 40 or 50 more than we got before," said Red Kramer. "So let's go again next Sunday."

The next week there were more than 600. After the 3rd week, there were 2000 people every Sunday. The Kramers wanted to open up Friday and Saturday nights also. Alan proposed to Gilbert that he tape the last 2 hours of his 7-to-midnight broadcasts Friday and Saturday nights so he could be at the Casino.

"I see nothing wrong with that," said Gilbert.

Then Jim Christie came to him. Contract for the remote broadcasts he'd been doing in the lobby of Club 182 was ending. Did he think Walled Lake Casino would buy those remotes? They would and did. Then Club 182 decided to renew and found they couldn't. They slapped one suit on Lee Alan and another on the station for $100,000 each for "maliciously and with intent" destroying their business.

Lee Alan: "John Gilbert called me in and he had a suit and I had a suit, and he said, 'All right, since we're sued separately, you get your attorney and we'll get ours and they can work together on it.'

"Well, I hired a gentleman who was a judge in Dearborn. The station had George Trendle, Junior. My attorney had it thrown out of court before George Trendle, Jr. ever read the brief.

"So that's the story of how Walled Casino got started."

The Kramers had 7- to 10,000 people in their Casino every weekend. It made Lee Alan a super star. It was also his downfall—at least the first one.

A moving van was being loaded in front of the Striker home in Arcade, New York. The Strikers were moving to Buffalo and Fran Striker felt good.

"You go ahead so you'll be there for the utilities," he told Jan. "I'll wait until they get everything loaded and then I'll go."

The family drove off. Fran was pulling himself together at last, after 5 very difficult years. He had completed work on his book, *One More River*, and hoped for publication.

Not once in five years had Fran Striker imposed his sensitive sorrow, loss of confidence and apprehension upon his family.

Jan Striker: "He was good and kind and gentle—and interesting. He made life interesting for all of his family and all of the people he loved. If he did anything for you, he would do it in a special way."

Janet: "One of the Christmasses I remember the best is one in Arcade when we didn't have much money because things had been rough. Doctor bills for Donnie were rough. Dad went out to the farm and cut down a tiny little Christmas tree, just for me. He called it my sugar plum tree and he put a lot of thought into fixing this tree and tied on a lot of meaningful little things."

Bob Striker: "I remember one time after he left Trendle and was trying to find some outlet for script writing, he wrote to Rod Serling to find out if he was buying scripts, and he got an answer from Serling addressing him as 'Miss Striker.' It just cut him down completely that Serling was not aware who Fran Striker was. Things like that tore dad up something terrible."

Jan: "And everybody assumed he was wealthy. He might have accepted a job with the Buffalo station if they'd known he needed one."

But now, after five years, he was pulling himself together. That summer, 1962, he had been adopted into the Hawk Clan of the Tonowanda band of Seminole Indians during a 3-day powwow of the Iroquois Nation. Fran had made a little speech and was given the name Ha Yah-Donh, "the writer."

He had also raised money for the Chaffee Memorial Hospital in Springville, just west of Arcade, and they put him on the board of directors. He would continue with that.

He was teaching at the YMCA and the YWCA. But the big thing—he had conducted a seminar at the University of Buffalo and had received a letter congratulating him on his success from the Director of Creative Education, Sidney

J. Parnes. Fran Striker was going to college at last! As a teacher!

The van was loaded and rolled away. Fran went across the street to say good-bye to his mother who now lived in a nursing home.

"I'll miss you laddie boy," she cried.

"Buffalo isn't far," he said cheerfully. "I'll be out to visit you and you will come in and stay with us now and then."

He took a deep breath of damp September air as he left the nursing home. He packed the last few things into his car. The next door neighbor came out on her porch.

"Good-bye, Fran," she called.

He grinned and waved to her.

"See you later," he called, opening the door on the driver's side.

She said afterward, he looked very happy.

It started to rain as he drove from Arcade. Construction work made one stretch of road very dangerous. When Fran's family had driven over it earlier, two flagmen who should have been on the road were sitting on a stone wall drinking coffee.

"Somebody's going to get killed here," Jan Striker had said.

As the accident was reconstructed afterward, Fran Striker was driving fast, as usual, came to a rise too late to see the cars lined up in front of him in time to stop. He pulled over to the left to avoid hitting them and ran head-on into a car coming in the opposite direction. He was killed. September 4, 1962.

The media called WXYZ for information. The station knew nothing about him, called Dick Osgood.

"Did you ever hear of a Fran Striker?"

The writer of Milestones, obituary column in *Time* Magazine, wrote of Fran: "...author of the saga of the masked rider of the plains, *The Lone Ranger*, by far the most enduring of all western radio heroes.... Striker first conceived of the straight-shooting law man in 1930, and the first episode was broadcast by Detroit radio station WXYZ in 1933."

The Milestone editor later received this letter from Fran Striker's son:

> Dear Sir:
>
> "I understand that my father's claim to origination of *The Lone Ranger* has again been challenged by Mr. G.W. Trendle. Perhaps this is a question which will never be completely answered; frankly, it hardly seems important. As a matter of fact, correspondence dating back to 1932 indicates rather strongly that *The Lone Ranger* was not *created* but *evolved*. The central character, the horse, the silver bullets, etc., were framed in the background of *Covered Wagon Days*, one of dad's first syndicated radio plays. Although the program was scheduled and broadcast as early as January 30, 1933, it was not until May, 1934, that a formal agreement was signed wherein dad relinquished all rights to *The Lone Ranger*, *Manhunters*, and *Thrills of the Secret Service*.

Dad once said that 'only God can create.' I'm sure you believe this, too. The important and indisputable fact is that Dad was able to build...a characterization which lived and was idolized by countless millions of American children. *The Lone Ranger* exemplified the traits of character with which every parent hopes to see his children mature. He brought a familiar touch of home to GI's in the far corners of the earth via Armed Forces Radio. He represented the ideals of the United States throughout the world.

The Lone Ranger could not have existed without financial backing. Mr. Trendle provided this backing...Dad gave life to the ideal for more than 25 years.

Please let me know if I can be of any assistance in burying Mr. Trendle's bone of contention.

Sincerely
Robert H. Striker

Bob Striker received this reply from *Time* Magazine:

Dear Mr. Striker:

Thank you for writing apropos of *Time's* September 14 Milestone item noting the death of your father, Francis H. Striker.

Our reference to the fact that your father was the author of the saga of *The Lone Ranger* and that he first conceived the character in 1930 was based on careful research. Since the item appeared, we have also received a letter from Mr. Trendle and it appears that the point is in controversy between you. As we wrote Mr. Trendle, it is not within our province to offer any solution in the matter, but we are glad to have the information you sent us and hope you will keep us informed of any further developments.

George W. Trendle sent a blanket of orchids to be draped over Fran Striker's casket. The surviving Strikers refused to let it be used.

49
Caught in the Swinging Doors

Dick Femmel stood tall in the doorway of John Pival's office. His suit hung snugly from broad shoulders; his hair was neat except for one crop that refused to lie down at the back of the part.

"Sit down," Pival invited. It was morning and Pival was clean and fresh, but cherry-eyed. A cold December sun sparkled on the crust of snow outside Pival's windows, and naked black limbs of the apple trees were frosted. Pival leaned back in his chair and put his feet on his desk. "Did you see Osgood's interview with Lenore Romney?"

George Romney was running for Governor and Osgood had interviewed his wife in their home for Channel 7.

"I thought it was quite good," Femmel admitted.

"That old bastard keeps going." Then Pival came to the point. "I've got a new job for you. We're going into editorials. ABC is going on with 'em in New York, and probably all the O-and-Os will get into it. We want an editorial director and we want you to take the job." Pival gave a warm, friendly little smirk as he added, "At an increase in pay, of course."

Femmel smiled.

"That makes you happy." Pival grinned openly. "We'll let Jack Powers' buddy take over TV news, Tom A'Hearn can handle radio, and we'll pull you out of the news room and find you an office downstairs here where you can work with Gilbert and me."

"Who'll do these editorials on the air?"

"I was going to have Ed McKenzie do 'em, but ABC wants the station managers on the air with 'em. I owe Ed a favor and I'm going to put him on with a late-night talk show."

"How long do you want these editorials?"

"Short. Not more'n two minutes, maybe a minute—depends on what we have to say. I'll do 'em for TV and Gilbert'll handle it in radio. You write for both of us, and if either of has to be out of town, you'll substitute."

"Will you both be doing the same editorials?"

"Not necessarily. I may have some ideas of what I want to do; Gilbert may have some ideas; and we'll expect you to come up with some. But one thing you gotta watch out for: if I take a stand *for* something, he can't come out *against* it; and I'll lay off anything I don't agree with him on. The station has to be consistent."

"How soon do you want to start this?"

"Let's say the 1st of the year." Pival lifted his feet off the desk and sat up. "Gilbert's waiting to talk to you."

Femmel found the dynamic Gilbert pushing Bob Koch out of his office with a friendly hand on the sales manager's shoulder.

"Fantastic! Fantastic! Cookie, you're the greatest!"

As he saw Femmel enter the reception room, Gilbert's free hand shot out to grab him. He slapped Koch on his way and drew Femmel into his office. His handsome features were alight with energy and bounce as he offered Femmel a cigarette.

"Pival said you wanted to talk to me about editorials," said Femmel.

"We had to find someone who has contact with the news makers, a guy who can write and who has good judgment, and we looked around. Hal Neal is in on this, of course. And we said, who has all this? Couldn't be anybody else! Dick Femmel!"

"How far can we go on these editorials?"

"Hal says confine it to our own area—go into national or international events *only* if they have some sort of link to what's happening here."

"In other words," Femmel reasoned, "if there's a bill in Congress that would affect the automotive industry we can take a position on it. Otherwise, no."

"You've got it! Great!"

"Or maybe the housing situation."

"You handle it. What do I know about all that stuff?"

Femmel's long face lit up with a grin. "You know what I like about this? I won't have to edit Lou Gordon any more."

"Oh, Jesus!" Gilbert threw up his hands and laughed. Then his hands banged down on the blotter as another happy thought struck him, and he leaned forward. "And what do you think is on deck for our big ham, Jack Powers?"

Femmel's laugh was arrested.

"Hal Neal wants him in New York."

Femmel's jutting chin dropped in disbelief.

"New *York*?" he managed.

"*New York*!" exclaimed Gilbert. "Hal said, 'When can I have him?' I said '*Yesterday*!"

Dick Femmel moved out of the news room into a small office without windows on the first floor. WABC in New York and WXYZ in Detroit were the first ABC stations to get on the air with editorials. After Trendle's *Town Talk* and Femmel's aborted effort with Osgood, the first WXYZ editorial was broadcast on January 2, 1963. John Pival approached the cameras with the flair of a showman, but Femmel had problems with John Gilbert. Curiously the dynamic extrovert carried little conviction before a microphone.

For the first time, managerial interest in news was aroused. Channel 7 purchased its first station wagons for camera and sound crews and cruising reporters. New York pushed the button that revved up journalistic activity, and in radio at least, the finger on the button belonged to Harold L. Neal, Jr.

This dedicated one-time announcer had done what no one else had been able to accomplish: he had turned WABC-Radio around. From the weakest individual station owned by ABC, he had built it to one of the strongest. To get the sound he had made a success of in Detroit, the kid from the sticks had had to make changes in well-established New York personnel. He had had to dig up Martin Block by the roots, and the roots were deep. But under Neal's direction, WABC would finally attain a position where it could reasonably use the slogan "Most Listened to Station in the Nation." Hal Neal was promoted to network vice president in charge of all Owned-and-Operated stations.

His contact and influence at his old alma mater, WXYZ, was stronger than ever. "Get more actualities on your newscasts," he commanded. Tom A'Hearn was no longer young. His hair and bushy eyebrows were salted with grey. He was tired; he was sick; and he didn't respond to the new requirement. After nearly a quarter of a century with the station he had to be replaced.

The new man was Ed Hardy from WBIM, Jackson, Michigan—blond, bland and bullish. He installed the gimmick of paying listeners for phoning in useable news tips. He and Gilbert nearly burst with joy when Hal Neal took the roly-poly neurotic Jack Powers to ABC-New York where he lasted but a short time. His pal was replaced on the TV side by a former Washington reporter, George Cheeley, whose stay at WXYZ was brief.

Ed Hardy imported a student of journalism from out-state Michigan, a professorial youth named Bob Hardt. Hardy was said to have housed him in his basement until the pedantic, heavy-voiced young man could get on his feet. Hardy also hired a pleasant, capable fellow from a small Detroit station named Joe Vaughn. Both worked first as announcers; both announced Osgood's newscasts.

Osgood was now over 60. Bob Baker called him "the grand old man of radio;" office boys addressed him as "mister" instead of "Dick;" and occasional newspaper pieces referred to him as "the veteran." He felt vaguely uncomfortable with Bob Hardt and Joe Vaughn, as though they were straws blowing in who-knew-what direction; and he instinctively distrusted Ed Hardy and felt unaccountably embarrassed in his presence. His instinct proved to be right.

There was now the same sense of turbulence in the WXYZ news rooms as had first been felt 10 years before when live programming was gradually phased out and replaced by records. Straight broadcast reporting was giving way to "electronic journalism"—a peculiar combination of ego and showmanship imposed upon facts.

The American Broadcasting Corporation was shuffling men from city to city and from title to title. A series of such changes put two new men on the Detroit payroll. One would be the most controversial character to exert influence over WXYZ in all its history.

David Klemm had full lips, wide-set eyes and a way with the ladies. He was keen and ambitious, loved the theater and the arts, and was in his early 20s. He was a farm boy from Indiana. His first work with radio had been with WOWO in Fort Wayne. Then college, service in the army, and work as a salesman for Harry and Barney Balaban at KWIL in St. Louis. The Balabans owned KWIL, KBOX in Dallas and WRIT in Milwaukee. David Klemm was made promotion director of all 3 stations.

As Awards Chairman of the Broadcast Promotion Association, he solicited entries from stations all over the country and came into contact with a number of people at ABC. WXYZ-Radio was "big on promotion; its salesmen always had outstanding materials to present"; and in 1963 the man most responsible, after quite a parade of promotion men, was one of the best—a short man with salt-and-pepper curls, Leo Collins.

As was happening now with a frequency that was sad for WXYZ, New York skimmed off the cream and Leo Collins soon rose to the top. A corporate shuffle produced a vacancy in New York and Collins was drawn into Manhattan to work with Ron Sack. Somebody was needed for Detroit.

Collins suggested David Klemm.

"I offered him my job in Chicago when they brought me here from WLS," said Sack. "He wouldn't leave St. Louis."

"He might now."

"Sound him out."

David Klemm was thrilled. WXYZ was to him "a fabulous radio station." It was the most profitable of the ABC-owned stations. It took very little pressure from Collins and Sack, plus the approving nod of John Gilbert, to move David Klemm from St. Louis.

John Gilbert's nod was made without much thought. He had other things on his mind. He himself had been caught in the swinging doors. No less than Simon B. Siegel, Executive Vice President of ABC, had offered Gilbert the position of General Manager of WABC-TV, the network's key station. John Pival, with his record of success, might have been considered first in line for such a promotion, but many things were now working against him, including a determined enemy at corporation headquarters.

John O. Gilbert leaped without hesitation into the higher branches of the corporate tree and never once looked down—or back.

WXYZ was now in the market for a new manager. Hal Neal was actually shopping for two managers, for when he was promoted to vice president, he needed a manager to replace himself with WABC-Radio, New York. The first man he thought of was his old friend, Detroit representative of the John Blair company, Charles D. Fritz. He offered Fritz either station, WXYZ in Detroit or WABC in New York. Fritz was definitely not interested in New York and that position went to a Wally Schwartz. Fritz took his own time deciding about WXYZ.

Hal Neal flew out to Detroit and met Fritz in the mixed grille at the Oakland Hills Country Club.

"You turned us down before," Neal reminded him, "because you said it was exciting to be out on the street. Still exciting?"

"No," Fritz admitted. "Selling is a young man's game and I'm 35 now."

"Good age for a station manager."

Fritz ignored the comment, picked a hair from the sleeve of his suit coat, and, half to himself, continued, "No, I can't picture myself trying to compete with these younger guys much longer. You do a lot of entertaining; you're out a lot; you have station people in from all over the country. I have to admit I've just really got kind o' tired o' that."

"Chuck, you once advised me to get out on the field with ABC, and I took your advice. Now I'm advising you. There's a lot of interest in radio in New York. There's going to be a lot of expansion—news, FM—. Fly into New York with me and talk with Si Siegel."

Fritz was fond of the Blair company. John Blair had set up a generous profit-sharing system with his employees. At the end of each year he gave Fritz a bonus of half a share of stock in the company which was worth $20,000 a share and paying 33 1/3 dividend. The bonus thus was worth $10,000. The dividend of $3,300 served as down payment on the balance and Fritz would have a year to pay the remaining $6,700. Stock in the Blair company was not yet on the market; it went only to employees; and it was phenomenally successful, splitting 2-for-1, 3-for-1; once even 50-for-1. Fritz was fully invested, had been with Blair for 12 years, and was a wealthy man.

He flew to New York with Hal Neal and talked with Si Siegel. Simon B. Siegel had been head accountant of Paramount Theaters. He was serious and shy but had tremendous influence with the ABC organization.

"How's my old friend, John Pival?" he asked as Fritz sat down.

"John's doin' all right," Fritz told him. Fritz actually had not been at all close with Pival and didn't intend to be in the future. "Hal tells me a lot of interest is going to focus on radio now."

"He's right. It is," said Siegel. He beamed on Neal. "This man has done a lot for us. Do you know that when we brought Hal into New York, WABC was losing a million-and-a-half a year? WLS was losing. KABC out in Los Angeles wasn't making any money when we bought it. KGO tried everything and couldn't get out of the red. WXYZ was the only profitable station we had."

"You're talkin' to an old Blair man, Mr. Siegel," Fritz smiled. "We knew what was goin' on."

"WXYZ was a profitable station when we bought it from Trendle; Jimmy Riddell kept it that way; and Neal made it more so."

"I'm trying to convince Chuck he can keep it that way," said Neal. "May do even better than I did."

"You from Michigan?" Siegel asked.

"I've been in Michigan all my life and if I go with WXYZ I won't be using it for a stepping stone for anywhere else."

Siegel looked from the slow-talking, steady-eyed Fritz to the vice president in charge of O-and-O stations. This might not be good. If Fritz delivered, they might want him in New York. Neal knew what was in Siegel's mind.

"We don't need to talk about stepping stones at this point," he said. "The problem is Detroit. It's been operating for a month without a manager."

"I happened to be talking to Jimmy Riddell on the phone from Los Angeles," said Siegel, turning back to Fritz. "He says you're doing a good job with Blair. And he says millionaires have left the Blair company because of the profit-sharing John has in there. Are you a millionaire?"

"No. But I'm not hurtin'."

Siegel studied the quiet smile, the cold steely eyes, and the tailored appearance, then turned back to Neal.

"Well, I trust your judgment, Hal."

"Let's talk a deal," said Neal.

"We can talk salary," Fritz told him, "but over and above that, I want a percentage—an X-amount of the profits above a certain point. And—there's John Pival."

"What about Pival?" Siegel demanded.

"According to Johnny Gilbert, Pival has been settin' up some very screwy deals for radio," Fritz went on quietly. "He's been usin' radio as a kind o' fringe benefit in sellin' TV, and radio's been losing money on that. Now, if I went with WXYZ, John Pival would have nothin' to do with radio. I would answer only to Hal Neal."

Neal agreed. "We don't mess in with television and television should leave us alone."

"I always liked John," said Siegel. "And he did a fine job out there putting Channel 7 on its feet. But I know Hal has no great affection for him, and Ted Shaker here in New York hates his guts, wants him out o' there."

Ted Shaker was to the owned ABC-TV stations what Neal was to radio; he was John Pival's enemy at headquarters. Siegel was studying Fritz. On the surface, Fritz seemed soft. His speech was slow and his forehead wrinkled as his eyebrows lifted and fell in jerks as though he were nervous or shy. But underneath he was firm, and the man had class.

"Okay," said Si Siegel. "Let's talk a deal."

In July, 1963, Charles D. Fritz became the 7th general manager of WXYZ-Radio.

"How many people do I have working for me? How many departments are there? How many in this department? How many in that department? What does this department cost me? What did it cost last year? How are we doin' this year? How do the union contracts read? What does this guy do? What is her job? Who are the important people? What kind o' guy is this one? That one? How many salesmen we got? Who brings in the most business? Do we have to have that many engineers? How long has this been going on? Got any suggestions on that?"

Some of the answers came from Bob Baker, some from Bob Koch, but most came from Bud Reynolds. Reynolds was impressed with the new manager, found him a "commanding-type figure."

The day he had come on the job, the staff had been summoned to one of the large TV studios to meet him—news men, jockeys, traffic department, production men. Hal Neal had flown in from New York to introduce him and Fritz had made a little speech, "nervous as hell," his eyebrows jumping up and down.

Now it was up to him to take hold. He had been locked in his office for several days—incommunicado. Some things he already knew from his experience with the Blair company. Television, for instance, got the higher dollar from the national advertisers, but with radio it was the other way around, local dollars were higher than national. The thing for radio to do was to go after local accounts.

Sales were down on the Fred Wolf show but that was understandable; his ratings had dropped, too; still the books showed the station was still making a million-and-a-half just opening up his show in the morning. Fritz knew that Wolf was threatening to quit, and he recalled that John Gilbert had said, "I pity the poor sonofabitch that's at this station when Fred Wolf leaves because he's an institution." But Neal was right. Wolf's style and the new music were incompatible. More with the times was Lee Alan.

Even so, Wolf was bringing in a hell of a lot of money. Maybe Bob Baker should press down a little harder on him. How long had Baker been with the station? Thirteen years. He seemed to be protecting his old friends.

Although Gilbert had told him of Pival's use of radio in selling television, Fritz was amazed to discover in the books the number of radio accounts that were not paying established rates; and he made a note of the expiration date of every contract and determined that when each of those contracts expired, if it were renewed, it would be on radio's terms and radio's alone.

There was another thing—the Friday luncheons at Palombo's. Baker, Baldrica, Davenport, Hengstebeck, Osgood, Koch, Jim Christie, Dick Wozniak, salesman—they and others gathered at the bar, drank, moved to a big round table, drank some more, ate a leisurely, sociable luncheon, and finally got back to the studio. He joined them two or three times in the beginning but—those luncheons were bad business.

Charles D. Fritz, head of WXYZ Radio.

All in all, however, Charles D. Fritz was satisfied that he had come into a good thing. He would not meddle with the format—maybe a little refinement here and there, and organizational improvement. But altogether, WXYZ-Radio was everything he had hoped for.

50
Battle of the Giants

1963 was quite a year. Five U.S. helicopters were shot down in Vietnam as the Viet Cong scored victories; astronaut Cooper splashed down on target in the Pacific after making 22 orbits; 1,200 Cuban refugees arrived in Florida in exchange for food and drugs; John F. Kennedy was assassinated and Lyndon B. Johnson became the 36th President of the United States.

The Lone Ranger and *The Green Hornet* were being syndicated by transcription but the actors were receiving no more residuals. Charles D. Livingstone had left California in 1956 following a breakdown, had been producing slide film and commercial movies with the Ford Motor Company in Dearborn, and now switched to the Jam Handy organization. Edythe Fern Melrose had now built and sold her third "house of charm" and was chosen one of the Top Ten Women in Detroit.

Charles D. Fritz and David Klemm soon realized that the most phenomenally successful element in the station's programming was the response to Lee Alan and his record hops at the Walled Lake Casino.

Dick Kernen remembered it well: "Lee understood instinctively what was goin' on and he knew how to use it. He knew what to say on the air, what records to play. He had a tremendous feeling for what was going on."

Alan had showmanship and a flair for quality. He made the Kramers install the best possible sound equipment in their Casino. He had 2 turn-tables installed, and a sound mixer. The operation had class. His manner of build-up on his broadcasts, and presentation in person at the Casino were made to coincide with the interests of the high schoolers and college crowd. He built up the success to 4 hops a week.

Kernen was all enthusiasm: "It was just a very exciting thing. I remember a number of people that Lee brought out to the Casino that just drew crowds beyond belief. He brought in Fabian and I remember Lee and Red Kramer and Leo Kramer pacing up and down waiting for Fabian to get there and the crowd inside like 5,000 people going bah, bah, bah, where is he? And he finally showed up—just a super colossal event.

"Lee was instrumental in Stevie Wonder's success. Lee brought Stevie Wonder out to the Casino, promoted it, told the story about—you know—Stevie Wonder is a blind kid—staged it in such a way that Stevie Wonder at the Casino was probably one of the all-time entertainment events I have ever seen. It was not only because Stevie was good—obviously, you can't palm off a turkey—but the build-up that Lee was able to do on the air and the presentation at the Casino, the suspense, the showmanship that he used in presenting Stevie Wonder was just incredible."

Another event was Chuck Berry. Berry had not made any new records for a couple of years, but jockeys were playing the old ones—especially Lee Alan in Detroit. He liked Berry's music because "the words were simple and the songs communicated." Chuck Berry's old records had been made by Leonard and Phil Chess in Chicago. Alan called to see if he could get Berry for an appearance.

"You know he's in jail," said Chess Records.

"Well, when is he going to get out of jail?"

"We're not certain. He may be paroled. We've heard what you've done over there in Detroit, because we have to ship in the records that still sell there, thanks to you."

"Well, look," said Alan. "Why don't you let us know if he ever gets out?"

Berry had occupied a cell in Terre Haute, Indiana, for approximately 2 years as the result of a border incident involving a young girl. This is why there had been no new records.

Lee Alan had a telephone call one Tuesday, was told that Berry would be out of jail on Wednesday, and could appear at Walled Lake that Friday and Saturday. Alan grabbed him at $1,000 a night. He had only 2 days to build up the appearance on his radio show. Late Friday afternoon, Alan drove to a motel to pick up Berry and take him to the Casino. He had never met Berry, was impressed by his height, and found him humble but suspicious.

"You're lucky," the Chess man told him. "Chuck won't do any interviews or see any reporters. He's your exclusive."

"I appreciate it."

"I'd like to see my thousand dollars," said Berry.

"Fine," Alan said. "The Kramers will pay you when we get out to the Casino."

"No," the musician said. "We can't have a check."

Alan packed Chuck Berry, his guitar, an amplifier and himself into his little Corvette and they drove to a bank. Alan got the cash and gave it to Berry who counted it as they started to Walled Lake. It was now almost 6 o'clock.

"By the way," Berry said, squirming around in the tightly packed little car to get the cash into his pocket. "I have to buy a tube for my amplifier. Could we stop at a drug store?"

Even though Berry had been out of circulation for 2 years, the clerk in the store recognized him. By the time they got to the freeway, it was choked. Traffic crawled.

"What's goin' on?" the recording star wanted to know.

"I don't know. The afternoon rush from downtown should be over. What did you do in jail?"

"Oh, I wrote a lot o' songs. I'm going' to record one next week."

"What is it?"

Berry pulled an old crumpled envelope from his breast pocket. He said, "I don't have it done yet."

He had written the words but not the music. He read some of the lyrics: "As I got on a city bus and found a vacant seat I thought I saw my future bride walkin' down the street."

The Swingin' Kind
WEDNESDAY, AUGUST 11
7:00 PM WXYZ-TV
7
LEE ALAN presents, in person,
LESLEY GORE • STEVIE WONDER
THE COASTERS • THE SHANGRI-LAS
The "Detroit Sound" taped "LIVE" from Metropolitan Beach

"What are you going to call it?"

"*Nadine*," said Berry.

They discovered that most of the traffic which slowed them down was bound for the Walled Lake Casino. It was jammed both nights.

Kernen was filled with superlatives: "He went on and just absolutely killed the place. I mean—*just—killed*—'em. I was out there the second night and he counted out $1,000, mostly singles, before he would go on, because he didn't trust *anybody.*"

The next week, Chuck Berry wrote the music for *Nadine* and recorded the song. In 6 weeks it sold 1,000,000 copies.

Lee Alan arranged for portions of his hops to be broadcast. The brothers who owned the amusement park across the street watched what was happening at the Casino, went to WXYZ, bought time, and arranged for their own hops to be hosted by WXYZ jockeys. Lee Alan's best friend, Dave Prince, alternated with him between the Casino and the smaller building in the park which they named The Band Box.

Kernen's memory was explicit: "It got to a point where people would stream out to Walled Lake and you would have Red Kramer sitting on his front porch with a bull horn yelling, 'Come to the Casino tonight. Tonight, appearing—!—and he'd run down all the guests. It was just bizarre. He would send his two kids across the street to check out the crowd at the Band Box. And then, after it was all over, we'd all go out and have dinner."

In August, 1963, after 14 years, Dick Osgood was removed from the morning newscasts. One long-time sponsor, The Automobile Club of Michigan, threatened to cancel but was saved by a promise that Osgood would continue to do its commercials as he had done. Osgood bore the embarrassment, but after a few weeks, the Auto Club cancelled anyway.

Osgood's place was taken on the newscasts by the two men who had been announcing him: Joe Vaughn and Bob Hardt.

Bob Baker stood listening to Dave Prince. He stuck a thick middle finger into one ear and joggled his hand up and down. He had trouble with that ear and it drained. He snuffed loudly and vigorously.

"Wa-al!" he cried when Prince was finished. "You have to be out of your mind! What can *they* do? Nobody can hear the damn station. They could have Jesus Christ on there and nobody'd know it."

That was that as far as Baker was concerned. Alan, Prince and Joel Sebastian went to Fritz. They had received their information from record pushers, and as usual the pushers were right. Mike Joseph, station analyst, was holed up in a Detroit motel mapping out an attack on WXYZ.

"Dave Klemm has told me about it," Fritz told them. "But I'm not losing any sleep over it."

"Mike Joseph's a pretty smart guy," Sebastian warned.

"I know who Mike Joseph is," Fritz said. "He was with ABC and worked with Hal Neal in New York when they turned that station around."

"So he knows how," Prince pointed out. "He knows how to turn WKMH around."

Fritz smiled at the jockeys as though they were little boys hopping up and down over a garter snake.

"He's been listening to all the stations in the area and he's done a lot of research—a lot o' market research—when factories have their shifts, when the kids get out o' school—a lot o' things."

"He's done a survey to find out exactly what this market wants."

"WKMH is going to be an entirely different station—a new station. They're even going to change the call letters. They're gonna play music, lots o' music."

"He'll kill us with our Breakfast Club from Chicago."

"And our hour-and-20 minutes of talk between 6 and 7:30 at night."

"Mike Joseph is going to take advantage of us and we know how and we know when he's going to kick it off."

"It's going to happen in October."

"His only problem now is to rid this town of WKMH, wipe it right off the map and never have it appear again. He's got that all planned, too."

"We could beat 'em to it. We can beat 'em to it by kickin' off the Breakfast Club *now* and all that news and featurette stuff in the evening—*now*.

"And go all music, and I mean—rock."

Charles D. Fritz was quiet, poised and pleasant. He had let the boys have their say.

"I know Mike Joseph and Dave Klemm knows Mike Joseph," he said. "And we don't think he can hurt WXYZ. They have a bad signal over there. You can't hear 'em from here to Grosse Pointe."

"I know," said Lee Alan. "All these years I felt the same way: WXYZ's an impenetrable powerhouse. But Mike Joseph is going to do it. This will change radio in this market."

"I like your interest and all that," said Fritz. "But we just don't think Nellie Knorr's little WKMH can compete with us."

At midnight on Hallowe'en, 1963, WKMH ceased to exist when a voice came on the air on that same wave-length, saying, "This is WKNR, new radio 13, where in exactly 23 hours, 22 minutes and 8 seconds, the battle of the giants begins."

For 23 hours, 22 minutes and 8 seconds, the "new" station broadcast no commercials and no music. It played ghost and detective stories, teasing the public with announcements of the coming "battle of the giants." And 23 hours, 22 minutes and 8 seconds after the first announcement of it, the battle began.

The world informed Lee Alan. "Friday night when I went to Walled Lake Casino, all the kids were talking about 'that new radio station playing ghost stories.' Then—the whole new format began. And it happened. In 60 days, WKNR was Number One in about 3 time slots."

Charles Fritz was shaken: "They made a hell of a run at us and they were beating us on the Hooper reports."

Within 3 months, WXYZ Radio had fallen from its pedestal, like Humpty

Dumpty from the great wall; and all the king's men and all the king's horses—tried!

"Lee."

Leon McNew looked up. It was the new weather girl.

"I have a feeling I'm not getting all the money I should be getting," she said. She had a beer sponsor.

"What makes you think so?"

"The agency tells me that they're paying *so* much, and I'm not getting that much."

"The old story."

"I just saw the light on down in John Pival's office. I'm going down and find out why they're not paying me that money."

"Do me a favor. Postpone it."

"What?"

"Go in during the daytime, during office hours."

"Why?"

"Don't go down there tonight."

"But why?"

"Well—John is apt to be stewed."

"I don't care."

She left the news room and McNew went back to the preparation of his 11 P.M. news. A few minutes later, the girl was back, crying like a baby.

"What happned?"

"You—can't—*believe*—"

McNew knew what had happened, but she told him anyway.

"I walked into that office—and—the first thing Mr. Pival did was—get up from his desk—stumble 3 or 4 times, walk over, and pull my blouse right up over my head and—start—fondling me, before I could even say a word."

"Don't say that I didn't warn you."

In New York, Charles Fritz was returning with Hal Neal to Neal's office after a talk with Si Siegel. As usual, Siegel had asked, "How's my friend? How's John Pival?"

"Hal," said Fritz as they walked along the polished corridor to Neal's office, "one of these days, if Si asks me that, I'm going to ask him if he really wants to know, and if he does, then I'm going to tell him."

John Pival was going through hell. Alcohol had been a habit before; now it was an escape. His enemy in New York has been variously described by those who knew him as "a very dapper lookin' feller; good smile; a real sales-type person, "urbane; a good, tough administrator; very effective" and "the Madison Avenue type; up-and-coming."

Bess Ashton knew about the feud between Pival and "the man in New York." To her it was "all personal. Personality conflict. That's all there was to it."

Some people say he wanted to get John Pival out of WXYZ.

Ashton said: "He wasn't the only one who wanted to get Pival out. Ted—Ted—what the deuce was his name? Any—oh Lord—yes."

Ted Shaker had left CBS to shape up the spot sales division of ABC. He had done an outstanding job and was made President of the owned TV stations, still retaining leadership in spot sales. In both capacities, he had come in contact with John Pival. The two had disliked each other immediately.

Shaker formed his own organization within the corporation and you were either on his team or you were not. Obviously John Pival was not. Nor, as it turned out, were men in Pival's position on other ABC-owned stations around the country.

John Pival had been a favorite of Si Siegel and Leonard Goldenson, but he no longer found their doors open; he had to report to Ted Shaker, and he met nothing but frustration. One who has full knowledge of the situation has said, "No question about it, Ted Shaker never let an opportunity go by to zing John."

He wanted to buy a package of theater film. He was not granted the budget. He wanted to carry a golf tournament.

"No. You can't spend the money."

He was running out of storage space for scenery and props and wanted a new building.

"I've been out there. I've seen it. You have plenty of room."

"God damn it! You come out here for 2 days. You don't know what the hell we need. I'm here every day and I say we have to have more space."

He was denied the money.

Where once he had been free to create shows and run his station, he now felt as though bound by ropes. He became suspicious of those around him, trusted only his brother-in-law, Ed Stock, who worked there, and Harold Christian. He knew he was unpopular with many of his employees. A technician was caught snitching food from the Charm Kitchen and was fired; this made Pival unpopular with the engineers even though he had sent two of them who were alcoholic to the Brighton clinic and held their jobs for them while they "dried out."

There was also pressure at home. Virginia Pival had stopped drinking, but there was a strain. She knew he was "playing around"; communication between the two was at a minimum; but she sympathized with what he was going through.

"Before Shaker got in there, John had no problems," she remembered. "I think every single idea Johnny ever got or ever promoted made money. Then they bring in *this* guy and John has to go through *him*! Well! John isn't going to stand for that. There's a lot of it I don't understand myself. John couldn't talk about it. I know now that he did receive a communication from Shaker once—not unfriendly—warning him that he should straighten up and stop drinking. But he'd come home and he'd walk in that door and he'd mutter, 'That God damn—.' *You* know he'd had it. He had just *had* it!"

Ginny thought up things for him to do to work his mind loose from its obsession.

"He enjoyed making shelves for a closet or changing draw pulls or repairing a table."

Late every summer, when the new season's network programming was set and before the season got underway, vice presidents from the network and managers of the owned-and-operated stations would get together for a preview of upcoming features. There would be press parties and meetings, and in Detroit it was natural that there would be a gathering in the home of John Pival.

Among those present one clear, cool August night, were Elton Rule, then head of the Los Angeles station, Red Quinlan, vice president in charge of ABC, Chicago, and the man they hated—Ted Shaker. As usual, John Pival drank too much. He disappeared and nobody missed him.

Suddenly he emerged from his basement, shirt removed, flexing his muscle, and carrying an axe. With a mighty swing Hercules brought the axe down on a coffee table, splitting it in two. Then he extended his index finger toward Ted Shaker and growled, "You're next."

Shaker escaped from the room onto a neighboring golf course. Pival chased after him. He was finally subdued. Ted Shaker had every right to be angry.

"I'm going to get you out of that station, if it's the last thing I do," a witness heard him say.

Virginia Pival who was not there that night, declares no such incident could have occurred; if her coffee table had been split, she says, she'd have known it. But she, herself, told of his repairing a table and Leon McNew swears he was there and saw the whole thing happen.

James G. Riddell was sympathetic and understanding: "There was a lot of pressure put on John by Ted Shaker. Pival loved to work, loved to be at the office. He would stay there and get thinking, get working. Well, while he was in the office, he drank. He could drink; he was no amateur; I mean, he could pour it in. But there came a time when he couldn't.

The next year, 1965, was to be that time.

One Sunday, two years after the death of Fran Striker, *The Detroit Free Press* printed the following complaint in a general information column, without comment:

> Do you suppose we can put an end to the oft-repeated fable that Fran Striker 'created' *The Lone Ranger*? The story had been tested on WXYZ a number of times before Striker was hired as script writer but, for the record, *The Lone Ranger* was my brain-child.

It was signed George W. Trendle.

By contrast:

James G. Riddell: "Johnny Pival was the one who found Soupy, you know. No one else. People say to me, 'You discovered Soupy.' I had nothing to do with it. John found him."

The new numbers were in and WXYZ-Radio was losing the battle of the giants. Charles D. Fritz sat at his desk listening to Hal Neal on the phone. David Klemm, looking urbane in striped grey-and-black slacks, a dark, tailored jacket and a bright figured tie, relaxed on the couch. He had warned Fritz that Mike Joseph would lead

WKNR to victory and now that the losers were on a rampage, he could sit relaxed and guiltless. Bob Baker, on the other hand, fidgeted in a chair that looked too small for him, snuffing drainage from his ear noisily. Big to begin with, he was developing the additional bloat of a man who would rather drink than eat, and his suit was tight. But the little blue eyes were keen and they watched Fritz sharply as Fritz listened to Neal. Fritz' eyes turned from the wall and concentrated on Baker. Then he hung up.

"Hal says we've got to tighten up some more," he said. "Bob, you've got to clamp down on those guys."

"How are you going to clamp down on Fred Wolf?"

"Don't give him so much choice in records."

"Wa-a-al!" Baker's voice was thin but strident. "He doesn't get that much choice."

"And Hal wants more actualities in the news, and he wants everything tightened up—less talk and more music."

"Why doesn't Hal look at the business we're doing?"

"But Keener is grabbing our audience."

Everybody was referring to WKNR as Keener now.

"What concerns me," said David Klemm, "is promotion. I knew the contests they'd hit with; I knew the amounts of money they'd give away. Even though they have a crappy facility, they've got the audience."

"They still don't have the advertisers," Baker muttered.

"Not yet," Fritz admitted.

"That follows," said Klemm. "You first have to have the audience. But you also have to have a degree of believability and acceptance for the advertisers, and XYZ still has a great acceptance."

"And what gives us that acceptance and believability?" Baker demanded. "It's the Fred Wolfs and the Dick Osgoods and the Lou Gordons and the Lee Alans and the Paul Winters."

Paul Winter's 40th birthday was a milestone. He was balancing WXYZ against an offer from Station WEEI in Boston. The degree of panic to which Wyxie had succumbed was indicated to him when Bob Baker came to him and almost apologetically told him "they" felt he sounded too highbrow. The important thing was that the kick had gone out of performing. No longer did he get off the air with a feeling of having done the job. The WEEI offer looked good. He went to Fritz.

"I just had my 40th birthday," he said. "And I thought to myself, Saul, you're getting too old to do this rock and roll stuff."

"You'd better find out what they want you to do in Boston."

"I have devised a rule for myself, Chuck. I will never again do anything that I myself wouldn't listen to. I admit it's kind of a silly rule; a man has to make a living. But—I want to try to live up to it and I'm tired of being a rock and roll disc jockey."

"Paul, I don't think it's going to work." Fritz was speaking of the offer from Boston. "I don't care how much talent you have, and I admit you have it, but if you don't know what their expectations are, you are going to be in a hell of a lot of trouble in Boston."

"They want me to talk. And I like to talk."

Paul and his family moved to Wellesley, Mass., and he worked at WEEI. Fritz was right; it was a disaster. The Winters returned to Detroit. Paul went into educational broadcasting.

Late one night in April, there were still traces of ice on the road. A huge semi-trailer truck zoomed along fast and heavy, a small car trailing behind it. Suddenly the truck stopped. The little car swerved to the left to miss it and slid onto ice. The driver was unable to straighten out. The car hit a ditch, went end over end, then rolled over horizontally three times. When it came to a stop it was upside down with a telephone pole on top of it. The gas tank exploded. Somehow the driver managed to crawl out of the car. But his back was broken. It was Lee Alan.

Fan letters were delivered to the hospital in sacks. He went home with an orthopedic brace from neck to waist. While he recuperated he did some hard thinking. Part of the cause of the accident had been exhaustion. He had worked at the station late that night, writing a script for a television show he was doing with Sebastian. He was working 10 hours a day, 7 days a week. Too much. He was making a lot of money but there were tax problems. And since he had put his work first and his home second, his family was on the verge of breaking up.

In June, he went to the boss.

"Chuck, I'm going to ask that you take me off 7 to midnight. Everybody else in town, all the contemporary stations are running the night-time disc jockeys 7 to 10 instead of 7 to midnight. So I'm going to ask that you give me 7 to 10, and that you don't make me do any more remotes."

"What about Walled Lake Casino?"

"I will forego that."

"What will Walled Lake Casino do?"

"Well, I—Look, I don't want to cut off the revenue for the station. I'll do the broadcasts at the Casino if you want me to; and maybe they can hire Dave Prince for the hops. I'm having difficulties with my family."

"Okay. Let me think it over."

That night, Lee Alan did the hop at Walled Lake. The next morning, Saturday, Dave Prince cornered Dick Kernen.

"Lee told the Kramers last night he wants more money or he'll have the remotes pulled."

That was the story Cleo Kramer repeated to Jim Christie. The salesman called Fritz at his home.

"Get Bob Koch and join me at lunch at Oakland Hills," said Fritz.

On Monday morning, Fritz went out to Walled Lake with Christie and talked with Cleo Kramer. As soon as he reached his office, he called Bob Baker, and Baker phoned Lee Alan.

"You better get your tail in here. Fritz wants to see you."

Alan got Prince on the phone.

"Dave, they want to see me at the station and the way Baker talked sounds to me like something fishy's up. Could you come to the station in case I need you to come into the meeting?"

"Sure."

Alan hurried to Baker's office. Baker was seething.

"Downstairs!" was all he would say.

Fritz closed the door to his office.

"Lee," he said calmly, "I'm going to fire you today."

"Why?"

Fritz' lips parted in a skeptical half-grin as though Alan had to be putting him on.

"Gosh. We're Number One," Alan continued innocently. "We're carrying 28 to 33 share in every Hooper.

"Well," said Fritz, "the reason is that after our discussion last Friday, you told Walled Lake Casino that unless they double your salary, you'd see to it that not only would you not do their dance hops, but there wouldn't be any remotes; the station just wouldn't be out there. Lee, we can't have that and I'm firing you."

"Would you like to hear my side of the story?"

Fritz leaned back, patiently. Alan spoke earnestly without raising his voice.

"After the dance Friday night, I went to Red and Cleo Kramer and I said, 'I want you to know that I've asked Chuck Fritz to release me from doing 10 to midnight and I told him that I needed more time at home and—.' Now let me stop there for a minute. I'm getting ahead of myself."

Alan lit a cigarette and continued.

"About 2 weeks before my accident, Walled Lake came to me and they said, 'Lee, we're going to do some remodelling here and we want to know if you can take a cut in your fee for a month or two.'

"And I said, 'What kind of a cut?'

"And they said, 'Well, could we pay ya like $250?'

"And I said, 'Fine. Just do it.' I was getting a flat fee of $350 a dance. So I agreed to $100 less. They said, 'What do you want to make it—60 days?'

"And I said, 'Yeah. Sixty days. Fine.'

Alan flicked an ash.

"Now we'll go back to the other point," he continued. "Last Friday night. I told them I'd gone to you and asked you to release me from doing 10 to midnight. And I told them I wouldn't do their dances any more and they could hire Dave Prince to do 'em. And I said—this is after the 60 days—I said, 'There is one other thing that I want you to know and that is: the fact that you have not increased my salary back to where it should be has nothing to do with this'."

Fritz was skeptical.

"I went out there this morning and talked to this gal," he said, "and that's not the way she told it."

"Well that's the way it was," Alan insisted. "Listen. Dave Prince is upstairs. He knows all about it."

Alan jumped up, opened the door, and asked the secretary to get Prince. Prince was not to be found.

Lee Alan was fired that day. He appeared briefly on other stations, opened a radio school, eventually dropped that, went back to school and passed his bar examination.

He was replaced at WXYZ by a strange person from Texas who called himself, appropriately, the "Weird Beard." Dave Prince took over the Walled Lake Casino dances, but they were on the decline. The era of the record hop was passing.

51
Black Friday

Black Friday came in August, 1964. It began when David Klemm returned from having lunch with Ed Hardy and found the message from Nancy Gallagher. Nancy was his secretary—tall, swingy and pleasantly aggressive. Mr. Fritz wanted to see him.

"Dave," Fritz said, "I am firing Bob Baker this afternoon and naming you Director of Operations."

Klemm was speechless.

"I am calling a meeting of the staff," Fritz continued. "And I want you prepared to tell the people what we are going to do. We've discussed it and you know."

Klemm's face had turned scarlet.

"You'll have my support in anything you want to do," said Fritz. "The only thing we'll stay away from, Dave, is Fred Wolf."

It was not easy for Charles D. Fritz to fire people, and Bob Baker had been with the company a long time.

"Bob," he began, "I know it has been tough for you. Those people up there are all your friends. But Hal—"

"I know. Hal has become disenchanted with me because I have not gone along 100% with the formula. I have wanted to cushion it with Lou Gordon, with Dick Osgood, with the newscasts, with Wolf, with the things that I think should be done."

"You have to leave," Fritz announced quietly. "You know that, Bob."

"Of course I do, Chuck. *You* know—I've been working up to this for 6 months."

Baker was stunned nevertheless. The first one he told was Bill Hengstebeck. Hengstebeck told Dick Kernen.

"Dave Klemm is taking his place."

"*Dave Klemm*!!!" Kernen exclaimed.

"That's the rumor."

"Dave *Klemm*?? Anyone else in the entire *world*! But *Dave Klemm*!"

Several others were fired that day with Baker. Within two hours they were all out of the building and the remaining staff, in a state of shock, was summoned to the board room. Fritz announced that Dave Klemm was now Director of Operations—a new title. He had high praise for Klemm and his ideas, a large number of which Klemm aired at the meeting.

"We are going to attempt to rev up, in terms of tightness, in terms of music selected, in terms of promotion, in terms of production, in terms of the overall attitude of everybody on the staff so that we convey a more enthusiastic, a more contemporary attitude. We will try to convey a new WXYZ. We are going to shake away all the cobwebs and put outselves into a more competitive sound."

By Monday morning, David Klemm was installed in Baker's old office. Nobody would have recognized it.

The Baker office had been a joke, as Dick Kernen remembered: "Everybody had put Baker on: why don't you do something about this office, Baker? Looks like a public toilet. And Baker said, 'Go to hell. I don't care about that'. "

Kernen: "Then Klemm comes in. We-e-ell now—stand back! Wall-to-wall carpeting, panelling, soft lights, a new desk, a fake lemon tree—the whole smash!"

Long-legged, black-eyed Nancy Gallagher moved upstairs with him.

Adding to the confusion and an overall sense of insecurity was the installation of new equipment in studios rebuilt by Jack Tierney. Disc jockeys now had some 20 knobs and buttons to get used to—buttons that started and stopped turntables, buttons to tie in with a loud speaker, knobs for this, knobs for that.

The new equipment did away with the "producers." Don Davenport was saved by Fred Wolf; John Dew was given work in the record library; Bill Hengstebeck was let go. The saddest fate of all befell Dick Kernen. He was banished to the traffic department.

Only one person at WXYZ kept in touch with Lee Alan after he was fired, and that was Milton DeLoche. Milt was a black man who held a unique position. He replaced light fixtures, drove VIP's to and from the airport, provided female companions for visitors who desired them, and often took John Pival home and put him to bed. He was powerfully built, like Pival, and the two were quite close.

DeLoche often called on Lee Alan while he was convalescing from the auto accident, and he continued to call occasionally after Lee was fired. During one visit, he said, "Lee, I want you to call Pete Strand. There's going to be an opening on the announcers' staff at Channel 7."

"No, Milt. Pete Strand won't talk with me. I've tried."

"I want you to know from the highest authority," DeLoche insisted. "You should call Pete Strand."

The opening to which DeLoche referred was created by the firing of John Slagle. Johnny was one of the first to discover that when a performer is dead in television, he's dead everywhere. To keep him working, Pival had given him the announcing job, but John's voice deteriorated and they finally had to let him go.

Lee Alan was so embarrassed to return to the station that he entered the building as inconspicuously as possible through a door into master control hidden under the executives' carport. It was night. He went up the backstairs and auditioned. A week later he had a call from Pete Strand.

"We'd like you to come back for another audition."

"Don't give up," Milton DeLoche advised him afterwards on the phone. "Just don't give up."

The station was testing his humility. He was called a third time and asked to read news on camera. More days went by before Pete Strand called him again.

"We have a job for a staff booth announcer," he said. "You would have to open up the station at six o'clock in the morning and you would be through by noon. Do you want it?"

"I'll take it." The super star of radio became a booth announcer in television.

Winter came and a big snow storm kept Rita Bell from her morning show; she was hostess of a morning movie. Lee Alan did the show for her. That night, John Pival, "pretty well in the bag" sought him out.

"I've got a television show we're going to do," he said. "I want to do one at Metropolitan Beach, and I want to do one at Edgewater Park, and I want to do one downtown by the Michigan Consolidated Gas Building, and—I want you to do it."

The clean brown eyes of Lee Alan came alive.

"Will you do it?" Pival asked.

"Yes sir. I'd be glad to do it."

The show was called *The Swingin' Kind*. It was for and about young people. Pival wanted it syndicated through ABC but, although it was successful in Detroit, such was his relationship now with the New York hierarchy that Detroit was as far as it went.

Lee Alan: "I had a lot o' respect for John Pival and it turned out that he was my benefactor at that station."

John Pival was loyal to many people, but he was under continual pressure from the towering, top-heavy corporate superstructure above him. Ted Shaker took a dislike to Bill Hendricks who was now salesmanager of TV. Shaker wanted Pival to replace him with a man of Shaker's own choice, a tall San Franciscan named Donald Keck. Pival refused. Shaker won. Hendricks was out and Keck was in.

Another victim of the New York pressure was Larry McCann. Pival told him he was up against a wall and would have to give him his notice.

"However," Pival promised. "I can work something out."

The solution was not altogether satisfactory, as McCann remembered: "He worked it out on paper and it came close to $10,000 if I'd quit. So I worked out a deal where they would give me half of it in '64 and the other half in '65 for income tax purposes. I had to sign something the size of a telephone book telling me I had no recourse until the end of the world."

McCann left WXYZ in 1964, worked for an advertising company, opened his own agency, went bankrupt, joined WINZ in Miami, moved to Mutual and rose gradually to Mutual headquarters in Washington.

Charles D. Fritz: "The first year I was at the station I had a bit of luck. There was a newspaper strike. But even aside from that we just went through the roof."

There is no denying that the newspaper strike helped. On July 14, the press men and paper handlers of the *Detroit Free Press* and the *Detroit News* walked out and there were no newspapers until the end of November. Advertisers, unable to get into print, rushed to the broadcast media and created a bonanza.

As of the first week of September, Channel 7 alone had telecast more than 400 news "extras" to keep viewers informed. On the radio side, AM greatly expanded its news coverage and Fritz re-activated FM which served with continuous news. Printed "WXYZ Radio News Briefs" were distributed in restaurants and hotels—one-page summaries of ten top stories such as the orbiting of a new weather satellite, the death of Gracie Allen, the travels of Hurricane Cleo, bargaining talks between the UAW and Chrysler, and the activities of President Johnson, Governor Romney and Mayor Cavanagh.

Fred Wolf and Marc Avery finished a game of golf together at the Lochmoor club in Grosse Pointe and were dressing in the locker room. Avery, a short man with a mop of straight hair, was morning disc jockey at WJBK.

"How'd you like my job when I retire?" Wolf asked.

"You? Retire?"

"I'm thinkin' of it. Want my job?"

"Sure."

Avery laughed. But Wolf was serious. He went to work on it. Toward the end of 1964, Marc Avery was given drive-home time in the afternoon on WXYZ, replacing Joel Sebastian who went to WINS, New York.

One of the casualties of Klemm's tightening-up process was *Hollywood Headlines*. The United-Detroit Theaters had sponsored it for 13 years, and still wanted it. Marie Meyer who handled advertising for the theaters was indignant.

"We've been spending around $80,000 a year on WXYZ," she complained. "And other theaters and movie distributors spend a lot of money on WXYZ because of Dick's programs. He's known nationally in the trade and that means business for WXYZ."

Her argument fell upon deaf ears. United-Detroit gained one concession: Osgood would transcribe a one-minute spot—a 60-seconds guide to what was playing in 12 theaters. It was soon cancelled.

With the rise of the disc jockey, the need for announcers in radio diminished. Fred Weiss and Don Zee saved themselves by becoming disc jockeys; Joe Vaughn, Porter Bourne, Gene Avram and Bob Hardt went into news. Don Karle crossed over to TV where booth announcers were still needed. By 1964, there was not one announcer on the radio staff.

A problem arose. The station was still obliged to carry Don McNeil's *Breakfast Club*. This required someone locally to make station breaks and to do commercial announcements between the network segments of the program. It was felt that the disc jockeys were too important for this. Osgood was not doing anything in the morning now; so he was pushed into duty as local announcer on the *Breakfast Club*. *Show World* was still on, and he ended the year with a new half-hour program, once a week, called *Call Board*. It combined "fun tips," show world features such as interviews, reviews and recommendations, and recorded music from current stage, movie and night club attractions.

"Keep everything very short, and use only up-tempo music," Klemm ordered.

"You mean I can't use Mancini's *Moon River* from *Breakfast at Tiffany's* or that haunting theme from *Ship of Fools* by Ernest Gold?"

WXYZ
WXYZ RADIO DETROIT
PERSONALITY PLUS
MARC AVERY
PAT MURPHY
DAVE PRINCE
LEE ALAN
DETROIT SOUND
JOEY REYNOLDS
JIM HAMPTON
DANNY TAYLOR
1270 ON YOUR DIAL
Hear it First on WXYZ Radio

"Nope. And make your movie reviews 30 seconds or less."

"That's absolutely impossible."

"Well, keep 'em short."

One afternoon Lee Alan was standing by for Don Karle who was out freelancing for some extra income. The phone rang. It was David Klemm.

"Hi! Lee! Would you come over to my office, please?"

As soon as he was relieved, Lee Alan visited the radio section of the building for the first time since he had been fired from it. Klemm was warm and friendly.

"How'd you like to come back to radio?"

"What do you mean? Are you serious?"

"The night ratings are way down and the Weird Beard is no longer with us. Have you been listening?"

"No, Dave. I haven't."

"Well, I've had a disc jockey in from England for two weeks—7 to 10—as a stunt—until we could find a permanent replacement. And for 2 weeks, we've been announcing a big surprise for tonight."

His face wore an expression of mischief as he added, "And at this point nobody knows what that surprise is going to be, least of all myself or Chuck."

Klemm chuckled.

"We've found a funny little man named Pat Murphy, Madcap Murphy, and we might put him on but—how'd you like to come back?"

"Tonight?"

Klemm looked at the clock on his desk, smiling.

"In about an hour. Seven o'clock. Chuck wants to see you."

Fritz shook hands with him and his smile was friendly.

"Lee, you've done your penance and we'd like to have you back. Frankly, I'm not too sold on Madcap Murphy. We'd rather make *you* the surprise."

"Well I would be just fooling anybody to hedge," said Alan. "Pride is something I don't live with now. Of course! I'll be glad to be the surprise."

He accepted $100 less then he'd been paid in radio before.

"There is one thing I really have to do first," he said. "I have to go down the hall and talk with Mr. Pival."

"I'll wait here for you," said Klemm.

Lee found John Pival sitting at his desk. It was 6:30. When Alan told him what had happened, Pival extracted a bottle from the lower right-hand drawer of his desk, poured a drink into a glass and held it out to Alan.

"I don't think I'd better," said Alan.

Pival downed the liquor.

"I wondered how long it would take 'em to get smart enough to hire you back," he said. "But I'll tell you something, Lee. You're not getting out of *The Swingin' Kind* and I'm not letting you out of my booth up there until we find a replacement."

Klemm was waiting for him in Fritz' office with a look of impish glee.

"Where do I sign?" said Alan. It was now less than 10 minutes before air time.

"I'd better get upstairs and pick some records."

"They're all picked," said Klemm, grinning. "And I've got Jim Hampton, one of our new jockeys, standing by to run 'em for you because we've installed new equipment since you were here. Jim doesn't know to this minute who he'll be running them for. *Nobody* knows. At this moment, Chuck and you and I are the only ones who know what's going on."

They left Fritz' office and started for the studios.

"You can thank Hal Neal for this," Klemm told him. "He called a couple o' days ago and he said, 'Look, you're running all over the country looking for somebody and you've got somebody right there in the TV booth'."

It was a dramatic re-entry, as Alan remembered: "I was an emotional wreck. I walked into the program area and as I walked in there, still out o' sight, somebody handed the young feller that was spinning the records a cartridge. It had no label on it. And they put it in the machine and Klemm told him to run it, and when he ran it, it was me and my old theme song. Jim Hampton's face was white, couldn't believe it. And I walked into the studio and looked him in the eye and he still couldn't believe it. There were people comin' in from the hallways. The phones were ringing. The girl downstairs was having trouble with the switchboard. People I hadn't seen for 3 months were all coming into the studio and talking. I was getting calls from everybody and it was like somebody hit me over the head and it never happened."

Madcap Murphy followed in Lee Alan's wake from 10 to 1 A.M.

Gene Avram was doing news on weekends. A page of WXYZ slogans was inserted in the book containing commercial copy, and newscasters and jockeys were to use those slogans at intervals stated on the log. One Saturday morning, there was no page of slogans. Avram used slogans by memory from the day before. The phone rang. It was Klemm. Klemm was sharp.

"Why aren't you using the slogans?"

"There aren't any in the book, Dave. I used one from memory."

"That was yesterday's slogan."

Stewing over this, Avram bobbled on his next newscast. It was a bad weekend. He had to write his own stuff, make phone calls, tape actualities, go out in the station wagon and have engineers back at the studio record what he got newsmakers to say into a telephone. Nothing went right.

On Monday, driving to Biff's for his lunch, he heard Ed Hardy do a newscast, substituting for his protege, Bob Hardt, who was away on personal business. Obviously, Ed Hardy had not prepared anything. He was reading newswire copy at sight and ad libbing around it. It was a miserable little broadcast.

When Avram returned to the studio, Hardy called him into his office.

"Klemm says you didn't use the slogans on Saturday."

"Ed, they were not in the book. I used some from memory to be safe."

"They were not the right slogans. I heard you, Gene, and your newscasts just were not good. Can you give me a reason?"

Avram had a sly sense of humor.

"I don't know, Ed. Maybe it's in the horoscope. Who knows? It's just something that happens to all of us. For example, did you think your before noon broadcast today was an example of perfection?"

That stopped Hardy. But Avram went to John Pival.

"John, I want a job in television. I want to escape from a jackass?"

"Which one?" Pival asked.

Lee Alan continued his morning TV announcing until April. Then it was decided to let him out and put Avram in. It was a happy move all around.

David Klemm was replaced in the promotion department by a warm, straight-haired young dynamo from Brooklyn. Martin Greenberg had progressed from an independent New York station to ABC and ABC sent him to Detroit. He was as excited about joining WXYZ as many others had been before him, not because of the reputation gained from the Trendle dramas but because of its more recent fame as the most profitable of the ABC-owned stations and the one which had produced James G. Riddell, John Gilbert and Harold L. Neal, Jr.

Pival and the television crowd impressed Marty as alcoholic savages. "Hard," is how he described them; and he felt he would never make it with such a group. Radio, he thought, was different; but he felt uncomfortable with David Klemm, as did almost everybody else.

On the same Saturday when Greenberg reported to the station, Klemm found Bruce Miller. Miller came from Norfolk where he had been running a small advertising agency of which he was part owner. Before that, he had been talent and program director on station WNOR in Norfolk. Now he wanted to move into a larger market. Through a chain of acquaintances and friends, he heard that WXYZ was making changes. He made up an audition tape giving his background and examples of the commercials he had written, produced and performed in many different voices; and sent the tape to Klemm. Klemm hired him.

Bruce carried one shoulder two or three inches higher than the other because of a football injury. It was difficult to imagine him playing football. He was not large; and he was gentle. His pale skin was finely textured; his brown hair was curly; he wore spectacles and looked collegiate—much younger than his 35 years.

He was hired for very little base pay but was promised extra income as he worked with the salesmen, Bob Baldrica, Bob Koch, Jim Christie and a new one who bore some resemblance to Rock Hudson, Bartley Walsh. Miller's job would be to create, write and produce commercials. He found the prospect exciting. He felt ready to conquer Detroit. He figured he could double his salary within a few weeks. But although he had the ball and was ready to run, he was shoved off side by Fritz and Klemm whose zest for station promotion outweighed their desire for commercials. They were not interested in making money at the moment; they wanted his help in the battle with WKNR for an audience.

"Tighten up. Close the gaps. Make up some promotional spots. Create some clever introductions for the disc jockeys and put them on tape."

They kept him under pressure.

"Where are those promos? Have you got an idea yet for a campaign to promote

Murphy? Do it now."

He had never before been under such pressure for non-commercial in-put. And the meetings! Interminable. Non-productive. Everything had to be talked about, and the talk was foreign to him, "sort of Madison Avenue." He felt he had moved into a Never-Never Land.

Ten years before, Bill Morgan had said he thought talent lived in a dream world. Now Miller felt the same way about executives.

An area next to the record library was glassed off to create an office for him, and a booth the size of a large closet was equipped to serve as his "studio."

In December, 1964, Harold Christian retired. He had a summer place in northern Michigan and a winter home in Florida, and he seemed happy at his retirement party.

"In my 33 years with WXYZ," he reminisced, "we merchandised over 5,000 products in the grocery, drug and department stores. We demonstrated food products in over 10,000 stores. This type of service has never been duplicated on any other station throughout the country.

"John Pival and I tried to convince New York that they should continue our merchandising set-up, but to no avail. They continually brought up the fact that I was unique in this field and they couldn't expect to find a man with my experience and following with the food industry, which is a lot of bull in my book. They just were not interested in spending the money to train somebody for 6 months."

An attempt was made to persuade Bob Baldrica to continue the department, but Baldrica agreed with everyone else that the department could not succeed without Harold Christian, and he turned it down.

James G. Riddell agreed: "There was no way of finding a replacement. For years and years we had tried to find somebody that had his connections and his knowledge of the food business and his friendships and associations with brokers and supermarket operators. There just wasn't anybody. When he was through, the department had to be through."

Although the station had a profitable year, 1964 was a year of turmoil and defeat for both sides of the Wyxie fence. WKNR's victory hit radio hard; Ted Shaker's pressure on John Pival was unrelenting; and David Klemm was discovering that his worst enemy—was David Klemm.

52

"Blessed are the Dead"

Ash Wednesday, 1965, was raw, damp. Still, people lined the street outside the Lutheran church on Seven Mile just east of Woodward, collars up, hands in gloves or pockets, staring at the comings and goings. The sanctuary was filled to overflowing. Don Karle sang a solo. Sitting down front were the pallbearers: Fred Flowerday, Bill Kennedy, Dick Osgood, John Pival, Jim Sherman and Ernie Winstanley. There were 41 honorary pallbearers including Sonny Kresge, Raymond Meurer, Dr.

Molner, George W. Trendle, Nate Shapero of the drug stores, Paul Sutton, Congressman William S. Broomfield, H. Allen Campbell, Major Robert Borg, Jim Fletcher, Paul Hughes and J. Edgar Hoover.

"Blessed are the dead which die in the Lord, they rest from their labors, and their works do follow them." Rev. 14, 13. The words appeared on the 3rd page of the funeral program and seemed appropriate. The man had "died in the Lord" on several counts, not the least of which was his victory over alcoholism. He died sober.

He had been playing cards with his wife and two friends when at 12:10 A.M., Monday, March 1st, his heart stopped beating. He left behind a widow, 4 children, race horses he had been preparing for the tracks, and a partly developed subdivision. Also a thousand memories. The 6-foot-3 carcass in the casket was all that remained on earth of Brace Beemer.

Eleven years had passed since he had played the Lone Ranger, but to his death, he maintained the image of the character he had portrayed for 13 years. He helped children, befriended stray animals, visited hospitals, and answered fan mail. Only the week before, he had granted an interview which was published the day after the funeral.

John Pival returned from a Florida vacation with a big catch of fish. The cafeteria closed in the afternoon, but Pival arranged for someone to fry the fish and someone to mix the drinks, and the party was on. Among those drinking was Pival's arch enemy from Manhattan, Ted Shaker.

Leon McNew joined the party after his 6 o'clock newscast. Shaker put his arm around McNew and said, "Lee, as far as I'm concerned you are the best newscaster in the city of Detroit."

At 3 o'clock the very next afternoon, Pival called McNew to his office and pushed a memo across his desk for Lee to read. It was signed by Ted Shaker.

"Fire Lee McNew," it said. "We want younger men on the news. Put on Bill Bonds."

"This will pass," Pival predicted. "Ride it out with me."

"Ride it out where?" McNew grinned. "He wants me off."

"Go into the announce booth and pull a regular shift. I will pay you the same salary."

"Whatever you say, John."

"I know what Shaker is trying to do and I'll get that son-of-a-bitch before he gets me. I've got enough on him to—hell! You hang in here with me. This will pass."

Bessie Ashton: "We all felt, we all knew, we—New York was, was trying to get in. And that's one thing that people have to give Pival credit for: he kept out New York as long as he was there."

Bess herself knew Pival's protection. When Franz Werner retired, John gave Bess the remains of his job. It was a hold-over from the days when radio had a vast library of musical scores and arrangements, and required one to keep it in order and report the use of numbers for the American Society of Composers, Authors and Publishers for collection of royalties. The job had shrunk and been lost sight of.

Bess Ashton: "John told me, he says, 'I have a sinecure here,' he said. And uh—so—so he was trying to figure, you know, somebody to give it to when Franz retired, besides which—uh—uh—I had been getting—I didn't care too much about Keck and very few people cared much about him."

So Bess fell heir to the sinecure.

"That job paid about $30 more a week than I was getting in traffic. I kid you not!" she says.

The office she moved into was a windowless closet off the corridor of the radio area. She sat there reading and doing crossword puzzles, and began to wither a bit, the red of her hair fading, her face losing its color. But she maintained her pride, eating with others in the cafeteria and doing a brave, swingy kind of walk down the long corridor to her hide-away, sometimes singing a bit to herself to keep her spirits up.

But with Pival, matters were not improving.

Nor were they with David Klemm. He turned from a charming, enthusiastic promotion man into an irascible martinet. The air was filled with hostilities. Such epithets as "unmitigated ass" and "little Hitler" flew in his direction from behind.

"Because I fired a number of people, because I changed the operation of long standing," Klemm said defensively, "a lot o' people looked at me with a calloused eye. Keep in mind that XYZ had been the big money maker, and suddenly it is no longer the big money maker in the ABC group. The good old days were gone."

After losing its audience to WKNR, WXYZ was now losing business. Klemm was under constant pressure. He fought the engineers, the jockeys, the news department. And he fought Fred Wolf.

Wolf was outraged: "I am fighting a losing battle. Not only are our ratings going down but our commercial load is going down, too,—and I mean—going—*down!*"

Even so, Don McNeil's *Breakfast Club* was taped when it came over the network at 9 o'clock and played back to Detroit at 10, so that Wolf's time could be extended one hour. He was now on from 6 to 10 instead of from 6 to 9—and he hated every minute of it.

"Fred, you talked longer than 20 seconds between records this morning," said Klemm.

"I had something to say."

"It's a rule now, Fred—no longer than 20 seconds between records."

"Who's rule?"

"My rule, Fred."

"Am I supposed to cut off in the middle of a sentence?"

Klemm was turning fiery red.

"If you can't finish the sentence before the 20 seconds are up, don't start it."

"How do I know how long it's goin' to take me to say something?"

"Either you follow the rules—or else."

"Else-*what*?"

"Or *else*."

Fred Wolf steamed down to Fritz.

"Gotta talk to ya."

"Well, I'm having a meeting here in just a few—"

"Thirty seconds. Chuck this is gonna take 30 seconds. Thirty! Not 20."

"Come in." But Fritz did not close the door.

"You either get that clown out of that office upstairs or I'm through—*today*!"

"Oh now, Fred. He's only trying to do a job. Try to get along with him."

"I have, Chuck, for 3 months, and I'm through. I can't go him—*No* mo'."

"What happened?"

"He gave me this bit: either you do this—or *else*!. So either he goes or I go, period. Now, what do you want?"

"Fred," Fritz said finally, "if you're really serious about this, you've got to give me time to find a replacement. Stick it out for another month."

"One month," Wolf agreed. "But you tell that clown upstairs to lay off."

Dick Femmel and John Pival drove downtown to attend a dinner for the Detroit Symphony. They braced themselves for the civic event in a bar.

"Bill Fyfe wants another mobile unit for the news room," Pival announced.

"He should have it," said Femmel.

"New York won't give it to me."

Femmel sat realizing that this man was losing his autonomy. People were being hired over his head; his budget was no longer his own; he could not even produce shows. Yet, as his position weakened, his appearance improved. His suits were tailored; his coats were cashmere. "A pig in mink" one old friend called him with affection. Pival seemed to know he was losing. That night, in the bar, he said to Femmel, "Dick, if you ever come across anything for this old man to do—like manage an educational station—something like that—I want to get out."

Dick Femmel himself left WXYZ in 1965 to join George Pierrot whose *World Adventure* series had jumped channels from 7 to 4, and was still going strong as a travel-film-lecture series at the Detroit Institute of Arts. The work of preparing editorials for Pival and Fritz went to Joe Vaughn.

On the 30th of June, Fred Wolf walked into the office of Charles D. Fritz and said, "Chuck, I am all through July 30."

Fritz leaned back in his chair and tried to smile. Wolf was not steamed up this time; he was quiet and firm.

"I want out for two reasons. One, this music's killin' me! I hate to go to work. It's like lockin' myself in a lion's den for four hours. I hate goin' in and I can't wait to get out o' there. *Plus*, I'm just tired o' gettin' up so damn early in the mornin'."

"Has Dave Klemm been after you again?"

"Chuck, I have in my basement at least 50 tapes that Klemm has done o' my show which he thinks I don't know was bein' made. Every engineer in the joint warns me. They say, 'Wolf, I have to tape your next segment for Klemm, so be ready'."

But Wolf waved Klemm out of the argument with one hand.

"Reason Number Two," he said. "Twelve years ago, I set myself up with my life savings into my bowling business. It was a deferred payment type thing. Well, I've bought out two of my partners and I'm major stockholder. I can't defer anything: I have to accept every damn cent that comes in now—and I'm payin' Uncle Sam 75%. Chuck, I want out."

"You're serious."

"Y—up. I've talked it over with Em. She's been waitin' 5 years for me to make up my own mind."

"If you would just stick it out for another 3 months. I'm trying to get a line on a team from Cleveland."

"A *team*! You gotta be outa your mind." But Fred let that pass. He repeated, "I'm tired o' gettin' up at 4:30 in the mornin'. I'm retiring, Friday, July 30."

Jerry Liebman was a tough little kid in a tough little town—Kittanning, Pennsylvania—a brick center surrounded by coal mines, gas fields, and clay. Jerry was head of his gang when he was eight.

Toward the end of July in 1934, there was a family picnic. Jerry was into everything until he dropped. They thought he was just tired; then they thought it was the summer grippe; but after 6 weeks, when his left side was atrophied and his knees came up and he couldn't straighten his legs, they knew it must be something else. It was polio.

"He will never walk again," said Dr. Crowe.

Mrs. Liebman refused to believe it. She massaged. She scolded. She made him try. And when she summoned the doctor and Jerry walked from the couch to a chair, Dr. Crowe had tears running down his cheeks. It was a miracle—a mother miracle. And the power of a kid's will.

Jerry went back to the gang. He didn't get any sympathy and he didn't want any; but he got justice. They let him play basketball and baseball. He even ran in the foot races. They'd give him a handicap of 20 paces. Then he got so good they only gave him 15.

"They kept backing me up, and suddenly—hey!" said Jerry, "the world isn't all bad."

He went to Allegheny College in Meadville, Pennsylvania, earned 30 hours of credits for law; then, in his senior year, switched over to get a Bachelor degree in speech and dramatics. A friend offered him a job announcing on a little radio station in Sharon, Pennsylvania, and he took it. He found the girl in Sharon whom he married. One of his pals moved to WTAM in Cleveland and got Jerry a staff announcer job there in 1954.

In 1956, WTAM became involved in a situation unique in the history of radio. NBC which owned WTAM in Cleveland, had no outlet in Philadelphia and wanted one, offered to buy KYW there from Westinghouse. Westinghouse didn't want to sell, liked Philadelphia. NBC got tough, threatened to build a new station in the Friendly City and grab off its own network shows. NBC offered an alternative: it would give

Westinghouse WTAM plus cash in exchange for KYW. Westinghouse agreed to the swap but complained to the Federal Communcations Commission.

Moving to Cleveland, Westinghouse changed the call letters of WTAM to KYW, dumped NBC network service, and went independent. It decided to go to records and invited the eight announcers inherited in the swap to make up auditions. With one exception, the staid, conservative NBC announcers declined to debase themselves. The exception was Jerry Liebman.

"What have you got to lose?" his wife reasoned.

So Jerry put some records together, knowing nothing whatever about music, taped a show, turned it in, and started looking for a job.

The new Westinghouse management was so impressed with the guts of any young man who could submit such a terrible show it promptly put him on the air an hour and a half a day, Monday through Friday, and gave him special shows to do on Saturday (*Pops of the Past*) and Sunday (*Breakfast in Bed*). And it changed his name. It got "Specs" from the fact that he wore glasses, and "Howard" from the phone book. Jerry Liebman was now Specs Howard.

While waiting for action from the FCC, KYW's ratings plummeted down the cellar steps and hid in a dark corner. Management decided it couldn't wait; something had to be done. William Kaland, national program director for Westinghouse, had created a very successful two-man team in New York known as Klavan and Finch. So it was decided to put together another two-man team in Cleveland. Specs Howard, already established, would be the straight man and the search began for a comic.

Two men were brought in, each matched against Specs like swatches of cloth to see if one would go with the other. The likelier of the two was a man from San Diego. He had wit, could fire off one-line zingers like unravelled firecrackers, and was a brilliant story teller. Like Specs, he had been a radio single for 8 years. Like Specs he was married and had 4 children. He had sung with the Robert Shaw Chorale but you'd never guess it from the frantic rock and roll style he had developed. They called him Happy Hare. The name was Harry Martin.

The rock and roll format was too fast, too tight, for ad libs; so they looked around for a writer. Luckily they found Ray Koeppen, a prolific writer, an intellectual and a man who would rather rap with a cop than smirk over a martini at a hostess.

Martin & Howard hit the air in July, 1962. It took 6 months to settle in, to "get the feel" of each other. Ken Draper, program director, oversaw the development of the team, personally, advised, criticized, helped them build. The show began to climb, and as it did, it dragged the station out of the coal bin and up the cellar steps into the light of day.

Then, 9 years after Westinghouse filed its complaint with the FCC, the Commission decided Westinghouse was right. NBC was obliged to relinquish Philadelphia and return to Cleveland. After much stalling around, on June 18, 1965, NBC signed off in Philadelphia and Westinghouse signed off in Cleveland, and the two switched places. The next morning, Westinghouse was in Philadelphia, taking the KYW call letters with it and NBC resumed management of the Cleveland station whose letters it changed to the closest it could get to the KYW sound, WKYC.

The conservative senior network now found itself in repossession of a station that was rockin' and rollin' like a lobster pot in a hurricane. But the station was Number One in Cleveland and the men who had made it Number One, radio-wise, were Specs Howard, Harry Martin and Ray Koeppen.

However, the new NBC management began to meddle. Little suggestions were made to take the "sharp edges" off the Martin & Howard show, to slow it down a bit, to cool it. This threatened to undo what had painstakingly been accomplished over a 3-year period; so when Specs Howard received a telephone call one morning, he limped off to a motel on the edge of Cleveland to meet the man who had phoned. It was the tall, aristocratic Charles D. Fritz, and with him was the short, ever-smiling Bob Baldrica.

"We've been listening to your show," said Fritz.

"Word got to us that we were being air checked," Howard admitted. "WJW has been taping us off the air and sending our stuff to New York."

"Yeah. Hal Neal has been listening to you, too," said Fritz. "I don't mind tellin' you that we want you in Detroit. We're kind o' draggin' in the morning. I like the sound of your show and I like the way you tie in with the community. We could use some o' that."

"Is the *Congo Kurt* thing your own?" Baldrica asked.

"Oh hey! Yes. Everything we do is ours. *Congo Kurt* is our non-hero. We made an album of some of those skits and sold 2,000 down here."

"Of *Congo Kurt*?"

"Right."

"How many commercials do you run in your 4 hours?" Fritz asked.

"I don't know how many we had this morning."

"I counted 65," said Baldrica.

"Would you consider moving to Detroit?"

"It doesn't do any harm to talk about it," Specs observed with a bright little grin.

It got to be a very wet session, but a firm offer came out of it: $30,000 for Howard, $30,000 for Martin and $15,000 for Koeppen.

The three drove up to Detroit with their wives to look the place over. They were met by a mild litle man named Bruce Still who had been hired by Dave Klemm to work as a production man with Bruce Miller. Still drove the visitors around Detroit and gave them a tour of Broadcast House. They were a little disappointed in the physical lay-out of the radio area but liked the sound of the station. They decided to move to Detroit.

But there would be a delay. Their attorney pointed out that although their contract had been signed with Westinghouse, NBC had agreed to honor all Westinghouse contracts, therefore they were under contract to NBC.

"You're the greatest asset NBC has here in radio," he argued, "and I believe they could prove that by ducking out of your contract you hurt them. They could sue you and in my opinion they could win the case."

Unaware of this, Fritz was trying to reason with Fred Wolf.

"Fred, I'm getting these two guys from Cleveland. Just stick with it until we work out a deal."

"No way," said Wolf. "I want *out*!"

"You have to give me time to get these guys in here."

"Chuck, it's dragged on over a year. You got a man right here on your staff that could fall into my seat tomorrow morning and you'll never know I left."

"On staff? Who's that?"

"Marc Avery. Christ, he was a morning man for 15 years before you hired him and put him on in the afternoon."

"These two guys from Cleveland—"

"Chuck! You gotta be outa your mind. You're goin' pay those guys a fortune; you can have Avery for peanuts."

"Well—give me one day."

Charles Fritz called Cleveland—and learned that Martin & Howard had just signed a new 5-year contract with NBC!

"That was the very last day I'll give ya," said Wolf.

"All right. Anything. You tell Avery."

Fred Wolf's retirement was given a big publicity build-up, not to mention his replacement by Avery. Wolf's final broadcast, ending 29 years in radio, was an event. Lee Alan read some poetry, Marc Avery gave him a pair of golf shoes, and as Wolf states it modestly, "that was that."

There was another change. Don Zee left the station and was replaced by a young man from Texas, Steve Lundy.

One Friday morning, the veteran engineer Al Schaub, was slow raising the gain on a newscast when it first went on the air. A few words later he heard Klemm's heels clacking down the passageway from his office to the control room. Klemm's face was flushed.

"Goddam it, Al, you old-line guys are nothing but trouble. All you have to do is sit here and turn those knobs. Couldn't you hear that news when it came on? That was the whole thing: you *didn't* hear it. *Nobody* could hear it. Nobody could hear it because you couldn't lift your hand to turn a knob; you just sat there! Now get with it! Watch those knobs!"

Klemm spent the rest of the day with a migraine headache. He brooded over his flare-up with Schaub all weekend.

"I had never shouted at anyone in my life. I literally found myself shouting at people. I was in a job that I didn't want. I loved the money and I loved the challenge but I wasn't happy. I felt tremendous pressure."

He was living, eating and breathing WXYZ 18 and 20 hours a day. It had to end. He went to Fritz on Monday.

"Chuck, I have to leave."

"Okay. So you want a vacation."

"No, Chuck. I'm not myself. I don't have any life of my own. I'm going to Europe for 3 or 4 months."

Fred Wolf

"Okay. A leave of absence."

"Well—that's very kind but—I'm resigning."

Later in the day, Klemm had a call from New York. It was Hal Neal.

"Listen, Dave, we know what the pressure has been. It's been a grind. Take a leave of absence. Go to Europe. Then, when you're ready—"

"No, Hal, I have a very definite feeling that if I took a leave of absence and came back, I would be coming back to the same situation. I feel that to sever the relationship would be to everybody's best interest."

Klemm spent 3 months in Europe, returned, worked for a time in Florida, then joined the John Blair company in New York, married, bought an historic old house in the country, and found happiness. He also recovered the charm he had had before Black Friday.

He was replaced at WXYZ by the gentle, mild, non-combative little man he had hired to "produce" and who, so far, had producing nothing—Bruce Still.

53
House Divided

The "magazine of New York advertising," *Madison Avenue,* quoted Charles D. Fritz in 1965:

"We may seem inconsistent. We have no formulas. Our station reflects a different personality as we go through the broadcast day. From the *Breakfast Club* to the hard-hitting news, to contemporary music—the vitality of our program concept is proved by the size and responsiveness of our audience. And that's why we can say that 'WXYZ Radio is the influential sound of Detroit'."

Brave words. The kind of dream world talk that confused Bruce Miller at staff meetings. The fact was Charles D. Fritz was facing a major problem. His dilemma was indicated in the comment of an agency man upon returning to his office after a tour of Broadcast House.

"Chuck, I can't understand how you people afford all of this, playing the junk that you do."

"Apparently no one cares for the junk but the people," Fritz fended, smiling.

He had to use constant pressure to convince advertisers that adults would listen to a station that played such frantic music. The men up front in his battle for listeners now were Marc Avery, David Prince, Lee Alan, Pat Murphy and Steve Lundy. But the arguments he used behind the lines to reassure advertisers were Lou Gordon, Bruce Miller, Dave Diles, Dick Osgood—and the newsroom.

A new, separate newsroom had been built for WXYZ-Radio in space previously occupied by part of the control room and a studio. Radio now had its own newswire services and bragged of its own 6-man operation around the clock, and of its Sigma Delta Chi Award for editorials.

One legitimate excuse for the word "influential" in the slogan "influential sound of Detroit" was Lou Gordon. With his commentaries, he roused the Detroit Common Council to an investigation of a shady operation of Detroit's Civic

Center.The president of the Pressman's Local went to him for suggestions as to how the newspaper strike might be ended, and Gordon arranged for him to meet with Walter Reuther of the United Auto Workers for help which paved the way for a settlement. Generally credited with influencing many of the votes which had put Jerome P. Cavanagh into the Mayor's office, Lou had now begun to sour on Mr. Cavanagh and there would be repercussions from the Mayor. Charles Fritz was spending too much time smoothing feathers ruffled by the Gordon broadcasts, but there was no denying Lou was an influence.

As for Dick Osgood, the station bragged that in one year he had hosted 30,000 listeners at 10 exclusive WXYZ movie preview parties. In 1965, he went on one junket to Hollywood and on another to Cairo. In July, sponsored by the University of Detroit, he hosted a Green Room series on "Inside Show Business." This was the idea of an imaginative public relations man named Bill Rabe. On the 3rd of September, in a single edition of a paper, his reviews were quoted in the ads of four different theaters. And in November, he was master of ceremonies at the annual civic event called "Detroit Aglow" held at Grand Circus Park in downtown Detroit. Combined choruses from all the high schools sang as the Mayor pulled a switch which presumably lighted all the Christmas decorations of the city. This was telecast.

Even so, in January, 1966, the station cancelled *Show World*. It had been on the air 30 years. Osgood was now left with *Marathon Fun Tips, Call Board* Sunday nights, and the humiliating chore of doing station breaks on the *Breakfast Club*. He began to check up on Social Security and what he could count on from the pension fund of The American Federation of TV and Radio Artists. He was 65.

"How's my friend John Pival?"

"Si," said Fritz, "you've asked me that many times. Do you really want to know?"

"I never ask a question in a meaningless way," said Siegel.

"All right," said Fritz. "I'll tell ya. He's drunk a lot of the time and he's abusive. He gets kicked out of our club about once a month, and he beat down the door of another place. He keeps people hanging around the station until 3 or 4 o'clock in the morning. One of my salesmen, Johnny Lyons, is in hock to the station for $2,000, an' I swear half of it he lost to John Pival, rollin' craps. So here's radio givin' Lyons a damn loan to pay his bills an' he's out gamblin' with Pival. John has no responsibility to people this way. I can't understand, Si, why ABC just doesn't fire him. 'You're through.' Fire him."

"Well, we can't forget that John did a good job out there. Besides, he has a good friend on the Board. But I will take care of the situation."

The "good friend" on the Board was James G. Riddell. He understood Pival, and he understood Shaker. In his estimation, Shaker was young, sharp—a business-executive type—and a little devious. Pival was none of this; Pival was blunt and open, even when doing wrong. Riddell knew how to handle Pival; Shaker couldn't be bothered. Shaker just wanted him out. But Riddell went to bat for his old friend, with

Simon B. Siegel and with Leonard Goldenson.

Sometimes Virginia Pival, who managed to get into Broadcast House now occasionally, would find her husband with his head in his arms on his desk, or asleep on the office couch. At home, he sought release in his basement workshop.

Virginia: "Sometimes I wonder if this pounding down there that he used to do wasn't just giving vent to his anger."

To add to the vague sense of insecurity that already permeated WXYZ, in December, 1965, the boards of directors of American Broadcasting Company, Inc. and the International Telephone and Telegraph Corporation agreed to a merger.

One who had become too busy to dwell on insecurity was smooth-skinned Bruce Miller. A series he did for Knorr Soups won an award from the Radio Advertising Bureau. He conceived a campaign which turned a cheap jewelry chain into plush stores with improved merchandise. For a local group of small department stores he created two characters, Rodney and Uncle Morris, both of whom he played himself. He wrote and taped 3 episodes for audition, and then sent them via salesman to the agency. The agency thought it was bad.

"Just play it for the client," Miller begged.

"I wouldn't insult him," said the agent. "*You* play it for the client."

Miller and the salesman went directly to the client who fell in love with it and ran the series for 3 years. It won another award.

However, the radio programming of WXYZ was virtually without leadership, and at no time of the day or night in 1965, was WXYZ-Radio Number One in the ratings. From early morning until 3 P.M. WJR with its conservative "old-fashioned" programming was the top station. J.P. McCarthy of JR had the town by the ears from 6 A.M. to 9 A.M. From 3 P.M. to midnight, Pulse rated WKNR the leader. WXYZ sank from second place in the morning to 4th, and stayed there except for 3 hours in the afternoon when it managed to lift itself to 3rd position.

Bruce Still was no champion for the battle. He was "a very nice little man way over his head." He emerged from the think tank of staff meetings nearly drowned in vague theories. Glad as everyone had been to see Dave Klemm leave, his departure had created a vacuum. One man saw this and began to move in.

"If you'll just take your stripes off," Lee Alan said, grinning eagerly, "I come to you with some ideas."

Charles D. Fritz had some amused reservations as to the private and personal character of Lee Alan, but he respected the young man's verve and was ready to listen.

"I saw some of the ratings," said Alan, "and it looks to me like the whole station could be in trouble."

"What are your ideas?"

"I've got a lot of 'em. A helicopter over the city during drive-time in the morning, reporting where there are traffic tie-ups so a guy driving to work could avoid 'em. And they're doing something in Chicago we should buy."

"In Chicago?"

"Joel Sebastian's back there and I was talking to him. They've got a thing over there called *Chicken Man*—2-minute skits. Joel says it's very funny and very successful, and the guys who do it—staff guys—are hungry and would tape 'em and sell'em. We could spot 'em in the deejay shows. The station could stand a laugh."

"Look into it," said Fritz. "What else?"

"The jingles we're using. We've had the same musical jingles on the station for a couple o' years and they're old and they're scratchy. We should have new ones. And there's still a lot o' deadwood on the station that I'd knock out."

His brown eyes were commanding, his face shone with eagerness.

"Chuck, I know we can make this the Number One station."

Fritz was impressed, and after a few meetings, he announced that Lee Alan was Program Director. Bruce Still kept the title of Operations Director, but he was relegated mainly to the area of public service.

Lee Alan wrote new jingles, and recorded them with Anita Kerr and her singers in Chicago. He arranged for the use of *Chicken Man* in Detroit. He hired a man with a helicopter, Barney Stutesman. Lee now did the *Whirlybird Traffic Reports* from 6 to 9 A.M., worked a full day as Program Director, and went on the air with his record show at night. Something had to give. He decided to hire a replacement for his air show and looked for someone who could challenge the leadership of WKNR.

One day his face was a living mask of triumph when he said to Dick Kernen, "I am hiring somebody to take over my night show and when Keener finds out who I have hired, when they *hear*—"

His eyes and grin indicated complete annihilation of WKNR.

"There's only one guy in the country that could send Keener into total fits," said Kernen at once. "It has to be Joey Reynolds."

"Right," said Lee Alan.

"Far out," Kernen agreed.

Joey Reynolds had been Number One in Buffalo. He had moved to a small station in Cleveland and in a short time boosted it to top position for his time period. Trade papers had more than once named him Number One Disc Jockey in national polls. Lee Alan had gone to Cleveland, sat in a motel room and listened to him. He returned to WXYZ with tapes of his show, and Fritz was impressed.

"Go back and talk to him," he said. "And while you're there, listen to Martin & Howard."

An old friend of Fritz from Blair days was manager of the Buffalo station Reynolds had left. Fritz called him.

"The guy's got fantastic ratings," the man told him. "But you have to watch him."

"He's wild," Alan admitted. "He owes all kinds of money. But he's got a manager now and he assures me that we'll have no problems."

"Okay," said Fritz. "We'll use him."

They had to find a place for him to live, temporarily. Fritz called Fred Wolf who was a member of the Detroit Athletic Club. They put him up at the D.A.C. Within 2 weeks, Wolf was on the phone.

"What have you guys done to me? Get him outta here. I'm in disgrace."

Reynolds was moved into rooms over the cafeteria at Broadcast House. He turned them into a marijuana pad and a love-in. Also, a wreck with broken doors, burned upholstery and dried vomit.

"It was one bizarre thing after another," Dick Kernen remembered. "The thing just got progressively worse."

During the Reynolds tenure, Dick Osgood was moved from television news, where he had been left behind, to a desk in a medium-sized office shared with four disc jockeys.

Osgood could get along with most of these people: "These jocks were a whole new breed. Marc Avery had balance and maturity. Danny Taylor and Steve Lundy were pleasant enough. Pat Murphy was absolutely without class, but affable. I liked Prince. But this Joey Reynolds was a freak. He would smash in, make senseless long-distance calls and talk loud by the half-hour—about nothing. I was appalled by his ego, his gall."

Reynolds soon began to be late to meetings. Then Kernen complained that he failed to fill out and initial the Logs. Then Fritz learned of his bacchanalian activities and became disenchanted. Finally, he missed one meeting altogether, went to a recording studio instead to engage in some activity which was strictly against the ABC policy he had been told about. Lee Alan fired him.

One Sunday, Charles Fritz was called at his home by a Mr. O'Rourke who said he was from New York and wanted to speak with him in private away from the station. They met in the back bar at Topinka's Country House.

"I'm here to investigate John Pival," said O'Rourke.

The detective trailed Pival for a week, and talked with a number of people. Pival knew he was being followed but he was not subtle and continued to live as he pleased. Mr. O'Rourke departed; and nothing happened.

Late in 1965, the corporate gods of ABC met with managers of radio stations on the network as was the annual custom. The meeting was held in Puerto Rico and combined a business-like exchange of ideas and information with the pleasures of beach, tennis, golf and card tables.

Si Siegel sat alone at the bar before dinner, enjoying a highball. Charles D. Fritz, freshly showered following tennis, joined him. After the usual pleasantries, a silence fell between them, broken finally by Fritz.

"Si," he said quietly. "I'm disappointed in you."

Siegel's face was without expression as it turned in Chuck's direction. He said nothing.

"You told me you were going to take care of the situation in Detroit and get John Pival out o' there. And—nothing has been done."

Siegel turned back to his drink.

"When Jimmy Riddell was in Detroit," Fritz continued, "we had such a wonderful reputation. Everywhere you went, people who met Jimmy had this great—feeling about him. WXYZ had just a marvelous reputation, and it still does, but Pival is destroying all this. When you see a guy—well—I don't want to go into it."

Fritz knew Siegel was hurt. He still said nothing.

"This guy O'Rourke was in," Fritz concluded. "He made his investigation and I know the kind o' stuff the company has knowledge of. And it just seems like nobody is doin' anything."

Siegel took another sip of his drink, held the glass in his hand a moment, looking at it; then put it down on the bar.

"I'll take care of it," he said. And walked away.

Jimmy Riddell was walking through the lobby on his way to the golf course the next morning when somebody took him by the arm. It was Siegel.

"Jimmy, we've—we've finally got to get rid of John."

Riddell grinned and started to object, but he was stopped.

"Jimmy, we've finally got to do it."

"Well, Si, you know he's been with us a long time and he's done a hell of a job. Don't forget, he and I put that station on the air together and—really he had more to do with the programming at the beginning, and—uh—despite the fact that he and Ted are not compatible, he's entitled to some kind of a break."

"Well, what do you think?"

"Well, let me try to work out a formula of some kind."

"You want to call him? Or do you want me to call him? You can work out the deal."

"I'll call him," said Riddell.

"You tell Pival he's got 2 weeks."

Recollections of how, when or whether John Pival was confronted with O'Rourke's little black book, and by whom, are conflicting. One says Pival was faced by Shaker himself. What, if anything, John Pival knew that might be damaging to Shaker also is unknown. Virginia Pival discovered some taped telephone conversations but wouldn't discuss them.

Riddell: "I don't know if he had anything on Shaker. There was no discussion, absolutely none at the time, of John having Ted under the hammer. Nothing. Nothing."

On January 26, 1966, the employees of WXYZ, Inc. received the following communication:

TO ALL MY FRIENDS AT WXYZ, INC.

> It is with regret that I announce my resignation from the Company. It will take effect February 11th.
>
> Next month marks the end of 20 years with the station. This anniversary seems to be a proper time for me to go to other endeavors. There are many things I have wanted to do other than management. I will now have the opportunity to do them.
>
> I'd like each and every one of you to know it's been a genuine pleasure to work with you. We have had slim days and we have had some wonderfully successful days together. I give you my sincere gratitude for the teamwork

you have applied to help make this the proud station it is today.

The company has not yet selected a replacement, but I am certain a selection will be made soon.

Again, best wishes—please keep up the good work—many thanks—and goodbye.

Sincerely,

John F. Pival
President

A publicity release announced that Pival would form Pival Productions and produce travel programs for national syndication. It also included a comment by Theodore Shaker, President of the Owned Television Stations Division of ABC, which said, in part, "it is with regret" etc., and "we wish him well in his new venture," etc. and "ABC looks forward to programming the Pival Productions."

The day after employees received Pival's note of resignation, another communication was received from him:

> I am very pleased to announce that John E. Campbell, currently General Sales Manager of WBKB-TV, has been appointed General Manger of WXYZ-TV, effective February 11. John has been with ABC for 11 years...was a fighter pilot in the Marine Corps during World War II.... I am sure that under the guidance, WXYZ—TV will continue to grow and prosper.

James Riddell worked out a deal whereby Pival would receive almost full salary for 5 years. Pival also owned 2,000 shares of ABC stock worth at the time close to $200,000. More important, the family situation was improving. John reduced alcoholic intake to normal amounts and started building a beautiful home outside Naples, Florida. Together he and Virginia produced a film documentary whose subject was lumbering. It has never been telecast.

Thin-faced Pete Strand with the piercing black eyes and the little moustache had been director and program director under Pival from the beginning.

"Your program is to go off the air," he said.

Edythe Fern Melrose was momentarily speechless.

"But we want you to continue making color fashion specials for us," he told her. "Every 6 weeks—a one-hour special."

"The new manager?"

Strand's smile was fixed. He was non committal. Edythe attempted a recovery.

"Doesn't he know I have all the wives of all the men who buy time on WXYZ?" she asked. "I have Ernie Jones' wife. I have ever so many wives from General Motors. I have the wives of Campbell-Ewald. I have the head man of Chevrolet. His wife is glued to the television set. She sends me a Christmas card, so does he, every year."

But Pete Strand was obviously under orders from above. It was indicative of the great change that Edythe Fern Melrose was talking with Pete Strand and not with the new manager.

"The decision makers are watching my show," she argued, uselessly. "Not the masses, maybe. But the decision-makers. And that is important to WXYZ."

She knew that after 25 years of ruling the roost, she was no longer even in the barnyard; and there was nothing Peter Strand could do about it. He, too, had lost his power.

"You know not what you do," she declared, not as much to him as to the new man in the other office and the impersonal corporation above him. "You knoweth *not*—what you do!"

Edythe stood up to go and faced Pete Strand.

"They do not know what they do to people," she said. "They do not know human beings. They know numbers. They do not know the human heart."

"The fashion show—"

"Oh bosh, Pete!" The high chime of Edythe's laugh had lost its ring. "You know that isn't going to happen."

It didn't. After 25 years with the station, Edythe Fern Melrose left it.

Pete Strand left soon after to join the staff of an educational station in Chicago.

54
The Soup Was on the Ceiling

Charles D. Fritz sat low in his leather chair, turned sideways, his feet resting on the corner of his desk. He held the phone to his ear. As he listened to Hal Neal, the long fingers of one hand pushed at Lou Gordon's script lying before him on the desk.

"I'm sitting here spending more damn time on this man than on probably anything else at the station," he said. "Cavanagh says Lou interjects himself too much; Lou says the mayor has ignored him too much. They've been at each other ever since they got into the scrap over the election money."

He listened to Hal's response, sat up, and shoved the script into a drawer.

"No, Hal, I don't want to do that. We need him. I feel, in fairness, that Lou gives the station a dimension that we need. People do listen and talk about him and it's a merchandisable situation. But I got people from all over town comin' in my office all the time."

He leaned back in his chair again and put his feet on the desk, listening.

"Every script is cleared through here," he said. "All the scripts come in here and we meet around 4:30 or 5 in the afternoon."

The conversation ended on a lighter note, with smiles, and Fritz hung up. Immediately, his phone rang.

"The doctor is here," Nancy Gallagher announced. She was now Fritz' secretary.

The doctor was shown into the office with his lawyer. Lou Gordon had broadcast an attack on the doctor's small hospital and had called his nurses "phony." The lawyer threatened to sue.

"I have talked to Lou," said Fritz coolly, "and he tells me the nurses are unlicensed. You don't have a registered nurse there."

"But he used the word 'phony'," the doctor complained.

"Do you want him to say 'unlicensed'?" Fritz asked.

"We object to his method," said the lawyer, not denying that the nurses were unlicensed.

"I'm not going to defend that at all," said Fritz. "But I have never encountered one situation where Lou was wrong. If Lou starts in on somebody, you better believe that he has his information. In fact, he has information about you and about your hospital that he has not used. You know what that is. Do you want him to go on and really tell?"

The lawyer turned from the icy blue eyes of Fritz to his client. He did not know about the additional information. The doctor looked flushed and frightened. After a moment, the doctor shrugged his shoulders, and the lawyer shrugged his, too, and the fight was over. The threat was dropped and the visitors left.

Lou Gordon had moved his TV activities to Channel 50, the Kaiser Broadcasting Company; but he continued on WXYZ radio.

One morning in the spring of 1966, on WKYC in Cleveland, the Martin & Howard show was even slicker than usual. It was fast and it was funny. As soon as they were off the air, they had a call.

"This is Chuck Fritz."

"Oh, hey, yes! ABC, Detroit."

"Right. I'm over at the Hollenden House with Hal Neal. How about joining us for breakfast?"

At breakfast, Neal and Fritz had nice things to say about Martin & Howard's hold on the Cleveland audience.

"I think our success is due to Mr. Khruschev," said Harry Martin.

"You have to know that Cleveland is divided in two," Specs Howard explained. "East and west, like Berlin, and the people from one side of the city never visit the people on the other side of the city—I mean, literally! So way back when we were floundering, we decided we'd conduct tours, and if that didn't work, we'd build a wall, but we oughta get some advice."

Harry Martin spread his hands. "And who knows better about building a wall than Mr. Khruschev?"

"So we placed a call to the Kremlin."

Charles Fritz looked up from his eggs in disbelief.

Specs continued, "And hey, would you believe it? We got through! We got through to Khruschev's male secretary who spoke impeccable English and was just beautiful. He had to know it was a put-on."

Now Neal looked up from his eggs.

"How we got through, nobody knows," Specs was grinning.

"But we did," Martin smiled.

"You actually got through to Russia?" Fritz asked.

"Oh, yes," Specs assured him. "We have the tape. We got through to the Kremlin. We-ell! Don't ask."

"It was World War Three," Martin chuckled.

"Ken Draper, program manager, just got reamed every direction," Specs laughed. "Didn't he know what was going on the air? Why did he let us do it? Clear it first! How could he let us do this!"

"Ray Koeppen, our writer," said Martin, laughing, "suffers from lack of confidence anyway. He thought he was going to be fired. For the first 3 months he was with us, he never took off his galoshes. You can imagine the state *he* was in."

"Well, Draper sat us down in his ofice—as white as your shirt. Looked—terrible!" Specs bubbled.

Martin continued, impersonating Draper, "He says, 'All right. I want to tell you one thing. It's not your show. It's not your show! It's *our* show'."

"Only he sprinkled it with profanity very nicely all the way through," said Specs.

"And he said, 'From here on in, I want a script on my desk by 5 o'clock in the afternoon of every single thing that you put on the air. We're going to find out what's right and what's wrong—on involvement.' Yeah," Martin concluded, "we owe our success to Draper—and the Kremlin."

"Draper finally took us and held us together, watched us," Specs said, turning pleasantly serious. "What bothers me more than anything else is the new people here not understanding the two-man concept. It just isn't the same as two singles being put together."

Martin and Howard went back to their breakfasts. Fritz pushed his plate back a little and sat back, tall and straight.

"I guess you know why I called you," he said quietly.

Martin and Howard ate silently, waiting. It came.

"We still want you guys to come to Detroit."

Neal reached into his pocket and brought out a pack of cigarettes, extracted one, and felt in the side pocket of his blue flannel suit jacket for a match.

"How much do you want, to come up?" Fritz asked.

"Six figures," said Specs Howard.

Neal's hand caught in his pocket. He looked at Fritz, Fritz was steely, self-contained, but jolted.

"And a 5-year contract," Howard added.

After a pause, Fritz said, easily. "Six figures is quite a broad jump from the $75,000 we talked about a year ago, isn't it?"

Neal lit his cigarette. He said, calmly, "This is something that can be negotiated, if you want to go to Detroit. What we want to know this morning is—are

Specs Howard and Harry Martin in Cleveland, 1964, receiving Red Cross award for work on a blood drive.

Specs Howard and Harry Martin with Cleveland Browns in Municipal Stadium, Cleveland, 1964

you interested?"

"Yes," said Harry Martin. He wanted to get out of Cleveland and matrimony, with his mistress.

"We'll talk it over with our attorney," Specs decided.

Their new attorney was a very sharp man named Edward Stillman. He had just won a million-dollar case and was riding high when the four met—Stillman, Specs Howard, Harry Martin and Ray Koeppen.

There was no question in Martin's mind. Detroit offered escape. Ray Koeppen had two reasons for wanting to change stations. First, he had been writing for the Martin & Howard show for 4 years and he was afraid of going dry; material could be repeated in Detroit. Second, he was a sports buff, Detroit was a great sports town, and he wanted to work the Tigers, Red Wings, Lions and Pistons into the show.

But there was the contract with NBC.

"Tenants break leases by becoming obnoxious neighbors," Stillman pointed out, smiling blandly.

"No," Specs objected. "I don't even want to open it up unless we can get out clean from NBC."

"That contract has three and a half years to go," Martin moaned.

"You go to a new market," Koeppen said, "you grow. You expand."

"We've been here since '62," Martin agreed.

"Four and a half years in this business, hammering out the same stuff," Koeppen added.

"Okay," said Howard. "You vote to go. I vote *not* to go unless we can get out of NBC clean and unless WXYZ comes in on the money. And I want a 5-year firm contract in Detroit."

"How much money you asking for?" Stillman asked.

"Six figures," said Howard. "On the way over to the Hollenden House I figured that if somebody comes back to you a second time, they really want you. When they called us the second time, I thought, well hey! We gotta be worth 25 or 30 thousand more. Besides I love Cleveland and if I go anywhere else, I want the same feeling of security."

"I'll see what I can do," said Stillman.

Then the pressure cooker exploded and the soup was on the ceiling. Bill Barrett, writer of a radio and TV column in the *Cleveland Press*, had seen Martin and Howard with Hal Neal from New York and Charles D. Fritz from Detroit, and had done some checking around.

Specs Howard: "He wrote a big story about Martin & Howard going to Detroit. This was big, bold type in his paper. *Well*!"

Charles D. Fritz received a telegram advising him that NBC would sue ABC for "willful intent to breach contract" if he did not refrain from negotiating with Martin & Howard. ABC's lawyer warned Fritz not to even talk with them.

Specs and Harry, both frightened by the unwanted publicity, sneaked up to Detroit for a discussion of the situation, but nobody would see them. Fritz ducked out of the station to avoid them. In Cleveland, they were summoned with their attorney to a meeting with the top executives of WKYC. The station was determined to hold them to their contract.

"How are you going to make sure they give you 100% efficiency?" Stillman asked. "The first time they don't, even if it is not for the reason that they want to leave, you're going to think it is. It isn't a very healthy situation, is it?"

It wasn't and WKYC knew it. At last, the station agreed to release them from their contract if they would remain until February, 1967.

"It's fine with me, except for my personal problem," said Harry Martin. "I want to tell you, if my wife comes in the front door with her attorney, I'm going to disappear out the back."

The station stipulated that there would be absolutely no publicity about their leaving, because it would hinder sales for the new year. Meanwhile, WKYC began to search for a replacement, and Stillman, who was receiving a $4,000 fee from each, carried on secret negotiation with Fritz. It was like a spy story.

The 6-figure cost of the package was agreed to. Howard and Martin would each receive $48,000, and Ray Koeppen would get $15,000—a total of $111,000 the first year with escalation after that. Specs Howard was still insisting on a firm 5-year contract.

Because of the cost involved, Fritz had to get an okay from New York. He did a great selling job to the Board. He explained how Martin & Howard was right in line with his policy to counteract the rock and roll music with jockeys who could capture accounts of adult appeal. It was significant, he pointed out, that every commercial airline was advertising on the Martin & Howard show, proving they could capture adults.

"Our ratings have just gone to hell," he argued. "And we just have to spend this money to get them back. For $111,000 we better damn well be right."

Si Siegel looked up at him, smiling, and said, "It's your decision. Where do you get this '*we*' shit?"

But he got his okay. One thing he could not sell, partly because he himself was not enthusiastic about it, was the firm 5-year contract. Specs Howard had to give in. After all, they had agreed on the money and they were getting out of NBC "clean."

"We need a 3-year firm," he insisted, however.

"Specs," Fritz pleaded. "I can't getcha three. You got a 2-year firm. None of us are that smart in the business to know how things are goin' to work out. Two years and you guys are going to make it. We're going to be behind promotion and build-up. You won't have any problem."

"Take 2 legs off the contract," Howard argued. "Just give us a 3-year firm. At the end of 3 years, if you want us, we'll talk again; and if you don't, you're not obliged for severance or anything else—nobody owes anybody anything."

So it was agreed: Martin and Howard would be in Detroit January 16, 1967, to prepare to hit the air on WXYZ on January 31st.

Then the pressure cooker exploded again and new soup was on the ceiling. What Harry Martin was afraid of, happened. Mrs. Martin came in the front door with her attorney. Martin flew out the back—all the way to San Diego, California. Two months before they were contracted to appear in Detroit, the team WXYZ had signed up—ceased to exist!

Howard: "Harry wrote me a beautiful letter, apologizing. It was just spine-tingling. November 4 of 1966—that was the end of the Martin & Howard show in Cleveland."

Specs Howard was under pressure from Mrs. Martin who blamed him for Harry's escape. He was under pressure from WKYC: was he sure he wanted to go to Detroit to team up again with a man like that who would run out and leave him?

Howard: "I hated to go to work because I knew what I was going to go through every day. I hated to get off the air. 'You want a 5-year firm? We'll give it to you. It isn't going to work in Detroit. We know ABC'."

WKYC hired another team to replace Martin & Howard and moved Specs from morning drive to 9-to-noon.

John and Ginny Pival were enjoying life and each other once again. They had built a beautiful home in Florida, had just moved in and were working together on furnishings and decoration. They were sober and congenial and happy. One beautiful day in November, John had been "up town" to buy rope and rigging to install a bait pound at the end of their pier. Their home was on the ocean and they loved to fish. Before installing their new bait box, they did a little casting and John caught a fine spotted sea trout.

"I'll take it up and clean it and put it in the fridge and we'll have it for dinner," said Ginny.

"Good. Come on back when you're done."

Ginny stepped lightly up the path to their new house. The sun was shining in her life again and she was filled with vitality. She cleaned the fish, her actions swift and definite. Her short skirt swung out as she spun around from sink to refrigerator. She went out to join John on the pier, but he was nowhere in sight. Back in Detroit, she would have worried, but things were different now. He had probably gone next door to talk with the neighbor.

But he hadn't. They found his body in a few feet of water at the end of the pier. In installing his rig for the bait trap, he had somehow got caught in the rope and the weight of the box had held him down. John Pival drowned on Friday, November 11, 1966.

In Detroit, Marc Avery received a telephone call from an old friend in Cleveland.

"Sorry to hear about your bad luck."

"What bad luck?" Avery asked.

"Losing your morning show. Are they putting you in another spot?"

"Another spot? What are you talking about?"

"You mean to tell me you didn't know Martin & Howard here are set to replace you in January?"

Five minutes after that call, Marc Avery walked into the office of Charles D. Fritz.

"What's this about Martin and Howard?" he demanded.

"Who's Martin & Howard?"

"Chuck! Don't hand me that shit. Okay. Look. My father-in-law is very ill in Cincinnati. My wife and I are goin' down there—today. You figure out what you owe me, and pay me. I am through here—as of the show I just finished."

"To this day, Fritz doesn't believe that bit about my father-in-law," Avery remembered. "But it was true. He died."

Avery went back to WJBK. Later, he became one of the top personalities at WJR.

Meanwhile, with a temporary substitute, the Wyxie ratings sank even lower. And in Cleveland, Specs Howard continued to mark time. WKYC relented, told him if he'd work through New Year's Day, he could leave. This brought relief to all who had agreed on the starting date in Detroit. Harry Martin assured Specs on the phone that he would keep the rendezvous at WXYZ on January 16. The team would be reunited. All seemed well.

But another ingredient was added to the evil brew. Shortly after Bruce Miller joined WXYZ, he had predicted, "If that sleeping giant in Canada ever wakes up, he can zap this market within a year."

The sleeping giant was CKLW with 50,000 watts. CKLW woke up. And began zapping.

55
"Fire Burn and Cauldron Bubble"

Shakespeare's witches who disposed of Macbeth and his Lady must have been in invisible session around Broadcast House toward the end of 1966. Poisoned entrails, sweltered venom, fenny snake and wool of bat were being gleefully dumped into the steaming cauldron to give W-HEX—Y—Z another troubled year.

The efforts of ABC and ITT to get married still dragged on. Wall Street saw it as a very good match. *Time* hailed the union as a "new colossus" which would even outrank the mighty RCA. There were many signs of potential congeniality. ITT made record players; ABC made records. Both companies had dovetailing interests abroad. Each stood to benefit. Harold S. Geneen, President of ITT, wanted prestige. Nobody in America up to that time seemed to have heard of the giant company he headed; through ownership of a national radio and television network, the initials "ITT" would be known in every household in the country. ABC, on the other hand, stood to receive an immediate dowry of $50 million with which to improve facilities and programming and lift it from 3rd position among the networks. With all that money, ABC could move immediately into full color telecasting.

The stockholders seemed eager, but the Federal Communications Commission had to bestow its blessing before there could be a union, and the FCC stalled for 12 months. Reportedly, it had been ready to vote *for* the merger as early as February, 1966; but there had been dissenting voices and apparently a minority had insisted on hearings which were delayed until late in the year. The matter was considered in December. After only 2 days of hearings, the FCC voted 4 to 3 to approve the merger.

So at the end of 1966, the wedding was on and the cake was in the oven. There was more gall of groat in it than sugar.

In 1966, for the 3rd time, WXYZ put its creative minds to work on the problem of separate programming for FM. The FCC ruled that 50% of FM broadcasting had to be independent of AM. Charles Fritz and Lee Alan collaborated on a new idea which, unfortunately, was several years ahead of its time.

The music would be conservative pop: Sinatra, Mancini, Belafonte, Peggy Lee, Les Elgart, Count Basie, the Tijuana Brass, Jackie Gleason, Roger Miller, Ferrante and Teicher, Ella Fitzgerald, the Ray Coniff Singers.

There would be personalities hired to ride the records with an aura of mystery and class. The jockeys would be known as "Casey J," "Shanahan," "Donovan," "Sebastian" and "Michelle." Care was taken to give these jobs to the right people: Dave Leukhardt (simplified to Lockhart), Ken Lawrence, David Manheimer, Bob Lane (son of an old Lone Ranger actor, Bob Liggett) and Bruce Miller's wife, Jean.

Again, happy Jack Tierney bustled about building a new studio—this one for FM only.

Into the witches' brew went nose of Turk and Tartar's lips, finger of birth-strangled babe.

There were other changes. Ed Hardy went to ABC in New York and was replaced at WXYZ by a man from WJR. Frank Tomlinson had a weighty personality and a heavy voice. He was a good newsman and easy to get along with. Bob Baldrica took his permanent smile to WJBK and Bob Koch was now sales manager of WXYZ-Radio. Marty Greenberg took over national sales for the station; and handling publicity now for radio was a nervous Texan named Bill Axtell. Axtell had been in radio news in Dallas when John F. Kennedy was assassinated and had written a book about it which was so hot nobody would publish it.

Bruce Still was increasingly ineffective. A public service program he had attempted was taken off the air, and more time was given to Osgood's *Fun Tips* which was renewed by Marathon Oil.

Not only WXYZ was tasting the witches' brew. CKLW had "zapped" the market, and not only had WKNR "gone to hell in a handbasket," WXYZ had in the tipsy whirl of the brew, staggered far, far behind. Credit for the upset was given to a station consultant named Bill Drake.

Well known in the broadcast craft as creator of "the Drake format," he had quietly slipped into Detroit and sat listening to WKNR while that station fell under the heady influence of adder's fork and blind worm's sting. Drunk with high ratings, Keener grabbed every dollar that came its way, and too many came. As soon as the station was up to "21 units" an hour—meaning commercial interruptions of the music—Bill Drake had CKLW kill all its commercials and go solid music—5, 6, 7 records in a row, without interruption. Jockeys screamed a minimum of words between segments. As one observer describes the results, "listeners, instead of going over to CKLW in little spurts, went over in droves." Keener simply fell apart.

In the dizzy whirl caused by taste of stale sea shark and root of hemlock digg'di' the dark, Charles D. Fritz found himself tipsy on a bar stool in the Oakland Hills Golf Club challenging JR's J.P. McCarthy. As the gossip goes, Fritz said, "J.P., you gotta know we're going to knock you and JR off your high horse. We got a couple guys comin' in here named Martin and Howard that'll knock you out of top spot in 6 months."

It is said that J.P. McCarthy responded quietly and soberly, "In the first place, you're not a gentleman to say these things here, Mr. Fritz. In the second place, I don't believe you're going to do that."

Obviously Chuck Fritz was not himself. He was trying hard to shake off the brain-blurring effect of steam from lizard's leg and howlet's wing. He began to look around for a weapon with which to clobber his enemies. He now had two instead of one.

The weapon had to be music and personalities. CKLW, WKNR and WXYZ were all playing the same music, and CKLW had the watts. He decided Wyxie'd have to change its music. The whole sound of the station had to be changed. Fritz was not a musician; he could not put a mental finger on the type of music he wanted. WNEW in New York had a sound he liked and it had popular personalities. One station in Detroit, WCAR, came near the sound, musically, but it had only one personality—a fellow named Bacarella.

Joe Bacarella grew up in Monroe, Michigan. Two brothers and an uncle had a band and rehearsed in his house. He decided he wanted to be a musician, too. His uncle put him on trumpet. Joe had his own band when he was in junior high school and earned spending money with it all through high.

From school he went to work as an expediter with the Monroe Auto Equipment Company, manufacturers of shock absorbers. The company owned radio station WMIC in Monroe. Business with the auto equipment company went slack and Joe faced unemployment.

"What'll I do?" he asked.

"Go downstairs and see if they have some other job for you," his boss suggested. "We're terminating a lot o' jobs but we don't like to lay off people."

There was nothing, and Joe was "a little bent out o' shape." But two days later, he had a call from WMIC.

"Come on out. We have a job for you."

The closest Joe Bacarella had been to broadcasting was a PA system in a dance hall. The afternoon he went to work at WMIC he went on the air from a shopping center in Monroe and did a one-hour remote. It was a total disaster. But they were patient and he eventually was moved to morning drive time. He also did some selling.

Three years after he joined the station, WMIC was sold to Dick Jones, Ross Mulholland and the Brink family. The new owners installed a studio in the Brink Building behind Hudson's department store in Detroit, and hired veterans Eddie Chase, Ed McKenzie and Ralph Binge. They moved Bacarella to Detroit with the

Joe Bacarella

station, and gave it new call letters—WQTE.

It was a heavy payroll and business was not as good as the new owners had hoped; so they changed to rock music and hired a popular young disc jockey named Tom Clay. And they kept Joe on. Rock music didn't help. Overnight, the format was changed again to "good" music, lush arrangements. Bacarella was bored.

"A disc jockey had nothing to do," Bacarella remembered. "I would send the secretary out and have her buy me some paint and I would paint the walls while I'm waitin' for the record to end."

Finally, he called Barney Morris who was working at WCAR. They had worked together in Monroe.

"Boy, I gotta get outa here," Joe told him. "I can't stand the smell o' this paint."

"I think there may be an opening here," said Barney.

There was. Bacarella was hired to do a show from 7 to midnight. On his first day at WCAR, they wanted to change his name.

"It's too long. It's melodious but it's too long."

"Okay. What'll I call myself?"

"Ted Stewart."

"Ted Stewart? Okay—if that's what you want."

Half an hour before air time, they called him again.

"Don't use Ted Stewart."

"Well, okay. What'll I use?"

"We haven't decided yet. Don't use anything."

He went on the air at WCAR with no name. The next day, they decided to call him Art Cooper. He used that name for a year, then they called him again.

"Joe, how'd you like to use your own name?"

"I'd love it a lot."

"Okay. Starting tonight, you're Joe Bacarella."

When he went on the air that night, he said, "This is Joe Bacarella doing the Art Cooper show." After 2 or 3 days, he dropped all reference to Art Cooper and was himself. A listener called.

"I don't know who this Joe Bacarella is but he sure does not sound like Art Cooper."

"It's the same guy," she was told.

"Oh don't tell me that!" the woman complained. "What do you take me for? He doesn't sound at *all* like Art Cooper."

Joe Bacarella worked at WCAR 5 years and 3 months. Hi Levenson, his boss, was a little like George W. Trendle; he liked to become involved in every phase of the station's business. Joe felt this hampered certain departments; many things were not done which Joe felt should be done to make WCAR a bigger and better operation. One of his friends, a girl named Artie Lindsay, was with an advertising agency.

"They might be interested in you at WXYZ," she offered. "Keith McLellan is a salesman on Channel 7. He's a friend of mine. I'll talk to him."

She did. TV's McLellan talked with radio's Bob Koch. Koch talked with Fritz. Keith McLellan called Bacarella.

"Joe, Chuck Fritz wants to talk with you."

Fritz and Bacarella met at Berman's Chop House. Bacarella was impressed with Fritz immediately. As Joe commented afterwards, "Fritz was just reekin' with class." Fritz noticed that Bacarella turned down a cocktail.

"You don't drink?"

"No. My dad—one of his prides is making his wine. I used to love to watch him and help him, but I couldn't stand the smell of it. And then, playing in the band, I saw so many people get smashed."

"Joe, who picks the music at WCAR?"

"I do."

"You do a good job."

"Well—*you* know—I really dig what I'm doing."

The little man who had cut potatoes so thin at Berman's was gone and Fritz missed the rapid rhythm of the knife on the cutting board.

"We've got a problem at WXYZ," he told Bacarella. "Keener, CKLW and WXYZ are splitting the rock audience 3 ways. We're just hangin' on which is tough when you've been up there and performed well. These other people make runs at you. You haven't got any place to go but down. We've done everything we can do as a rocker."

"Well, you can keep it going without rockin'," Joe pointed out. "At CAR we don't scream but we do run as tight a show as we can. We keep it going, only with a different sound than the Top 40."

"We have found we can't run a good Top 40 with ABC," Fritz explained. "ABC insists we do editorials. ABC insists we carry *The Breakfast Club*. And insists on a lot o' things that slow us down."

"There has to be a compromise," said Bacarella, savoring his steak. "There's a marriage taking place between good music and rock and roll. This could turn out to be a very healthy situation."

"I may call you again," said Fritz.

Joe Bacarella didn't drink but when he left Fritz he had a hangover of hope. He tried to cool it, but he thought, "Jeez, I'd love to work with that guy." Being in 3rd place at the moment didn't mean anything. WXYZ was where the biggies were. Anybody who was anybody, Joe thought, was, or had been, at WXYZ—the Fred Wolfs, the Ed McKenzies. And that was where *The Lone Ranger* started. He didn't really go back to that but he was aware of it. What did interest him was that so many people had left other stations to go "up" to WXYZ.

Joe Bacarella cooled it for 2 weeks. His second luncheon at Berman's was with Chuck Fritz, Marty Greenberg and Bob Koch. They all fired questions at him.

"Why do you want to leave WCAR?"

"There are many things I want to do as program director that I'm not allowed to do."

"Do you pick your music from the trade magazines?"

"No. I pick music because I listen to it and like it."

"We don't want to sound like WCAR," said Charles Fritz. "We wouldn't mind sounding more like WNEW in New York. They have Klavan and Finch in the morning, and we've got a team coming in next month, Martin & Howard—2 funny guys. We're going to go heavy on personalities."

"What do you think of ratings?" Greenberg asked.

"Well, I guess it's the only way to go," said Bacarella. "Personally, I don't care about the ratings if the station sounds good."

"The last time we talked," said Fritz, "you spoke about a marriage of good music and rock and roll. What did you mean?"

"The Beatles have written some good music. On the other hand some of the good singers are beginning to take a crack at rock and roll songs, taking the sharp edges off. They let you hear the lyrics and the lyrics are good."

"That may be what we get into," said Fritz. "Sort of middle of the road."

"Good music doesn't mean WJR necessarily," Bacarella explained. "Or what you hear in a shopping center. You have to listen to it and pick what you want to play."

"I like country music," said Koch.

"Country music can be good," Bacarella agreed.

"How are you in a lion's den?" asked Marty Greenberg.

Bob Koch laughed.

"A lion's den?" Bacarella laughed.

"He means upstairs at Broadcast House," Fritz said quietly, smiling. "We got a lot o' wild people up on the 2nd floor in the studios. We call it the lion's den. What we need is a lion tamer."

"Lee Alan can't handle it," said Greenberg.

"Lee Alan can't handle anything," said Koch.

"Lee has alienated the salesmen," Fritz explained. "And he's alienated most of the talent. I've got a real feud on my hands with Alan and Dave Prince. And we have a feller named Bruce Still."

"A sweet guy," said Greenberg. "A real sweet man, but over his head."

"I've got to let him go," said Fritz.

"If you hired me, what would I go in as?" Bacarella asked.

"You would go in as program director," said Fritz.

"Lion tamer," Marty grinned.

Joe Bacarella was hired.

The steam from the cauldron smelled good. Make the gruel thick and slab, cool it in the baboon's blood, then the charm is firm and good. Peace? The charm's wound up.

56

Joe in the Lion's Den

Joe Bacarella, Specs Howard, Harry Martin and Ray Koeppen all reported for work at WXYZ on the same day—January 16, 1967. The station plunged into a heavy campaign to promote the new team. Bruce Miller, instead of creating commercials for the salesmen, made up billboards advertising Martin & Howard. Ray Koeppen

Dave Diles

wrote short skits and Specs taped them by telephone with Phyllis Diller, Mike Douglas and other celebrities. These were used on the air as promotional teasers. Fritz observed that Harry Martin did not appear on these blurbs, and learned he had skipped off to Mexico with his girl. Mr. Fritz was not pleased. Specs Howard, on the other hand, was not happy with his build-up. It was too much. They were being oversold.

The day after they reported for work, he and Bacarella were flown to New York with Fritz and Marty Greenberg to meet the brass.

"Dress real nice, and please—no boots," said Fritz.

It didn't occur to Joe Bacarella that the order was directed at him. He was a fussy dresser. Six feet tall and built like a full back, he wore clothes that proved it. His Beatle Boots went well with what he wore and he was proud of them. But when Fritz saw him at the airport, he said, "Come on, now. You don't go in with boots. It just doesn't look right."

"What am I going to do?" Joe asked Greenberg on the flight to New York. "I have to have my shoes made for me. I wear a 13 double-A."

The next morning, before reporting at ABC, Joe and Marty went to 6 stores before they found shoes. And Bacarella had to borrow the money to pay for them.

On the 40th floor of the ABC building, Fritz introduced his new operations supervisor and half of his new morning team. ABC brass and men from the New York office of Blair, including Dave Klemm, seemed pleased. But Specs Howard received a jolt: Fritz announced that WXYZ was going to move into something new—a new format. The fuse of a bomb had been lit.

Joe Bacarella had no office. Bruce Still occupied the one used previously by Baker and Klemm, and Lee Alan had the small room between that and the hallway. Bacarella spent his first weeks sitting in the visitor's chair in Still's office, feeling transient. Nobody knew who the boss was—Still, Alan or "the new man."

A friendly record pusher advised Bacarella, "The first thing you have to do is get rid of Lee Alan."

"Look. I'm sorry," said Joe. "Give the guy a chance. He and I might work well together, and if we do he's got a job."

Finally, David Klemm found an opening for Bruce Still in Cincinnati, and the little man escaped from the den. The first lion to feel the eye-to-eye domination of the new tamer was Lee Alan. At a meeting, the disc jockeys, as usual, recommeded records they wanted to play and suggested others to be dropped from the list. Lee Alan was lenient.

"Well—okay—but."

"No 'but'!" Bacaralla declared flatly. "No more disc jockeys pickin' records. That'll be in one pair o' hands from now on—mine."

Bacarella established a new policy with record pushers. They were to talk with no one else at the station except himself, and he gave them less time than they were used to. He listened to their pitch with occasional blunt comments, auditioned the records they left behind, and threw most of them away.

The feeling of insecurity at Broadcast House shortened tempers, and one day Bacarella heard Lee Alan yell at a jockey. He called Lee into his office.

"Don't *ever* chew out a guy like that again," he said. "And another thing. Why is it I never see you here until 10 or 11 in the morning?"

"I work late at night."

"I never see you here after 5:30."

"I work at home."

"Look. We have to set an example. So let's have you get in here at 9 in the morning."

"I can't. I'm monitoring the station."

"Your office is loaded with radios. You can close the door and monitor all you want—right here."

"I won't work that way," said Alan. He walked out.

Another lion was Dave Diles. Dave was friendly with the tamer but put him to the test occasionally. He now had a telephone show, *Dial Dave Diles*, on which he and a guest sports celebrity talked on the phone with listeners. Diles had strong likes and dislikes in the sports world and broadcast poisoned darts now and then. Once, he sent a dart into the air which Bacarella felt was bad. He called Diles into his office the next day and they had an argument. Bacarella ended it with: "You're one of the highest paid guys on the station an' you're a big name, but I'm going to take you head on, man, because when all is said and done, I'm drivin' the boat."

Harry Martin managed during his brief visit to Mexico to divorce Mrs. Martin and marry his girl. Carol, the new Mrs. Martin, was open, friendly, blonde and pretty and a lady of considerable charm.

Martin & Howard went on the air January 31, 1967. They were under a hex from the beginning. In spite of the mammoth build-up, they were told to "hold off" doing their thing. Fritz and Bacarella were working on a "new sound."

"Let us do our stuff," Specs pleaded. "Let us get our stuff out o' the crates and *go*."

"No. No. Just be a couple o' nice guys for a while."

"Well, fellers," Specs said privately to Koeppen and Martin, "I have to say I told you so. They just don't know what to do with us. They don't know how to handle a two-man show."

In February, 1967, the day before the ITT-ABC marriage was scheduled to occur, the Justice Department arose to object to the wedding. The FCC was driven to "reconsider." The FCC was irritated, pointing out that the Justice Department could have expressed its objections earlier, but agreed 5-to-2 to postpone the wedding. The Justice Department declared the FCC had batched up the wedding cake too hastily. The FCC gave the Justice Department "two weeks" to introduce any evidence that could stop the wedding.

A word and an expression new to radio crept into conversations at Broadcast

House and were introduced in New York by Charles D. Fritz at an important meeting. He had to get an okay for the radical changes he had in mind from Ralph Beaudin, overall program director of the ABC-owned stations. He, Beaudin and Hal Neal lounged around a coffee table in a hotel suite kept by ABC in Manhattan. Fritz found it difficult to explain exactly what he had in mind, but felt he had a good starting point: WXYZ-Radio had reported $650,000 profit for the year just ended.

"Let's get into the demographics of the thing," he said. "Demographics" was the new word. He continued, "CKLW has the numbers, but the rock fans are in their teens. WJR has high ratings, but it appeals to listeners 50 or older. It's the young marrieds who are buying the furniture and refrigerators and the stoves and the houses. They're the ones we want to go after at WXYZ—the 18s to 34s. I don't care if we're Number Three in the ratings if we're Number One with this age group."

"So how do you propose to do this?" Beaudin asked.

"With the right personalities and the right music," said Fritz. "We've got the personalities. What we have to do is change the music, the whole sound of the station."

"What kind of music?"

"Well, it's kind of hard to explain," said Fritz, wishing he had brought Bacarella with him. "It can be rock—but not hard rock—soft rock—bright—but gentler than what CKLW goes with. On the other hand, it would be more—contemporary—than what they use at WJR. I guess you'd call it middle of the road."

That was a new expression: Middle of the Road. It would be referred to in the trade as MOR.

"It means tossing out all the teen business," Fritz added.

The room was silent.

"We want to be the WNEW of Detroit," said Fritz.

Still no reaction from Beaudin.

"I think this station's long-term progress will be better as—middle of the road, rather than the rocker," Fritz concluded, waiting.

At last Beaudin put down his cup and asked, "When are you going to make the change?"

"Well—I want to take some time to plan it out."

"Don't stall around with it, Chuck. You know what you want to do. Do it. Now."

"What—uh—is your idea of now?"

"Next month."

Fritz' eyebrows jerked up and down. "Well—"

"I'll get on it with you," said Hal Neal. "I'll come out to Detroit. And we'll get Klemm out there. He'll have some ideas.

On the last Friday in February, one month after Martin & Howard went on the air, the staff was called to a meeting and the bomb was exploded.

"We are leaving rock and going Middle of the Road."

It was not as simple as it sounded. No longer could records be chosen from lists. More than ever, Joe Bacarella depended on his gut feeling in choosing what was to be played.

Dick Kernen: "Joe understood early that something was going on musically. There were things happening in rock music—a soft sound. It was rock music, but it was nice. Melodic. So they went into this and it was very tough because Joe was fighting a lot of tradition."

Charles D. Fritz: "No one knew precisely how to put a list together for this. Rock—you could just look at the charts. Now you get into this broad spectrum of Middle of the Road and there are no criteria. Hal had his ideas. Dave Klemm had different ones. Dave would come out here and tell us our music was bad. I had my ideas. Bacarella had his."

Bacarella: "I knew a marriage of good music and rock and roll was taking place and some day it would be a healthy situation."

Kernen: "What is good music? The radio definition of, quote, 'good music,' unquote, is Patti Page, Jerry Vale, Tony Bennett, Sinatra, Sergio Mendez, Herb Alpert—this is MOR, Middle of the Road music. *But*—so is a lot of other stuff. So is Simon and Garfunkel. So is *Chicago*. The Carpenters. A lot of stuff. There was beginning to be a thing happening that didn't necessarily mean that the smartest thing to do was to play Jerry Vale just because he had a new record, or Tony Bennett just because he happened to have a new record. So Joe starts to build this format. Well—it was just niff-nawing all the way. Why are you playing this? Why are you playing that?"

Bacarella: "We tried the best of the rockers and the best of the good music, and put 'em together."

Specs Howard: "It was the ruination of our show. We were directed to get on the air and just be two nice guys talking. Had they told us what they were going to do I think we would not have come. We had proven material and we were not allowed to use it. We were told what *not* to do, never what *to* do.

"Another problem was to whom to report. Lee Alan was program manager. Bacarella was operations supervisor. If you have a problem, who do you go to? Lee used to call us in and say, 'Hey. Look now. I'm program manager. If you have a problem come to *me*.' Bacarella took offense at our going to Lee."

The demographic span was narrowed to 25 to 35. This not only knocked out the teen-age listeners but all commercials appealing to that age group were pulled off the station. This upset the sales department. Chuck Fritz called Bacarella and Lee Alan into conference.

"I want you both to write a presentation," he told them. "We've got to sell what we're doing to the salesmen. They're about to take a bath as far as money is concerned, because they have to go to their pimple-cream clients and tell 'em we're changing our format; so we have to throw this money off the station unless they want to change their spots."

Joe Bacarella put a lot of thought into his presentation. It filled 12 to 14 pages. Lee Alan did about 1 page.

Alan liked less and less what was happening. He had openly disapproved of hiring Martin & Howard, said it meant altering the sound of the station to fit personalities, and they had gotten away from that. He didn't like MOR either. He became less and less co-operative. Finally, Bacarella called him into his office, handed him his last pay check and said, "I'm sorry."

Lee Alan had been fired again.

As manager of national sales for WXYZ, Marty Greenberg was doing a strong job. The Detroit Chapter of American Women in Radio and Television pointed to him in 1967 as one of the "Five to Watch" and the Gallagher Report picked him as one of the 7 top sales managers in the country.

Marty cornered Bacarella one day and said, "Now that you've let Alan go, you're going to have to have a new assistant. Why don't you consider Dick Kernen?"

Greenberg, Bacarella and Kernen were congenial and often had lunch together. Bacarella knew Kernen only as the man in traffic, but was impressed by his knowledge of the studio, its people—and music. He judged from Kernen's conversations that he would take orders and get things done, and he knew from traffic that he was a good detail man. So Bacarella rescued him from his boredom and gave him Lee Alan's office.

Kernen was replaced in traffic by Jim Neil, thin, short and hungry-looking.

"Chuck, we're in trouble."

Hal Neal was calling from New York.

"Mayor Cavanagh wants Lou Gordon off the air. He has called Jimmy Riddell in California. So have some of his friends—Cavanagh's friends and Jimmy's friends. So we've set up a meeting out there at the University Club next Wednesday."

"Here? Who set up the meeting?" Fritz demanded.

"I don't know. It just came about. We want you there."

"Do you want Lou?"

"No. God no."

The Mayor's chief objection seemed to be that Gordon was telling people the Cavanagh marriage was on the rocks. Cavanagh maintained this was a lie.

"You never heard Lou Gordon mention anything like this on the air," Fritz declared. "I read every damn script before he broadcasts, and there has never been anything about you havin' any trouble with your wife."

It was true. But the Mayor objected to Gordon's talking about his troubles privately.

"Chuck, why don't you take him off the air?" Riddell asked.

"I just don't feel under the circumstances it would be the right thing to do."

"Can't you get him to back off Jerry?" asked one of Cavanagh's friends.

"I won't make any deal that he can't talk about the Mayor," Fritz insisted. "Detroit is the 5th largest market in the country and the Mayor is involved; he is a very active person. Lou's got to talk about this. He's got to be able to do it."

Riddell was pleasant, seeking an alternative. "Can't you report his activities on the straight newscasts?"

"We could get along without his commentaries," said Neal.

Dick Kernen—December 1971

Lou Gordon—1964

"Lou Gordon means a lot to our station," Fritz persisted. "He has a hell of a lot of listeners and a hell of a lot of influence. I want him on."

It was the best they could get from Chuck Fritz.

"We'll watch it in New York," Neal promised Cavanagh. "We'll have our lawyers go over the scripts."

From that time on, the New York lawyers were on Fritz' back constantly. Every time Cavanagh was mentioned on a Lou Gordon commentary, a red flag went up in New York and Fritz received a call.

Lou Gordon proved to be right; the Cavanagh marriage did break up. Meanwhile, Gordon was peeved because he had not been invited to the meeting, and Fritz was angry because the meeting had been called over his head.

"Who the hell is runnin' this radio station?" he demanded on the phone to Si Siegel. "If you want me to run it, you don't plan meetings and make commitments without me."

The ITT-ABC engagement which wore on and on contributed to the confusion and uncertainty, even on the lower levels. A second series of hearings began in Washington in April. They ran for 3 weeks. The main issue was what the merger would do to the independence of ABC.

In July, 16 months after the merger had first been announced and 6 months after the Justice Department had halted the wedding, *Newsweek* reported that the FCC, in a bitter 4-to-3 vote, had again given its blessing to the joining of ABC and ITT. The majority saw the benefits in the marriage for ABC.

The minority, however, maintained that ITT control would slant ABC news, and in 131 pages expressing their viewpoint, the 3 who voted against the merger presented evidence that ITT public relations men had even managed to color ABC's coverage of the hearings.

Nevertheless, the majority ruled, the proposed wedding had the official blessing of the FCC, and ABC fell heir to an advance of $25 million. However, the wedding had to be postponed again. The Justice Department threatened to go to the U.S. Court of Appeals. Once more the cake was removed from the oven.

1967 was the year of Detroit's second race riot. It began in the middle of the night of July 23, when police raided a blind pig in a black neighborhood and arrested 73 after-hours drinkers. Within hours whole blocks of the city were ablaze. In 5 days of havoc and hatred, 1,383 buildings were burned, 1,700 stores looted, 5,000 people arrested and 43 people dead.

WXYZ newsman Tom McIntyre was broadcasting from a station car in one center of conflict. The car got caught in a crossfire. McIntyre dropped to the floor and his "eyewitness report" consisted of "My God, let's get out of here. Oh, this is awful. They're shooting at us. Let's get out of here."

Such was the confusion of the times at WXYZ that this was held to be great reporting and when Frank Tomlinson left the station to join the network, Tom was made head of WXYZ radio news.

The departure of Lee Alan brought an end to *Chicken Man.* Specs Howard did his best to persuade the station to let him haul *Congo Kurt* from the crate. Fritz was afraid of it; one of the main characters was black.

"But hey! He's the hero!" Howard argued. "*Congo Kurt* always gets into a jam and it's *Shango*, the native guide, who gets him out."

Fritz might have been softening a little before the race riot, but *Congo Kurt* and *Shango* remained in the crate. Ray Koeppen went around Broadcast House looking as though he were having labor pains. He gave birth to comedy bits for Martin & Howard but it was a losing battle for all three. The station was moving more and more away from their style.

The riots also killed *Whirlybird Traffic Reports.*

Bruce Miller: "They couldn't make any money with it. Detroit's traffic pattern is not that complicated; there are not that many alternate routes. You can say there's a bottleneck at Ford and Lodge expressways—well, it bottlenecks every day; so what's new? It was another idea that didn't work, that's all."

Lou Gordon was completely disenchanted with WXYZ. He knew New York wanted him out. His TV commentaries on Channel 50 were widely watched, but the radio side of his endeavors had gone sour.

Also souring was Pat Murphy. He remembered he had been on the verge of something good when, at the last moment, Lee Alan had displaced him. Now he resented Martin & Howard.

"I could be funnier than them on my worst day," he said. He began to let down, developed a habit of arriving a minute or two before air time—or even a minute or two late. The tamer cracked his whip; another lion left the den; Pat Murphy was fired.

Added to the hex was FM. Donovan, Michelle and the others cost money; advertisers were not interested in FM. The operation was ahead of its time. The main thing then was, it was losing money.

So was AM. Martin & Howard, stymied as they were, just couldn't get moving. Hard as they tried, the numbers just lay there. WXYZ Radio turned a profit of not more than $15,000 in 1967.

"It was a miserable year," confessed Fritz. "I was practically out of my gourd. It was all of this adjustment. We told the guys, 'You have got to read, you've got to be knowledgeable, you gotta get *Variety*, and all of this. You're goin' to be Mr. Show Biz'."

This stepped on Dick Osgood; he had been Detroit's Mr. Show Biz for 29 years. *Show World* was on again. Marty Greenberg sold it to the Metropolitan Exhibitors of Detroit on a solid 52-week contract. But now the jockeys were getting into his act. He resented it and became more glum than ever.

The ear of Hal Neal was in constant attendance. In New York, he had a direct line to WXYZ and could sit in his Manhattan office and listen to the station. Once Fritz was paged in a New York restaurant. Neal was on the phone. He'd been listening to Detroit.

"Do you know," he said, "that Dave Prince has not given the call letters in the last 8 minutes?"

"Hal," Fritz moaned. "I'm not in Detroit. What do you want me to do—slice my throat?"

Bacarella felt the tension: "It was just niff-nawing all the way. Why are you playing this? Why are you playing that? We got questions from all over the place. From the deejays. I got questions from Fritz. There were times when Hal Neal was in town and wanted to know why we were playin' certain records. I just flew by the seat of my pants on a lot of music. We were in the ball park but we weren't doing precisely what we wanted to do."

Fritz was displeased: "1967 was a disaster. Everything! I was president of the Michigan Association of Broadcasters. I went up to the convention in Hidden Valley and I made a speech and I said, 'I'll tell you what's happened in the year since I've been president.' And I listed all of these things. The Mayor was in trouble with his wife; there's scandal in the police department; the riots—I listed all of these things. One was just worse than the other. And I said, 'You know what will happen? Al Kaline will blow the pennant for the Tigers on the last day. You watch and see.'

"And he did! Oh, that was an abominable year."

Somehow it was not as easy to turn a rock station into Middle of the Road as it had been the other way around. Time and time again, good music stations had switched to rock and won the numbers within weeks. It was a downhill slide, but for some reason to abandon rock and fight for a MOR audience was a long, long climb back up the hill."

Bacarella: "We were getting ratings books after rating books. Nothin! We were getting flack from agencies, listeners, and it was just—the station was going further downhill."

Back in 1933, WXYZ had saved itself with *The Lone Ranger*. Twenty years later, Riddell had rescued the station again with his audience-flow theory and the hiring of Wolf, McKenzie, Surrell and Paul Winter. A decade before this "abominable year" Hal Neal had begun his boost to a new peak with his tight adaptation of the McLendon sound and the Top 40 format.

What could be done this time—if anything?

57
New Sounds and Old Echoes

On the morning of New Year's Day, 1968, 17 directors, called to a meeting by Harold S. Geneen, settled into deep blue chairs around a big oval table in the New York headquarters of International Tel and Tel. According to the agreement with the American Broadcasting Company, this was the last day in which they could bow out of the wedding. President Geneen and the 17 directors concluded that morning that whatever decisions government agencies might eventually come to, the bride was not worth the dowry, and broke the engagement. The wedding was off.

The network must now return $25 million it had accepted in advance from ITT. Leonard Goldenson issued orders for strict economy.

On Friday afternoon, January 5, Elton Rule, manager of KABC-TV in Los Angeles, boarded a plane for New York. He had been summoned to headquarters by Mr. Goldenson who offered him the job of network president.

Elton Rule, cool, experienced, and not easily demoralized, considered the offer for 2 days. He had found ABC New York headquarters in a state of shock. Several promising young executives had already left to work elsewhere. The network had marked time, waiting for the wedding, and had fallen so far behind the other networks in ratings that sponsors were in a position to dictate policy. Although there were only 3 major television networks, ABC was now jokingly said to rank 4th.

Elton Rule's hesitation, however, was due not so much to the emotional disorder at ABC as to the position in it of one man. For 6 years he had had to report to that man as his superior and like all other managers of ABC-owned TV stations, including John Pival, he had found that man overbearing.

"I will accept the presidency," he told Goldenson at last, "but only under the condition that I take no direct orders from Ted Shaker."

It was agreed. The corporate set-up was reorganized so there would be 4 different group presidents. Mr. Rule and Mr. Shaker headed different groups. Each would be accountable only to Simon B. Siegel.

The same month, John Campbell who had followed John Pival as TV's top man in Detroit, was hauled off by the network to replace Elton Rule as manager of KABC-TV in Los Angeles. John MacMahon was brought in from WLS, Chicago, to manage WXYZ-TV.

ABC-Radio was divided into 4 separate networks: 'Information," "Contemporary ," "Entertainment" and FM. In Detroit, WXYZ became part of the Entertainment network. ABC Radio now owned or had affiliate agreements with 740 stations in 200 markets.

At WXYZ, FM had become a major problem. To save money spent on disc jockeys, it was decided to convert to an automatic operation. Two were transferred to AM. Dave Lockhart took the segment following *Dial Dave Diles* at night. The other, assigned to the all-night trick, was tall, aggressive young David Manheimer. He had wanted to be on WXYZ from the time he was 12, had performed on teen shows on the west coast, played the lead in a Greek drama at the University of Michigan, worked as Dave Mann on WYSI in Ypsilanti, and had been "Donovan" on FM at Wyxie. His name was now changed again—to Johnny Randall.

Happy Jack Tierney who had not changed an ounce or gained a wrinkle in all the years, now bounced about installing $50,000 worth of automatic equipment. It was all too complicated for that other veteran engineer, Al Schaub. After 38 years with the station, he retired.

With the new automatic FM equipment, Dick Kernen became Mr. Computer, putting together long-running tapes of solid music. Station breaks, jingles, slogans, weather reports and commercial announcements (when there were any) were taped separately, and at pre-determined moments, the long musical tape would turn off, the spoken stuff would click in, then click off, and the music tape would start running again.

"It was a like a giant juke box," Kernen remembered. "I picked the music and began to deal with a lot of things that I saw in the trade papers and I said what is all this? Weird things like the West Coast Pop Art Experimental Band and the Jefferson Airplane—what is this? So I began to play the records—and *wow* this stuff is pretty good. So we started playin' that on this automated thing which had a lot of technological hang-ups. It was a bastard! It was a disaster!"

The equipment got jammed and stopped running. But at least FM was on the air and separate from AM half the time as required by the FCC, and costing less money as ordered by Leonard Goldenson.

AM continued its battle for listeners, and changed sales managers. Bob Koch left the station and purchased a country music station, and a jolly gent with sparkling blue eyes and a very red face was sent in from the east to replace him. His name was Lee Vanden-Handel, and wrapped inside his expansive personality was concealed a small core of inadequacy.

Meanwhile, in its effort to regain an audience, AM fell back on the ancient gimmick of bribing people to listen with a new campaign named *Cash Call.* Names and numbers were picked "at random" from telephone directories. People were called at frequent intervals and, if they were listening, were given whatever sum of money was in the "jackpot." Each time a person was contacted who was not listening—which was embarrassingly often—more money was tossed into the pot.

A refreshingly unprofessional blonde with long legs, long hair, beautiful teeth and an Irish sense of humor combed the regional telephone directories for numbers according to a rigidly honest system devised by Joe Bacarella. Her name was Patricia Pendergast but everybody called her Myra to distinguish her from the other Patricia in the office. Elaine Baker had left to get married, Nancy Gallagher moved downstairs to work for Fritz, and Pattie Robertson had become Bacarella's secretary. Pattie Robertson was simply in love with the whole thing. If WXYZ was in trouble, she wouldn't believe it. She was devoted to Bacarella's job and stayed overtime and got in early to help him with it. The two Patties were rosy oases in a dark brown desert.

On the 19th of January, 1968, Lou Gordon quit WXYZ to devote full time to Channel 50. His commentaries and interviews were syndicated to Kaiser stations in other cities.

Saturday night, January 20, at 6:05 P.M. a 25-minute documentary was presented on a new series called *Close-up,* produced by WXYZ-Radio news. It was the result of an experiment by The Motion Picture Council of Greater Detroit. Students from the Seaholm High School in suburban Birmingham were invited to an exclusive screening of a film called *Don't Look Back* which explored the work and personality of Bob Dylan. Dick Osgood was asked to moderate a discussion of the film and went up and down one aisle with a portable microphone after the picture was shown. In the opposite aisle, with mike, was Ross Caccavale, manager of the theater in which the film was screened. Comments were taped, and the gap was wide between those of the young people from Seaholm High and the ladies of the Council. The older people had a hard time understanding Dylan and what he was all about, but the students considered him a poet-philosopher and "dug" him all the way. Osgood edited the tape to bring out the interesting and sometimes amusing contrast in points of view, and the tape was used on *Close-up.*

The following September, the Motion Picture Council started monthly showings of films for selected high schools students, followed by discussions moderated by different personalities. The series started with City of Detroit schools only, but within five years began to spread to include schools within a 30-mile area. It became an institution (not broadcast) called Youth Film Forum.

Head of WXYZ *television* news in 1968 was William Fyfe, and working for him as "assignment editor" was Frank Benesh. Television now had units cruising around the city in constant communication with news headquarters at Broadcast House. Its own film developing laboratory had been installed on the first floor. Anchor man was Joe Bacarella's old friend from Monroe and WCAR, Barney Morris, a big young man with a fresh, round face. Working with him was another large fellow, a natural clown who tried hard to be serious, Bill Bonds. Another newsman who specialized in feature stuff was a thin, grim-faced humorist and frustrated actor, Jim Herrington. Osgood was put on the news Friday evenings with reviews of plays and pictures. For him, this became the high spot of the week.

Channel 7 news was fighting hard for an audience, but found it impossible to alter the long-established viewing habits of Detroit which kept most eyes and ears receiving from Channels 4 and 2 at news time.

The dishes had been cleared, their coffee had been "hotted up," and the two men sat smoking at their table in Topinka's Country House. One was Dick Purtan of WKNR. His little brown eyes observed life through spectacles with cold amusement, and the thin lips under his comical black moustache curled with wry cynicism when he smiled. Both men were in their 30s.

"I can remember one record by Al Martino," Joe Bacarella was saying. "It was a million-selling smash. Now this artist comes to town and tried to get it on the rock stations and they wouldn't play it. Yet it was a smash record."

Bacarella's jacket hugged his fullback shoulders smoothly, the collar of his jacket fit snugly around his shirt collar which rose rather high on his big neck, and the shirt cuffs extended properly beyond his jacket sleeves. His tie was full. His wide-lensed, wire-rim spectacles were the "new thing." Opposite him, Dick Purtan looked dowdy.

"Now it's totally different," Joe continued. "The best example would be—Perry Como. Look at his hit *It's Impossible.* The rockers are playing it. A few years ago they would not have touched it, I don't care if it sold a million copies."

"Yeah. Right." Purtan muttered.

"On the other hand, look at the *Easy Listening* list in *Billboard*. Jerry Vale can come out with a record and not sell copy number one of that record and still come out Number One on the *Easy Listening* list, because that list doesn't take sales into consideration. It only takes into consideration how many MOR stations are playing this record. So every good music station in the country can be playing Jerry Vale and he doesn't have to sell one copy. They're playin' it only because it's Jerry Vale's latest. Well—we don't do that, because it may be a lousy record. But people come to me and say, 'Why aren't you playin' Jerry Vale?' "

The two men smoked in silence.

"How'd you like to join us at WXYZ?" Bacarella asked.

Purtan feigned a little surprise. "WXYZ?"

"Afternoon drive," said Joe.

Purtan made a face.

"I'll tell you what's happening," Bacarella offered. "Martin & Howard are not doin' it. The ratings just aren't there; and there's no business. We just gotta make a change."

"But you say *afternoon* drive."

"Well, they have a big contract. We got to go slow."

"How does Fritz feel about it?"

"Chuck wants to make sure and move lightly because there's such a large investment. But Dick, they've just got to go."

"Why afternoon drive?"

"Come over; get used to us; and when we open up the morning you go in."

"What's this do to Dave Prince?"

Bacarella didn't hedge. His Italian brown eyes were steady. His manner was forthright.

"Dave Prince has been on thin ice for a long time," he said.

"What happened?"

"It was something before I got there. I think it was the Walled Lake thing. Dave is a good man, a very good man. He's good in the afternoon, but he's not our idea for the morning."

"Gee, I don't want to do afternoon drive, Joe. I've had an offer from Baltimore—WBAL—and—I just don't want to do afternoon drive. I'm not an afternoon personality."

"Well, if you want the job, you've got it."

"I'll let you know but I think I may take the Baltimore job."

He did. About a month later, he called Bacarella.

"They got a governor down here named Spiro Agnew and with Maryland taxes halfway to the moon, I called the guy *Spiral* Agnew and got chewed out. I use the word "guts" and I'm told "guts" is not a word that you use in Baltimore. Is your offer still good?"

It was. Purtan returned to Detroit and replaced Dave Prince on afternoon drive in May 1968.

Bacarella understood that Martin & Howard were not his lions but he began to

Dick Purtan

hammer away at Fritz. One of three rating services indicated that Martin & Howard were making a little progress. Fritz felt if there was any sign of picking up numbers, they'd better stay with the team. Bacarella kept hammering.

If Martin & Howard did not know Dick Purtan was now hanging over their troubled heads, they soon learned. A record pusher phoned Specs Howard at his home.

"I don't want to shake you up, Specs, but the talk is WXYZ is going to buy up your contract and put Purtan on in the morning."

"I've heard nothing about it. Who told you?"

"I can't tell you. It came from New York."

Tough little Jerry Liebman laughed.

"Hey," he bubbled. "We got a 2-year firm, that's all I can tell ya. If they buy me out, I'm not going to starve. And if they don't, there's still a chance for us to prove ourselves."

Before Martin and Howard went on vacation in August, they had a meeting with Fritz. Bacarella was present. There had been no word of a contract renewal, and Specs Howard knew it would be a fight.

"Look," he proposed, "we're willing to prove some things to ourselves and to you. Renew the contract and bring us down from $48,000 to $30,000—on a six months thing. You know—the 6 we have left plus another 6 which will give us a year. Just give us time to tie this together."

"Go on your vacation and don't worry," said Fritz.

As Specs Howard limped down the hall after that meeting, he said to Harry Martin, "Well, Hare, the handwriting's on the wall."

The Howards spent 10 days in the Catskills; they went to Gettysburg and Washington; and they stopped off in New York City. Lee Vanden-Handel, the jolly sales manager, knowing nothing of what was happening, had asked Specs to make some commercial calls with him to drum up business.

"Sure," said Specs. "I want the kids to see the Bowery and the Statue of Liberty anyway."

Meanwhile, the tamer kept working on a lion named Fritz. One day, the ice-cold eyes of the lion reflected the Italian fire of Joe Bacarella. Fritz smiled.

"Let's go to lunch."

"If you don't mind," said Bacarella, "I'd rather *not* go to lunch. I'd rather solve this problem."

The discussion went on for 4 hours without interruption. The next day Fritz called Bacarella.

"We could have saved ourselves all that hassle yesterday," he said. "The new ratings are in."

"And?"

"They're down again."

That settled it. The Howards returned to Detroit on a Friday. Specs had a call from Nancy Gallagher; Mr. Fritz would like to see him. Putting down the phone, the spunky little Jerry Liebman turned to his wife, smiling.

"Honey, I don't know how fast we're going to be let go, but I think this is it."

It was. Fritz and Bacarella were waiting for him.

Specs Howard: "It was like, uh—you know—'I don't know how to tell you this' type thing. And—we sat, and—were very nice. I remember, the scene was a very tender scene. I think that they respected the fact that we had worked hard and tried."

They did. Fritz admits today the two had been oversold in the advance build-up, and that the change of format before they got off the ground ruined what chance they might have had.

Bacarella left the meeting to tell Dick Purtan his move to the morning drive was definite. When he had gone, Fritz offered Specs Howard afternoon drive at $30,000. Specs turned it down.

Specs Howard, Harry Martin and Ray Koeppen—never great personal friends—met for the last time in Koeppen's little office at Broadcast House. Koeppen was relieved; Detroit had been a traumatic experience for him. He returned to Cleveland and managed an FM station which he purchased with a friend. Harry Martin drove a U-haul truck to San Diego, followed by Carol in their car. Martin resumed the career he had dropped as a single disc jockey. Specs Howard went to Cleveland and worked for a time at the station they had left. He eventually returned to Detroit and established the Specs Howard School of Broadcasting Arts, Inc. Johnny Randall had told him the school was for sale and he bought it from Lee Alan.

Dick Purtan's afternoon drive-time was taken over by a heavy-set jockey named Mike Sherman.

A faint echo of the Lone Ranger's call was heard in the mid-'60s. Fred Flowerday put together a special *Lone Ranger* album which he felt would be a collector's item.

"I had the idea that the sponsor of this should be General Mills," Flowerday remembered, "'cause they had been associatied with the program so long. Ray Meurer and I flew up to Minneapolis—and incidentally, saw Russ Neff. We thought it should be a premium offer—a box-top thing—one dollar. I thought sure General Mills would do this, but a whole new group had come in there, and they didn't have the feeling for it."

Flowerday returned to Detroit and sold his idea to Dodge dealers who became "the good guys in the white hats." It was a successful promotion.

In August, 1968—the same month Martin & Howard broke up in Detroit—John Campbell was moved from Los Angeles to New York, and John MacMahon was moved from WXYZ to replace him at KABC. Asked what his contribution had been during his 8 months as top man of WXYZ-TV, one associate thought hard for a moment and said, "Well, he put down a blue carpet."

Dean McCarthy who had followed Pete Strand as program manager, was promoted to the job previously held by John Pival, Campbell and MacMahon.

In March, 1968, Dick Osgood was asked by Milton H. London, President of the National Association of Theater Owners of Michigan, to be the master of ceremonies

Dave Prince's housewife program segment has responsive audience.

at a "celebrity luncheon" at which the national president of the association, producer Julian S. Rifkin, was principal speaker, and Phyllis Diller received a plaque as "Star of the Year."

Osgood himself was surprised with an award in whose metal are etched the words: "The Show World's Award of Appreciation to Dick Osgood for distinguished service to the world of entertainment, presented by the National Association of Theater Owners of Michigan, Inc., Milton H. London, President, March 28, 1968——."

In June, Marty Greenberg arranged a surprise cocktail bash for Osgood in one of the large TV studios. All the town's top theater men came. They presented the veteran with a handsome watch with a gold mesh wrist-band; and Charles D. Fritz gave him an oil portrait of himself, saying, "This is not a hint that Dick is about to, or should, retire."

"The only way WXYZ will ever get rid of Dick Osgood," Marty Greenberg commented, "will be to scrape him off the walls with a putty knife."

In August, WXYZ—Radio lost Marty Greenberg to ABC's newly acquired station in Houston, Texas—KXYZ. He enjoyed Texas and liked Houston, but there was not enough action at the station for him and he was glad to return to WXYZ in March, 1971, replacing Lee Vanden-Handel as radio sales manager.

Osgood was now 67 years old, not feeling his age so much as the great changes that had occurred in the 34 years he had been with the station. For the first time, people were growing old in the broadcasting industry.

Dick Osgood: "It bothered me that nobody knew anything of the past of the station. Only a few at the surprise party had ever heard of George Trendle."

Osgood made an appointment with Trendle and the two met again on July 8, 1968, 4 days after Trendle's 84th birthday. He had not changed much in outward appearance, although he complained of a painful jaw condition and was recuperating from a broken leg. He was still going to his office in the Mutual Building every day.

Forty-five minutes of the station's history was taped with the man under whose leadership it had attained its call letters, won international recognition, and lifted Detroit to 4th position as a production center—the man who was being forgotten. The tape was edited to fit the 25-minute format of *Close-up*, and given to the news department.

Mr. Trendle began to write rather pathetic follow-up letters. When was his interview going to be on the air? He would like to hear it first and okay it.

"My leg is getting better all the time," he wrote, "and I'm beginning to drive further and further.... I'm still messing with my jaw, but three trips to Ford Hospital haven't developed anything either.... People are suggesting that I go up to Mayo's but I think it's useless. Maybe if I got out to WXYZ and somebody hit me a punch in the nose, it will take care of the whole situation."

When he "went out to the station" to hear the tape, nobody hit him on the nose. In fact, nobody even bothered to say hello to him, except Osgood. Isabelle Stouffer who was in charge of personnel did come into the board room to ask if he'd like coffee. He had a pitcher of lemonade and listened to the tape. He approved it, except for one thing.

"I didn't say anything about Ray Meurer. When you put that on the air, you say something nice about Ray."

It didn't go on the air. Nobody was interested.

58
Triad

More and more the three divisions of the station—AM, FM and TV—became separate entities, each competing for an audience not so much against one another as with opposition elsewhere. Where once there had been a half dozen radio stations in all Detroit, there were now more than 30. There were 7 television outlets, and FM radio was developing like crab grass.

WXYZ television, under Dean McCarthy, had no Soupy Sales, Johnny Ginger, Marv Welch, Johnny Scat Davis nor Betty Clooney. No Don Ameches nor Herbert Marshalls were being imported to host the movies as in Pival's time. But weekdays stepped off on the right foot with easy-going, affable Bob Hynes and the *Morning Show*. Hynes came from Boston by way of New Haven and his set was remarkably relaxed and congenial. The morning movies were doing well with Rita Bell. One of the new personalities was Don Haney, tall and black with a keen mind. He had switched from pre-med to broadcasting. While announcing at WJR he had cracked television with a show of his own once a week in Canada. WXYZ-TV hired him to do news, and he had his own talk show on weekends, called *Haney's People*.

TV's main effort now was expended in the news room. Personalities came and went; only a computer could keep track of it. William Fyfe left and Frank Benesh moved up to take his place. Bill Bonds went off to Hollywood and Barney Morris became anchor man. Later Barney would go west and Bonds would come back. Roy Allred abandoned the weather to Jim Smith. There were Ken Thomas, thin-faced, smiling and smug, and Solon Grey, masculine, handsome and unhappy at WXYZ from the beginning; and Larry Adderly, Dave Diles' likable assistant from WWJ. There were Ven Marshal, older than the others, grey-haired, short and heavy-set; Jim Herrington with the reddish hair and tight-lipped smile; and Erik Smith who had the cherubic look of an old-fashioned stock-company juvenile. Thrown into the battle now were cameramen, sound men, film editors, writers and expediters. It was a busy room and had all the makings of a Number One news operation, but somehow—no one could figure why—it wasn't winning.

In July, 1969, there was another change at the top. Dean McCarthy was summoned to Manhattan to be vice president in charge of programming for owned-and-operated TV stations, working under John Campbell. He was followed into Pival's old office by Donald Kech. Some wondered why it had not happened before because they suspected he had contributed substantially to Pival's downfall. Kech made a good appearance on the editorials, and under his leadership, television moved along smoothly enough without conspicuous innovation.

Dick Purtan gave AM the spring it needed to loosen its flippers and start a slow float upward. He introduced an element of practical joking to the early morning hours which caught attention. He would call people by secret arrangement with their friends, and "put them on." He would call, for example, a small counter lunch stand and put in a mammoth take-out order so full of complications and changes that the poor man on the other end was beside himself. "Thirty with mustard and 10 with piccalili, 5 with piccalili and chopped onion—no, make that 6 with picalili and chopped onion. Wait a minute. What's that Sam? Sam doesn't want mustard so that's 29 with mustard and 6 of those with chopped onion. Now how many does that make with relish?" And so on. It was amusing to the audience but catastrophe for the victim. Some of the calls seemed in bad taste, but anything for listeners, and Purtan was getting them.

Dick Kernen had said that there was something about WXYZ that attracted people, but now—

Kernen: "The only time I've known this particular phenomenon to wane was around 1969 when you did not have people knocking down doors to work there."

Three years before, the *Detroit Free Press* did a feature on Detroit disc jockeys. Aside from those already at WXYZ, two others had since moved over: Joe Bacarella from WCAR and Dick Purtan, WKNR's "odd ball." Now came the 3rd, Tom Shannon. CKLW dropped him because it wanted all its jocks to sound the same. Tom Shannon was a compact little figure, a lot less naive than he seemed. It was his act to seem confused by it all. He loved horses, bought a farm, and attracted a stable of women.

Tom McIntyre ended his tenure as head newsman in AM and moved on. Joe Vaughn replaced him, and a solid Michigander with the Bostonian name of Lowell Newton, became editorial director.

George W. Trendle became anxious about the airing of the interview Osgood had made with him. He wrote an inquiring letter:

> Dear Friend:
>
> About two years ago you were all hot and bothered about some kind of a story you were planning for which I supplied some information at your request incurring a trip to Broadcast House.
>
> Since then I have not heard a thing about it. Is the whole thing dead or do I meet with a surprise one of these days? I'd be interested in hearing from you.
>
> Sincerely,
> George W. Trendle.

The reply:

> Dear Mr. Trendle:
>
> I am sorry about the documentary. It has been in the hands of the news

department since your visit but I do not have the power to insist that it be used. I hope it will help you understand the situation if I tell you what happened the other morning.

The men in the newsroom were watching secretaries arrive for work. One of them commented, "Gosh, there's such a turnover I don't know half of them."

I said, "It bothers you that you don't know them. What gripes *me* is that I've been here more than 30 years and *they* don't know who the hell *I* am."

Cordially,
Dick Osgood

Trendle's reply was filled with understanding:

Friend:

I really enjoyed reading your letter and I can appreciate how things have changed at WXYZ!

I did not know whether they had thrown all that material into the waste basket and had forgotten about it or whether it was being deferred until it could be used for my obituary or something like that.

Best wishes, always.

Sincerely,
George.

In addition to the lift Dick Purtan was giving AM radio, Fritz and Bacarella with help from Neal and Klemm were gradually creating a format which was original and which would one day serve as a model for other stations. It was based on the demographic appeal to young marrieds 25 to 34 years old. They also had help from a tall, thin, rather bloodless man named Bob Henaberry. Henaberry became a sort of advisor to the program directors of the ABC-owned stations. Eventually he borrowed the demographic format, created and refined in Detroit, for other stations. New York was on top of everything that was happening at WXYZ. At times, too much so.

Charles D. Fritz: "The biggest decisions are the easiest ones to sell at ABC. When I tell 'em that our transmitter has got to be replaced and it's a hundred thousand dollar item, they say, 'Sure. Let's find out which is the best transmitter.' But I can't get a 2-buck raise for the gal that types up the contract."

It was a combination of the demographic approach and Bruce Miller's unique department that helped Charles Fritz wrest a big account from the smug embrace of WJR.

Fritz met Bob Dale at a restaurant in suburban Detroit. Dale was merchandise manager for Sears Roebuck. Both men were graduates of the University of Michigan. They were not close friends although they had socialized occasionally. Fritz, in his easy, direct but rather hesitant manner, tried to explain what he was attempting to do at WXYZ.

WXYZ softball team around 1964

"We are goin' to position ourselves so that our particular audience is goin' to be the best audience for you to buy, because the 25-to-34-year-olds are the people with the new homes and the young children and they're in the age of acquisition. They're buying everything. I just don't think WJR is delivering that audience to you."

"WJR is not delivering much of anything," Dale conceded. "They won't give us our spots where we want them, besides which there are billing foul-ups."

Fritz smiled and continued, "Our figures show that their audience is generally older and when you're over 50 or 60, you slow down on refrigerators. We have information that shows what kind of merchandise you're selling and it matches the demographic make-up of our audience."

"We're already placing some spots with you."

"You buy a number o' stations, but you only buy one station as an anchor and give that station a firm one-year contract."

"Well—WJR seemed to be the big one."

"WXYZ will put your spots on when they'll do you some good," Fritz promised. "And there'll be no billing foul-ups."

Fritz smiled confidently. Dale was silent, thinking.

"Now the other thing that we'll do," Fritz continued. "You have terrible commercials and I've got a guy at the station and we will produce commercials for you to use on all stations at a package rate that is just scale. Anybody else is goin' to charge you 3 or 4 times this. We'll toss in a $10,000 production package if you give us a year's contract."

"You make it sound good," Dale admitted. "But I can't go over the agency's head like this."

"You'll hear from your agency," Fritz said, picking up the check. "This lunch never happened."

Fritz called Bartley Walsh, the young salesman.

"Bart, I have a hunch you should go after the Sears account. Renewal time is coming up at WJR. Why don't you contact the Werner agency and you and Bruce Miller work out some idea for them for Sears?"

Bruce Miller outdid himself with new, fresh ideas for Sears Roebuck. Walsh took the tapes and a deal he had worked out with Fritz to the Werner agency, and Werner went to Bob Dale. Dale, smiling a little strangely, listened to the tapes and the deal.

"Sounds good," he responded. "I don't see how we can turn this down, except—these tapes are trying to sell Sears' name and Sears doesn't need to have its name sold."

The criticism got back to Bruce Miller.

"Some of the best things I've ever done," he moaned, "and they'll never be heard."

"Dale says to just name the items and prices—bang, bang," the agency man explained.

Bruce was frustrated but he did add polish and a little class to the bang, bang. It was an important sale and, aside from the money, a victory for the station.

Fritz' steel-blue eyes looked up at Kernen and a faint smile played along his lips.

While AM was slowly surfacing and TV was coasting with Kech, FM fought a revolution. The first shot was fired at the annual managers' meeting by two young men from Chicago. Alan Shaw and George Wahraes pointed out that a quarter to a third of the 100 best-selling record albums listed in *Billboard*, the trade magazine, were made up of so-called "underground" music, or "progressive rock." The point Shaw and Wahraes made was that although these albums were selling millions of copies in the stores, they were not being used on radio. They proposed that the ABC FM stations which had been experimenting with one format after another without success latch onto this music.

Hal Neal and Ralph Beaudin dug into a study of the situation and decided to install a format using this music at all ABC-owned FM stations. Rather than have each station program its own selections, they decided to produce and tape the shows in New York. It would be cheaper and, of course, it would diminish the possibility of local program people straying from what they wanted.

For their jockey they chose a Lutheran minister with a rich voice named John Rydgren. They called him "Brother John" and the whole package was labelled *Love*. Each day's programming was sent on tape to each station. The results were disappointing.

"We did not make the inroads that we had hoped," Dick Kernen complained.

"I think Hal Neal finally was the guy who realized that you have to get back into the local market," Charles Fritz remembered.

"We discovered that the automation type of approach just would not work with this format," Hal Neal conceded.

The real cause of its not working in Detroit, aside from mechanical problems, was a little FM station called WABX. WABX-FM tied into this same underground music and was programming albums by *Original R & B,* Bill Dorgett, Hank Ballard, Buddy Holly, Ike and Tina Turner and all the others.

Dick Kernen: "WABX did not knock the market on its ear, but they began paying the bills."

Hal Neal: "Some competitors developed in Detroit, doing this kind of music on a non-automatic basis, and we were coming off a poor second."

"Brother John" was sent back to more heavenly work, automation was junked, and Dick Kernen got a promotion he did not want.

Kernen: "In the Fall of '69, it became evident to me that there were a lot of serious failings here, on a number of levels; and because my function was a bureaucrat, I said to myself, Joe is overall responsible but FM is my deal. Now if they find out that this thing is screwed up, they're going to come to old Dicky Kernen and say, 'Why is this screwed up, Dick?' And I better have some answers. So I wrote this long memo, and I said this is the problem, blah, blah, blah. We're not logging correctly, it's not being sold, just on and on and on."

Fritz read the memo and sent for Kernen and Bacarella. He was re-reading the memo when they stepped in.

"*Oh*-oh," said Kernen.

"If you're so smart why don't you run it?" he asked.

"Well, Chuck," said Kernen, "I think really you're missing the point. I'm not down here for myself. What we need here is a salesman. We don't need someone to run—I can run it as Joe's assistant and that's fine with me."

"I've got my hands full with AM," said Joe.

"Hal wants someone to run it," Fritz told them.

"Look," Kernen insisted. "People tend to view me because I do certain things competently as, you know, here is another guy going places. I don't want to go places."

Bacarella feigned good-natured disgust, "Come on, Dick."

Kernen ignored him. "Making a living to me is exactly that and nothing more. All I want to do is to work and do the job, and when it's over, I go home and turn it off. I have—you know—very close relationship with my wife and my brothers and my mother and father—which sounds kind of weird."

"You apologizing for that?" Fritz demanded.

"No. And I'm not criticizing people who take a career seriously. I just can't be that interested. That is not what I consider my life. Work is something I do because I have to do it."

"Okay," said Fritz, grinning at him. "You have to do it. FM is what you have to do. You run it and you hire a salesman."

"Well—I don't think that's the whole answer. What we need—"

"What you need—get. You're running FM."

"Okay," said Kernen. He felt miserable. But it was exciting. The first thing he did was remarkable: for his salesman he took a young engineer off the control board. His name was Jim French. Together, they discovered a whole new world.

Kernen: "The FM stations that were showing growth were the, quote 'Underground Free Form,' unquote, radio stations that were playing not only what has now come to be called progressive rock. But they were playing Wagner; they were playing Czechoslovakian folk music; they were playing just an incredible gamut of music. And they were talking a lot. They were playing 20- and 30-minute segments from philosophers like Marcuse. In some instances, it was extremely creative, but no one can be extremely creative 24 hours a day 7 days a week. The bursts of creativity were few and far between but they were brilliant when they occurred."

FM radio was giving expression to what was happening in the heart of youth. As Bobby Dylan sang, something was happening in America. The ladies of the Motion Picture Council had not seen it or felt it, but the high school students had felt it, understood it, and were in tune with it.

Youth was questioning the Establishment and the Establishment's answer was tear gas and rifle shots. Haight-Asbury was the mecca; the inner city ghetto was home. Many divergent groups found themselves moving defiantly in one direction by different routes. ABC-FM became the voice of this revolution.

Perfectly suited to be its champion at WXYZ was Richard Kernen. His hitch in the Navy had been served with disdain; he had no respect for the ordinary process of formal education; and he was far from eager to move into the role of an executive.

The first live talent he engaged was "a fine fellow who had great credibility in the so-called hip community," Harvey O'Shinsky. O'Shinsky went to Broadcast House late on Thursday nights when no one was around, and taped an hour of "comment," later—two hours.

Kernen: "We began to hire disc jockeys later who were very hip and groovy—I mean people who were freaks in the finest sense of the word. I mean they all had shoulder-length hair and they all smoked dope and were supporting the revolution and that was where it was at."

Charles Fritz may have had some apprehension as to the direction in which FM was moving, but did nothing to change its course. Other conservative members of the staff disapproved openly. Porter Bourne who was now in charge of "actualities" in the radio news room next door to the FM studio, complained loud and clear about the smell of marijuana that permeated the area on weekends.

Dick Osgood was concerned with more than the smell. *Call Board* was simulcast, and on FM it followed a Sunday night talk show which was pure anti-Establishment propaganda. One Sunday was devoted to the Chicago Seven who had led in the disruption of the 1968 Democratic Convention. No opposing view was presented. The next Sunday, another group championed the cause of draft dodgers who were running off to Canada rather than serve in Vietnam. The third Sunday, young Americans who had just returned from Cuba where they had gone, presumably, "to help bring in the sugar crop," devoted the half hour to telling how much better everything was in Cuba than in the United States.

Osgood wrote a memo addressed to Fritz demanding that *Call Board* be taken off FM, but he didn't send it; he was afraid *Call Board* might be cancelled on AM too. He justified his cowardly passivity with the argument that a bright program of show music, and interviews with people of some accomplishment might be an antidote for all the destructive propaganda that preceded it.

Dick Kernen and the FM people could not have cared less about the reactions of aging members of the Establishment such as Porter Bourne and Dick Osgood.

Dick Kernen: "Within a year, ABC FM stations became the hippiest group of radio stations in the country. I mean, we were doing just fantastically weird things.

59
Strike!

The midnight news was ending and the all-night disc jockey was standing by. Joe Bacarella stuck his head in the studio door.

"I'm not going home tonight," he said. "I'll be in my office; I'm going to try to get some sleep. Just knock on my door when you're called to go out on strike."

Bacarella cleared off the top of his desk and tried to lie down on it but he was too tall and big. He stretched out on the carpeted floor. His mind went spinning back through the past few incredible days.

He had sat in on negotiations and kept his mouth shut most of the time.

Negotiating was foreign to his nature. Fritz, Keck and the ABC lawyer had carried the ball against Forsythe and AFTRA's national executive secretary, Sanford I. Wolff. Richard Forsythe, now executive secretary of the Detroit Local, was hardly a union type; he was a gentleman with a dignified humor. He was getting old, and often, as he drooped over the big table in the board room, he appeared to be going to sleep, but he had a mind like a steel trap and when something against the interests of AFTRA fell into it,he snapped up and was surprisingly forceful.

Chuck Fritz handled the situation admirably. They would retire to one of the offices, and Fritz would ask for an explanation of something the union had brought up and how it worked "upstairs"; then he'd go back and tackle the opposition head-on.

Joe Bacarella found the floor uncomfortable. He got up and slouched down in his chair with his feet on the desk and closed his eyes.

Negotiations had dragged on since the middle of April; the union had voted on the 4th of June to strike; and it was now the 17th; and Joe thought, "This is too bad. Just when we're startin', just when the station is doin' fine, it has to have something like this to upset it."

"Asleep, Joe?" The disc jockey was at the door of his office.

"No."

"Okay. I have to go now."

"What time is it?"

"4:30"

"All right. I'll see ya."

For the first time since he had left WCAR three and a half years before, Joe Bacarella went on the air. He was a disc jockey again.

All executives spent most of the night in Broadcast House planning strategy. TV was prepared. Sales manager Steve Mathis would handle Bob Hynes' *Morning Show*. Frank Benesh himself would do the news, and the handsomest woman on staff, Jan Coil, account executive, would be the weather girl. The announcing would be handled by secretaries.

"Read the commercials in the same low, breathless tones you use on the telephones and the sponsors will love it," Donald Keck told them.

For FM, Dick Kernen found some old *Love* tapes and got busy editing out Brother John. He and Jim French would substitute—live.

In AM, Andy Driscoll, publicity man and amateur actor, would get his days of glory on the air; Joe Vaughn would broadcast the news; Bacarella would do the Dick Purtan show. In fact, that first day, Joe Bacarella was on the air eight and a half hours, waiting for relief to fly in from New York.

The strike began at 4:30 A.M., June 17, 1970. By 8 o'clock most of the 34 striking air personalities were at the end of the driveway to start their picketing. Later, they would picket in pairs or fours, but the union wanted all the "names" to be out in full force in the first morning. Newspaper reporters and photographers were there. There were hot coffee and sweet rolls; everybody was smiling; it seemed like an early morning picnic.

Behind the happy facade of the strikers, emotions were mixed. There were defiance and determination; there was embarrassment; and there were jokes.

The veteran news man Ven Marshall was unenthusiastic about the strike, predicted it would be lost, yet was a willing supporter who retained a grim sense of humor.

"I've gone over my financial situation," he said, "and I figure I can only hold out about—ten years."

Others who expected to gain nothing in the strike—Bob Hynes, Dave Diles, Dick Osgood, Jim Smith—marched around in a tight little circle where the driveway parted from Ten Mile Road. The bitter monotony ahead was not yet realized; egos were buoyed this first morning by photographers' requests for pictures. They especially wanted to get Barney Morris who had broken his ankle and was marching with a cast on one leg.

Dick Purtan stood apart talking in a low voice with a reporter, his thin lips curled in their cynical smile as his little brown eyes watched the pickets as though they were engaged in a comic pantomime.

Shortly before 9 o'clock, nearly 250 non-striking employees began to arrive. Some seemed surprised to see the pickets; others had been listening to the station and knew something was wrong although mentioning the strike on the air was forbidden. Some drove through the line waving and smiling to their friends. Others kept their eyes fixed on the driveway trying not to see anybody.

Bruce Miller was a member of AFTRA only because he had to be. He and his wife, Jean, had attended meetings and taken such a positive stand against striking that some wanted them barred from the meetings, but they paid dues, and to keep them out would be unconstitutional. An agreement was reached with the station that Bruce could work at his writing as long as he did not go on the air. He drove through the lines.

Bess Ashton turned off Ten Mile into the drive without slowing down. As she stepped inside Broadcast House, she found a little group gathered.

"Isn't this awful?" one was saying with a long face.

Bess gave a derisive hoot.

"I almost ran Jim Herrington down. On purpose!" she anncunced happily. "I shaved him."

She laughed again and went swinging gaily down the hallway.

Patty Prendergast slowed down in her Volkswagen. She thought the strike was stupid, but she didn't want to knock down a picket. Her jaw dropped low as she saw one of the men standing there, and her voice filled with astonishment as though she had caught her grandfather on the porch of a whore house.

"*Mis*ter *Os*good! What are *you* doing here!"

It was a question he asked himself later in the day as he marched back and forth for his allotted 2 hours. During negotiations, the union had asked members to appear at meetings in the board room as observers. They were not to say anything, but it was felt that their presence alone would have a psychological effect on management.

One morning, Osgood had arranged to leave early for a screening downtown, and dropped off to "observe." By coincidence, during the 10 minutes he was there, he heard ABC's New York lawyer propose that when staff performers reached the age of 65, they be forced to retire without separation pay. Osgood was now 69 and had been with WXYZ almost 35 years. His separation would be substantial. He remained at the meeting only long enough to hear Charles D. Fritz half agree in principle but suggest such a new rule, if agreed to, not apply "to present employees." Fritz won his point—the only point that would benefit Osgood. So why *was* he there on the picket line.

Dick Osgood: "First, I considered it my duty as a member of AFTRA. Second, I was Detroit's senior broadcaster and I wanted to set an example. Besides, I owed the union several raises in base pay through the years, sizable talent fees on a semi-freelance basis, separation pay if I was fired, and a decent pension when I retired.

"There were other reasons. I was demonstrating against the inhumanity of a giant corporation and its ruthless remote control, symbolized by that New York lawyer. I was walking there because of what George W. Trendle did to Fran Striker. I was protesting the ignorance of the disc jockeys who didn't know what the preceding generation had done for them. I was marching in memory of all those pioneers who had made it good for us: Gwen DeLany, Earl Graser, Ruth Dean Rickaby, Jay Michael—all dead now. And the living—Al Hodge, Ernie Winstanley, Ted Robertson—who had put their jobs on the line during the Great Depression when it was impossible to find jobs. They had gambled, not only to make things better for themselves, but for these disc jockeys three decades later. I was picketing in tribute to the pioneers and doing my bit for those who would come after me."

What were the feelings of those inside, working?

Bruce Miller: "Sick. Sick—about the whole situation. Nobody would talk about it. They hated to have the guys out there. They hated the whole situation. Nobody believed it. But all of a sudden one morning everybody's out there and they're raving maniacs. People were kind of amazed that it could come to this. They wanted somebody to say, 'Come on! Get the hell off the road and come back in to work and let's be friends again. Why did this have to happen? Why! Look at this beautiful land, this beautiful building. We got the best working conditions in the world and—they want something. What do they want? It was just amazing."

What they wanted was fees over and above their staff salaries for on-the-air appearances, equal to those paid by other ABC-owned stations in New York, Los Angeles, Chicago, San Francisco and Pittsburgh. Fritz and Keck said no, WXYZ's competition was not in San Francisco or Pittsburgh, it was in Detroit, and no other station in Detroit was paying such additional fees.

Strike headquarters were set up in a neighboring Howard Johnson motel. Elected strike chairman was a 24-year-old black newscaster, Paul Berry, who had better reason than most to be active. Paul had been hired by Frank Benesh as a "trainee" in the TV news room. He began doing things on the air and his co-workers thought he should be paid scale. Osgood, president again of the Local, took the matter up with the Board and the Board stood firm; Paul Berry must be paid. He was not fired, as some feared; he was paid AFTRA base which meant a substantial raise. Paul devoted himself whole-heartedly to the strikers' cause even though his mother was in surgery and he himself was not well.

Least enthusiastic among the strikers were 3 men who had started work at the station just in time to find themselves outside at the end of the driveway with little understanding of the reasons for it. TV's Solon Gray had been there a little longer than the others and was finding it tough to win acceptance by the management; participation in a strike was not going to help his situation. Arthur Penhallow had been hired by Dick Kernen to produce for FM, but because he was also on the air, like Bruce Miller, he had been forced to join AFTRA. Although he was part of the "revolution," he wanted no part in the strike. He was short, scraggy, shabby and smelled of marijuana. The 3rd was Jack Hayes, a strong personality in his early 30s, hired from New York by Bacarella who had heard him on WNBC and knew he would add power to the AM line-up.

On the 20th of June, 3 days after the strike began, Tom Jones was booked for an appearance at Olympia. The Tom Jones television series was being carried on Channel 7. A group of pickets, including Ven Marshall, Barney Morris and Don Haney, went to Olympia and carried signs which read: "Tom and I work for ABC." It was hoped that the English star might refuse to perform. He performed.

Two nights later, at the opening of *Hair*, members of the audience were handed a mimeographed message as they filed into the theater. It read:

> CONGRATULATIONS———to the management of WXYZ for its generosity in supporting the Detroit premiere of *Hair*. As you know, the proceeds will be used to help the United Nations World Youth Assembly.
>
> We would hope that the management of WXYZ would be as generous in considering the proposals of their 34 AFTRA performers currently on strike at the station.
>
> We hope you enjoy the show.

All the fun went out of it after a few days. People inside Broadcast House worked long and hard. Executives flew in to help from other ABC stations across the country.

Charlie Kocher earned a new nickname. Stage hands, electricians, camera and sound men threatened to stay out with the performers. Kocher literally "tough-talked" them across the line and kept them on the job. He had several angry confrontations with the business agent of the engineers' union. Kocher was so full of fight that the agent complained to New York, New York got to Keck, and Keck called Kocher to his office.

"Charlie, take it easy."

Kocher had a peculiar way of laughing; he chuckled on the exhale and laughed on the inhale, and the laugh was a little grating. Now, he laughed.

"Well!" he managed. "I've finally become as tough as Riddell wanted me to be back in *his* time. I finally made it."

The technicians stayed on the job and the executives began to call Kocher "The

Rock."

On the 1st of July, Detroit was in the middle of a hot spell. Page 1 of the *Free Press* was dominated by the picture of a wilted polar bear in the Detroit Zoo. Relief was promised on the 4th. A top headline shouted, "Nixon Claims Success in Cambodia; Senate Votes to Bar More Invasions." *The Detroit News* announced the expectation that President Nixon would reveal his peace plans during a 1-hour live televised "conversation" with correspondents of the 3 networks beginning at 9 P.M. In London, American Vice President Spiro T. Agnew had declared that some people in the United States should "be separated and discarded" from society. Wall Street was going through what the experts called "a liquidity crisis." Profits of the 2nd quarter would probably show little improvement over the disastrous 1st. The House had okayed a $3.1 billion bill to help cities fight crime; and Congress had overridden a presidential veto for the 1st time in 10 years by passing a $2.7 billion hospital construction bill.

But neither newspaper reported that the WXYZ call letters were celebrating their 40th anniversary that day. Inside Broadcast House in a drawer in Joe Vaughn's desk was the story of how those letters were obtained from the Army and the Navy, told by the man who had won them—a 25-minute documentary that would serve well in a time of need. Nobody thought to use it. Outside, the pickets trudged up and down, holding their signs on poles or wearing them suspended over their shoulders.

"I have always said the sandwich man is the lowest form of show biz," Osgood remarked, "and here, after 50 years and 3 months, I've made it."

Inside, Keck and Fritz faced other worries. Profits were down. The Federal Communications Commission had ordered the networks to give up 30 minutes of prime time in the early evening to be filled by local stations. And the government was going to ban all cigarette advertising. This last was the blow.

As for the strike, there was one gleam of hope. The union could keep no secret. Everything it thought and did was reported to the management. And word came in that there was a split in the ranks of AFTRA.

60
Stray Thoughts on a Picket Line

The polarization began to assume its final definition when other unions muscled in. They felt that Aftrans were too gentle. Eight days after the strike began, A.L. McLain of the *Detroit News* wrote a piece headlined: "TV 'Hits the Bricks' with Show Biz Style." He described the "carefree setting," the lawn chairs, sandwiches and bright blue outhouse. He told of the people who continued to work inside and waved and exchanged greetings with the pickets as they drove through the lines.

"Labor leaders from other unions," he noted, "stop by daily to lend moral support—and leave shaking their heads."

These included Douglas Frazer, UAW Vice-President; Gus Scholle then President of the Michigan AFL-CIO; Charles O'Brien, International Representative of the Teamsters Union; Myra Wolfgang, International Vice President for Hotel and Restaurant Workers, and Tom Turney, Metropolitan Detroit AFL-CIO President. They said it would be bad for all unions if AFTRA lost this strike and asked permission for their unions to join in and show muscle. The striking Aftrans called a special meeting to consider the offer.

It was 10 P.M. and the night was hot. Ven Marshall and Dick Osgood had finished their 2 hours of picket duty but there was no relief in sight.

"You go along. I'll wait," said Osgood.

He walked up and down by himself. As he watched the newsman's stubby, vigorous figure disappear in the darkness, his mind filled with stray thoughts about the new "electronic journalism."

It was smug, egotistical and dangerous. For an hour and a half a network could capture the rapt attention of mesmerized viewers with a "documentary" on some controversial issue. The presentation was invariably one-sided, usually these days, left-sided. Before "electronic journalism" a sensible newscaster and his writers could distinguish between political axes and the plain facts of what happened. What happened was reported. Political statements were usually ignored. Today, so much was political hot air! It was fanned by reporters into a wind. A good critic would say, at least, "It's a bad comedy but the people laughed." A poor critic would simply say, "It's a bad comedy." Journalism had slipped to the edge of bad criticism.

Dick Femmel was right when he said, "We do not need advocates in journalism unless they're doing editorial work. What we need is someone who is going to be a reporter, who is going to give you the truth, who is going to be discerning enough to say, these are the facts, you make up your own mind."

And the good advice of Robert Novak: "Tell it like it is, not the way you think it should be."

Radio news was obsessed with actualities. Anything to get as many different voices crammed into a newscast as possible. Never mind developing a story; get the voices. "In depth" reports by one station lasted 90 seconds by the stop watch.

And the power! The whole gist of a man's speech could be controlled, even reversed by editing the tapes. George Romney was crucified by the "journalistic," out-of-context blow-up of his declaration to Lou Gordon that he had been brainwashed in Vietnam. As it proved later, he was right; there was brain washing in Vietnam, and in Washington as well; but the confirmation came too late to help Romney.

A President can make a speech, and before the viewer or listener can digest it, a commentator plunges in to tell him what he heard. And there was the engineer on a network who objected to the way a commentator was directing him to hack up the speech of a leader until the whole meaning was altered. The engineer was "chewed out" and removed from the job.

And the lack of taste! Going into blatant, up-beat, happy commercial jingles on the heels of serious, important or saddening news—death followed by Faygo, hundreds killed in battle, so use liquid eye-shadow.

The negative approach. The news "analyst" asserts that a certain government experiment "has only been partially successful." Couldn't he be positive instead, and say, "has been partially successful"? In addition to asking, "What is the danger zone in this government action?" could he not also point out what might be a zone of hope?

"Why don't you sit down for a while?"

Stray thoughts scampered off at the sound of the voice. It was the guard. WXYZ hired guards to protect its property from the strikers. Whenever there were pickets, there were guards. Even so, someone had managed to cover the driveway with nails and there had been an epidemic of flat tires at Broadcast House.

The old guard got up from his chair on the grass and joined the aging picket who paused to adjust the sign hung over his shoulders and to comb his hair out of the sweat on his forehead.

"I'm a retiree and I live alone with my dog," the guard said, wanting conversation. "I just do this to make a little extra."

The guards had been instructed to write down the license numbers of all cars whose occupants stopped to chat with the pickets. That was against orders. The strikers had persuaded one of the guards to take down the numbers of Donald Keck and Charles D. Fritz, and turn those in too.

"It's easier for me if I keep walking," said the senior picket. "I feel tired when I stop walking."

He resumed his march. The old guard slowly returned to his chair and sat down.

At the emergency meeting to consider the offer of help from other unions, Jack Hayes made an angry speech calling for an end to the strike and a return to work. Bud Haggart, who was president of the Local during this difficult time, called for opinions. Two or 3 fiery speeches demanded that AFTRA accept help from the tougher unions. Moderates maintained that performers were not teamsters, that AFTRA was not that kind of a union, and that nothing would be gained by smashing windshields, hauling scabs out of their cars, and having 100 pickets on the line armed with clubs.

Someone proposed that AFTRA hold off affiliation with any other union until Bud Wolfe, the national Executive Secretary, could return and give advice, maybe try a new approach to management. This view was supported by Richard Forsythe who had left the annual national convention of AFTRA to attend this meeting.

The voters voted to postpone acceptance of the offer from other unions.

We're soft, the old man thought, as he walked up and down at the end of the driveway. We wouldn't have to get violent if we were only united and firm. But we're not. Our union is a collection of egos. It's amazing that actors, opera stars, vaudevillians, night club artists, and now us, ever formed their several unions in the first place. I'm the only one on this strike who remembers the risk taken to get AFTRA started, and the struggle for that first contract, and what it cost to raise the net of base pay under today's high-flying disc jockeys. They think they've done it on their own.

"Hey! Want some help?"

A carload of high schoolers had come to a stop across Ten Mile. A boy and a girl, recognizing the picket, had crossed the street.

"How's it going?"

Everybody was all smiles.

"Give us signs and we'll walk with you."

The young man's offer was turned down but appreciated. The three talked a bit. The boy and girl returned to their car; all waved; and the kids drove off.

The picket resumed his walk, 20 paces west, 20 paces east, 20 paces west, 20 paces east—45 minutes to go.

The morning after the turndown of help from other unions, all hell broke loose. Welcome or not, help came to AFTRA. As the *Detroit News* reported, "chain-swinging, rock-throwing members of several local unions joined pickets at strikebound WXYZ." According to the *Free Press*, Myra Wolfgang of the hotel and restaurant workers was "slightly injured" when a car swept past her. It was said to have been the car of Donald Keck, but Keck denied it. The car of Charles D. Fritz was damaged. Hot coffee was thrown at Frank Benesh. Police were called and one man was arrested. The strike had become ugly.

"May I speak with somebody on FM, please?"

"I'll give you Mr. Kernen, the manager."

"You the station manager," the man asked. "Right?"

"Well—yes—I guess," said Kernen.

"Well, I was listening to your station last night and it's really nice you're live, but—the guy you got on at night is not really that great."

"Oh?"

"Yeah. I'd like to see you about a job."

"The guy that was on last night," said Dick, "was me."

Back and forth under the sweltering sun. Paul Winter, now with an educational TV station, drove by with 2 children. They had been collecting specimens to plant in his wild flower garden. He saw the picket, parked beyond, and ploughed back on foot "to say hello to Richie." Paul had grown a wiry grey beard to match his hair and looked like a forceful character from the time of Moses. He even took a couple of turns with a sign, for kicks.

Now the old man was in the middle of his 2-hour walk—alone again. Paul had gone, and the disc jockey who had been scheduled to picket with him had failed to show up. Again his mind went wandering.

The art of radio, its very special art, was lost. The art had stimulated imagination more intimate than the eyes of a vampire in close-up. No other art could draw one so completely into the scene itself. Reaction to a book was slower and required more concentration. There was a relaxed immediacy of identification for the radio listener. His mind drew the picture automatically. He laughed because Jack Benny's Maxwell and the crazy people of Allen's Alley and the avalanche of possessions tumbling from Fibber McGee's closet existed in his own imagination.

He was thrilled because the squeaking door of *Inner Sanctum* and the galloping Silver lived in his very own images. In his heart, he carried his own Amos and Andy, and Marge of *Easy Aces*, and Ma Perkins and the members of *One Man's Family*. they were all his own, evoked by the suggestive artistry of creative radio.

Artistry today was all on the records; creativity went into commercials. More and more in television the best entertainment was the old movies, or new ones "made for TV." Radio was a juke box with news; TV, movies and news. In passing from scientist to showman, from showman to salesman, and now, they predicted, from salesman to auditor, broadcasting had abandoned its potential nobility.

Way back in 1928, in the days of George Harrison Phelps, Henry B. Joy, early president of the Packard Motor Company, said, "When the brotherhood of man shall have reached the degree of perfection that national boundaries are no more, some historian of that distant period will record that radio was the first medium to disseminate the mode of thought that wiped narrowness from the minds of mankind."

On the contrary, today's radio and television seemed to stir up apprehension and hatred among races, nations and generations. They don't build up; they expose and tear down. The end impression is one of destruction rather than creation, of prejudice, pollution and decay, of selfishness and savage violence. Man seems to be plunging into a new Dark Age and the medium that inspired idealistic hope in the mind of Henry B. Joy was doing nothing to stop the plunge, was instead expediting it.

I'm getting too old for all the sound and the fury, the old man thought. I miss the lack of wit and wisdom and good manners. God forbid that radio or television return to the past. Let the dead past bury the dead. This time and this generation will pass, and so will the next; then, maybe, some bright young man will come up with a brand new idea out of which will be born a 21st century Vic and Sade, or Myrt and Marge, or Clara, Lu and Em—and radio will be simple and fun again.

The picket jumped back as a car came to a stop at his elbow.

"Getting your daily exercise?" the driver grinned.

Taking license numbers had not stopped the stopping.

"Pretty rough out here this morning," said the friend from inside. "It really shook 'em up. I hear your man from New York is in there now talking with Fritz and Keck. Hey, I dropped by your desk for your mail."

He handed the picket his letters and drove away. A familiar name was on the envelope of one letter, and the picket read:

> Dear Friend,
> A lot of water has gone over the dam since I saw you last, and I was wondering whether WXYZ was ever going to do anything with that record you made of certain things I did which were to be included in some kind of broadcast. I'm curious about it.
>
> Sincerely,
> George W. Trendle

The first man in Detroit to have signed a contract with AFTRA had just passed his 86th birthday. Nobody remembered him; the tape would never be used.

Yeah, the picket sighed, we enter; we do our thing; and we exit.

He wiped the sweat from his face and looked at his watch. Another half hour before relief was scheduled. He tossed the rest of his mail, unopened, onto a camp chair, folded Trendle's note, put it in his hip pocket, and resumed his march.

The strike ended on July 9. Newsmen won a raise and an 8-hour day; and time to reach the top of the pay scale was reduced from 10 years to 2; but they and the announcers did not win their talent fee structure. However, both sides expressed satisfaction with the new contract.

Joe Bacarella celebrated by taking his disc jockeys and their wives to dinner.

Bacarella: "We just felt so good that this thing was over with! There was just a—a great feeling of brotherhood after the strike."

Bruce Miller: "Afterwards, when it was all over, there was such an up-tight feeling. People had a very hard time even speaking to each other."

For some weeks the atmosphere of Broadcast House, especially in the cafeteria, sparked with little animosities, but eventually, if not forgotten, they were swept under the rugs of the mind, and normality seemed to return to the station.

Five years went by. Bob Hynes changed channels, Dave Diles and Paul Berry, Bruce Miller and Dick Kernen left Wyxie, and Keck was gone. Howard Pierce died, and Harold True, Floyd Tracy, Raymond Meurer, Rollon Parker, Ruth Dean Rickaby and Bea Leiblee. Dead, too, was Harry Golder. And dead was George W. Trendle.

July 16, 1975.

The weather co-operated with elaborate plans for a memorable occasion—an invitational party for several hundred people on the front lawn of Broadcast House. Elton Rule, President of the American Broadcasting Company, and Harold L. Neal, Jr., President of ABC Radio, flew in from New York with Fred Foy for the occasion; and James G. Riddell, virtually retired but serving the George B. Storer interests as consultant, flew in from California. The 50th anniversary of WGHP—WXYZ Radio was being celebrated—2 1/2 months early.

On TV's *Morning Show* that day a very short version of *The Lone Ranger* had been performed with Hal Neal in the title role, Fred Foy announcing and playing a character, and Dick Osgood as Tonto and doubling as an "old timer." The morning host did sound effects. He later interviewed Dick Purtan and Lee Alan, and there were hints that Alan, now an agent, would return to a radio mike.

As the day grew dark, live music filled the outside air and there were lights in the apple trees. Champagne flowed freely and bars were set up to serve cocktails. Ladies remarked that the plates were not paper. Food was served on china, knives and forks were not plastic, and red and white wine was poured into glassware. Each table was covered with cloth, set for ten, and decorated with flowers.

One columnist had written that he looked forward to meeting John Slagle.

WXYZ-Radio 50th Anniversary party, July, 1975. (Actual anniversary, October 10, 1925. Left to right: Lee Alan, Don Davenport, Bill Hengstebeck, Ernie Winstanley, Fred Flowerday, Jack Tierney, Paul Hughes, Joan Vitez, Jack Rellis, Fred Wolf, Mickey Shorr, Specs Howard and Dick Osgood. At the podium, Fred Foy.

Johnny had been dead for years. There was one passing word in a speech for George Harrison Phelps, and Jimmy Riddell announced in a short speech that George W. Trendle's widow was present. She was over 90, carried a walker, and was accompanied by a gentleman so stiff and cold he might have been her butler, and his wife who could be mistaken for a housekeeper. Mrs. Trendle was as bright as a silver bullet.

A number of personalities from the past were introduced briefly, including Bill Hengstebeck, Fred Flowerday, Mickey Shorr, Paul Winter, Fred Wolf, Joan Vitez (Heidt), Porter Bourne who was still working, and Dick Osgood, who was not, and another now hot in several cities, Lou Gordon. There was a polite smattering of applause.

Following these introductions came what was planned to be the surprise piece de resistance. Fred Foy, in his most dramatic style, over the exciting music of the *William Tell Overture,* delivered the famous lines introducing the Lone Ranger, winding up with a thunderous, "The Lone Ranger RIDES A—G—A—I—N!!"

With that, a saggy-backed, once-white horse sauntered silently across the lawn with all the speed of a summer sunset. Aboard him, as uncomprehending as the nag, rode an unmasked man in overalls. With some difficulty he slid off the horse and lifted a box containing stubs of invitations for a give-away car to a platform.

DickPurtan took over and introduced each current disc jockey. Each drew a stub and through a process of elimination an agency lady won the car.

A few days after the affair which was in excellent taste and reflected the character of Charles D. Fritz, James Jewell died in Chicago. Obituaries there, and in other cities referred to him as "creator of *The Lone Ranger.*" A few days before the celebration, Ted Robertson died in California at the age of 55.

Dick Osgood left the party in a wistful mood, brooding over the transitory nature of life. So many characters had come and gone; so many would come and go in the future. But the station, through all the changes, remained. It had weathered the Great Depression, the loss of a network, 3 wars, Top 40 and the onslaught of hard rock music. It would outlive the current problems of inflation, unemployment, dimishing sources of energy, a continuing battle between Congress and President Ford—and demographics. It might even outlive the auditors.

When Anne Osgood joined her husband in their car, she carried a floral display she had stolen; and she kept the flowers alive for weeks.

Marty Greenberg, President of ABC-owned FM stations

Ben Hoberman, President of ABC Radio

Bibliography

"Wyxie Wonderland"

Additional Sources:

As stated in the foreword, most of the material in this book, including the dialogue, was supplied by the characters in it with whom we taped. Many dates were confirmed by referring to the author's scrapbooks, and to the column he wrote for *The Detroit Free Press,* 1951-1958.

Other sources of information are as follows:

On WGHP, George Harrison Phelps, the Dodge brothers and their widows, Walter Chrysler, and the disposition of the Dodge Motor Company:

American Automobile Manufacturers — Chilton
The Automobile Industy — E.D. Kennedy
The Gasoline Age — Glasscock
Men, Money and Motors — MacManus & Beasley

Automobile Topics:	54: 790, June 21, 1919
	August 7, 1920
	October 8, 1921
	February 4, 1922
	April 8, 1922
	April 1, 1923
	February 16, 1924
	January 17, 1925
	September 25, 1926
Detroit Free Press:	April 8, 1922
	October 10, 1925
	October 4, 1925
The Detroit News:	April 8, 1922
	April 9, 1922
	September 17, 1945
Radio Broadcasting:	May, 1925, p. 24
	June, 1926, p. 56
Detroit City Directory	1927-1928
	1928-1929
Saturday Night:	February 9, 1929

On the "opening" of WXYZ:

Detroit Free Press:	Sun. June 29, 1930
	July 1, 1930
	July 6, 1930
Detroit News:	July 1, 1930
Detroit Times:	July 1, 1930

On the Great Depression:

History of the American Economy — Robertson
A History of American Life, Vol. XIII: *The Age of the Great Depression* — Dixon Wechter

On Indian Village: Louis Mendelssohn and the Mendelssohn Mansion

Men Who Made Michigan—	Pipp
The Detroit News:	February 16, 1970

The Detroiter: October 22, 1922, pp. 6 & 14
October 3, 1927, p. 5
May 21, 1928, p. 7
May 27, 1928, p. 7

On Change of the Blue Network (NBC) to ABC:

Printers Ink, December 5, 1941, p. 82
Business Week, Dec. 13, 1941, p. 58
August 7, 1943, p. 78
August 21, 1943
Sept. 30, 1944, p. 93
Newsweek, May 12, 1941
August 9, 1943
Time, August 9, 1943, p. 62
January 3, 1944, p. 71
New York Times Index 1941, p. 1609

On the merger of ABC and the Paramount Theaters:

Time, June 4, 1951
Business Week, April 7, 1951
January 24, 1953, p. 31
February 14, 1953, p. 31

On ASCAP vs. BMI, Inc.:

A History of Popular Music in America — Spaeth
Radio and Television — Chester & Garrison
Saturday Review, Feb. 23, 1957, p. 15
Saturday Review, March 2, 1957, p. 10

On Todd Storz, Gordon McLendon, Bill Stewart and Top 40:

Variety Mar. 12, 1958, p. 53
Dec. 10, 1958, p. 57
Coronet May, 1961, pp. 128-132
Business Week, Sep. 9, 1961, p. 124-6
Billboard, Dec. 30, 1972
Dec. 23, 1972
Jan. 6, 1973
Jan. 13, 1973

On "payola":

Variety, March 12, 1958, p. 1
July 23, 1958, p. 7
Nov. 25, 1959, p.55
Billboard, March 17, 1958, p. 2
March 24, 1958, p. 2
October 6, 1958, p. 2
Time: Sep. 1, 1958, p. 38
Sep. 8, 1958
Oct. 6, 1958, p. 50

Life:	Sep. 15, 1958, p. 24
Down Beat,	Oct. 15, 1959, p. 16
	Nov. 12, 1959, p. 14
Newsweek:	Nov. 16, 1959, p. 66-68
U.S. News and World Report:	Nov. 16, 1959, p. 41
	Dec. 21, 1959, p. 82
Life:	Nov. 23, 1959, p. 45

On the disk jockey conventions:

Billboard:	Mar. 10, 1958, p. 3
	Mar. 17, 1958, p. 2

On Rock and Roll, the above, plus:

Billboard:	Apr. 21, 1958
	May 5, 1958
	May 12, 1958
	Dec. 15, 1958
	July 20, 1959

On Top 40 (in addition to sources listed under Gordon McLendon, etc:

Variety,	Mar. 12, 1958, p. 53
	Dec. 10, 1958, p. 57
	Apr. 20, 1960

On The Listeners Lobby:

The New York Times,	Feb. 21, 1960
Saturday Review,	Apr. 21, 1962
The Detroit Free Press,	Dec. 10, 1961

On Ted (Theodore) Shaker and Elton Rule:

Television, the Business Behind the Box — Brown

On the NBC-Westinghouse exchange:

Variety	Aug. 26, 1964
	Sep. 2, 1964
Broadcasting:	68:83, June 14, 1965

On the ABC-ITT merger:

Business Week,	April 7, 1951
Newsweek,	Dec. 13, 1965, p. 82
	July 3, 1967, p. 61
	Jan. 15, 1968, p. 66
Time,	Dec. 17, 1965, p. 53
	Dec. 30, 1966, p. 53
	Feb. 10, 1967

	Jan. 12, 1968, p. 57
	July 26, 1968, p. 72
New Republic,	Oct. 15, 1966, p. 6
	Feb. 25, 1967, p. 8
	July 8, 1967, p. 17
The Nation,	Mar. 20, 1967, p. 361

Other sources are mentioned in the text.

INDEX

B

D

G

H

L

MC

O

P

Q

R

S

T